Lecture Notes in Artificial Intelligence 1068

Edited by J. G. Carbonell and J

Subseries of Lecture Notes

T0230319

Henrik Schärfe   Pascal Hitzler
Peter Øhrstrøm (Eds.)

Conceptual Structures: Inspiration and Application

14th International Conference
on Conceptual Structures, ICCS 2006
Aalborg, Denmark, July 16-21, 2006
Proceedings

 Springer

Series Editors

Jaime G. Carbonell, Carnegie Mellon University, Pittsburgh, PA, USA
Jörg Siekmann, University of Saarland, Saarbrücken, Germany

Volume Editors

Henrik Schärfe
Peter Øhrstrøm
Aalborg University, Department of Communication
Kroghstraede 3, 9220 Aalborg East, Denmark
E-mail: {scharfe, poe}@hum.aau.dk

Pascal Hitzler
University of Karlsruhe, Institute AIFB
76128 Karlsruhe, Germany
E-mail: hitzler@aifb.uni-karlsruhe.de

Library of Congress Control Number: 2006928040

CR Subject Classification (1998): I.2, G.2.2, F.4.1, F.2.1, H.4

LNCS Sublibrary: SL 7 – Artificial Intelligence

ISSN   0302-9743
ISBN-10   3-540-35893-5 Springer Berlin Heidelberg New York
ISBN-13   978-3-540-35893-0 Springer Berlin Heidelberg New York

Springer is a part of Springer Science+Business Media

springer.com

© Springer-Verlag Berlin Heidelberg 2006
Printed in Germany

Typesetting: Camera-ready by author, data conversion by Scientific Publishing Services, Chennai, India
Printed on acid-free paper   SPIN: 11787181   06/3142   5 4 3 2 1 0

Preface

The 14th International Conference on Conceptual Structures (ICCS 2006) was held in Aalborg, Denmark during July 16 – 21, 2006.

Responding to the Call for Papers, we received 62 papers from 20 different countries, representing six different continents. This clearly indicates the international nature of the ICCS community as well as the widespread interest which was spawned by the previous conferences. By a thorough review process, 24 papers were selected to be included in this volume. In addition, six invited speakers made contributions which can be found in the first section of this volume.

The theme of ICCS 2006—Conceptual Structures: Inspiration and Application—points to a dual focus of interest that is also reflected in the constellation of papers. From the beginning of the planning of this conference, we focused on inspirational sources that have led to the current state of research in our community, by tracing important historical influences which daily effect work in representing knowledge and in handling representations of conceptual structures. At the same time, we also focused on ways in which these legacies are employed to further advance theory and practice in the field of knowledge representation and processing. With this volume, we believe that a valuable contribution to both aspects of this field is being made.

We wish to express our appreciation to all the authors of submitted papers, to the members of the Editorial Board and the Program Committee for all their work and valuable comments.

More information regarding the details of the conference can be found on the conference homepage at http://iccs-06.hum.aau.dk.

July 2006

Henrik Schärfe
Pascal Hitzler
Peter Øhrstrøm

Organization

The International Conference on Conceptual Structures is the annual conference and principal research forum in the theory and practice of conceptual structures. Previous ICCS conferences were held at the Université Laval (Quebec City, 1993), at the University of Maryland (1994), at the University of California (Santa Cruuz, 1995), in Sidney, 1996), at the University of Washington (Seattle, 1997), in Montpellier (1998), at Virginia Tech (Blacksburg, 1999), at Darmstadt University of Technology (2000), at Stanford University (2001), at Borovets, Bulgaria (2002), at Dresden University of Technology (2003), at the University of Alabama (Huntsville, 2004), and at the University of Kassel (2005).

General Chair

Peter Øhrstrøm   Aalborg University, Denmark

Program Chairs

Henrik Schärfe   Aalborg University, Denmark
Pascal Hitzler   University of Karlsruhe, Germany

Editorial Board

Galia Angelova (Bulgaria)
Michel Chein (France)
Frithjof Dau (Germany)
Aldo de Moor (Belgium)
Harry Delugach (USA)
Peter Eklund (Australia)
Bernhard Ganter (Germany)
Mary Keeler (USA)
Sergei Kuznetsov (Russia)
Wilfried Lex (Germany)

Guy Mineau (Canada)
Bernard Moulin (Canada)
Marie-Laure Mugnier (France)
Peter Øhrstrøm (Denmark)
Heather Pfeiffer (USA)
Uta Priss (UK)
John Sowa (USA)
Gerd Stumme (Germany)
Rudolf Wille (Germany)
Karl Erich Wolff (Germany)

Program Committee

Radim Bělohlávek (Czech Republic)
Anne Berry (France)
Tru Cao (Vietnam)
Dan Corbett (Australia)
Pavlin Dobrev (Bulgaria)
David Genest (France)
Ollivier Haemmerlé (France)
Udo Hebisch (Germany)
Joachim Hereth Correia (Germany)
Richard Hill (UK)
Andreas Hotho (Germany)
Christian Jacquelinet (France)
Adil Kabbaj (Marocco)
Pavel Kocura (UK)
Yannis Kalfoglou (UK)
Robert Kremer (Canada)
Markus Krötzsch (Germany)
Leonhard Kwuida (Switzerland)
Michel Leclère (France)

Robert Levinson (USA)
Michel Liquière (France)
Carsten Lutz (Germany)
Philippe Martin (Australia)
Claudio Masolo (Italy)
Engelbert Mephu Nguifo (France)
Jørgen Fischer Nilsson (Denmark)
Sergei Obiedkov (South Africa)
Ulrik Petersen (Denmark)
Simon Polovina (UK)
Anne-Marie Rassinoux (Switzerland)
Gary Richmond (USA)
Olivier Ridoux (France)
Sebastian Rudolph (Germany)
Éric Salvat (France)
Janos Sarbo (The Netherlands)
William Tepfenhart (USA)
Guo-Qiang Zhang (USA)

Table of Contents

Formal Ontology, Knowledge Representation and Conceptual Modelling: Old Inspirations, Still Unsolved Problems

Nicola Guarino

Laboratory for Applied Ontology, ISTC-CNR, Trento, Italy
`guarino@@loa-cnr.it`

Abstract. According to the theme of ICCS 2006, I will revisit the old inspirations behind the development of modern knowledge representation and conceptual modelling techniques, showing how the recent results of formal ontological analysis can help addressing still unsolved problems, such as semantic interoperability and cognitive transparency.

H. Schärfe, P. Hitzler, and P. Øhrstrøm (Eds.): ICCS 2006, LNAI 4068, p. 1, 2006.
© Springer-Verlag Berlin Heidelberg 2006

The Persuasive Expansion - Rhetoric, Information Architecture, and Conceptual Structure

Per F.V. Hasle

Department of Communication - Aalborg University
phasle@hum.aau.dk

1 Introduction

Conceptual structures are, as a rule, approached from logical perspectives in a broad sense. However, since Antiquity there has been another approach to conceptual structures in thought and language, namely the rhetorical tradition. The relationship between these two grand traditions of Western Thought, Logic and Rhetoric, is complicated and sometimes uneasy – and yet, both are indispensable, as it would seem. Certainly, a (supposedly) practical field such as Information Architecture bears witness to the fact that for those who actually strive to work out IT systems conceptually congenial to human users, rhetorical and logical considerations intertwine in an almost inextricable manner.

While this paper shows that Rhetoric forms an obvious communication theory for Information Architecture, it will not deal with the questions of how to utilize this insight in concrete practise. The focus is on how Information Architecture (IA) and Rhetoric meet in what is in essence a common conceptual structure. I shall describe the basic concepts of classical rhetoric and then proceed to show how these fit most closely to the main concepts of Information Architecture. Specifically, the "Information Architecture Iceberg" model of Morville and Rosenfeld can be shown to have a predecessor in Cicero's considerations on *oratio* (speeches). Then an important current development, in this paper called the *Persuasive Expansion,* is examined with an emphasis on its implications with respect to IA and Rhetoric. Finally, and most strikingly of all, perhaps, it is suggested how the "hard" computer science paradigm of object orientation is rooted in the Topics of Rhetoric. The paper is concluded by a brief discussion of implications for Conceptual Structures and raising a vision of a *Computer Rhetoric.*

In discussing Rhetoric I shall follow what has become standard usage in textbooks on classical rhetoric and use both Greek and Latin terms. This is partly to make the terms more readily recognisable, but partly also because in some cases the Greek terms cover the concept in question slightly better than the Latin terms, and sometimes vice versa.

2 Core Concepts of Rhetoric

What is Rhetoric about? Classical rhetoric is as a rule associated primarily with giving speeches (in Latin: oratio) whose aim is persuasion (in Latin: persuasio). However, while this is not entirely wrong, it is amputated to the point of being misleading, even

H. Schärfe, P. Hitzler, and P. Øhrstrøm (Eds.): ICCS 2006, LNAI 4068, pp. 2–21, 2006.

when only classical rhetoric is considered. There are good historical and cultural reasons why classical rhetoric indeed gave its attention to speeches rather than other media, but even in the classical apparatus there is nothing at all which necessitates a limitation of the field of Rhetoric to speeches, or even to words, spoken or written. Rather, the concepts of Rhetoric have to do with how to present a subject matter with a specific purpose – in general, how to achieve effective or efficient communication. In this connection presentation should also be thought of as more than simply the question of how the exposition is couched in words and other expressive means. The notion of exposition is inherent in the rhetorical notion of presentation – thus the logical and temporal structure of the delivery is part of the presentation, and in fact, part of the relevant subject matter. The great Roman rhetorician Quintilian (ca. 35-100 A.D.) clearly dispels any idea of limiting Rhetoric to a matter of outward style or persuasion only:

> Accordingly as to the material of oratory, some have said that it is speech, an opinion which Gorgias in Plato is represented as holding. If this be understood in such a way that a discourse, composed on any subject, is to be termed a speech, it is not the material, but the work, as the statue is the work of a statuary, for speeches, like statues, are produced by art. But if by this term we understand mere words, words are of no effect without matter. Some have said that the material of oratory is persuasive arguments, which indeed are part of its business and are the produce of art, but require material for their composition (Quintilian, IO, 2,21,1-2).

What Quintilian is saying here (in a perhaps somewhat complicated manner) is in essence that rhetorical work is really not on words, but on a subject matter; however the work consists in giving the subject matter an appropriate expression through words (or any other relevant expressive means). This passage thereby also states another fundamental tenet of Rhetoric, which we have already touched upon: the idea that form and content are inseparable. Any change in form implies a change in content – however small – and any change in content necessitates a change in form. That is why presentation is not merely about expressive means and their delivery, but inevitably also about conceptual structure.

Indeed, we here begin to deal with nothing less than the contours of a rhetorical epistemology, and a rhetorical perspective on conceptual structures, however lacking it still is in detail. So this is probably the place to pause for a few but important precautions. Rhetoric began in ancient Greece about 500 BC. Since then this important tradition of Western thought has been developed further till this very day. This fact makes for both historical depth and great systematic refinement of Rhetoric, but it also introduces a complication – the simple fact that various thinkers and epochs have conceived of Rhetoric differently, have emphasised different aspects and so forth. In particular, there was and is an approach to Rhetoric which sees it mainly as a set of communicative techniques with no or little philosophical import (to which I would count, for instance, the classical standard work Corbett 1999/1965).[1] Indeed, one of the greatest contributors to Rhetoric, Aristotle (384-322 B.C.), is sometimes understood this way (again, Corbett is an example of this). It is quite clear that a discussion of the arguments for or against this approach as opposed to a more philosophically inclined understanding of Rhetoric is quite beyond this paper. Nevertheless, decency demands that it be made clear here and now that this paper is based on the assumptions of what we could suitably call epistemic rhetoric (following Scott 1967). More precisely, the conception presented here is based on the works of in particular Karl Otto Apel (1963), Robert Scott (1967 and later), Ernesto Grassi (1980), Michael Billig (1996), and – in some ways - most of all Karsten Hvidtfelt Nielsen (1995).[2] However, this reservation does not imply any reservations with respect to what I have to say about the basic meaning of rhetorical terms – such as oratio and persuasio, and a number of other ones to follow – explications which will be readily recognised by all professionals of Rhetoric.[3]

So, we should now be ready for a fuller picture of Rhetoric and its epistemology. Rhetorical work sets out by a kind of question, or theme, or issue, which is perceived as problematic – the Latin term for this is *quaestio*:

> The question in its more general sense is taken to mean everything on which two or more plausible opinions may be advanced (Quintilian: 3,11,1).
> To a rhetorician, all issues present themselves under the aspect of a quaestio or causa ambigiendi, that is a sort of "issue in doubt"... In rhetoric, a case

---

[1] This is particularly evident in the manner in which Corbett repeatedly stresses that argumentation and human understanding should proceed on the basis of pure logos: 'Ideally, people should be able to conduct a discussion or argument exclusively on the level of reason [i.e. logos]. But the rhetoricians were realistic enough to recognize that people are creatures of passion and of will as well as of intellect. We have to deal with people as they are, not as they should be.' (Corbett: 71-72). Thereby cognitively cogent thought is associated with pure logic, whereas the remaining rhetoric concerns must be banned from philosophical epistemology, although they may still be relevant to how human cognition actually works. But as pointed out by epistemic rhetoricians and not least Robert Scott, rhetoric really calls for a notion of human rationality, wherein full human rationally rests on ethos and pathos as well as logos. Advances in neuroscience such as Antonio Damasio's works (e.g. 2000) seem to provide actual empirical underpinnings of this ancient notion - traceable in Gorgias, Protagoras, Cicero and Quintilian to mention some.

[2] Unfortunately, Hvidtfelt Nielsen is ambiguous in this matter. The ambition underlying his (initially) epistemological reading of rhetoric is the dissolution of epistemology – in essence, a post-modern contention that makes content disappear. But we may disregard these grand ambitions and stick with his otherwise excellent examination of possible epistemological consequences of above all Cicero's rhetoric.

[3] Moreover, these references to modern thinkers hopefully make it clear that this paper's focus on classical rhetoric is not meant primarily as an historical exercise. The fact that for instance New Rhetoric is not discussed is simply due to the fact that the core concepts of classical rhetoric are fully sufficient to demonstrate the points of this paper. Since New Rhetoric is mainly an extension and adaptation of classical rhetoric, a demonstration of the systematic relevance of the latter is a fortiori a demonstration of the relevance former. It may be added, however, that the difference between classical and modernised rhetoric is smaller than often assumed, as shown by e.g. Lunsford and Ede (1994).

constitutes a question with as many angles and sides as there are competent or
imaginative orators to represent them (Nielsen 1995: 61-62)

Rhetorical work is aimed at reaching a presentation and a concomitant understanding
of the subject matter. This process is directed by an intention implicit in the quaestio –   .
for instance the intention of presenting a convincing case for the acquittal of a defen-
dant, or the intention of finding out whether violent computer games affect children
adversely, and so on. The process initiated by quaestio is divided into five phases, the
Partes Rhetorices, or the five canons of Rhetoric:

- *Inventio* – in this phase the subject matter is determined and delimited, that is,
  a number of potentially relevant elements are selected (and others passed by,
  i.e. omitted). The selection is governed partly by the intention and partly by
  relations between the elements selected.
- *Dispositio* – the arrangement of the selected elements, for instance into argu-
  mentative sequences or conceptual hierachies.
- *Elocutio* – in this phase the style of presentation is chosen and suitable means
  of expression selected – words and terms, of course, but all kinds of expressive means may come under this phase (pictures etc.). Thus the presentation is
  given its final or almost final form.
- *Memoria* – the presentation is gone over and memorised as much as possible
  (in classical times, the presentation was often learned by heart; even so, the
  speaker should also be able to improvise).
- *Actio* – the delivery, i.e. the time and place when the presentation meets its
  audience (hearers, receivers, users).

We thus have in view an entire process, initiated by quaestio and its associated intention and leading to a presentation. But we need to determine a little more closely how
to conceive of the subject matter, and how the process operates upon it. It is fair, I
hope, to say of this paper, that its subject matter is Rhetoric and Information Architecture – with an affinity to conceptual structures. But it is immediately clear that this
description opens up a huge domain of possible topics that could result in very many
very different papers. We should therefore say that the subject matter (Latin res)
roughly circumscribes a large domain of possibly relevant elements. This goes also
for much more narrowly defined matters. For instance, the presentation of a case
before court may be seen as concerned with, say, guilt or non-guilt of a person with
respect to an alleged crime. Even so, the preparation of the defence may lead the in-
vestigator into realms not immediately within the scope of the matter – for instance,
statistics, laws of acceleration of cars, developmental psychology etc. etc. – often
topics not even thought of at the beginning of investigation. Therefore, we shall say
that the process operates on a loosely delimited domain of elements. The elements we
call doxa, following Greek tradition. Doxa means facts, loosely speaking, but not the
kind of hard facts envisaged in referential semantics (like the building stones of the
world in Wittgenstein's Tractatus). Doxa are plausible facts, arguable tenets, and
commonly held opinions.

The phase of inventio searches for these facts and selects among them. The selection is governed by the intention, of course, but also by relevance criteria. Relevance criteria partly stem from the elements themselves – for example, how one
element relates to another one in a possible conceptual hierarchy. But it is also most

significant that the selection of some facts (at the cost of others) by itself forms a momentum attributing for the further selection higher relevance to some facts and less to others. The following picture may serve as illustration (the term Topica will be explained later):

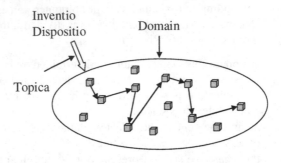

Fig. 1

The chain of arrows indicates a series of consecutive selection of facts from which to build the presentation. Thus there is really an element of dispositio present already in this phase. Element is added to element – juxtaposed, as it were – thus forming a pattern, which, even if it does not determine the presentation definitively, surely anticipates structures and sequence of the presentation to follow. Moreover, the working through the elements, the forming of a pattern, is also the forming of one's understanding of the case. The crucial epistemological consequence of this picture is that the investigation, the understanding, and the presentation of a subject matter are mutually dependent, indeed inseparably interwoven. The question as well as the intention of the process has to do with the need of a presentation/exposition of the subject matter – governing already the first probing steps of inventio. And the way one comes to see and understand the matter is obviously dependent upon what one has selected (and de-selected) and which patterns have been formed. We may picture it thus:

Fig. 2

Hopefully, the idea that form and content are inseparable may become more tangible through these observations.

We shall now examine some more specific concepts of Rhetoric, and show how very closely these fit with the very modern discipline known as Information Architecture (IA). Later, we shall see how the rhetorical understanding of IA's core model can be embedded in the foregoing general outline of Rhetoric and what some of the possible consequences are.

3 The Aptum Model and the Information Architecture Iceberg

Classical rhetoric identifies three fundamental parameters for effective and felicitous presentation, known as logos, ethos, and pathos. The presentation must make an appeal to the rationality of the receivers (logos), it must establish the plausibility of the sender as well as the presentation itself (ethos), and it should also appeal to and involve the receivers' emotions (pathos). These concepts are laid out in Aristotle's Rhetoric, and later developed by Cicero (106-43 B.C.) into the idea of the three duties (officia) of the rhetor: he must inform his audience (docere), he must "delight" it (delectare), and he must stir the audience's feelings (movere). As for delectare this does not merely mean "entertain", but rather creating a personal rapport with the audience such that the good will and honesty of the speaker are perceived. Nevertheless, the idea that a communication is there not just to inform and to achieve a goal but also to create as much pleasure, or joy, as possible is very characteristic of Rhetoric (Cicero's, in the very least).

It is not sufficient, however, simply to be aware of these functions or duties. The decisive point - and one of the pieces of really hard work for any presenter – is to bring them into the proper balance according to the situation. In a lecture, informing should play the primary (but not exclusive) role; and in a birthday talk, the aim of delighting should (ordinarily) play the greater part. When the right balance is found it bestows upon the speaker a dignity, decorum, which can greatly contribute to the success of the communication in question. To achieve the right balance, however, it is necessary to consider some more concrete parameters of communication. Already Aristotle was aware of the triad sender, content, and receiver (or speaker, message, and hearer) as constitutive entities of communication. This insight has since been the starting point of most communication theory. Cicero added to this triad two further parameters, namely expressive means (verba), and the context of the communication in question (situatio). The full list then becomes:

- *Orator*, that is speaker/sender.
- *Scena* (or *auditores*, *spectatores*), that is audience/hearers/receivers.
- *Res*, that is the subject matter to be investigated and presented – and hence the theme, respectively content, of the presentation. Cicero also calls this *causa*, the reason why a presentation is called for, respectively the cause which has initiated rhetorical discourse.
- *Verba*, that is the style, choice and deployment of expressive means, in a broad sense the form of the presentation.
- *Situatio*, that is the circumstances surrounding the presentation. This of course applies to the direct context of the presentation itself, but also to the wider setting in which it is given. As already mentioned a lecture is one kind of situation, and a talk at a birthday another one, each setting different felicity conditions for the presentation act to be performed.

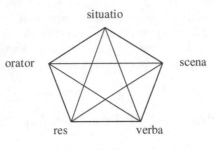

Fig. 3

In Scandinavia, it has become commonplace to set these parameters up in a figure called the rhetorical pentagon:[4]

The lines connecting all parameters directly are there not just for decorative purposes. They are to be taken seriously as emphasising the fact that there are relations between all of them. Just to exemplify this, let me suggest some of the possible factors to take into consideration with respect to the relations between *orator* and the four other parameters:

- *Orator – res*: whether the sender is an expert on the subject matter, does he have special interests in it (such as possible personal gain), etc.
- *Orator – verba*: what kind of expressions, which words are befitting this orator – is he or she young and in a position to use trendy expressions, or is the orator an older person who should refrain from such youngish language, etc.
- *Orator – scena*: is the orator a person in a position to make demands of the audience, being for instance its commander or its lecturer, or is he or she rather a supplicant, say a salesperson or an attorney appealing to a jury.
- *Orator – situatio*: is the occasion festive or grave, is the presentation ordinary as a lecture which is just the part of a pre-planned course, or is it extraordinary as a lecture given in the honour of a recently deceased colleague, etc.

Ideally, the task of the rhetorician is to bring these five parameters of presentation into their optimal balance. In practise, one must often be satisfied when a reasonably good balance is achieved, and when this happens the presentation is apt (*aptum*) – obviously, the better the balance, the more apt. Where a high degree of good balance is achieved the presenter, as well as the presentation, achieves *decorum*. The good balance will by implication also be a good balance of *logos*, *ethos*, and *pathos*. Hence the task of the rhetorician is really not located at the orator-parameter, as one might expect at first

---

[4] The figure is 'unauthorised' in the sense that classical rhetoric did not avail itself of graphical figures such as this one. Therefore some purists find it at least anachronistic to use it. Be that as it may, in this paper I am not out to argue an interpretation as historically correct in all details as possible, but rather in seeing how fundamental concepts of classical rhetoric are applicable to information technology. Of course, it is for my endeavour still a crucial point that these parameters of communication can be clearly documented in Cicero's thought and writings.

glance, but rather at the centre of the model – as the professional who is to work out and ensure the optimal adaptation between the five parameters of communication.[5]

We now make a leap of two millennia and turn to Morville and Rosenfeld's *Information Architecture Iceberg* (Morville and Rosenfeld 2002: 258):

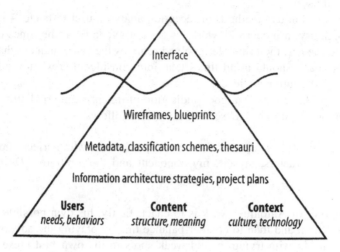

Fig. 4

The similarity with the Aptum Model is in many ways striking. The bottom line of the triangle, or the iceberg, contains three concepts which obviously correspond to the Aptum-Model. The same thing applies to the top of the iceberg, its visible part. In general, we have the following correspondences:

- *Users* corresponds to *scena*
- *Content* corresponds to *res*
- *Context* corresponds to *situatio*
- *Interface* corresponds to *verba*

While this correspondence is strikingly systematic we should not think of the relations in question as relations of identity. Rather, the IA-Iceberg is a special case of the Aptum-Model, calling for nuances and closer determination. This is obvious, when one considers the middle of the iceberg, with concepts such as *wireframes, meta-data, project plans* etc. These indicate principles and methods of organization which clearly have to do with the specific IT context considered in IA. Even so, the affinities are in my opinion obvious, also when one goes into greater depth with the conceptual relations at stake here. For instance, *scena* is in Cicero's work to be thought of not only as the audience present at the time of delivery. Rather, it is a

---

[5] In fact, Ciceronian rhetoric stresses that the persona of the orator is – just like verba - to be shaped and adapted according to the overall objective of achieving aptum. Hvidtfelt Nielsen lays out this point convincingly and with an emphasis on the fact that the shaping of one's persona in a rhetorical context has nothing to do with "make-believe or trickery" (Nielsen 1995, p. 31).

factor to be considered at all times of rhetorical work – and, moreover, the ultimate touchstone of the quality of the presentation, no matter how diligent the rhetorician has been in his work, and no matter which good arguments the rhetorician himself could give for the choices made (a point not irrelevant to the relation between IT developers and users).

To be true, present day methods of reception analysis, user tests etc. – highly relevant and obligatory in modern IT work - were not available at the times of classical rhetoric. Nevertheless, the principle involved remains the same, namely that the presentation to be made should at all times take into consideration whatever knowledge one can have of the future audience.

In *De Oratore* Cicero gives a good indication of this principle by letting one of his figures, the attorney Antonius, describe his work as follows:

> ...when [my client] has departed, in my own person and with perfect impartiality I play three characters, myself, my opponent and the arbitrator (*De Oratore*: 2,120).

The starting point of Antonius' work is to enact, firstly, his own possible role as the client's spokesman, and then to counter-balance this by playing the adversary, which of course has to do with trying to find weak spots in his own first presentation. He then proceeds to consider how a special kind of audience, namely the judge, would probably react to the two previous competing presentations. At which point the whole process can be repeated to improve the first presentation, or alternatively, the case could be abandoned.[6] In all of this the principle of imitation (*imitatio*) is involved – Antonius imitates to himself a possible adversary and a possible judge. (In fact, he even imitates himself to himself.) Surely IA-workers, while availing themselves of as much solid information as can be had about future users, still are doing the same thing constantly in the course of their work – trying to imagine how future users will react to various features of the system to be developed.

For all these similarities, there also is a thought-provoking difference. In the Iceberg, there is no sender-instance analogous to *orator* in the Aptum Model. Surely this does not mean that there is no sender at all, but rather that this component has become much more complex than in classical rhetoric. The sender-parameter comprises several entities with complicated mutual relations – such as an organization commissioning the production of its website, a web-site company developing it, the information architects working in the process, and so on. Nevertheless, the continuing importance of sender-receiver relations is clearly indicated by the following remark:[7]

> The choice of organization and labelling systems can have a big impact on how users of the site perceive the company, its departments, and its products (Morville and Rosenfeld 2002: 54-55).

---

[6] In fact, if the process shows the case for the client to be untenable, it is Cicero's advice to decline taking the case – for professional as well as ethical reasons.

[7] In fact, this applies to all the relations of the Aptum Model.

The same perspective is emphasised by Jesse James Garrett:

> In the minds of your users, an impression about your organization is inevitably created by their interaction with your site. You must choose whether that impression happens by accident or as a result of conscious choices you have made in designing your site (Garrett 2003: 42).

This very difference between the Aptum Model and the IA Iceberg focuses attention on the special conditions and features of IT-based communication. At the same time, the Aptum Model is a useful reminder to the IA-worker that he or she should carefully bear in mind the relation between sender(s), receivers (i.e. users) and the information architecture itself.

Morville and Rosenfeld in their 2002 book put much care and energy into determining the work of the information architect as precisely as possible. To my mind, the end result is in fact a picture of the IA worker very much akin to the classical idea of rhetorical work: as the person whose task it is to ensure the optimal adaptation of the various components to each other – databases to interfaces, form and function to content, senders' intentions to users' needs, etc. The information architect is not a specialist in programming, graphical design, user tests or other specific IT-disciplines. He or she is a specialist in relating all these areas to each other in a manner striving for their optimal balance, or to use classical terms: to develop an optimal fit between form and content. Speaking of IA in the year 2006 this kind of work must of course take heed of the conditions and features specifically appertaining to the IT-medium.

4 The Persuasive Expansion

The concluding quotes of the previous section lead us straight into an emerging and apparently rapid development in IT, perhaps most poignantly characterised by the idea of *Persuasive Design* (PD). The most important individual contributor to this notion is B.J. Fogg, whose book *Persuasive Technology* (2003) described persuasive uses of technology, in particular computers, in greater detail than had been done before. Fogg himself calls the field "Captology", an acronym for "Computers as Persuasive Technologies", but I shall stick to the term *Persuasive Design* (which also seems to have achieved wider acceptance in IT- communities). Fogg defines persuasion as

> ... an attempt to change attitudes or behaviours or both (without using coercion or deception) (Fogg 2003: 15).

Furthermore, Fogg describes PD as a field and/or discipline by saying that it

> .. focuses on the design, research and analysis of interactive computing products created for the purpose of changing people's attitudes or behaviour (Fogg 2003: 5).

The decisive insight underlying Fogg's work is the fact that software is increasingly being used with the conscious aim of *influencing* people in various ways. In a narrower sense, it is used for *persuading* people – to buy a product, to join a party, to support a cause, to become a good leader or a considerate driver. Indeed, in this respect a momentous and general development is going on in IT. Even good old-fashioned information systems such as, say, library portals are increasingly given an overlay of persuasion. It has often been noted that the computer began its history rather as a super-calculator. With the personal computers in the 80'es its scope was broadened a good deal, a development which was brought further by the growth of the Internet in the 90'es. Altogether a development which made the use of the computer as *information system* more prominent than its use as calculator.[8] The idea of PD, however, indicates yet another expansion of the scope of the computer, which may be briefly characterised as its expansion from *information system* into *communication system*. I call this development the Persuasive Expansion, to emphasise two points: that it is a development which expands rather than supersedes the customary use of computers for information purposes, and this expansion is guided by an increasing emphasis on persuasive purposes. As explained in the first section on rhetorical concepts, persuasio, and the goal of Rhetoric in general, should not be seen narrowly as the attempt to get one's way, but rather as the purposeful use of communication to achieve a goal (which may very well be idealistic and in the best interest of the receiver). We may illustrate the Persuasive Expansion as follows:

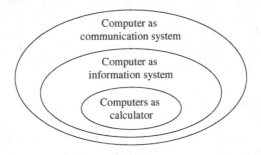

Fig. 5

While Fogg focuses on persuasion in a somewhat narrower sense than is done here, it is clear that his proviso "without coercion or deception" is meant to indicate a purpose more flexible than brute persuasion. Significantly, it seems to me, (Fogg 2003) in at least one place focuses on the wish to influence in a completely general sense:

> As I see it, if someone didn't want to influence others in some way, he or she would not take the time or energy to set up a website (Fogg 2003: 147).

The original core task of IA was to organize and categorize knowledge in a way that would optimally support users in their attempts to find information for which they were looking:

---

[8] Of course, the use of computer as calculator has not disappeared. On the contrary, it is more crucial than ever to financial and technological development in general. However, the number of people (respectively the amount of time spent) interacting with computers as information systems vastly outstrips the corresponding numbers of interacting with computers for obviously calculatory purposes.

As information-architects, we organize information so that people can find the right answers to their questions (Morville and Rosenfeld 2002: 50).

It is clear that this task is still most necessary, but it is also clear that as it is stated here, it associates IA with the computer as an information system. PD takes the step of subsuming this IA core task under more general communicative purposes. Nevertheless, the observation previously quoted that *"The choice of organization and labelling systems can have a big impact on how users of the site perceive the company..."* shows that *Morville and Rosenfeld are quite aware of the fact that categorization is not just a logical endeavour. It also creates the setting for achieving effective communication and has an impact on the image of the company or organization inevitably projected by the website. There is, however, more to this than image projection. The communicative design – that is, the choice of rhetorical conceptual structure – determines the very understandability of the website, as noted by Jesse James Garrett (a contributor to IA with an especially keen eye for communicative aspects):*

If your site consists mainly of what we Web types call 'content' - that is, information - then one of the main goals of your site is to communicate that information as effectively as possible. It's not enough just to put it out there. It has to be presented in a way that helps people absorb it and understand it (Garrett 2003: 14).

Coming back to Fogg (2003), this work itself explicitly points to Rhetoric as at least part of its background (e.g. p. 24). Already at the beginning of his development of PD, Fogg wrote:[9]

For example, Aristotle certainly did not have computers in mind when he wrote about persuasion, but the ancient field of rhetoric can apply to captology in interesting ways (Fogg 1998: 230-231).

On the other hand, Fogg 2003 does not purport to be a scientific theory, to the best of my comprehension. It is a presentation of concepts and guidelines which are useful for developing persuasive software. But it is also clear that at a general level it shares central concerns with classical rhetoric. In fact it also has many interesting connections with rhetorical concepts even at the level of detail. Especially the concept of *Credibility* has a strong connection with Rhetoric, which deserves to be mentioned. Aristotle's concept of *ethos* is determined more precisely by identifying three components of *ethos*. These are

- *Phronesis*, approximately the same thing as 'competence'
- *Eunoia*, approximately the same thing as 'benevolence'
- *Arete*, approximately the same thing as 'honesty' or 'trustworthiness'.

---

[9] The development of PD and Fogg's work was investigated by Sine Gregersen in her Master's Thesis (2005). Herein she also points out that Fogg – as described in Fogg 1998 – met informally with other researchers interested in the intersection of persuasion and computing technology at the CHI - Human Factors in Computing – conference in 1997. I am indebted to her work for leading me to the quote used here.

To discuss the full philosophical meaning of these concepts certainly requires a deeper knowledge of Aristotle as well as ancient Greek culture than the present author commands. But from a rhetorical perspective we can have a quite satisfactory picture by saying that to exhibit, or achieve, *ethos*, the speaker must show relevant competence, good will towards his audience, and honesty. Indeed, *ethos* was arguably recognized as the most important factor in communication already in ancient Rhetoric. This corresponds entirely with the importance attributed to credibility by Fogg. Fogg describes credibility as consisting of two components, namely perceived expertise and perceived trustworthiness, using this figure (Fogg 2003:123):

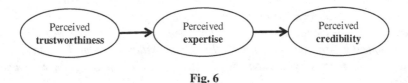

Fig. 6

It is evident that *Phronesis* roughly corresponds to expertise and *Arete* roughly corresponds to trustworthiness. The fact that Fogg qualifies these components as 'perceived' raises some interesting aspects. Cicero argued that the speaker must mean what he says (in a telling contradiction to the bad reputation which is sometimes given to Rhetoric) – thus it would not be sufficient to simulate emotions or values or convictions not really held:

> ...I never tried, by means of a speech, to arouse either indignation or compassion, either ill-will or hatred, in the minds of a tribunal, without being really stirred myself... by the very feelings to which I was seeking to prompt them... (181) ...the power of those reflections and commonplaces, handled in a speech, is great enough to dispense with all make-believe and trickery... (*De Oratore*: II, 191).

While this passage deals especially with *pathos*, the demand for personal honesty is arguably generalizable also to *ethos* and *logos* within the thought of Cicero. Nevertheless, it may be that we have here found one of those points where classical rhetoric cannot be applied without modification to the field of IT. The place of *eunoia* (closely related to Cicero's idea of *delectare*) and the call for sincerity stated above may depend on the special situation of face-to-face communication characteristic of the classical speech. Without doubt computer-based communication also bears some imprint of the 'authors' behind it, but it is not necessary and in fact hardly possible to establish that kind of personal rapport between speaker and hearer which is so central to classical rhetoric. Thus the omission of *eunoia* in Fogg's determination of credibility is really comparable to the omission of the *orator* from the IA Iceberg (if 'omission' I may call it).[10] Nevertheless, the comparison between classical and present-day determinations should inspire consciousness about what has been changed, and why. Put

---

[10] In this connection it ought to be mentioned that Fogg devotes a whole chapter to a discussion of 'The Ethics of Persuasive Technology' – thus an aspect which is surely an integral part of his work. The point here is simply that these ethical deliberations have nothing or little to do with eunoia or arete for that matter.

negatively, it is certainly still the case that a website whose authors are obviously *not* benevolent towards their users will lose its persuasive power immediately.

5 Object Orientation, Categorization and Topica

In my experience, the observation that persuasive concerns in the uses of IT are on the rise is not lost on practitioners of IA and other IT-professionals – in many cases, their daily work support this point, often quite strongly. So, in these communities that observation is apparently a persuasive and convincing argument for the relevance of Rhetoric, once its concepts are explained a little more closely. Nevertheless, the relevance of Rhetoric to IA is rooted at a much deeper level, having to do not only with the development of IT and the goals of IA, but also the very foundation of IA. One fairly straightforward way of showing this is, I believe, by focussing on the systems development paradigm of Object-Orientation (OO). OO is at the same time a programming paradigm. In the following, by OO I mean OO in its entirety.

The most characteristic feature of all in OO is that it is designed for categorization, especially for the construction of conceptual hierarchies.[11] In OO, these are

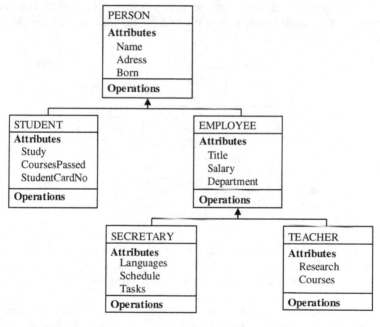

Fig. 7

---

[11] From a computer science point of view, the crucial technical effects of these constructions are inheritance and what is known as virtual operations (or methods), which makes for code-sharing. It is however clear that these properties derive from categorization at the conceptual level of OO (as in object-oriented analysis and design) as well as its linguistic level (i.e. object-oriented programming languages).

subdivided into two major kinds, generalization structures and aggreration structures (cf. Mathiassen et al. 2000: 69 ff.) – also known as "is-a" hierarchies and "is-part-of" hierarchies. Here is an example of a generalization structure in UML-notation (UML is a standard within OO):

This diagram expresses immediate hierarchical relations: a Secretary is an Employee, and an Employee is a Person. This of course also very directly explains why this is called an "is-a" hierarchy. In other words, Secretary is a sub-concept (subclass) of Employee, and Employee is a sub-concept of Person.

The introduction of a common superclass for two or more classes is called generalization within OO. Person is a superclass of Student as well as of Employee, and so forth. If you want to know what the properties of objects from a specific class are – say some Student – you read off the attributes stated in the Student class and then move upward in the tree to the next immediate superclass, the attributes of which must be added – and so on until you reach the top, i.e. a class with no superclass above it. So in the case of Student the relevant properties are:

```
{StudentCardNumber, CoursesPassed, Study, Born, Ad-
dress, Name}.
```

In other words, objects from a subclass inherit the properties of all its superclasses.[12] In OO, the differentiation of a class into two or more subclasses is called specialization.

An example of an aggregation structure could be this one, taken from (Mathiassen et al. 2000: 76):

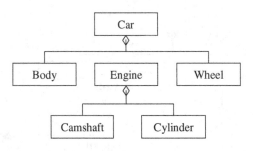

Fig. 8

This diagram indicates for instance that a Camshaft is part of an Engine, and a Cylinder is part of an Engine, an Engine itself is part of a Car, etc. Thus an aggregation structure can describe how an object from a class is composed of objects from other classes.

It is a remarkable fact that OO-languages contain constructions which are direct formal expressions of generalization structures. Thus in JAVA, for instance, we might introduce some class representing persons like this:

```
Public class Person {....}
```

- assuming, of course, that {...} fills in the relevant content of the class.

---

[12] Indeed, this is exactly the procedure which leads to the determination of Human Being in the famous arbor Porphyrii – see also the brief remark on Porhyrios below.

With this definition, the language construct named extends can be used for directly creating a subclass of Person, with inheritance and other resulting computational consequences:

```
Public class Student extends Person {....}
```

Apart from being a very powerful programming construction, it is a remarkable fact that we have here a crucial programming language construct which is there primarily in order to be able to express the *conceptual structure* of the relevant domain. The importance attributed to OO-programming language constructs' ability not just to achieve computational advantages (such as code-sharing) is evident in many places in   OO-literature. The effect of having conceptual hierarchies directly in the language is to aid programmers in grasping the information structure of the domain as well as bridging the gap between domain understanding and computational modelling:

> To program is to understand: The development of an information system is not just a matter of writing a program that does the job. It is of the utmost importance that development of this program has revealed an in-depth understanding of the application domain. (Madsen et al. 1993:3)[13]

As for aggregation structures, these are expressed differently and partly on the basis of choice in various OO-languages, but they too are generally simple to program, at least when you have the corresponding UML-diagrams. Thus for instance, assuming that we have definitions of

```
Public class Body {....},
Public class Engine {....},
Public class Wheel {....},
```

the aggregation expressed in Figure 8 could be programmed in JAVA as follows:

```
Public class Car
{
        Private Body myBody;
        Private Engine myEngine;
        Private Wheel rightFront, rightRear, leftFront,
leftRear;
}
```

These concepts and the uses of these constructions derive directly from Aristotle's work on *Categories*, further developed by the Greek philosopher Porphyrios

---

[13] Madsen et al. 1993 is in my opinion a culmination in the emphasis on the conceptual benefits of OO which is characteristic of the Scandinavian approach to OO. This emphasis is explicit and explained at length in the work quoted, which is also a textbook on the BETA programming language.

(ca. 232-304 A.D.) and, of course, refined more over the centuries.[14] In his *Rhetoric* Aristotle placed these concepts in their constructive and communicative context. Classical rhetoric contains a whole subfield called *the Topics (Topica)*. Roughly and very briefly, these may be described as sets of strategies and general questions which support the investigation of the domain – especially the phase of *inventio*, but also *dispositio* in the manner described in the first section. Among the Topics the most important of all is known as *Definition*. In Rhetoric, *Definition* is normally subdivided into

- *Genus/species*, which in effect means the establishment of conceptual hierarchies relevant to the domain,
- *Divisio*, which means the investigation into and the working out of relevant "part-of" hierarchies.

As the ideal form of definition Aristotle favoured *genus/species*, but he was aware that often the precise hierarchical specification is exceedingly difficult to establish, in which case we may resort to defining a concept by enumerating components of which it consists, or which it has as parts. (On the other hand, it can also be argued that definition in terms of composition is not merely something used for want of a better definition, but a conceptualization in its own right.) Be that as it may, *genus/species* and *divisio* are crucial concepts and methods for the overall rhetorical process, especially inventio. It must here be noted that categorization in IA, and in general, is more than what is contained in the topic of Definition. More flexible and nuanced approaches, such as grids or faceted categorization are clearly needed. But it is fair to say that historically and systematically (and arguably even cognitively) the forms contained in Definition are most fundamental. Moreover, in software matters programming remains the basis of everything, even if IA-practitioners usually work with Content Management Systems and similar tools, which make work considerably easier than programming proper. Now OO is the only programming paradigm, at least with massive real-world usage, which is systematically built for categorization. As is hopefully evident by now, *genus/species* directly corresponds to the OO notions of generalization and specialization, while *divisio* corresponds to aggregation. But clearly, conceptual hierarchies and whatever else is found in *inventio* and *dispositio* is in the end reflected in the presentation (cf. Figure 2).

Now the full importance of this goes far beyond the mere demonstration of the philosophical and also rhetorical roots of OO and categorization. For the very point of epistemic rhetoric is that the early choice of conceptualization (including categorizations) is inevitably reflected in the final presentation; and conversely, the goals inherent in *quaestio* (i.e. the presentation to be made and its purposes) inevitably direct the whole conceptualization, including the categorization which takes place even at the earliest stages of the process.[15] Some of the quotes from Garrett as well as Morville and Rosenfeld have already suggested this kind of connection – the importance of

---

[14] A remark which also applies to OO itself, since OO is in a computational refinement of the classical concepts – that is, an operationalization.

[15] For OO, the consequences of some very similar observations are taken in (Soegaard 2005). In particular, the more classical OO-notion of the domain as a pre-existing "referent system" to be modelled is replaced by a notion of "programming for the future", i.e. not a pre-existent but a future referent system.

categorization etc. for the company's image etc. – and indeed, these IA-theorists have also recognized that categorization is important for the meaning which is created for users:

> The way we organize, label, and relate information influences the way people comprehend that information (Morville and Rosenfeld 2002: 50)

At what appears to me to be an even more fundamental level, Garrett (2005) observed how the combination of information elements (what Rhetoric would call *doxa*) and the manner in which they are selected and juxtaposed is at the root of the creation of meaning. Having first noted how individual data in isolation fail to create meaning as well as the fact that "Humans are pattern makers", Garrett's argument culminates in this determination of IA:

> Information Architecture is the juxtaposition of information in order to convey meaning (Garrett 2005: dias 29).

Hoping that this will not be felt as an imposition, it seems obvious to me that this determination could have been straight out of classical rhetoric.

6 Conclusion: Computer Rhetoric

In this paper, a special focus has been placed on the field and discipline of Information Architecture. Part of the reason for this, I gladly admit, is that this author has a special interest in IA beforehand. But more importantly, the very term and idea of Information Architecture has much in common with Conceptual Structures (CS). Depending on one's epistemology, one may say that an information architecture is, or *represents*, or *induces* a conceptual structure. But while there are thus different views on how the relation is to be conceived of, it seems wholly unreasonable to deny that it is there. The idea of conceptual structure is part of the idea of IA, whether implicitly or explicitly.

Similarly, Rhetoric offers an idea of CS. While this is somewhat different from established studies of CS, the issues and problems dealt with in Rhetoric quite obviously have to do with the conceptualization of subject matters or problem domains. But while this difference may have something to do with epistemological assumptions, it also has to do with different concerns. Roughly, classical work on CS has been oriented towards logical issues, whereas Rhetoric has concerned itself more with style, and how to structure presentations for specific communicative purposes. These two concerns as such are in no way contradicting each other – in fact they need each other.

But it is true that epistemic rhetoric departs from classical foundationalist notions of CS – whether these are mentalist as Chomsky's deep structures or referential like Montague's universal grammar. It should be carefully noted however, that epistemic rhetoric is different from postmodern thought. The first quote of this paper (Quintilian on the material of oratory) says how: in Rhetoric, content (the material of rhetoric) does not disappear as it does in, say, Rorty's postmodern conception. That is, in the latter there seems to be just different vocabularies which can be played off against each other, but cannot be measured against any kind of "external" standards or criteria. And here we find postmodernism in the flattest contradiction of Rhetoric: the emphasis upon the need for the rhetorician to be well instructed in the subject matter (= content) is unequivocal. Probably even worse to the postmodenist, Rhetoric admits of no doubt that certain styles and deliveries really *are* better than others. Thus is it almost objectivistic on a point which is often left over to mere subjectivity even by thinkers who are otherwise hard-nosed realists. So, as I see it Rhetoric by no means rejects a notion of content or for that matter truth. But it is relativistic in the sense that it abandons the ideal of definitive decidability underlying Chomsky's deep structures, Montague's universal algebra and lots of other kindred approaches – which means a great part of classical Western objectivism.

These remarks are meant mainly to stimulate future discussion on the possible place of Rhetoric in work on Conceptual Structures. But however interesting these epistemological questions are, we need not be faced with stark choices when it comes to more practical matters. The *usefulness* of rhetorical thought is strongly indicated by its relation to IA as well as PD (even if this study has not dealt much with practise). In turn, much of the sometimes confusing field of IA (together with PD) can be systematically determined within rhetorical theory.

But, to be true, there also is a still more encompassing vision at stake here, that of *Computer Rhetoric*. Thinking of the domain in terms of *doxa* to be selected and patterns to be formed rather than a set of objective facts to be pictured really does make for a "rhetorical turn" not just in IA but in systems development in general. The same goes for thinking of the modelling function "as programming for the future" (cf. footnote 15) rather than picturing a referent system. In short, epistemic rhetoric leads to a novel conception of these matters, a conception for which I suggest the term *Computer Rhetoric*.

While the practical implications of this idea need to be worked out in more detail, I think the general outlook is clear enough and may be suggested by a simple comparison. Anybody who has been taught to program in a computer science department has also been taught some basics of the relevant mathematics – automata theory, formal logic etc. But in fact, when actually programming one very rarely uses any of this knowledge directly. So why was it taught? The simple reason is that this mathematics is crucial for understanding what programming is (in its technical sense), and that understanding this stuff makes better programmers. The role of Rhetoric wrt. IA and PD, and Systems development in general is exactly the same. It is the theory of how to make all the communicative aspects of any computer-system work in relation to human users. When actually designing a system one may think only to a limited extent about *aptum, ethos, persusasio* etc., but the basic knowledge of these constitute the full understanding of what one is doing, and mastery of it makes better designers.

Note: Website for the M.Sc. study programme in Information Architecture at Aalborg University: www.infoark.aau.dk

References

Apel, Karl Otto (1963): Die Idee der Sprache in der Tradition des Humanismus von Dante bis Vico, Bouvier Verlag Herbert Grundmann, Bonn

Aristotle: The Categories, (Loeb edition)

Aristotle: The Rhetoric, (Loeb edition)

Billig, Michael (1996): Arguing and thinking – A Rhetorical Approach to Social Psychology. (2nd edition), Cambridge University Press

Cicero: De Oratore, (Loeb Edition)

Corbett, Edward P.J. and Connors, Robert J. (1999): Classical Rhetoric for the Modern Student, Fourth Edition, Oxford University Press, New York 1999

Damasio, Antonio (2000): Descartes' Error. Emotion, Reason, and the Human Brain, Harper-Collins/Quill, New York

Fogg, Brian J.(1998): Persuasive Computers – Perspectives and Research Directions, CHI98 Papers

Fogg, Brian J. (2003): Persuasive Technology - Using computers to change what we think and do, Morgan Kaufmann Publishers, San Francisco

Garrett, Jesse James (2003): The Elements of User Experience: User-centered design for the web, American Institute of Graphic Arts / New Riders

Garrett, Jesse James (2005): The Frontiers of User Experience. PowerPoint-presentation given at DF's second Conference on Information Architecture, Korsør, Denmark, 2005

Grassi, Ernesto (1980): Rhetoric as philosophy: The humanist tradition, Pennsylvania State University Press

Gregersen, Sine (2005): Persuasive Design – fra Praksis til Teori, Master's Thesis, University of Aalborg

Lunsford, Andrea A. and Ede, Lisa S. (1994): On Distinctions between Classical and Modern Rhetoric. Professing the New Rhetoric. Ed. Theresa Enos and Stuart C. Brown. Englewood Cliffs, New Jersey, Prentice Hall

Madsen, Ole Lehrmann, Petersen, Birger Møller and Nygaard, Christen (1993): Object-Oriented Programming in the BETA Programming Language, Addison-Wesley, New York

Mathiassen, Lars, et al. (2000): Object-Oriented Analysis and Design, Aalborg: Marko. (www.marko.dk).

Morville, Peter; Rosenfeld, Louis (2002): Information Architecture for the World Wide Web, O'Reilly & Associates, Inc 2002

Nielsen, Karsten Hvidtfelt (1995): An Ideal Critic: Ciceronian Rhetoric and Contemporary Criticism, Peter Lang Verlag, Bern 1995

Quintilian, Institutio Oratoriae, (Loeb Edition) (here quoted from www2.iastate.edu/~honeyl/quintilian/2/chapter21.html, retrieved on 25 March 2006.)

Scott, Robert L. (1967): On Viewing Rhetoric as Epistemic, Central States Speech Journal 18

Scott, Robert L. (1976): On Viewing Rhetoric as Epistemic: Ten Years Later, Central States Speech Journal 27

Scott, Robert L.(1990): Epistemic Rhetoric and Criticism: Where Barry Brummett Goes Wrong, The Quarterly Journal of Speech 76

Soegaard, Mads (2005): Object Orientation Redefined: From abstract to direct objects and toward a more understandable approach to understanding, Master's Thesis in Information Studies, University of Aarhus (Retrieved March 25, 2006 from Interaction-Design.org: www.interaction-design.org/mads/articles/object_orientation_redefined.html)

Revision Forever!

Benedikt Löwe*

Institute for Logic, Language and Computation, Universiteit van Amsterdam,
Plantage Muidergracht 24, 1018 TV Amsterdam, The Netherlands
bloewe@science.uva.nl

Abstract. Revision is a method to deal with non-monotonic processes.
It has been used in theory of truth as an answer to semantic paradoxes
such as the liar, but the idea is universal and resurfaces in many areas
of logic and applications of logic.

In this survey, we describe the general idea in the framework of pointer semantics and point out that beyond the formal semantics given by Gupta
and Belnap, the process of revision itself and its behaviour may be the
central features that allow us to model our intuitions about truth, and is
applicable to a lot of other areas like belief, rationality, and many more.

1   Paradoxes

Paradoxes have been around since the dawn of formal and informal logic, most
notably the **liar**'s paradox:

This sentence is false.

Obviously, it is impossible to assign one of the truth values *true* or *false* to the
liar's sentence without a contradiction. One of the most pertinacious urban leg-
ends about the liar's paradox and related *insolubilia* is that the problem is just
self-reference. But it cannot be so simple; a lot of self-referential sentences are
completely unproblematic ("This sentence has five words"), and others that formally look very similar to the liar, have a very different behaviour. For example,
look at the **truthteller**

This sentence is true.

As opposed to the liar, the truthteller can consistently take both the truth values
true and *false*, but it is still intuitively problematic: there is no way we can find
out whether the sentence is correctly or incorrectly asserting its own truth. The
same happens with the so-called **nested liars**:

The next sentence is false,
the previous sentence is false.

Here, the assumption that the first sentence is false and the second is true is
perfectly consistent, as is the assumption that the first sentence is true and the

---

* The author would like to thank Fabio Paglieri (Siena & Rome) for discussions about
belief revision and comments on an earlier version of the paper.

H. Schärfe, P. Hitzler, and P. Øhrstrøm (Eds.): ICCS 2006, LNAI 4068, pp. 22–36, 2006.

second false. If you mix the liar with a truthteller and let them refer to each other, you get the **nested mix**,

the next sentence is false,
the previous sentence is true,

which again does not allow a consistent truth value assignment.

Even though all of them are problematic, their status is subtly different and we get a rather clear picture of how and why they are different. Even more striking is the following **hemi-tautology**:

At least one of the next and this sentence is false,
both the previous and this sentence are false.

Here we get a unique consistent truth value assignment; the first sentence must be *true* and the second one *false*, and our intuition allows us to identify it accurately[1].

In this survey, we shall discuss structural approaches based on the concept of revision due to Herzberger [He82a,He82b] and Gupta and Belnap [GuBe93] called **revision theory**. We describe revision theory both as a partial truth predicate based on revision (this is the way Gupta and Belnap phrase it in their book) and as a conceptual method. We argue that the underlying ideas of revision theory are widely applicable; the formal semantics has been reinvented independently in many areas of logic (§ 6.1), and the conceptual framework of recurrence and stability describes a wide range of phenomena (§ 6.2).

2   Pointer Semantics

In § 3, we shall describe the semantics of Herzberger, Gupta and Belnap in the simple logical language of **pointer semantics** invented by Gaifman [Ga$_0$88,Ga$_0$92]. The presentation of the system in this section is taken from [Bo$_0$03, § 5].

We shall define a propositional language with pointers $\mathcal{L}$ with countably many propositional variables p_n and the usual connectives and constants of infinitary propositional logic ($\bigwedge$, $\bigvee$, $\neg$, $\top$, $\bot$). Our language will have **expressions** and **clauses**; clauses will be formed by numbers, expressions and a pointer symbol denoted by the colon : .

We recursively define the expressions of $\mathcal{L}$:

- Every p_n is an expression.
- $\bot$ and $\top$ are expressions.
- If E is an expression, then $\neg E$ is an expression.
- If the E_i are expressions (for $i \in \mathbb{N}$) and $X \subseteq \mathbb{N}$, then $\bigwedge_{i \in X} E_i$ and $\bigvee_{i \in X} E_i$ are expressions.
- Nothing else is an expression.

---

[1] For a critical discussion of reasoning of this type, *cf.* [Kr$_0$03, p. 331-332].

If E is an expression and n is a natural number, then $n\colon E$ is a **clause**. We intuitively interpret $n\colon E$ as "p_n states E". We can easily express all of the examples from § 1 as (sets of) clauses in this language. For instance, the liar is just the clause $0\colon \neg p_0$ ("the 0th proposition states the negation of the 0th proposition"). The truthteller is $0\colon p_0$, the nested liars are $\{0\colon \neg p_1, 1\colon \neg p_0\}$, the nested mix is $\{0\colon \neg p_1, 1\colon p_0\}$, and the hemi-tautology is $\{0\colon \neg p_0 \vee \neg p_1, 1\colon \neg p_0 \wedge \neg p_1\}$.

We now assign a semantics to our language $\mathcal{L}$. We say that an **interpretation** is a function $I\colon \mathbb{N} \to \{0, 1\}$ assigning truth values to propositional letters. Obviously, an interpretation extends naturally to all expressions in $\mathcal{L}$. Now, if $n\colon E$ is a clause and I is an interpretation, we say that I **respects** $n\colon E$ if $I(n) = I(E)$. We say that I respects a set of clauses if it respects all of its elements. Finally, we call a set of clauses **paradoxical** if there is no interpretation that respects it.

Proposition 1. *The liar $0\colon \neg p_0$, and the nested mix $\{0\colon \neg p_1, 1\colon p_0\}$ are paradoxical, the truthteller $0\colon p_0$, the nested liars $\{0\colon \neg p_1, 1\colon \neg p_0\}$ and the hemi-tautology $\{0\colon \neg p_0 \vee \neg p_1, 1\colon \neg p_0 \wedge \neg p_1\}$ are non-paradoxical.*

Proof. There are four relevant interpretations for the mentioned sets of clauses:

$$I_{00} \quad 0 \mapsto 0; 1 \mapsto 0$$
$$I_{01} \quad 0 \mapsto 0; 1 \mapsto 1$$
$$I_{10} \quad 0 \mapsto 1; 1 \mapsto 0$$
$$I_{11} \quad 0 \mapsto 1; 1 \mapsto 1$$

It is easy to check that none of these respects the liar and the nested mix. All four interpretations respect the truthteller, and the interpretations I_{01} and I_{10} respect the nested liars. In the case of the hemi-tautology, the only respecting interpretation is I_{10}.   q.e.d.

So, if the truthteller and the nested liars are non-paradoxical, does that mean that they are not problematic? Well, both I_{01} and I_{10} are interpretations of the nested liars, but the interpretations disagree about the truth values of both p_0 and p_1 and therefore do not allow any determination of truth. The situation is quite different for the hemi-tautology where there is exactly one respecting interpretation. We call a set of clauses Σ **determined** if there is a unique interpretation respecting Σ. With this notation, the truthteller and the nested liars are non-paradoxical but also non-determined, and the hemi-tautology is determined.

In [Boo02, §§ 5&6], Bolander investigates self-referentiality and paradoxicality in order to highlight that these two notions are related but there can be self-reference without paradox and paradox without self-reference. The framework of pointer semantics described so far is perfectly fit to making these claims precise.

Let Σ be a set of clauses. Then we can define the **dependency graph** of Σ by letting $\{n\,;\, p_n$ occurs in some clause in $\Sigma\}$ be the set of vertices and defining edges by

$$nEm \text{ if and only if } p_m \text{ occurs in } X \text{ for some } n\colon X \in \Sigma.$$

With this definition, we get the following dependency graphs for our five examples as depicted in Figure 1.

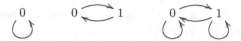

Fig. 1. Dependency graphs of our five examples from § 1: the first graph is the dependency graph for the liar and the truthteller, the second is the one for the two nested examples, and the third is the one for the hemi-tautology

We now call a set of clauses Σ **self-referential** if there is a loop in the dependency graph of Σ. With this definition, it is obvious that self-reference does not imply paradoxicality; the clause $0\colon p_0 \vee \neg p_0$ shares the simple loop as a dependency graph with the liar and the truthteller, but the interpretation $I(0) = 1$ respects it. Yablo [Ya93] gave an example for the converse of this simple fact:

Proposition 2 (Yablo). *Let* $E_n := \bigwedge_{i>n} \neg p_i$ *and* $\Upsilon := \{n\colon E_n \; ; \; n \in \mathbb{N}\}$. *Then* Υ *is not self-referential, but paradoxical.*

Proof. The dependency graph of Υ is $\langle \mathbb{N}, < \rangle$, so it does not contain any loops.

Let I be an interpretation respecting Υ. If for any $n \in \mathbb{N}$, we have $I(n) = 1$, then $1 = I(n) = I(\bigwedge_{i>n} \neg p_i)$, so we must have that $I(i) = 0$ for all $i > n$. That means that $0 = I(n+1) = I(\bigwedge_{i>n+1} \neg p_i)$, whence there must be some $i^* > n+1$ such that $I(i^*) = 1$. But this is a contradiction.

So, $I(n) = 0$ for all n. But then $I(E_0) = I(\bigwedge_{n>0} \neg p_n) = 1 \neq 0 = I(0)$. Contradiction.   q.e.d.

3   Revision

So far, our analysis did not involve revision at all – everything was solely based on the static picture given by the set of clauses. Revision theory now adds a rather natural idea of revision along the pointers established by the clauses. From now on, we shall assume that all sets of clauses Σ satisfy a simple **consistency condition**: If $n\colon E \in \Sigma$ and $n\colon F \in \Sigma$, then $E = F$. If Σ is a set of clauses, then we can define the **revision operator** on interpretations I by

$$\delta_\Sigma(I)(n) := I(E)$$

where E is the unique expression such that $n\colon E \in \Sigma$. This can now be used to recursively define a **revision sequence** of interpretations from an initial interpretation I (called "hypothesis" in revision theory) as

$$I^{\Sigma,0} := I$$
$$I^{\Sigma,n+1} := \delta_\Sigma(I^{\Sigma,n}).$$

We call an interpretation J Σ-**recurring** if there is some I such that there are infinitely many n with $J = I^{\Sigma,n}$ and we call it Σ-**stable** if there is some I and some n such that for all $k > n$, we have $J = I^{\Sigma,k}$.

Proposition 3. *Let Σ be a set of clauses and I an interpretation. Then I respects Σ if and only if I is Σ-stable.*

Proof. Obviously, "I respects Σ" is equivalent to $\delta_\Sigma(I) = I$.   q.e.d.

Let us check our examples from §1. For the liar and the truthteller, relevant interpretations are just one bit ($I(0) = 0$ and $I(0) = 1$). For the liar, both interpretations are recurring, but none of them is stable. For the truthteller, both are recurring and stable. For the two nested examples, we have four relevant interpretations whose revision sequences are as follows:

nested mix:	0: $\neg p_1$	0 1 1 0 0 $\cdots$	0 0 1 1 0 $\cdots$
	1: p_0	0 0 1 1 0 $\cdots$	1 0 0 1 1 $\cdots$
	0: $\neg p_1$	1 1 0 0 1 $\cdots$	1 0 0 1 1 $\cdots$
	1: p_0	0 1 1 0 0 $\cdots$	1 1 0 0 1 $\cdots$
nested liars:	0: $\neg p_1$	0 1 0 1 0 $\cdots$	0 0 0 0 0 $\cdots$
	1: $\neg p_0$	0 1 0 1 0 $\cdots$	1 1 1 1 1 $\cdots$
	0: $\neg p_1$	1 1 1 1 1 $\cdots$	1 0 1 0 1 $\cdots$
	1: $\neg p_0$	0 0 0 0 0 $\cdots$	1 0 1 0 1 $\cdots$

For the nested mix, all four interpretations are recurring, but none of them is stable; for the nested liars, all of them are recurring, but only 01 and 10 are stable.

Analysing the revision sequences for the hemi-tautology gives us a unique stable interpretation 10 and two more recurring interpretations 00 and 11 as described in Figure 2.

hemi-tautology:	0: $\neg p_0 \vee \neg p_1$	0 1 0 1 $\cdots$	0 1 1 1 $\cdots$
	1: $\neg p_0 \wedge \neg p_1$	0 1 0 1 $\cdots$	1 0 0 0 $\cdots$
	0: $\neg p_0 \vee \neg p_1$	1 1 1 1 $\cdots$	1 0 1 0 $\cdots$
	1: $\neg p_0 \wedge \neg p_1$	0 0 0 0 $\cdots$	1 0 1 0 $\cdots$

Fig. 2. The revision sequences for the hemi-tautology

All of this conforms with the analysis of §2, but does not add any new insights. However, the revision approach can add new insights in the case that there is no unique stable solution. For this, let us consider the following example that we shall call **nested liars with two observers**:

The second sentence is false,
the first sentence is false,
exactly one of the first two sentences is true,
exactly one of the first three sentences is true.

Intuition tells us that exactly one of the first two sentences should be true, and therefore the third sentence should be true and the fourth sentence should be false. (Again, we point the reader to Kremer's debate [Kr03, p. 331-332] concerning the dangers of applying ordinary reasoning to sets of sentences with self-reference.) The natural language sentences can be translated into a set of clauses as follows:

$$0: \neg p_1$$
$$1: \neg p_0$$
$$2: (p_0 \vee p_1) \wedge (\neg p_0 \vee \neg p_1)$$
$$3: \bigvee_{i \in 3} p_i \wedge \neg \bigvee_{\substack{i \neq j \\ i,j \in 3}} (p_i \wedge p_j)$$

They give rise to the revision sequences depicted in Figure 3, establishing 0110 and 1010 as the two stable interpretations, and 1100 and 0000 as recurring, yet unstable.

```
0│0 1 0 1 ··· │0 1 0 1 ··· │0 1 0 1 ··· │0 1 0 1 ···
1│0 1 0 1 ··· │0 1 0 1 ··· │0 1 0 1 ··· │0 1 0 1 ···
2│0 0 0 0 ··· │0 0 0 0 ··· │1 0 0 0 ··· │1 0 0 0 ···
3│0 0 0 0 ··· │1 0 0 0 ··· │0 1 0 0 ··· │1 1 0 0 ···
0│0 0 0 0 ··· │0 0 0 0 ··· │0 0 0 0 ··· │0 0 0 0 ···
1│1 1 1 1 ··· │1 1 1 1 ··· │1 1 1 1 ··· │1 1 1 1 ···
2│0 1 1 1 ··· │0 1 1 1 ··· │1 1 1 1 ··· │1 1 1 1 ···
3│0 1 0 0 ··· │1 1 0 0 ··· │0 0 0 0 ··· │1 0 0 0 ···
0│1 1 1 1 ··· │1 1 1 1 ··· │1 1 1 1 ··· │1 1 1 1 ···
1│0 0 0 0 ··· │0 0 0 0 ··· │0 0 0 0 ··· │0 0 0 0 ···
2│0 1 1 1 ··· │0 1 1 1 ··· │1 1 1 1 ··· │1 1 1 1 ···
3│0 1 0 0 ··· │1 1 0 0 ··· │0 0 0 0 ··· │1 0 0 0 ···
0│1 0 1 0 ··· │1 0 1 0 ··· │1 0 1 0 ··· │1 0 1 0 ···
1│1 0 1 0 ··· │1 0 1 0 ··· │1 0 1 0 ··· │1 0 1 0 ···
2│0 0 0 0 ··· │0 0 0 0 ··· │1 0 0 0 ··· │1 0 0 0 ···
3│0 0 0 0 ··· │1 0 0 0 ··· │0 0 0 0 ··· │1 0 0 0 ···
```

Fig. 3. The revision patters of nested liars with two observers

While the four recurring interpretations disagree about the truth values of p_0, p_1, and p_2, all of them agree that p_3 should receive value 0. Therefore, even in the absence of a unique solution, we can get information out of the revision procedure and define a partial truth predicate.

If Σ is a set of clauses and $n: X \in \Sigma$, then we say that p_n is **stably true** (**recurringly true**) if for every stable (recurring) interpretation I, we have $I(n) = 1$. Similarly, we define notions of being **stably false** and **recurringly false**. The difference between the stable partial truth predicate and the recurring partial truth predicate is roughly the difference between the Gupta-Belnap systems S_0 and $S_n{}^2$. Gupta and Belnap argue [GuBe93, Example 5A.17] that S_0 is not good enough to capture intuitions. The systems S^* and $S^{\#}$ proposed by Gupta and Belnap [GuBe93, p. 182 & 191] are refinements of these systems. The differences hardly matter for simple examples of the type that we are covering in this paper.

Proposition 4. *In the nested liars with two observers, the fourth sentence is recurringly false.*

Proposition 4 sounds like a success for the revision theoretic analysis of the concept of truth, as it gives a prediction or analysis for a truth value that coincides with the intuition. However, it is important to note that our reasoning used to intuitively determine the truth value of the fourth sentence used the fact that the third sentence seemed to be intuitively true. But the revision analysis is less informative about the third sentence: it is neither recurringly true nor recurringly false, but stably true. This phenomenon (with a different example) was the topic of the discussion between Cook and Kremer in the journal *Analysis* [Co02,Kr$_0$03,Co03] and will be discussed in detail in § 4.

4   Fully Revised Sequences and the Cook-Kremer Debate

In a dispute in the journal *Analysis* [Co02,Kr$_0$03,Co03], Roy Cook and Michael Kremer debated whether the revision-theoretic analysis of self-referential sentences yields intuitive or counterintuitive readings. Both Cook and Kremer focussed on what we called "recurring truth" in the last section.

The hemi-tautology from § 1 is a special case of the following set of clauses. Denote by $\binom{n}{k}$ the set of k-element subsets of $n = \{0, ..., n-1\}^3$. For every positive natural number n, the set Σ_n has the n clauses

$$k: \quad \bigvee_{X \in \binom{k+1}{n}} \bigwedge_{i \in X} \neg p_i$$

(for $k < n$), *i.e.*, "there are at least $k+1$ many false sentences". If n is odd, Σ_n is paradoxical, if n is even, then it has a unique respecting interpretation, *viz.* the one in which sentences $0, ..., \frac{n}{2}$ are true and the rest false. The original example in [Co02] is Σ_4, the hemi-tautology is the example used in [Kr$_0$03] and is Σ_2 in the above notation. Analysing the revision sequences in Figure 2, we get:

Proposition 5. *In the hemi-tautology, neither of the sentences receives a recurring truth value.*

Proof. The recurring interpretations are 10, 00 and 11, and so they agree on neither of the truth values.   q.e.d.

---

[2] *Cf.* [GuBe93, p. 123 & 147].

[3] The usual notation $\binom{n}{k}$ from finite combinatorics denotes the number of elements of the set that we call $\binom{n}{k}$. Of course, in most cases the set is not equal to its number of elements, but there is no risk of confusion in this paper.

Cook [Co02] contrasts the partial truth predicate of recurring truth as calculated Proposition 4 with our intuitive expectations of a favoured interpretation 10 for the hemi-tautology, and considers this a failure of the revision theoretic analysis.

It is surprising that neither Cook nor Kremer mention that this phenomenon has been observed by Gupta and Belnap. They discuss this in a slightly less transparent example [GuBe93, Example 6C.10]:

> The third sentence is true,
> It is true that the third sentence is false,
> One of the first two sentences is false,

formalized as

$$\{0\colon p_3, 1\colon \neg p_3, 2\colon p_1, 3\colon \neg p_0 \lor \neg p_2\},$$

where intuition suggests that 1001 should be the only solution. Analysing the revision sequences, we find that 1001 is the only stable interpretation, but 0101, 1011, and 1000 are recurring, and thus none of the four truth values is determined in the Gupta-Belnap revision semantics defined *via* recurring interpretations.

Gupta and Belnap deal with this situation with their notion of "fully varied" revision sequences. We extend the sequences from sequences indexed with natural numbers to transfinite sequences indexed with ordinal numbers[4]. Given a limit ordinal λ, we say that a revision sequence $s = \langle I_\xi \, ; \, \xi < \lambda \rangle$ **coheres** with an interpretation I if the following two conditions are met:

1. If for some $\xi < \lambda$ and all $\eta > \xi$, we have $s_\eta(n) = 1$, then $I(n) = 1$.
2. If for some $\xi < \lambda$ and all $\eta > \xi$, we have $s_\eta(n) = 0$, then $I(n) = 0$.

So, going back to the case of $\lambda = \omega$, if the value of n has stabilized after a finite number of revisions, then an interpretation must agree with this value in order to cohere. For those n that flip back and forth infinitely many times, the value of $I(n)$ can be both 0 or 1. Looking at the hemi-tautology as an example, we get four revision sequences as in Figure 2:

$$0\ 1\ 0\ 1 \cdots$$
$$0\ 1\ 0\ 1 \cdots$$

$$0\ 1\ 1\ 1 \cdots$$
$$1\ 0\ 0\ 0 \cdots$$

$$1\ 1\ 1\ 1 \cdots$$
$$0\ 0\ 0\ 0 \cdots$$

$$1\ 0\ 1\ 0 \cdots$$
$$1\ 0\ 1\ 0 \cdots$$

---

[4] The "forever" in the title of this paper is an allusion to this extension of the process of revision into the transfinite.

The ones starting with 01 and 10 stabilize on 10, and so only 10 is a coherent interpretation for them. The other two flip back and forth infinitely many times in both slots, and so every interpretation is coherent with those.

Using the notion of coherence, we can now define the notion of a transfinite revision sequence. If Σ is a set of clauses and δ_Σ is the revision operator derived from Σ in the sense of §3, then a sequence $s = \langle I_\xi ; \xi < \lambda \rangle$ of interpretations is called a **transfinite revision sequence** if $I_{\xi+1} = \delta_\Sigma(I_\xi)$ and I_ϱ coheres with $s{\restriction}\varrho$ for limit ordinals ϱ. Note that for a fixed interpretation I_0 there can be different transfinite revision sequences starting with I_0.

Gupta and Belnap call a transfinite revision sequence **fully varied** if every interpretation coherent with it occurs in it [GuBe93, p. 168]. For the hemi-tautology, the sequences starting with 01 and 10 are fully varied; the only coherent interpretation is 10 and it occurs in them. The other two sequences are not fully varied, as 01 and 10 cohere with them, but do not occur. However, we can transfinitely extend them to the four revision sequences

$$
\begin{array}{ll}
0\,1\,0\,1\cdots & 0\,1\,1\,1\cdots \\
0\,1\,0\,1\cdots & 1\,0\,0\,0\cdots \\[4pt]
0\,1\,0\,1\cdots & 1\,1\,1\,1\cdots \\
0\,1\,0\,1\cdots & 0\,0\,0\,0\cdots \\[4pt]
1\,0\,1\,0\cdots & 0\,1\,1\,1\cdots \\
1\,0\,1\,0\cdots & 1\,0\,0\,0\cdots \\[4pt]
1\,0\,1\,0\cdots & 1\,1\,1\,1\cdots \\
1\,0\,1\,0\cdots & 0\,0\,0\,0\cdots\,,
\end{array}
$$

characterized by their values at 0 and the ordinal ω as 00/01, 00/10, 11/01, and 11/10. All of these sequences (of length $\omega \cdot 2$) are fully varied, and together with the sequences starting with 01 and 10, they are essentially the only fully varied sequences.

We can now define a new notion of recurrence. Given a transfinite revision sequence s of length λ for a set of clauses Σ, we say that I is recurring in s if for all $\xi < \lambda$ there is some $\eta > \xi$ such that $s_\eta = I$. Based on this notion, we say that p_n is **transfinitely true** (**transfinitely false**) if for all fully varied transfinite revision sequences s and all interpretations I that are recurring in s, we have $I(n) = 1$ ($I(n) = 0$).

Proposition 6. *The first sentence of the hemi-tautology is transfinitely true, the second is transfinitely false.*

This alternative analysis arrives at the intuitive expectations by enforcing additional constraints on the notion of a revision sequence. Cook implicitly acknowledges this possible defense of the revision analysis when he says

> "The Revision Theorist might ... formulat[e] more complex revision rules than the straightforward one considered here, ones that judged the sentences [of the hemi-tautology] as non-pathological. [Co03, p. 257]"

The fact that there are so many different systems of revision theory, all with slightly different requirements on the sequences or variations of the semantic predicate, each of them with some other set of advantages and disadvantages, is raising a concern: we are trying to model a phenomenon as central as truth; if revision theory is a fundamental tool to understanding it, shouldn't it provide answers that do not depend on such minor details?

One possible way out of trouble would be to get rid of the idea that a theory of truth needs to define a partial truth predicate. Revision theory gives a rich analysis of what happens, yielding patterns of behaviour of truth values. Instead of superposing these patterns into a single (partial) interpretation as is done by the notions of "stable truth", "recurring truth" and "transfinite truth", we could understand the revision analysis as the description of what is going on:

The liar is problematic as there are no stable interpretations, the truthteller is because there are two conflicting ones. This difference explains how they are different types of problems for the theorist of truth – collapsing it into a uniform partial truth function (which would give the value "undefined" to both the liar and the truthteller) clouds a rather clear conceptual picture. We propose to think of the sequences and their behaviour as the real analysis of truth without the definition of a partial truth predicate; the fact that 10 is the only stable interpretation for the hemi-tautology is good enough to explain our intuitions with the set of sentences[5].

It is this approach to revision sequences that we believe to be a powerful tool for explaining intuitions with truth, much more than the different axiomatic systems proposed by various authors in order to deal with inadequacies of earlier definitions. We shall continue this discussion in § 6.2.

5   An Aside: "And What Is the Connection to Belief Revision?"

In the community of applied and philosophical logic, the word "revision" is much closer associated to the area of *belief revision* and *belief update* than to the revision theory described in § 3. In 2002, I gave a talk on the complexity of revision-theoretic definability at the annual meeting of the Pacific Division of the *American Philosophical Association* with the title "Where does the complexity of revision come from?"[6], and received questions from philosophical logicians asking about the complexity of *belief revision* in the style of [Li97,Li00].

Is the use of the phrase "revision" in both areas just an equivocation? Do the two underlying concepts of revision ("update of belief states in light of changing reality" and "update of truth value in a formal system") have nothing to do with each other?

---

[5] Note that by Proposition 3, this is equivalent to saying that 10 is the only interpretation that respects the hemi-tautology, so here the pointer semantics approach and the revision approach are just two different ways of looking at the same phenomenon.

[6] The results presented in this talk have in the meantime been published as [KüLöMöWe05].

In this section, we shall give a rough sketch of why revising belief states may be incorporated into the framework described in §§ 2 and 3. Since this is a side issue here, we cannot do justice to these questions here.

In belief revision and update, we have an ordinary propositional language and consider sets of formulae as **belief sets**. Based on new information about the true state of the world, we may get inconsistent intermediate stages of belief sets which we then have to update in order to reach a consistent belief set again. This is the main paradigm of an enormous amount of literature in philosophy, logic and artificial intelligence[7].

The most basic example is the following: an agent believes that p and $p \to q$ are true, but then learns that $\neg q$ is true. The belief set has to be updated to either $\{p, \neg q, \neg(p \to q)\}$ or $\{\neg p, \neg q, p \to q\}$. Of course, which one is the correct update will depend on the context.

We believe that revision theory as described in § 3 can provide a partial semantics for belief update procedures in general, but will only develop this idea for the simple examples given above here. Given a belief set Λ and some new fact represented by a propositional variable, we can assign a set of clauses in our language $\mathcal{L}$ as follows:

Let Λ^* be the set of propositional variables occurring in a formula in Λ and let $\pi : \Lambda^* \to \mathbb{N}$ be an injective function with coinfinite range. We can think of π as associating an $\mathcal{L}$-variable p_n to each element of Λ^*. Clearly, π naturally extends to all elements of Λ.

In a second step, we take an injective function $\pi^* : \Lambda \to \mathbb{N}$ such that $\mathrm{ran}(\pi) \cap \mathrm{ran}(\pi^*) = \varnothing$. If $n \in \mathrm{ran}(\pi) \cup \mathrm{ran}(\pi^*)$, we define a clause $n \colon E$ where

$$E := \begin{cases} p_n, & \text{if } n \in \mathrm{ran}(\pi), \\ \pi(\varphi), & \text{if } \varphi \in \Lambda \text{ and } \pi^*(\varphi) = n. \end{cases}$$

This defines the set Σ of $\mathcal{L}$-clauses associated to Λ.

In our given example, this would be

$$\{0 \colon p_0, 1 \colon p_1, 2 \colon p_0 \to p_1\}.$$

The dependency graph of our set of clauses is

$$0 \qquad\qquad 1.$$

The key difference between the setting of revision theory and that of belief update is that the new fact that triggers the update is given a special status: if the initial belief set is $\{p, p \to q\}$ and we learn $\neg q$ as a fact, then we do not want to disbelieve this fact in order to remedy the situation[8].

---

[7] As a token reference, we mention [Gä92], in particular the introduction.

[8] This is the traditional approach to belief revision. The AGM success postulate has been weakened in *non-prioritized belief revision*, as in [Ga₁92,Bo₁FrHa98,Ha₁99,Ha₁FeCaFa01].

We fix some $n \in \text{ran}(\pi)$ and some truth value $b \in \{0,1\}$ for this n, assuming that the new fact that we learned corresponds to p_n or $\neg p_n$. An $\langle n, b \rangle$-**interpretation** is a function $I : \mathbb{N} \to \{0,1\}$ that satisfies $I(n) = b$.

0: p_0	0 0 0 $\cdots$	0 0 0 $\cdots$
1: p_1	0 0 0 $\cdots$	0 0 0 $\cdots$
2: $p_0 \to p_1$	0 1 1 $\cdots$	1 1 1 $\cdots$
0: p_0	1 1 1 $\cdots$	1 1 1 $\cdots$
1: p_1	0 0 0 $\cdots$	0 0 0 $\cdots$
2: $p_0 \to p_1$	0 0 0 $\cdots$	1 0 0 $\cdots$

We see that 001 and 100 are the only stable interpretations. Taking our remarks at the end of § 4 seriously, we shall not use this to define a partial truth function (which would say that p_1 is recurringly false and the others have no assigned truth value), but instead look at the two stable interpretations and see that

$$\{p_0, \neg p_1, \neg(p_0 \to p_1)\} \text{ and } \{\neg p_0, \neg p_1, p_0 \to p_1\}$$

are the two possible outcomes for the belief set after belief update.

6   The Ubiquity of Revision

In the abstract, we mentioned that revision is a concept that is "universal and resurfaces in many areas of logic and applications of logic". It comes in two very different flavours as discussed at the end of § 4: as formal Gupta-Belnap semantics defining a partial truth predicate on the basis of revision sequences, and in the wider sense as a conceptual framework for analysing our intuitions about truth and circularity. So far, we have argued that revision plays a rôle in the analysis of paradoxes and *insolubilia*, and that the approach may be applied to belief revision. In this section, we shall lay out how the general ideas can be extended to yield applications in other areas. We split the discussion into applications of the Gupta-Belnap semantics and applications of the wider scope.

6.1   Independent Developments of Gupta-Belnap Semantics

The crucial mathematical element to the Gupta-Belnap truth predicate as defined in § 3 (as "recurring truth") is the following: we have a set of nonmonotonic processes assigning a function $I : \mathbb{N} \to \{0,1\}$ to each ordinal. While monotonic processes give rise to fixed points and thus allow us to talk about an "eventual value", nonmonotonicity forces us to be inventive here. The processes give rise to a notion of *recurrence*, and we can define

$$T_{\text{GB}}(n) := \begin{cases} 0 \text{ if for all recurrent } I, \text{ we have } I(n) = 0, \\ 1 \text{ if for all recurrent } I, \text{ we have } I(n) = 1, \\ \uparrow \text{ otherwise.} \end{cases}$$

This is a general idea to integrate the process of revision into a single definition, and Gupta and Belnap are not the only ones who came up with this idea. Essentially the same semantics was developed independently by Stephan Kreutzer in [Kr$_1$02] for his partial fixed point logics on infinite structures. Also Field's *revenge-immune solution* to the paradoxes from [Fi03] is based on ideas very similar to the Gupta-Belnap semantics[9].

Widening the scope to other types of transfinite processes, cognate ideas can be found in the limit behaviour of infinite time Turing machines as defined by Hamkins and Kidder [Ha$_0$Le00][10] and definitions of game labellings for nonmonotone procedures for game analyses in [Lö03].

6.2 The Wider Scope

Despite the fact that the general ideas have found applications in many places, there are several problems with Gupta-Belnap semantics as a theory of truth. As mentioned, there are many variants of formal systems with different properties, thus raising the question of how to choose between them. The Cook-Kremer debate discussed in § 4 is an indication for the problems generated by this. The revision-theoretic definitions are also relatively complicated, leading (in the language of arithmetic) to complete Π^1_2 sets, in the case of using fully revised sequences even Π^1_3 sets [We03a, Theorem 3.4]. This is too complicated for comfort, as is argued in [We01, p. 351] and [LöWe01, § 6].

As we have discussed in § 4, the conceptual idea of analysing the nonmonotonic process by looking at the behaviour of interpretations under revision rises above all this criticism. The problems associated with the arbitrariness and complexity of the Gupta-Belnap are related to the fact that the full analysis has to be condensed into one partial truth predicate. Allowing both 01 and 10 as stable solutions of the nested liars is much more informative than integrating these two solutions into undefined values.

This attitude towards allowing several possibilities as analyses should remind the reader of game-theoretic solution concepts. In game theory, Nash equilibria are not always unique. This connection between revision semantics and game theory has been observed by Chapuis who gives a sketch of a general theory of rationality in games based on revision analyses in his [Ch03]. We see Chapuis' work as an interesting approach compatible with the spirit of the analysis of belief update discussed in § 5, and would like to see more similar approaches to revision in various fields of formal modelling.

References

Bo$_0$02.   Thomas **Bolander**, Self-Reference and Logic, **Phi News** 1 (2002), p. 9-44

Bo$_0$03.   Thomas **Bolander**, Logical Theories for Agent Introspection, PhD thesis, Technical University of Denmark 2003

---

[9] Welch has proved in [We03b] that the set of *ultimately true* sentences in the sense of [Fi03] coincides with the set of stable truths in the sense of Herzberger.

[10] This similarity was pointed out by the present author in [Lö01] and used by Welch in [We03a] to solve the limit rule problem of revision theory. *Cf.* also [LöWe01].

Bo₁FrHa98.   Craig **Boutilier**, Nir **Friedman**, Joseph Y. **Halpern**, Belief revision with unreliable observations, *in:* Proceedings of the Fifteenth National Conference on Artificial Intelligence (AAAI-98), July 26-30, 1998, Madison, Wisconsin, Menlo Park 1998, p. 127-134

Ch03.   André **Chapuis**, An application of circular definitions: Rational Decision, *in:* Benedikt Löwe, Wolfgang Malzkorn, Thoralf Räsch (*eds.*), Foundations of the Formal Sciences II: Applications of Mathematical Logic in Philosophy and Linguistics, Rheinische Friedrich-Wilhelms-Universität Bonn, November 10-13, 2000, Dordrecht 2003 [Trends in Logic 17], p. 47-54

Co02.   Roy T. **Cook**, Counterintuitive consequences of the revision theory of truth, **Analysis** 62 (2002), p. 16-22

Co03.   Roy T. **Cook**, Still counterintuitive: a reply to M. Kremer, "Intuitive consequences of the revision theory of truth", **Analysis** 63 (2003), p. 257-261

Fi03.   Hartry **Field**, A revenge-immune solution to the semantic paradoxes, **Journal of Philosophical Logic** 32 (2003), p. 139-177

Ga₀88.   Haim **Gaifman**, Operational Pointer Semantics: Solution to Self-referential Puzzles I, *in:* Moshe Vardi (*ed.*), Proceedings of the 2nd Conference on Theoretical Aspects of Reasoning about Knowledge, Pacific Grove, CA, March 1988, Morgan Kaufmann, San Francisco 1988, p. 43–59

Ga₀92.   Haim **Gaifman**, Pointers to Truth, **Journal of Philosophy** 89 (1992), p. 223–261

Ga₁92.   Julia Rose **Galliers**, Autonomous belief revision and communication, *in:* [Gä92, p. 220-246]

Gä92.   Peter **Gärdenfors** (*ed.*), Belief revision, Cambridge University Press 1992 [Cambridge Tracts in Theoretical Computer Science 29]

GuBe93.   Anil **Gupta**, Nuel **Belnap**, The Revision Theory of Truth, Cambridge MA 1993

Ha₀Le00.   Joel David **Hamkins**, Andy **Lewis**, Infinite time Turing machines, **Journal of Symbolic Logic** 65 (2000), p. 567-604

Ha₁99.   Sven Ove **Hansson**, A survey on non-prioritized belief revision, **Erkenntnis** 50 (1999), p. 413-427

Ha₁FeCaFa01.   Sven Ove **Hansson**, Eduardo Leopoldo **Fermé**, John **Cantwell**, Marcelo Alejandro **Falappa**, Credibility limited revision, **Journal of Symbolic Logic** 66 (2001), p. 1581-1596

He82a.   Hans G. **Herzberger**, Naive Semantics and the Liar Paradox, **Journal of Philosophy** 79 (1982), p. 479–497

He82b.   Hans G. **Herzberger**, Notes on Naive Semantics, **Journal of Philosophical Logic** 11 (1982), p. 61–102

Kr₀03.   Michael **Kremer**, Intuitive consequences of the revision theory of truth, **Analysis** 62 (2002), p. 330-336

Kr₁02.   Stephan **Kreutzer**, Partial Fixed-Point Logic on Infinite Structures, *in:* Julian C. Bradfield (*ed.*), Computer Science Logic, 16th International Workshop, CSL 2002, 11th Annual Conference of the EACSL, Edinburgh, Scotland, UK, September 22-25, 2002, Proceedings, Berlin 2002 [Lecture Notes in Computer Science 2471], p. 337-351

KüLöMöWe05.   Kai-Uwe **Kühnberger**, Benedikt **Löwe**, Michael **Möllerfeld**, Philip **Welch**, Comparing inductive and circular definitions: parameters, complexities and games, **Studia Logica** 81 (2005), p. 79-98

Li97.   Paolo **Liberatore**, The complexity of iterated belief revision, *in:* Foto
             Afrati, Phokion Kolaitis (*eds.*), Database theory—ICDT '97, Proceed-
             ings of the 6th International Conference held in Delphi, January 8-10,
             1997, Springer-Verlag 1997 [Lecture Notes in Computer Science 1186],
             p. 276-290

Li00.   Paolo **Liberatore**, The complexity of belief update, **Artificial In-
             telligence** 119 (2000), p. 141-190

Lö01.   Benedikt **Löwe**, Revision sequences and computers with an infinite
             amount of time, **Journal of Logic and Computation** 11 (2001),
             p. 25-40; *also in:* Heinrich Wansing (*ed.*), Essays on Non-Classical Logic, Singa-
             pore 2001 [Advances in Logic 1], p. 37-59

Lö03.   Benedikt **Löwe**, Determinacy for infinite games with more than two
             players with preferences, **ILLC Publication Series** PP-2003-19

LöWe01.   Benedikt **Löwe**, Philip D. **Welch**, Set-Theoretic Absoluteness and
             the Revision Theory of Truth, **Studia Logica** 68 (2001), p. 21-41

We01.   Philip D. **Welch**, On Gupta-Belnap Revision Theories of Truth, Krip-
             kean fixed points, and the Next Stable Set, **Bulletin of Symbolic
             Logic** 7 (2001), p. 345-360

We03a.   Philip D. **Welch**, On Revision Operators, **Journal of Symbolic
             Logic** 68 (2003), p. 689–711

We03b.   Philip D. **Welch**, Ultimate Truth *vis à vis* stable truth, *preprint*,
             November 7, 2003

Ya93.   Stephen **Yablo**, Paradox without self-reference, **Analysis** 53 (1993),
             p. 251-252

Ontological Constitutions
for
Classes and Properties

Jørgen Fischer Nilsson

Informatics and Mathematical Modelling
Technical University of Denmark
jfn@imm.dtu.dk

Abstract. Formal ontologies model classes and their properties and relationships. This paper considers various choices for modelling of classes
and properties, and the interrelationship of these within a formal logical
framework. Unlike predicate logical usage with quantification over individuals only, in the applied metalogic classes and properties appear as
first class non-extensional objects. Using this framework miscellaneous
classification structures are examined ranging from mere partial orders
to distributive lattices. Moreover, we seek to capture notions such as intensionality of classes and properties ascribed to individuals and classes
in an coherent ontological framework. In this analytic framework we further present generative ontologies in which novel classes can be produced
systematically by means of given classes and properties.

1 Introduction

Ontology addresses the categorial structure of reality seeking answers to metaphysical questions such as: what is there

– in our entire common actual/imaginable world?
– in a naturalistic (or any other) view of the world?
– in a particular application domain?

How can what-there-is be adequately classified? Are there *a priori* general *classes*,
i.e. *categories*? As stated in the introduction in [20]: "Now to provide a complete
metaphysical theory is to provide a complete catalogue of the categories under
which things fall and to obtain the sort of relations that obtain among those
categories".

Formal ontologies in focus here serve to describe and systematize classes and
properties and their relationships in a formal (logical) language. Thus we have
to distinguish between classes and properties of entities in reality and on the
other hand modelling of classes and properties and their properties in turn, in
the formal logic.

1.1 Ontological Traditions

The philosophical ontological tradition as presented e.g. in [35, 38] has focussed on
ontologies as classifications, that is to say the identification of all encompassing

H. Schärfe, P. Hitzler, and P. Øhrstrøm (Eds.): ICCS 2006, LNAI 4068, pp. 37–53, 2006.

and appropriate universals, categories or classes, and the subsequent organisation of these into sub- and superclasses. The field of computer science dealing with data-, systems, and domain modelling has also addressed the identification of classes of entities stressing the various relationships which can be identified between entities as in the entity-relationship model [12]. Thus less emphasis has traditionally been put on classification as such, though, however, classification is considered central in object-oriented modelling.

The philosophical and the computer science approaches have begun merging into ontological engineering methodologies, which consider classification together with ascription of properties and relationships to entities in the identified classes as in conceptual graphs [34].

Elaboration of major formal ontologies as taking place e.g. in contemporary semantic web activities call for formal logical systems for the purpose of exact specification and computerised reasoning. This in turn appeals to the logical tradition examining properties and other intensional notions in analytical philosophy, as e.g. in [3, 13, 29, 37] to mention just a few disparate contributions.

However, logic tends to focus on mathematical existence (or absence) of objects within concrete symbolic systems or their associated purely abstract mathematical models, rather than the question of formation and existence of classes in the real world. There is moreover a inclination towards concern with ontological oddities and logical sophistry epitomized by the selfapplying properties giving rise to the Russell antinomy. By contrast contemporary ontologists, e.g. [6, 31], emphasises ontology building as being concerned with description of classes existing in reality.

There is yet a viewpoint to mention here, the linguistic one. The language aspect is prevalent in the field of terminology analysis, cf. e.g. [21], since methodologically the domain analysis there proceeds in a bottom-up fashion from the terms of the domain. In terminology analysis ontologies are understood as taxonomies serving to structure the specific terms and nomenclature of an application domain, typically a scientific domain such as medicine.

1.2   Predicate Logic as Metalogic

In this paper we abstract, compare and seek to reconcile ontological essentials by means of a metalogic apparatus. We encode ontological notions such as classes and properties as terms in first order predicate calculus on a par with individuals. This enables us to endow these ontological notions with appropriate (meta)properties of their own via introduction of suitable axioms. In this way methodologically we can tailor a chosen ontological framework to form a constitution prior to the elaboration of a specific domain ontology.

First order predicate logic (in the following just predicate logic) and its sublanguages and derivatives such as description logics and modal logics are commonly adopted tacitly or explicitly as the underlying logical medium for formal ontologies describing concepts and their relationships. This paper discusses metalogic use of predicate logic as "metaontologic" for design of ontological languages. This means that concepts (classes, kinds, properties) are conceived as terms contrasting the more conventional use of predicate logic, where only individuals are given as arguments to predicates. As such the metalogic approach bears resemblance to higher order logic, and indeed may be viewed as a simulation of such.

The metalogical approach favours an intensional, i.e. non-extensional, understanding of encoded predicates. Thereby it overcomes the extensionality of predicate calculus in traditional usage with its concomitant trivialisation of predicate denotations to sets of (tuples of) individuals. As such it competes with the possible worlds approach to intensionality as pursued in modal logics.

This paper endeavours at a clarification of the extensionality/intensionality dichotomy in ontologies in parallel with formalization of the various ontological notions. Therefore discussion of intensionality is conducted in a counterpoint manner interleaved with sections dealing with ontological issues. It is our key contention that the extension/intension distinction has not received the attention it deserves in ontology research. Thus, these notions are hardly mentioned in the volume [36].

Furthermore, the present paper advocates use of the metalogical medium for

- distinguishing and connecting linguistic and conceptual ontologies,
- stipulating appropriate classification structures
- introducing compound concept terms for producing novel subclasses.

The metalogic approach in this paper appeared in preliminary form in [25].

2 Logics for Ontologies

Until recently elaboration of ontologies often proceeded without resorting to any specific formal logic. However, the strive for large scale ontologies as e.g. in contemporary semantic web projects with ensuing problems of consistency maintenance calls for use of formal logic in forms amenable to computation.

2.1 Taxonomies with Inclusion Relation

At the most basic level an ontology consists of an organization of classes of entities into what is typically a hierarchy with the most comprehensive class, possibly the allembracing universal class, at the root. Such a tree structure, often referred to as a taxonomy, is spanned by the binary class inclusion or subsumption relation traditionally denoted *isa*, where *a isa b* expresses that the class *a* is a subclass of class *b*. This means that the individuals in the class *a* are bound to be members of class *b* as well by virtue of their possessing all the qualities possessed jointly by members of *b*. In philosophical parlance one would say that particulars belonging to the universal *a* also belongs to the universal *b*.

2.2 The Case of Individuals and Substances

In ontologies individuals (particulars) tend to be of less interest than the classes – with possible exception of distinguished particulars. These latter may, however, be accorded special treatment as singleton classes.

Sometimes the classes do not comprise individually distinguishable entities, e.g. in the case of chemical substances. Suppose one would state tentatively in an ontology that the class of vitamin comprises individuals vitaminA, vitaminB etc. However, vitaminB is itself in turn to be divided into vitamin B1, vitaminB2 etc. Therefore, preferably, substances are modelled as classes, with individual portions or molecules forming individual physical entities. A similar situation exists for classes of states, such as for instance a named disorder as opposed to the particular (occurrence) of a disease suffered by a particular patient.

2.3   Predicate Calculus as Candidate Ontologic

Returning to the logical formalization issue, predicate calculus – being the *lingua franca* of artificial intelligence – is the most obvious candidate for logical specification of ontologies. Trivially, the inclusion relationship stating that a is a subclass of b might be formalized as

$$\forall x \; a(x) \rightarrow b(x)$$

introducing 1-argument predicates for classes. However, this formalization approach suffers from the shortcoming that is focusses on the individuals belonging to the classes rather than the classes themselves. Thus it runs counter to the ontological focus on classes and class relations. Accordingly the above sentence is often replaced with the atomic factual sentence

$$isa(a, b)$$

where *isa* is a two-place predicate, and classes a and b logically reappear as individual constants. Trivial as it seems, this opens for quantification over classes and not just quantification over individuals as in standard predicate logic. Thus this is the initial step in a methodology where classes and properties become the objects dealt with within the logic in contrast to conventional use of predicate logic.

Most of the specifications in the following fall within the sublanguage of definite clauses as used in logic programming, with atomic formulae as a special case of these. All of the variables are here then distinguished by upper case letters as common in logic programming, they being then universally quantified, by default. We may therefore often drop explicit quantification. In those few cases where we need to go beyond definite clauses we resort to standard notation for predicate logic with explicit quantification. In order to account for negation, then, a few places we assume implicitly appropriate formation of the completion of the clauses for turning implication into double implication.

2.4   Description Logic as Candidate Ontologic

Description logic [2] has become a popular tool for elaboration of formal ontologies. Description logic in its basic form may be conceived as a fragment of predicate calculus supporting unary and binary predicates, only, and designed to meet computational desiderata such as decidability and preferably also tractability.

Moreover the language is shaped in the form of an extended relation algebraic logic, cf. the analysis in [7]. Accordingly it offers an inventory of predicate

formation operators such as conjunction and disjunction on classes conforming with meta-language definitions
$$\lambda x.(a(x) \wedge b(x)) \quad \text{and} \quad \lambda x.(a(x) \vee b(x))$$
and the so-called Peirce product
$$\lambda x.\exists y\ (r(x,y) \wedge b(y))$$
The latter one can be rephrased and understood as a first order predicate formation device avoiding λ-notation through the auxiliary
$$\forall x\ peirce_{rb}(x) \leftrightarrow \exists y\ r(x,y) \wedge b(y)$$
which derives a unary predicate from predicates r and b.

Sentences in the description logic are equations or inequations, for instance
$$a \leq b$$
meaning $\forall x\ a(x) \rightarrow b(x)$, that is, if stated alternatively as an algebraic equation, $a = a \wedge b$. In description logic the standard notation is $\sqsubseteq$ and $\equiv$.

The inequations are handy in particular for the specification of ontologies, cf. the above $isa(a,b)$, which becomes $a \leq b$. Moreover, the Peirce product serves to provide attribution with properties, which is crucial in ascribing properties to classes in ontologies as to be discussed.

The form of the language places it in the algebraic logical tradition dating back to Boole via Tarski (relation-algebraic logic) and Peirce. However, in the perspective of modern algebraic logic it may rather be seen as an extension of lattice algebras or Boolean algebras with relation algebraic operations as explicated in [7].

In spite of the undeniable appeal of description logic, especially from a computational point of view, the logic falls short of handling classes as quantifiable intensional objects, see however [22]. In addition and more fundamentally it takes for granted that the class ordering structure should be Boolean and therefore be a distributive lattice. This is in contrast to the more primitive pre-lattice orderings which can be provided in a metalogic approach and which may better reflect ontological practice.

2.5 Class Intensionality

Classes as used in ontologies are inherently intensional notions. That is to say, classes and properties in general cannot be reduced to the set of instances falling under them, let alone for the fact that one might well introduce two distinct classes in an ontology both having no known individuals and thus being co-extensional. On the other hand predicate logic is extensional in the sense that coextensional predicates are mutually substitutable *salve veritate*. This means that for all p and q if

$$\forall x(p(x) \leftrightarrow q(x))$$

then p and q can be substituted for each other without affecting logical consequences.

Description logic being in its basic form a algebraized fragment of predicate logic is *a fortiori* also extensional, and therefore tends to reducing concepts to their extension sets, cf. e.g. [26, 27]. However, [10] proposes a hybrid description language intended to overcome the extensionality of description logic. A desired non-extensionality is traditionally achieved in philosophical logic by resorting to higher order logic (type theory without extension axioms) or to modal logic with its accompanying notion of possible worlds. The latter notions may further facilitate specification of rigid properties and other essentialist notions as examined for the purpose of ontology building in [16]. We are going to advocate in favour of metalogic as an alternative for achieving intensionality in ontologies.

3   Fundamentals of the Class Inclusion Relation

The *isa* class inclusion relation, also known as subsumption, is commonly held to possess the properties of transitivity and reflexivity. Thus, for instance
$$isa(vitaminB1, vitamin)$$
by virtue of $isa(vitaminB1, vitaminB)$ and $isa(vitaminB, vitamin)$.

Moreover, there is reflexivity as in $isa(vitamin, vitamin)$.

In addition to being a preorder, class inclusion may also be considered to possess the property of antisymmetry, meaning that if $isa(a, b)$ and $isa(b, a)$, then the classes a and b are identified. With these three properties the inclusion relation thus becomes a partial order.

Philosophically inclined ontologists seem to prefer (if not insist) that the inclusion relation forms a hierarchy proper, that is a tree structure, cf. e.g. [32]. This requirement is fulfilled by imposing the condition: if $isa(a, b')$ as well as $isa(a, b'')$ then either $isa(b', b'')$ or $isa(b'', b')$. In other words cross categories are banned. However, this restriction is not endorsed by the object-oriented modelling tradition, where cross-categories are considered routinely, giving rise to the notion of multiple inheritance of class membership.

The inclusion relationship forming the backbone taxonomic structure of ontologies may be supplemented with another fundamental ontological relationship, the parthood relation for establishing partonomies. See e.g. [33] for a formalization proposal.

4   Metalogical Constitution of an Ontology

Adopting metalogic for classes and properties (jointly here called concepts) means that these ontological categories become represented as term encoded 1-ary predicates. Thereby they *prima facie* appear as arguments to predicates on a par with individual terms. Thus concepts can be quantified over. In this way the extensionality of predicate logic is neutralised for concepts.

The key principle in metalogic is to replace the atomic formula $p(t)$ with $\varepsilon(p', t)$ where the predicate ε expresses predication and p' is a novel constant term representing p. This principle is examined in [4] for reconstructing (to the extent possible) higher order type theory within predicate logic. We prefer here the metalogic reification-of-predicates-as-terms point of view in favour of the type-theoretical higher-order handling of predicates as arguments to predicates since the axiom of extensionality is not to be introduced and endorsed.

4.1 Ontological Constitution

The metalogic apparatus encourages a two phase *modus operandi*: The first phase sets up an *ontological constitution* specifying properties of classes and properties etc. We thereby commmit us to a meta-ontology for the ontological categories. This is to be used in the second phase for elaborating an actual ontology for the domain at hand. In the following sections we are discussing the various constitutional choices in the first phase.

The term encoded predicates constitute an object level for the target ontology, whereas the predicates proper form a meta level of universal ontological notions. In admitting quantification over classes and properties we recall Quine's dictum: to be is to be the value of a variable. This set-up proposed in our [25] is much in line with [5, 23]; however the emphasis is here put here on the class/property interplay and intensionality.

In [18] we describe a more comprehensive metalogic framework where not just classes but entire definite clauses with n-ary predicates and hence logic programs are encoded as terms. This framework is proposed for inductive synthesis of logic programs.

At the constitutional level of an ontology one introduces ontological notions like class, being instance of a class, property, class inclusion and class overlap etc. These general ontological notions are to be expressed by predicates $class()$, $inst(,)$, $prop()$ etc.

5 Intensional Versus Extensional Class Inclusion

Let us now reconsider formalization of inclusion *isa* in the adopted metalogic framework. Following sect. 2.3 there is a tentative definition

$$isa(p, q) =_{df} \forall x(p(x) \rightarrow q(x))$$

In our metalogic framework using predicate calculus proper this would become $\forall p \forall q \ isa(p, q) \leftrightarrow \forall x(inst(p, x) \rightarrow inst(q, x))$ for classes p and q, and where $inst(p,x)$ expresses that individual object x is an instance of p.

In our metalogic framework using predicate calculus proper we introduce, however, for *isa* only the weaker

$$\forall p \forall q \ isa(p, q) \rightarrow \forall x(inst(p, x) \rightarrow inst(q, x))$$

which might be referred to as intensional inclusion. This is because classes introduced in this way are not subject to the above-mentioned set trivialization making co-extensional classes collapse.

However, we also admit the former definition as the special so-called extensional subsumption relation

$$extisa(p, q) \leftrightarrow \forall x(inst(p, x) \rightarrow inst(q, x))$$

Accordingly, intensional inclusion implies extensional inclusion, but not *vice versa*: $isa(p, q) \not\vdash extisa(p, q)$. This conforms with the slogan that intensions determine extensions but not *vice versa*.

Using these axioms the intensional (non-extensional) inclusion *isa* cannot be verified by inspection of individuals – in accordance with the principle that individuals in general are less relevant to ontology construction. This is in contrast to the latter extensional inclusion which is verifiable by empirical observations. However, how is such a non-empirical class inclusion going to be established by the ontologist in the first place? This foundational aspect of intension/extension is discussed in connection with introduction of properties in ontologies in sect. 10.1. As next steps we consider first the name/concept distinction and then in sect. 7 various forms of class inclusion orders.

6   Linguistic and Conceptual Ontologies

At the basic methodological level we can distinguish between ontologies leaning towards relations between words versus relations between concepts. The latter kind of ontology is ideally language neutral, and probably one can only achieve such an ontology within "naturalistic" scientific realms. A classical example of this issue is met in the domain of colours where different cultures tend to partition the physical spectrum somewhat differently with respect to colour names.

The considered metalogical set up facilitates specification of the relationhip between a taxonomic word ontology and a proper conceptual ontology. A dualistic reconciliation of concepts vis-à-vis their linguistic manifestations may be provided by introducing a predicate lex, where

$word(W) \leftarrow lex(W, C)$
$class(C) \leftarrow lex(W, C)$

Thus the first argument of lex contains words, the second argument contains names of classes.

For instance in an ontology we may have entries

$lex(vitaminE, vitaminE)$
$lex(tocepherol, vitaminE)$

yielding two synonyms for a class. The relation lex may approach identity relation in practice. There may well be nodes (classes) in the ontology with no lexical counterpart.

Lexical semantic notions may now be formalized, e.g. synonymity between words X and Y

$syn(X, Y) \leftarrow lex(X, Z) \wedge lex(Y, Z) \wedge nonident(X, Y)$

and homonymity:

$hom(X) \leftarrow lex(X, U) \wedge lex(X, V) \wedge distinct(U, V)$

The lex relation thus may serve to rule out confusions in the conceptual ontology proper due to homonyms. The taxonomic relationships hypo/hypernomy are now distinguishable from subsumption via the tentative definition

$hyper(X, Y) \leftarrow lex(X, U) \wedge lex(Y, V) \wedge isa(U, V)$

Lexical translation from word X to word(s) Y between two languages (1) and (2) via a common ontology may be specified via intermediating concepts Z in the ontology with

$translate(X, Y) \leftarrow lex1(X, Z) \wedge lex2(Y, Z)$

The lex coupling may be extended from nouns to noun phrases as explained in [19] along the lines of sect. 11. In the following sections we assume for the sake of simplicity that lex is a one-one relation so that the distinction between conceptual and linguistic ontologies vanishes.

7 Forms of Ontological Classification

The philosophical ontological tradition dating back to Aristotle prefer categories which partition into disjoint i.e. non-overlapping classes. This leads to tree-shaped i.e. hierarchical orderings. The proto-typical hierarchical ordering is the Linnean biological classification into the levels of kingdom, phylum/division, class, order, genus (family), and species. The periodical table of elements exemplifies by contrast a paradigmatic non-hierarchical classification with its array-like organisation.

Let us assume that a classification is specified by means of an immediate subclass predicate, *sub*, as in the following sample ontology fragment:

    sub(vitaminA, vitamin)

    sub(vitaminB, vitamin)

    sub(vitaminB1, vitaminB)

    sub(vitaminC, vitamin)

    sub(vitaminE, vitamin)

Moreover, classes are introduced at the metalogic level with the ground atomic formula *class(vitaminA)* etc.

The following metalogic clauses then then contributes to the definition of the class inclusion relation in an ontological constitution

$$isa(X,Y) \leftarrow sub(X,Z) \land isa(Z,Y)$$
$$isa(X,Y) \leftarrow sub(X,Y)$$
$$isa(X,X)$$

with

$$class(X) \leftarrow isa(X,Y)$$
$$class(Y) \leftarrow isa(X,Y)$$

The inclusion *isa* is thus established as the reflexive and transitive closure of the *sub* relation. These clauses may serve in a logic program (indeed even in a DATALOG program) for browsing the ontology.

There may be introduced a distinguished *null* class without member instances: $\neg \exists x\ inst(null, x)$, and with

    $isa(null, C)$

Now for instance we may introduce *class(ufo)* claiming *sub(ufo,null)*. This class is distinct from though co-extensional with, say, an introduced empty class *unicorn*. Thus *extisa(unicorn, ufo)* and *vice versa*.

8 Hierarchies and Beyond

Although there are often norms for performing classification within a given scientific context, in general the ontologist faces the complication that classification can be done optionally according to different competing criteria. And in addition the partial ordering property of the inclusion relation does not favour hierarchical classifications *per se*, but admits non-hierarchial classifications as well.

As an example consider an extension of the above example with

    sub(vitaminC, antioxidant)

    sub(vitaminE, antioxidant)

If antioxidant is formally considered a class on a par with vitamin by way of a declaration *class(antioxidant)* this ontology is non-hierarchical, albeit still a partial ordering. The classes *vitaminC* and *vitaminE* are cross-categorial, then.

Alternatively and perhaps more intuitively antioxidant may be conceived as a property in case of which the above vitamin ontology remains hierarchical. This, however, calls for an additional ontological category of properties to be discussed further in sect. 9. Formally this may be instituted with *prop(antioxidant)*.

Consider, as another variant now, a sample a classification of substances into two additional classes *class(fatsoluble)* and *class(watersoluble)*. In combining this classification with the above vitamin ontology one obtains what is some times called a multi-hierarchy, that is two or more superimposed hierarchies. (We may assume the two latter classes being disjoint for the sake of the example.) However, instead of assuming "multiple hierarchies" one may prefer to conceive of the partial order as a non-hierarchical organisation, which may optionally possess some additional structural properties to be discussed next.

8.1   Lattices

As a general principle we endeavour to avoid reducing classes to sets. However, if we recognise that classes do have extensions in the form of individuals, then in principle we can perform set union, intersection and complementation on the classes.

As contributions to the formal ontological constitution this may give rise to existential assumptions forcing novel classes by way of compound terms with function symbols

$$isa(X, meet(Y, Z)) \leftarrow isa(X, Y) \wedge isa(X, Z)$$
$$isa(X, Y) \leftarrow isa(X, meet(Y, Z))$$
$$isa(X, Z) \leftarrow isa(X, meet(Y, Z))$$

Dually

$$isa(join(Y, Z), X) \leftarrow isa(Y, X) \wedge isa(Z, X)$$
$$isa(Y, X) \leftarrow isa(join(Y, Z), X)$$
$$isa(Z, X) \leftarrow isa(join(Y, Z), X)$$

These clauses, if added to an ontological constitution, posit existence of greatest lower bounds (*meet*, infimum, cf. conjunction) and least upper bounds (*join*, supremum, cf. disjunction) for all pairs of classes. The axioms turn the partial ordering into a lattice comprising in general a number of anonymous classes.

Here we have to distinguish existence in the logical model, vs. ontological existence vs. linguistic presence. It seems reasonable at the discretion of the ontologist to conceive of proper ontological class existence as being established somewhere between the over-crowded mathematical set model world and the sparse linguistic term world. Assuming underlying mathematical set models, the lattice models are bound to be distributive – implying even more classes than imposed by the above existential assumptions.

In the above trans-hierarchical example there is implicitly the class of vitamin-being-anti-oxidant having the subclasses vitaminE and vitaminC. Such ghost classes are mathematically brought into existence in the ontological constitution by the above lattice axioms, *in casu* as lattice meet of vitamin and anti-oxidant. Observe that this requirement does not insist that any two classes have a proper class overlap since the meet of two classes may degenerate to the distinguished empty *null* class. Accordingly, a tree may be conceived as a distinguished form of lattice by introducing the empty bottom class placed below all classes proper in the ontology. Thus the infimum of two classes on different paths is the empty class *null*.

As a next step one could argue that the ontological lattice should be specified as distributive since the underlying classes has extension sets which fulfill the usual rules for the set operations of union and intersection. In the algebraic theory of lattices, cf. [14], distributivity is achievable by introducing an axiom of distributivity known from Boolean algebra, cf. also the discussion in [8, 9]. Apropos, distributive lattices are implicitly set model basis for description logic.

A complement class *Cnon* of a class *C* is a class which do not overlap with *C* and such that for classical complement *C* together with *Cnon* forms the most general top class. If a distributive lattice is equipped with complement class for all classes present the resulting structure becomes a Boolean algebra. Alternatively it becomes a Heyting algebra in case that the complement *Cnon* is formed as the join of all classes being disjoint with *C*. This crowding of an ontology with not necesssarily useful ghost classes is sometimes referred to as "Booleanism". In particular the classes formed by set union and complementation of their extensions in most cases are useless ontological medleys, whereas on the other hand the additional classes coming about by set intersection may be empty.

One way of reconciling the set oriented claim of existence of swarms of class derivatives contra the wish to keep the ontology sparse is to recognise mathematical existence of the ghost classes but making the latter inaccessible via the *lex* relation. This means that only classes having been properly baptized are recognised as ontologically relevant. In the above example the ontologist may choose to introduce a cross-class *antioxidant_vitamin* comprising vitaminC and vitaminE.

The above considerations do not address the crucial question of how classes come about in the first place. This issue seems to be bound up with the notions of properties and property ascription being addressed in formal concept analysis (FCA) [15]. This is a mathematical technique for suggesting appropriate classes given properties for a population of individuals. FCA applies lattice theory and the classification structure resulting from applying the method forms a lattice. However, FCA relies on an extensional understanding of classes in contrast to the intensional view pursued here.

9   Classes and Properties

We conceive of the world as shaped not just by presence of individual material objects and events but by objects belonging to certain classes and therefore exhibiting certain characteristics. At the same time we classify objects based on perceived properties and established classes. Therefore the interplay between classes and properties is mutual and intricate.

In a linguistic view classes are expressed as common nouns (including nominalised verbs typically expressing activites or states), with individuals being denoted by proper nouns. The linguistic counterpart of properties is basically adjectives. In addition, prepositional phrases also serve to express property ascription, see e.g. [19]. An adjective qualifying a common noun may also serve to identify a subclass.

By contrast in a simplistic use of predicate logic both classes and properties are expressed as unary (one-argument) predicates with no *apriori* distinction between these notions. Such a unary predicate identifies a subset of the underlying model universe, irrespective of the predicate being (pre-)conceived by us as representing a class or as a property.

The data modelling approaches in computer science traditionally model individuals as frames or records comprising slots (ex. colour) into which property values (ex. blue) are placed. Thus this view tends right from the outset to recognise a distinction between an object and its properties in contrast to the simplistic predicate logical approach. The frames or records are organised into database relations or object classes (ex. car). Object classes form classification structures reflecting subclass/superclass relationships (ex. sportscar - – car – vehicle). Property slots are inherited as well as common properties for members of a class.

9.1   Properties in Conceptual Spaces

In the conceptual space view advanced in [17] in the context of cognitive science concepts (classes in our terminology) are conceived as coherent regions in an abstract space spanned by property dimensions. The property dimensions are typically physical measurements corresponding to sense data. The concept space is an abstract space, which may be visualised as, say, 3-dimensional Euklidean space. However, it may have any dimensionality reflecting the available properties. Concepts having cognitive import and/or linguistic representations are claimed to correspond to coherent or even convex regions in the space conforming with the notion of natural kinds, cf. [30]. A subconcept of a concept corresponds to a coherent subregion of the region of the concept. It is not quite clear how to endow such conceptual spaces with logical formalizations in the form of symbolic languages enabling computations. See [24] for an attempt using algebraic lattices.

In the conceptual space approach classes are distinguished from properties by their possessing a more complex structure due to their multi-dimensionality contrasting the one-dimensional property. In this approach intensionality may be claimed to be achieved by virtue of classes arising as situated point sets shaped and embraced relative to other classes and embedded in a property structure.

10   Property Ascription in Ontologies

The present metalogic approach is readily extended with properties, which like classes are conceived as term encoded predicates, in line with the property logics developed in [13, 3], and also with [37].

Given a design choice that being-antioxidant is modelled as a property rather than a class, the pure class ontology fragment from above

> *sub(vitaminC, vitamin)*
> *isa(vitaminE, vitamin)*
> *isa(vitaminC, antioxidant)*
> *isa(vitaminE, antioxidant)*

is accordingly replaced with the class/property ontology

> *class(vitaminA)* etc.
> *prop(antioxidant)*
> *isa(vitaminC, vitamin)*
> *isa(vitaminE, vitamin)*
> *hasprop(vitaminC, antioxidant)*
> *hasprop(vitaminE, antioxidant)*

Property ascription calls for extension of the ontological constitution.

10.1   Constitutions for Properties

An ontological constitution can now be further elaborated by means of metapredicates. First of all properties are to be inherited downwards in an ontology

> $hasprop(C, P) \leftarrow isa(C, C') \wedge hasprop(C', P)$
> $hasprop(X, P) \leftarrow inst(C, X) \wedge hasprop(C, P)$

Conversely all instances are to be "exherited" upwards

> $inst(C', X) \leftarrow isa(C, C') \wedge inst(C, X)$

In addition to the above properties possessed jointly by individuals in a class there are class properties such as

> *hasclassprop(eagle, endangeredspecies)*

with the sole inheritance axiom

> $hasclassprop(C, P) \leftarrow isa(C, C') \wedge hasclassprop(C', P)$

Thus class properties do not inherit to class instances in contrast to properties of individuals.

Further, there may be general metaclassifications

> $\forall c \; concept(c) \leftrightarrow class(c) \vee property(c)$

10.2   Intensionality Revisited with Properties

Consider the following pair of definitions forming basis for FCA

- The extension of a class is the set of individuals falling under the class as expressed by the predicate $inst(C, X)$.

- The intension of a class is the collection of properties possessed jointly by all the members of the extension of the class.

These intuitively appealing definitions unfortunately may lead to deflation of intensions to extensions in the sense that co-extensionality implies co-intensionality. This is noticed in [11] and further examined in [27, 26]. This is unfortunate since we would like to ensure that class inclusion be ontologically constituted by properties rather than instances, cf. sect. 5., by way of

$$\forall c1, c2\ isa(c1, c2) \leftrightarrow \forall p(hasprop(c2, p) \rightarrow hasprop(c1, p))$$

when all properties are recorded in the ontology. However, in the case of a pure class ontology where no properties are ascribed to classes, all classes are co-intensional. Therefore coextensional classes in particular becomes also co-intensional contrary to our intentions.

10.3   Property-Discernible Ontologies

However, reduction of intensions to extensions can be avoided in an ontology by requiering that all its classes are property discernible. This means that no two distinct classes possess the same properties, that is

$$\forall c1, c2\ identical(c1, c2) \leftarrow \forall p(hasprop(c2, p) \leftrightarrow hasprop(c1, P))$$

This may be achieved by insisting that any pair of sibling classes must possess a distinguishing property or trait. With this condition fulfilled recursively through the ontology we can maintain the above extension/intension definitions without risking Carnap-Oldager deflation of intensions. This issue is reminiscent of Leibniz' Identity of Indescernibles principle, with individuals, however, replaced by class differentiation here.

11   Generative Ontologies

Basically an ontology is understood as a fixed "chest-of-drawers" with a finite amount of classes. However, in sect. 8.1 we considered closed operations on classes leading to new classes coming about from a finite given collection. Now further operations combining classes with properties enable generation of useful further subclasses from given classes. This leads to potentially infinite ontologies spanned by a finite set of primitive concepts and operations on primitives. The crucial operation for forming relevant subclasses consists of conjunction of a class with an attributed property. Property attribution can be accomplished with the Peirce product from description logic mentioned in sect. 2.4. For instance the concept expressed by the phrase "lack with respect to vitaminB" can be achieved in the formal ontology as the derived class $meet(lack, peirce(wrt, vitaminB))$. This emerges as a subclass of a class $lack$ being itself a subclass of $state$. This process can be continued recursively so that for instance "disorders caused by lack with respect to vitamin B" becomes

$$meet(disorder, peirce(causedby, and(lack, peirce(wrt, vitaminB))))$$

being situated below the class of diseases. In the usual framelike notation this is recognised as the more readable $disorder[causedby : lack[wrt : vitaminB]]$.

The recursive formation of subclass terms gives rise to potentially infinite ontologies with evermore restricted subclasses along paths downwards in the ontology. The admissible attributes may be fetched from an inventory of case roles including e.g. causality, parthood etc. Obviously many of these potential nodes might be deemed useless if not senseless let alone for ontological reasons. For instance parthood might accept only material objects as parts of objects, and only states as part of states etc. This suggests development of a system of ontological typing constraints called ontological affinities in our [1,19], where in [1] BNF production rules are suggested as a simplified practical means of specifying generative ontologies.

In [19] the notion of generative ontologies is used to elaborate what is called an ontological semantics for noun phrases in which a generative ontology forms the semantic domain for noun phrases less their determiners as in the examples above. Prepositional phrases and adjectives accompanying nouns are considered property assignments which generate subclasses. As such this semantics extends the relation *lex* of sect. 6 in a principled compositional manner from common nouns to noun phrases less the determiner and disregarding cases with pronouns. The ontological meaning of a noun phrase is thus identified with a point in a generative ontology, which leaves room for phrases of unlimited syntactical complexity.

12 Summary and Conclusion

We have described a metalogic set-up in first order predicate logic for specifying ontological constitutions in a formal and principled manner. An ontological constitution primarily determines ordering principles for classes and operations on classes and properties and their ascription to classes and their metaproperties such as inheritance. In the course of this presentation we have discussed how to ensure non-extensionality of classes and properties.

The metalogical approach fits well into the logic programming paradigm in that many meta concepts can be readily expressed and computed within definite clause logic. As a question for further study we wish to capture the distinction between essential and contingent properties in the present framework.

Acknowledgement

I would like to thank my colleagues in the ONTOQUERY project [28] for many fruitful discussions. The ONTOQUERY project has been supported in part by a grant from the Danish National Science Boards.

References

1. Andreasen, T. & Nilsson, J. Fischer: Grammatical Specification of Domain Ontologies, *Data & Knowledge Engineering*, **48**, 2004.
2. Baader, F. *et al.*: *Description Logic Handbook*, Cambridge U.P., 2002.
3. Bealer, G.: *Quality and Concept*, Clarendon press, Oxford, 1982.

4. Benthem, J. v. & Doets, K.: Higher-Order Logic, in Gabbay, D. & Guenthner, F. (eds.), *Handbook of Philosophical Logic, Vol. 1*, Reidel, 1983.
5. Bittner, T., Donnelly, M., & Smith, B: Individuals, Universals, Collections: On the Foundational Relations of Ontology, in [36].
6. Bodenreider, O., Smith, B., & Burgun, A.: The Ontology-Epistemology Divide: A Case Study in Medical Terminology, in [36].
7. Brink, C., Britz, K. & Schmidt, R.A.: Peirce Algebras, *Formal Aspects of Computing*, **6**, 1994. pp. 339-358.
8. Bruun, H. & Gehrke, M.: Distributive lattice structured ontologies, in preparation.
9. Bruun, H., Gehrke, M. & Nilsson, J. Fischer: Lattice Structured Ontologies: An Ontological Account, in preparation.
10. Cappelli, A. & Mazzeranghi, D.: An intensional semantics for a hybrid language, *Data Knowledge Engineering*, **12**, 1994. pp. 31-62.
11. Carnap, R.: *Introduction to Symbolic Logic and its Applications*, Dover, 1958.
12. Chen, P.P-S.: The entity-relationship model: toward a unified view of data, *ACM Trans. on Database Systems*, **1:1**, pp.9-36, 1976.
13. Cocchiarella, N.: Properties as Individuals in Formal Ontology, *Noûs*, **6**, 1972.
14. Davey, B.A. & Priestley, H.A.: *Introduction to Lattices and Order*, Cambridge University Press, 1990.
15. Ganter, B. and Wille, R.: *Formal Concept Analysis: Mathematical Foundations.* Springer, Heidelberg, 1997.
16. Guarino, N. & Welty, C.: Supporting ontological analysis of taxonomic relationships, *Data & Knowledge Engineering*, **39**, 2001. pp. 51-74.
17. Gärdenfors, P.: *Conceptual Spaces, The Geometry of Thought*, MIT Press, 2004.
18. Hamfelt, A., Nilsson, J. Fischer, & Oldager, N.: Logic Program Synthesis as Problem Reduction using Combining Forms, *Automated Software Engineering*, **8**, 2001. pp. 167-193.
19. Jensen, P. A. & Nilsson, J. Fischer: Ontology-based Semantics for Prepositions, in: Patrick Saint-Dizier (ed.): *Syntax and Semantics of Prepositions*, Text, Speach & Language Technology, **29**, Springer, 2006. pp. 229-244.
20. Loux, M. J.: *Metaphysics, a contemporary introduction*, Routledge, 1998, 2002.
21. Madsen, B. Nistrup: *Terminologi, principper og metoder.* Gads forlag, 1999.
22. Motik, B.: On the Properties of Metamodeling in OWL, Proc. of the 4th Int. Semantic Web Conf. (ISWC 2005), Galway, Ireland, 2005, pp. 548-562.
23. Neuhaus, F., Grenon, P., & Smith, B: A Formal Theory of Substances, Qualities, and Universals, in [36].
24. Nilsson, J. Fischer: A Conceptual Space Logic, *Information Modelling and Knowledge bases XI*, Kawaguchi, E. (ed.), IOS Press, 2000. pp. 26-40.
25. Nilsson, J. Fischer, & Palomäki, J.: Towards Computing with Extensions and Intensions of Concepts, P.-J. Charrel *et al.* (eds.): *Information Modelling and Knowledge bases IX*, IOS Press, 1998.
26. Oldager, N.: Conceptual Knowledge Representation and Reasoning. PhD thesis, Informatics and Mathematical Modelling, Technical University of Denmark, 2003.
27. Oldager, N.: Intensional formalization of conceptual structures, in Proceedings of ICCS'2003, *Lecture Notes in Artificial Intelligence 2746*, Springer Verlag, 2003.
28. OntoQuery project net site: http://www.ontoquery.dk.
29. Prior, A. N.: Intentionality and Intensionality, *Papers in Logic and Ethics*, Geach, P.T. and Kenny, A.J.P (eds.), University of Massachusetts Press, 1976.
30. Quine, W.V.: Natural Kinds, in *Ontological Relativity and other essays*, Columbia U.P., 1969.

31. Smith, B.: Beyond Concepts: Ontology as Reality Representation, in [36].
32. Smith, B.: Ontology and Information Systems, draft 11.12.01.
33. Smith, B. & Rosse, C.: The Role of Foundational Relations in the Alignment of Biomedical Ontologies, Proceedings MedInfo 2004, San Francisco, CA., 2004.
34. Sowa, J.F.: *Knowledge Representation, Logical, Philosophical, and Computational Foundations*, Brooks/Cole Thomson Learning, 2000.
35. Thomasson, A.L.: Methods of categorization, in [36].
36. Varzi, A. C. & Vieu, L. (eds.): *Formal Ontology in Information Systems*, IOS Press, 2004.
37. Weingartner, P.: On the Characterization of Entities by means of Individuals and Properties, *Journal of Philosophical Logic*, **3**, 1974. pp. 323-336.
38. Øhrstrøm, P., Andersen, J., & Schärfe, H.: What has happened to Ontology, ICCS 2005, *Lecture Notes in Computer Science 3596*, 2005.

Peirce's Contributions to the 21st Century

John Sowa

VivoMind Intelligence, Inc.
sowa@vivomind.com

Abstract. Peirce was a precocious child, a 19th-century scientist who had an international reputation in both logic and physics, and a largely neglected philosopher in the 20th century. Peirce's research in logic, physics, mathematics, and lexicography made him uniquely qualified to appreciate the rigors of science, the nuances of language, and the semiotic processes that support both. Instead of using logic to understand language, the philosophers who began the analytic tradition — Frege, Russell, and Carnap — tried to replace language with a purified version of logic. As a result, they created an unbridgeable gap between themselves and the so-called Continental philosophers, they exacerbated the behaviorist tendency to reject any study of meaning, and they left semantics as an unexplored wilderness with only a few elegantly drawn, but incomplete maps based on Tarski's model theory and Kripke's possible worlds. This article reviews the ongoing efforts to construct a new foundation for 21st-century philosophy on the basis of Peirce's research and its potential for revolutionizing the study of meaning in cognitive science, especially in the fields of linguistics and artificial intelligence.

1 The Influence of Peirce and Frege

Charles Sanders Peirce is widely regarded as the most important philosopher born in America, and many of his followers consider him the first philosopher of the 21st century. An easy explanation for the neglect of his philosophy in the 20th century is that Peirce was "born before his time." A better approach is to ask what trends in the 20th century led to the split between analytic and Continental philosophy, and how Peirce's logic and philosophy relate to both sides of the split. The short answer is that his logic was adopted by the analytic philosophers, but the questions he addressed were closer to the concerns of the Continental philosophers. A longer answer is needed to show what Peirce's ideas can contribute to research and development projects in the 21st century.

Frege (1879) and Peirce (1880, 1885) independently developed logically equivalent notations for full first-order logic. Although Frege was first, nobody else adopted his notation, not even his most famous student, Rudolf Carnap. Schröder adopted Peirce's notation for his three-volume *Vorlesungen über die Algebra der Logik*, which became the primary textbook on logic from 1890 to 1910. Peano (1889) also adopted Peirce's notation, but he changed the logical symbols because he wanted to include mathematical symbols in the formulas; he gave full credit to Peirce and

H. Schärfe, P. Hitzler, and P. Øhrstrøm (Eds.): ICCS 2006, LNAI 4068, pp. 54–69, 2006.

Schröder and criticized Frege's notation as unreadable. Whitehead and Russell (1910) cited Frege, but they adopted Peirce-Schröder-Peano notation for the *Principia Mathematica*.

To illustrate the differences in notation, consider the English sentence *John is going to Boston by bus*, which could be expressed in Peirce's algebraic notation as

$$\Sigma_x \Sigma_y \, (\text{Go}(x) \bullet \text{Person(John)} \bullet \text{City(Boston)} \bullet \text{Bus}(y) \bullet$$
$$\text{Agnt}(x,\text{John}) \bullet \text{Dest}(x,\text{Boston}) \bullet \text{Inst}(x,y))$$

Since Boole treated disjunction as logical addition and conjunction as logical multiplication, Peirce represented the existential quantifier by Σ for repeated disjunction and the universal quantifier by Π for repeated conjunction. Peano began the practice of turning letters upside-down and backwards to form logical symbols. He represented existence by $\exists$, consequence by $\supset$, the Latin *vel* for disjunction by $\vee$, and conjunction by $\wedge$. With Peano's symbols, this formula would become

$$(\exists x)(\exists y)(\text{Go}(x) \wedge \text{Person(John)} \wedge \text{City(Boston)} \wedge \text{Bus}(y)$$
$$\wedge \, \text{Agnt}(x,\text{John}) \wedge \text{Dest}(x,\text{Boston}) \wedge \text{Inst}(x,y))$$

Figure 1 shows a conceptual graph that represents the same information.

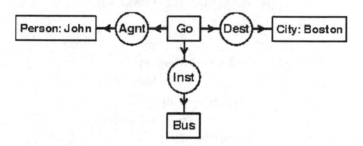

Fig. 1. Conceptual graph for *John is going to Boston by bus*

For his Begriffsschrift, Frege (1979) adopted a tree notation for first-order logic with only four operators: assertion (the "turnstile" operator), negation (a short vertical line), implication (a hook), and the universal quantifier (a cup containing the bound variable). Figure 2 shows the Begriffsschrift equivalent of Figure 1, and following is its translation to predicate calculus:

$$\sim(\forall x)(\forall y)(\text{Go}(x) \supset (\text{Person(John)} \supset (\text{City(Boston)} \supset (\text{Bus}(y) \supset$$
$$(\text{Agnt}(x,\text{John}) \supset (\text{Dest}(x,\text{Boston}) \supset \sim\text{Inst}(x,y)))))))$$

Frege's choice of operators simplified his rules of inference, but they led to awkward paraphrases: *It is false that for every x and y, if x is an instance of going then if John is a person then if Boston is a city then if y is a bus then if the agent of x is John then if the destination of x is Boston then the instrument of x is not y.*

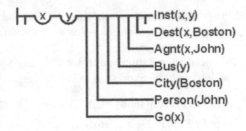

Fig. 2. Frege's Begriffsschrift for *John is going to Boston by bus*

Peirce began to experiment with *relational graphs* for representing logic as early as 1882, but he couldn't find a convenient representation for all the operators of his algebraic notation. Figure 3 shows a relational graph that expresses the same sentence as Figures 1 and 2. In that graph, an existential quantifier is represented by *a line of identity*, and conjunction is the default Boolean operator. Since Peirce's graphs did not distinguish proper names, the monadic predicates isJohn and isBoston may be used to represent names. Following is the algebraic notation for Figure 3:

$$\Sigma_x\Sigma_y\Sigma_z\Sigma_w \ (\text{Go}(x) \bullet \text{Person}(y) \bullet \text{isJohn}(y) \bullet \text{City}(z) \bullet \text{isBoston}(z) \bullet \text{Bus}(w) \bullet$$
$$\text{Agnt}(x,y) \bullet \text{Dest}(x,z) \bullet \text{Inst}(x,w))$$

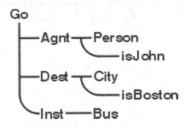

Fig. 3. Peirce's relational graph for *John is going to Boston by bus*

In 1896, Peirce discovered a simple convention that enabled him to represent full FOL: an oval enclosure that negated the entire graph or subgraph inside. He first applied that technique to his *entiative graphs* whose other operators were disjunction and the universal quantifier. In 1897, however, he switched to the dual form, the *existential graphs*, which consisted of the oval enclosure added to his earlier relational graphs. Peirce (1898) observed that metalevel relations could be attached to the oval to make further statements about the enclosed graphs. The most important innovation of the graphs was not the notation itself, but the rules of inference, which were an elegant and powerful generalization of *natural deduction* by Gentzen (1935).

Hilbert and Ackermann (1928) gave equal credit to Peirce and Frege, but later publications almost completely ignored Peirce. Frege was certainly a brilliant logician who deserves credit for the first publication of full FOL and for his high standards of rigor. Yet he had little influence on the technical development of logic, and mathematicians in the late 19th century were developing higher standards without any assistance from logicians. The historical footnotes have been amply documented by Putnam (1982), Quine (1995), Dipert (1995), and Hintikka (1997), but those studies don't explain why references to Peirce disappeared from the literature during most of the 20th century.

The primary reason for the focus on Frege at the expense of Peirce was not their logic, but their philosophy. Frege addressed narrow questions that could be expressed in logic; instead of broadening the scope of logic, many of his followers dismissed, attacked, or ridiculed attempts to address broader issues. In other areas of cognitive science, a similar emphasis on narrow technical questions led Watson (1913) to throw out the psyche from psychology by renaming the field *behaviorism*, and it led Bloomfield (1933) and Chomsky (1957) to throw out semantics from linguistics. Katz and Fodor (1963) reintroduced a tiny amount of semantics through a negative formula: "Language description minus grammar is semantics".

For linguistics and artificial intelligence, the narrow focus meant that the most important questions couldn't be asked, much less answered. The great linguist Roman Jakobson, whose career spanned most of the 20th century, countered Chomsky with the slogan "Syntax without semantics is meaningless." In AI, Winograd called his first book *Understanding Natural Language* (1972), but he abandoned a projected book on semantics when he realized that no existing semantic theory could explain how anyone, human or computer, could understand language. In a later book, coauthored with the philosopher Fernando Flores, Winograd (1986) abandoned the analytic foundations of his first book in favor of methods inspired by Heidegger's phenomenology. Winograd's disillusionment also affected many other AI researchers, who turned to the useful, but less ambitious problems of text mining, information retrieval, and user-interface design. Those techniques may be practical, but they won't solve the problems of understanding language, meaning, intelligence, or life.

After a century of narrow questions, it is time to examine the broader questions and ask how Peirce's methods might answer them. His first *rule of reason*, "Do not block the way of inquiry" (CP 1.135), implies that no question is illegitimate. Peirce applied that principle in criticizing Ernst Mach, the grandfather of logical positivism:

> Find a scientific man who proposes to get along without any metaphysics — not by any means every man who holds the ordinary reasonings of metaphysicians in scorn — and you have found one whose doctrines are thoroughly vitiated by the crude and uncriticized metaphysics with which they are packed. We must philosophize, said the great naturalist Aristotle — if only to avoid philoso-phizing. Every man of us has a metaphysics, and has to have one; and it will influence his life greatly. Far better, then, that that metaphysics should be criticized and not be allowed to run loose. (CP 1.129)

Whitehead and Gödel were two distinguished logicians who also considered metaphysics to be the heart of philosophy. The analytic philosophers cited them only for their contributions to logic, never for their philosophy. This article analyzes the origins of the extreme narrowness of analytic philosophy, Peirce's broader scope, and the potential of Peirce's semiotics to serve as the basis for reintroducing topics that the analytic philosophers deliberately rejected.

2 Logical Negativism

In his book *Beyond Analytic Philosophy*, Hao Wang, a former student of Quine and assistant to Gödel, classified philosophers by the terms *nothing else* and *something more*. The leaders of the analytic movement were mostly characterized by what they excluded: they chose a methodology that could address a limited range of topics and declared that nothing else was a legitimate matter of discussion. By applying logic to a narrow range of questions, they often achieved high levels of precision and clarity. But the philosophers who sought something more felt that the unclear questions were often the most significant, and they tried to broaden the inquiry to topics that the nothing-else philosophers rejected. Whitehead and Russell were two pioneers in logic who collaborated successfully on the *Principia Mathematica*, but were diametrically opposed in their attitudes toward philosophy. Whitehead (1929) constructed one of the largest and most ambitious metaphysical systems of the 20th century, but Russell was an outspoken critic of metaphysics. For the second edition of the *Principia*, Russell added a lengthy introduction based on his system of *logical atomism*, but Whitehead wrote a letter to *Mind* saying that he had taken no part in the revisions and he did not wish to be associated with any of the additions or modifications. Whitehead aptly characterized both of their philosophies in his introduction of Russell for the William James lectures at Harvard: "I am pleased to introduce my good friend Bertrand Russell. Bertie thinks that I am muddle-headed, but then, I think that he is simple-minded" (Lucas 1989, p. 111).

To describe the narrow scope, Wang (1986) coined the term *logical negativism* for the critical, but reductionist approach of his former thesis adviser:

> Quine merrily reduces mind to body, physical objects to (some of) the place-times, place-times to sets of sets of numbers, and numbers to sets. Hence, we arrive at a purified ontology which consists of sets only.... I believe I am not alone in feeling uncomfortable about these reductions. What common and garden consequences can we draw from such grand reductions? What hitherto concealed information do we get from them? Rather than being overwhelmed by the result, one is inclined to question the significance of the enterprise itself. (p. 146)

In support of this view, Wang quoted a personal letter from C. I. Lewis, the founder of the modern systems of modal logic, about the state of philosophy in 1960:

> It is so easy... to get impressive 'results' by replacing the vaguer concepts which convey real meaning by virtue of common usage by pseudo precise concepts which are manipulable by 'exact' methods — the trouble being that nobody any longer knows whether anything actual or of practical import is being discussed. (p. 116)

The negativism began with Frege (1879), who set out "to break the domination of the word over the human spirit by laying bare the misconceptions that through the use of language often almost unavoidably arise concerning the relations between concepts." His strength lay in the clarity of his distinctions, which Frege (1884) summarized in three fundamental principles:

1. "always to separate sharply the psychological from the logical, the subjective from the objective;"
2. "never to ask for the meaning of a word in isolation, but only in the context of a proposition;"
3. "never to lose sight of the distinction between concept and object."

These distinctions may sound good in isolation, but in practice the borderlines are not clear. Instead of trying to understand the reasons for the lack of clarity, Frege imposed arbitrary restrictions:

> In compliance with the first principle, I have used the word "idea" always in the psychological sense, and have distinguished ideas from concepts and from objects. If the second principle is not observed, one is almost forced to take as the meanings of words mental pictures or acts of the individual mind, and so to offend against the first principle as well.

With this interpretation, Frege made it impossible to formalize metalanguage as language about language because there are no physical objects that can serve as the referents of metalevel terms. In the *Tractatus*, Wittgenstein (1921) observed Frege's restrictions and defined all meaningful language in terms of references to physical objects and their relationships. Everything else, including his own analysis of language, had no legitimate reference: "My propositions are elucidatory in this way: he who understands me finally recognizes them as senseless" (6.54).

While reviewing Quine's *Word and Object*, Rescher (1962) was struck by the absence of any discussion of events, processes, actions, and change. He realized that Quine's static views were endemic in the analytic tradition: "The ontological doctrine whose too readily granted credentials I propose to revoke consists of several connected tenets, the first fundamental, the rest derivative:"

1. "The appropriate paradigm for ontological discussions is a *thing* (most properly a physical object) that exhibits *qualities* (most properly of a timeless — i.e., either an atemporal or a temporarily fixed — character)."
2. "Even *persons* and *agents* (i.e., "things" capable of action) are secondary and ontologically posterior to proper (i.e., inert or inertly regarded) *things*."
3. "Change, process, and perhaps even time itself are consequently to be downgraded in ontological considerations to the point where their unimportance is so blatant that such subordination hardly warrants explicit defense. They may, without gross impropriety, be given short shrift in or even omitted from ontological discussions."

> "It is this combination of views, which put the thing-quality paradigm at the center of the stage and relegate the concept of process to some remote and obscure corner of the ontological warehouse, that I here characterize as the 'Revolt against Process'."

Rescher found that the only analytic philosopher who bothered to defend the static view was Strawson (1959), who adopted *identity* and *independence* as the criteria for ontological priority: "whether there is reason to suppose that identification of particulars belonging to some categories is in fact dependent on the identification of particulars belonging to others, and whether there is any category of particulars that is basic in this respect" (pp. 40-41). By applying that principle, Strawson concluded that physical objects are "basic" because processes cannot be identified without first identifying the objects that participate in them. Rescher, however, found Strawson's arguments unconvincing and presented three rebuttals:

1. Since people are commonly identified by numbers, such as employee numbers or social-security numbers, Strawson should grant numbers ontological priority over people. Church (1958) observed that a similar argument could be made for the ontological priority of men over women because women are typically identified by the names of their fathers or husbands.
2. All physical things are generated by some process. Therefore, they owe their very existence to some process. Processes can generate other processes, but inert things cannot generate anything without some process.
3. The method of identifying an object is itself a process. Therefore, things cannot even be recognized as things without some process.

Undeterred by the rebuttals, Strawson (1992) published a textbook that he used to inculcate philosophy students with the thing-property doctrine. He mentioned *event semantics* as proposed by Davidson (1967), but dismissed it as "unrealistic" and "unnecessary." He took no notice of the rich and growing literature on event semantics in linguistics and artificial intelligence (Tenny & Pustejovsky 2000).

When the nothing-else philosophers turn their criticism on one another, they are left with nothing at all. In developing a semantics for a fragment of English, Montague (1967) stated his goal of reducing ontology to nothing but sets: "It has for fifteen years been possible for at least one philosopher (myself) to maintain that philosophy, at this stage in history, has as its proper theoretical framework set theory with individuals and the possible addition of empirical predicates." To disguise the emptiness of the foundations, Montague called the elements of his sets *possible worlds*, but the logician Peter Geach, who was strongly influenced by Frege, dismissed Montague's worlds as "Hollywood semantics" (Janik & Toulmin 1973). In his famous paper, "Two Dogmas of Empiricism," Quine turned his critical skills on the work of Carnap, his best friend and mentor. In the process, he destroyed the last positive claims of logical positivism. In his mature review of topics he covered during his career, Quine (1981) began with the reduction of ontology to sets, which Wang deplored; he then continued in chapter after chapter to criticize various attempts to add something more, such as modality, belief statements, or ethics. His conclusion was that precise, local, context-dependent statements could be made, but no formalized general-purpose system of logic, ontology, knowledge representation, or natural language semantics is possible. Quine's arguments would seem to justify Winograd in abandoning the quest for artificial intelligence. Yet people somehow manage to learn languages and use them successfully in their daily lives. Other animals are successful even without language. What is the secret of their success?

3   Peirce's Contributions to the Study of Meaning

Although Peirce had never read Quine's arguments, he wouldn't have been troubled by the negative conclusions. In fact, he would probably agree. Like Leibniz, Quine would agree that absolute certainty is possible only in mathematics and that all theories about the physical world are fallible and context dependent. Peirce went one step further: he even extended fallibilism to mathematics itself. A major difference between Peirce and Quine is that Peirce (1906) not only recognized context dependence, he even developed a notation for representing it in his existential graphs:

> The nature of the universe or universes of discourse (for several may be referred to in a single assertion) in the rather unusual cases in which such precision is required, is denoted either by using modifications of the heraldic tinctures, marked in something like the usual manner in pale ink upon the surface, or by scribing the graphs in colored inks.

Peirce's later writings are fragmentary, incomplete, and mostly unpublished, but they are no more fragmentary and incomplete than most modern publications about contexts. In fact, Peirce was more consistent in distinguishing the syntax (oval enclosures), the semantics ("the universe or universes of discourse"), and the pragmatics (the tinctures that "denote" the "nature" of those universes).

What is revolutionary about Peirce's logic is the explicit recognition of multiple universes of discourse, contexts for enclosing statements about them, and meta-language for talking about the contexts, how they relate to one another, and how they relate to the world and all its events, states, and inhabitants. That expressive power, which is essential for characterizing what people say in ordinary language, goes far beyond anything that Kripke or Montague, let alone Frege or Quine, ever proposed. As an example, the modal auxiliary *must* in the following dialog expresses a context-dependent necessity that is determined by the mother:

Mother:	You must clean up your room.
Child:	Why?
Mother:	Because I said so.

The necessity in the first sentence is explained by the mother's reply *I said so*, which is a context-dependent law that governs the situation. To clarify the dependencies, Dunn (1973) demonstrated two important points: first, the semantics of the modal operators can be defined in terms of laws and facts; second, the results are formally equivalent to the semantics defined in terms of possible worlds. For natural language semantics, Dunn's semantics can support methods of discourse analysis that can relate every modal or intentional verb to some proposition that has a law-like effect, to a context that is governed by that law, and to a lawgiver, which may be God, an official legislature, or the child's mother (Sowa 2003). Although Peirce could not have known the work of Kripke or Dunn, he anticipated many of the relationships among modality, laws, and lawgivers, and he recognized levels of authority from the absolute laws of logic or physics to more lenient rules, regulations, social mores, or even a single individual's habits and preferences.

Unlike Frege, Russell, and Carnap, Peirce did not avoid the challenge of characterizing the language people actually use by escaping to a purified realm of formal logic and ontology. He had been an associate editor of the *Century Dictionary*, for which he wrote, revised, or edited over 16,000 definitions. The combined influence of logic and lexicography is apparent in a letter he wrote to B. E. Smith, the editor of that dictionary:

> The task of classifying all the words of language, or what's the same thing, all the ideas that seek expression, is the most stupendous of logical tasks. Anybody but the most accomplished logician must break down in it utterly; and even for the strongest man, it is the severest possible tax on the logical equipment and faculty.

In this remark, Peirce equated the lexicon with the set of expressible ideas and declared logic as essential to the analysis of meaning. Yet he considered logic only one of the three major subdivisions of his theory of signs:

1. **Universal grammar** is first because it studies the structure of signs independent of their use. The syntax of a sentence, for example, can be analyzed without considering its meaning, reference, truth, or purpose within a larger context. In its full generality, universal grammar defines the types of signs and patterns of signs at every level of complexity in every sensory modality.
2. **Critical logic**, which Peirce defined as "the formal science of the conditions of the truth of representations" (CP 2.229), is second because truth depends on a dyadic correspondence between a representation and its object.
3. **Methodeutic** or **philosophical rhetoric** is third because it studies the principles that relate signs to each other and to the world: "Its task is to ascertain the laws by which in every scientific intelligence one sign gives birth to another, and especially one thought brings forth another" (CP 2.229). By "scientific intelligence," Peirce meant any intellect capable of learning from experience, among which he included dogs and parrots.

Many people talk as if logic is limited to deduction, but Peirce insisted that induction and abduction are just as important, since they are the branches of logic that derive the axioms from which deduction proceeds. Peirce also emphasized the importance of analogy, which is a very general method of reasoning that includes aspects of all three of the other methods of logic. In fact, analogy is essential to induction and abduction, and the method of *unification* used in deduction is a special case of analogy.

One of the pioneers of formal semantics, Barbara Partee (2005), admitted that the formalisms developed by Montague and his followers have not yet come to grips with the "intended meanings" of their abstract symbols and that lexical semantics and lexicography cover material that is very far from being formalized:

> In Montague's formal semantics the simple predicates of the language of intensional logic (IL), like *love*, *like*, *kiss*, *see*, etc., are regarded as symbols (similar to the "labels" of [predicate calculus]) which could have many possible interpretations in many different models, their "real meanings" being regarded as their interpretations in the "intended model". Formal semantics does not

pretend to give a complete characterization of this "intended model", neither in terms of the model structure representing the "worlds" nor in terms of the assignments of interpretations to the lexical constants. The present formalizations of model-theoretic semantics are undoubtedly still rather primitive compared to what is needed to capture many important semantic properties of natural languages.... There are other approaches to semantics that are concerned with other aspects of natural language, perhaps even cognitively "deeper" in some sense, but which we presently lack the tools to adequately formalize. (Lecture 4)

In Montague's terms, the intension of a sentence is a function from abstract sets (called possible worlds) to truth values, and the intensions of words are other abstract functions that can be combined to derive the function for a sentence. In lexical semantics and lexicography, words are decomposed into patterns of words or word-like signs, and any connection to logic or possible worlds is rarely discussed and often denounced as irrelevant. As Partee said, there are no known mathematical "tools" for mapping all the words and signs of lexical semantics to Montague-style functions. Even if the words could be mapped, an even greater challenge would be to map the relatively loose patterns of lexical semantics to Montague's strictly regimented functions of functions for combining the basic functions.

A more realistic way to bridge the gap between the formal and the informal is to recognize that loose informal patterns of signs are the foundation for perception and analogical reasoning by all mammals, including humans. Children learn language by mapping perceptual and motor patterns to verbal patterns, and for adults, there is a continuity between the informal patterns learned in childhood to the most highly disciplined patterns used in science, mathematics, and logic. The advantage of Peircean semiotics is that it firmly situates language and logic within the broader study of signs of all types. The highly disciplined patterns of mathematics and logic, important as they may be for science, lie on a continuum with the looser patterns of everyday speech and with the perceptual and motor patterns, which are organized on geometrical principles that are very different from the syntactic patterns of language or logic. Transferring the problems to a broader domain does not automatically solve them, but it provides a richer set of tools to address them.

4   Patterns of Symbols in Language and Logic

A semiotic view of language and logic gets to the heart of the philosophical controversies and their practical implications for linguistics, artificial intelligence, and related subjects. The analytic philosophers hoped that they could use logic to express facts with the utmost clarity and precision. Wang (1986) observed that Carnap, in particular, was "willing to exclude an exceptionally large range of things on the grounds that they are 'not clear,' or sometimes that 'everything he says is poetry.'" But the logicians Peirce and Whitehead and the poet Robert Frost recognized that clarity is often an oversimplification. Whitehead (1937) aptly characterized the problem:

Human knowledge is a process of approximation. In the focus of experience, there is comparative clarity. But the discrimination of this clarity leads into the penumbral background. There are always questions left over. The problem is to discriminate exactly what we know vaguely.

And Frost (1963) suggested the solution:

I've often said that every poem solves something for me in life. I go so far as to say that every poem is a momentary stay against the confusion of the world.... We rise out of disorder into order. And the poems I make are little bits of order.

Contrary to Carnap, poetry and logic are not at opposite extremes. They are complementary approaches to closely related problems: developing patterns of symbols that capture important aspects of life in a memorable form. Logic is limited to expressing factual content, but poetry can express aesthetic and ethical interpretations of the facts. Any particular interpretation of a poem can be asserted in logic, but a good poem can express a volume of possible interpretations in a single phrase.

The greatest strength of natural language is its flexibility in accommodating patterns ranging from poetry and cooking recipes to stock-market reports and scientific treatises. A very flexible syntactic theory, which is also psychologically realistic, is *Radical Construction Grammar* (RCG) by Croft (2001). Unlike theories that draw a sharp boundary between grammatical and ungrammatical sentences, RCG can accept any kind of construction that speakers of a language actually use, including different choices of constructions for different sublanguages:

Constructions, not categories or relations, are the basic, primitive units of syntactic representation.... the grammatical knowledge of a speaker is knowledge of constructions (as form-meaning pairings), words (also as form-meaning pairings), and the mappings between words and the constructions they fit in. (p. 46)

RCG makes it easy to borrow a word from another language, such as *connoisseur* from French or H_2SO_4 from chemistry, or to borrow an entire construction, such as *sine qua non* from Latin or $x^2+y^2=z^2$ from algebra. In the sublanguage of chemistry, the same meaning that is paired with H_2SO_4 can be paired with *sulfuric acid*, and the constructions of mathematical and chemical notations can be freely intermixed with the more common constructions of English syntax.

The form-meaning pairings of RCG are determined by language-specific or even sublanguage-specific *semantic maps* to a multidimensional *conceptual space*, which "represents conventional pragmatic or discourse-functional or information-structural or even stylistic or social dimensions" (Croft, p. 93). Although Croft has not developed a detailed theory of conceptual structures, there is no shortage of theories, ranging from those that avoid logic (Jackendoff 1990, 2002) to those that emphasize logic (Sowa 1984, 2000). The versions that avoid or emphasize logic represent stages along a continuum, which an individual could traverse from infancy to childhood to adulthood. Each stage adds new functionality to the earlier stages, which always remain available; even the most sophisticated adult can find common ground in a conversation with a three-year-old child. Following are the basic elements of logic, each of which builds on the previous elements:

1. Every natural language has basic constructions for expressing relational patterns with two or three arguments, and additional arguments can be added by constructions with prepositions or postpositions.
2. The three logical operators of conjunction, negation, and existence, which are universally available in all languages, are sufficient to support first-order logic.
3. Proper names, simple pronouns, and other indexicals are universal, but various languages differ in the selection of indexical markers.
4. Metalanguage is supported by every natural language, and it appears even in the speech of children. Metalanguage supports the introduction of new words, new syntax, and the mapping from the new features to older features and to extralinguistic referents.
5. Simple metalanguage can be used even without embedded structures, but the ability to encapsulate any expression as a single unit that can be embedded in other expressions provides enormous power.
6. When combined in all possible ways, the above features support the ability to define modal operators and all the intensional verbs and structures of English.

In addition to supporting any representation for logic, a general theory of intelligence must also support reasoning methods. The most primitive and the most general is analogy, which by itself supports case-based reasoning. Sowa and Majumdar (2003) showed how Peirce's three branches of logic — induction, deduction, and abduction — could be defined as highly disciplined special cases of analogy. Unlike the methods of logic, which are limited to language-like symbols, analogies can relate patterns of signs of any kind: they can support the metaphors described by Lakoff and Johnson (1980), they can link abstract symbols to image-like icons, and they can relate similar patterns of percepts across different sensory modalities.

5 Everything Is a Sign

In focusing their attention on tiny questions that could be answered with utmost clarity in their logic, the analytic philosophers ignored every aspect of life that was inexpressible in their logic. The Continental philosophers did address the unclear questions, but their prose was so opaque that few people could read it. Although Peirce invented the logic that the analytic philosophers adopted, he incorporated logic in a much broader theory of signs that accommodates every possible question, answer, perception, feeling, or intuition — clear, unclear, or even unconscious. With that approach, the border between analytic and Continental philosophy vanishes. In fact, all borders in cognitive science vanish, except for local borders created by differences in methodology.

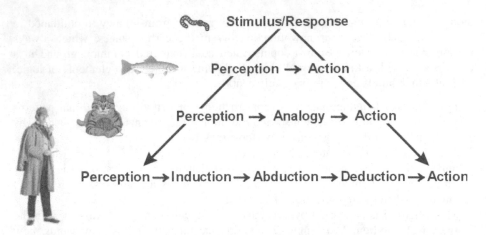

Fig. 4. Evolution of semiosis

To illustrate the generality of semiotics, the following examples show how Peirce's ideas can be applied to a wide range of topics:

- Figure 4 illustrates the evolution of cognitive systems according to the sophistication of their semiotic capabilities. For the worm, a sign that serves as a stimulus triggers a response with only a few intermediate levels of signs passed from neuron to neuron. The fish, however, has highly developed perceptual and motor mechanisms that depend on vastly more complex neural mechanisms. For the cat, the ball of string is a mouse analog, which can be used in exercises that build the cat's repository of learned sign patterns to be invoked when hunting prey. The human inherits all the capabilities of earlier levels and adds the symbol processing that supports language and logic.

- Peirce's fundamental assumption is that anything in the universe that can have a causal influence on anything else is a potential sign, independent of the presence of anything that can interpret signs. The big bang at the beginning of the universe, for example, could not be observed by any cognitive agent at the time, but astronomers today can observe its effects in the background microwave radiation.

- In the classification of signs, three basic categories are Mark, Token, and Type. A mark is an uninterpreted sign of any kind, a type is a pattern for classifying marks, and a token is the result of classifying a mark according to some type. For example, a pattern of green and yellow in the lawn is a mark, which could be interpreted according to the viewer's interests as a token of type Plant, Weed, Flower, SaladGreen, Dandelion, etc.

- A sign may be characterized by the way the mark determines the referent:
  1. **Icon:** according to some similarity of image, pattern, or structure.
  2. **Index:** according to some physical relationship; e.g., immediate presence, pointing to something remote, or causally indicating something not directly perceptible.
  3. **Symbol:** according to some convention; e.g., spoken words, written words, money, flag, uniform...

- Communication, memory, learning, and reasoning depend on signs — but most signs are not symbols. In Figure 4, organisms from the level of bacteria to worms respond to indexes. With larger brains and more complex sensory organs, animals from fish to mammals add icons. The human level of symbol processing supports the open-ended levels of complexity possible with logic and language.
- According to Peirce, the ability to respond to signs is characteristic of all living organisms. Since a virus cannot process signs, it is not alive. Instead, a virus is itself a sign, which a susceptible organism interprets by generating replicas.
- Pietarinen (2004) pointed out that Peirce had anticipated much of the modern work on speech acts, relevance, and conversational implicatures; although he hadn't listed the principles as conveniently as Grice (1975), he discussed and analyzed versions of them in many of his writings. Peirce had also anticipated Davidson's event semantics by insisting that actions and states were entities just as real as their participants, and he anticipated Perry's "Essential Indexical" by pointing out that every statement in logic requires at least one indexical to fix the referents of its variables.
- Although Peirce's graph logic is equivalent to his algebraic notation in expressive power, he developed an elegant set of rules of inference for the graphs, which have attractive computational properties. Ongoing research on graph-theoretic algorithms has demonstrated important improvements in methods for searching and finding relevant graphs during the reasoning processes (Majumdar et al. forthcoming).

The key to Peirce's modernity is his solid foundation in history. Unlike Frege and Russell, who made a sharp break with the Aristotelian and Scholastic work on logic, many of Peirce's innovations were based on insights he had derived from his studies of medieval logic. In fact, Peirce had boasted that he had the largest collection of medieval manuscripts on logic in the Boston area. In general, major breakthroughs are most likely to come from unpopular sources, either because they're so new that few people know them, so old that most people have forgotten them, or so unfashionable that nobody looks at them.

References

Bloomfield, Leonard (1933) *Language*, Holt, Rinehart, & Winston, New York.

Box, George E. P., J. Stuart Hunter, & William G. Hunter (2005) *Statistics for Experimenters: Design, Innovation, and Discovery*, 2nd Edition, Wiley-Interscience, New York.

Brunning, Jacqueline, & Paul Forster, eds. (1997) *The Rule of Reason: The Philosophy of Charles Sanders Peirce*, University of Toronto Press, Toronto.

Chomsky, Noam (1957) *Syntactic Structures*, Mouton, The Hague.

Church, Alonzo (1958) *The ontological status of women and abstract entities*, Lecture presented at Harvard University on April 18, 1958, available at http://www.jfsowa.com/church.htm

Croft, William (2001) *Radical Construction Grammar: Syntactic Theory in Typological Perspective*, Oxford University Press, Oxford.

Davidson, Donald (1967) "The logical form of action sentences," reprinted in D. Davidson (1980) *Essays on Actions and Events*, Clarendon Press, Oxford, pp. 105-148.

Dipert, Randall R. (1995) "Peirce's underestimated place in the history of logic," in Ketner (1995) pp. 32-58.

Dunn, J. Michael (1973) "A truth value semantics for modal logic," in H. Leblanc, ed., *Truth, Syntax and Modality*, North-Holland, Amsterdam, pp. 87-100.

Frege, Gottlob (1879) *Begriffsschrift*, English translation in J. van Heijenoort, ed. (1967) *From Frege to Gödel*, Harvard University Press, Cambridge, MA, pp. 1-82.

Frege, Gottlob (1884) *Die Grundlagen der Arithmetik*, tr. by J. L. Austin as *The Foundations of Arithmetic*, Blackwell, Oxford, 1953.

Gentzen, Gerhard (1935) "Untersuchungen über das logische Schließen," translated as "Investigations into logical deduction" in *The Collected Papers of Gerhard Gentzen*, ed. and translated by M. E. Szabo, North-Holland Publishing Co., Amsterdam, 1969, pp. 68-131.

Frost, Robert (1963) *A Lover's Quarrel with the World* (film), WGBH Educational Foundation, Boston.

Hilbert, David, & Wilhelm Ackermann (1928) *Grundzüge der theoretischen Logik*, translated as *Principles of Mathematical Logic*, Chelsea Publishing, New York, 1950.

Hintikka, Jaakko (1997) *The place of C. S. Peirce in the history of logical theory*, in Brunning & Forster (1997) pp. 13-33.

Jackendoff, Ray S. (1990) *Semantic Structures*, MIT Press, Cambridge, MA.

Jackendoff, Ray (2002) *Foundations of Language: Brain, Meaning, Grammar, Evolution*, Oxford University Press, Oxford.

Janik, Allan, & Stephen Toulmin (1973) *Wittgenstein's Vienna*, Elephant Paperbacks, Chicago, 1996.

Katz, Jerrold J., & Jerry A. Fodor (1963) "The structure of a semantic theory," *Language* **39**, 170-210.

Ketner, Kenneth Laine, ed. (1995) *Peirce and Contemporary Thought*, Fordham University Press, New York.

Lakoff, George, and Mark Johnson (1980) *Metaphors We Live By*, University of Chicago Press, Chicago.

Lucas, George R., Jr. (1989) *The Rehabilitation of Whitehead*, State University of New York Press, Albany.

Majumdar, Arun K., John F. Sowa, & Paul Tarau (forthcoming) "Graph-based algorithms for intelligent systems," in A. Nayak & I. Stojmenovic, eds., *Handbook of Applied Algorithms*, Wiley & Sons, New York.

Montague, Richard (1967) "On the nature of certain philosophical entities," revised version in R. Montague, *Formal Philosophy*, Yale University Press, New Haven, pp. 148-187.

Partee, Barbara H. (2005) "Formal Semantics," Lectures at a workshop in Moscow. http://people.umass.edu/partee/RGGU_2005/RGGU05_formal_semantics.htm

Peirce, Charles Sanders (1880) "On the algebra of logic," *American Journal of Mathematics* **3**, 15-57.

Peirce, Charles Sanders (1885) "On the algebra of logic," *American Journal of Mathematics* **7**, 180-202.

Peirce, Charles Sanders (1898) *Reasoning and the Logic of Things*, The Cambridge Conferences Lectures of 1898, ed. by K. L. Ketner, Harvard University Press, Cambridge, MA, 1992.

Peirce, Charles Sanders (1906) Manuscripts on existential graphs. Reprinted in Peirce (CP) 4.320-410.

Peirce, Charles Sanders (CP) *Collected Papers of C. S. Peirce*, ed. by C. Hartshorne, P. Weiss, & A. Burks, 8 vols., Harvard University Press, Cambridge, MA, 1931-1958.

Perry, John (1979) "The problem of the essential indexical," *Nous*, vol. 13, pp. 3-21.

Pietarinen, Ahti-Veikko (2004) "Grice in the wake of Peirce," *Pragmatics and Cognition* **12:2**, pp. 295-315.

Putnam, Hilary (1982) "Peirce the Logician" *Historia Mathematica* **9**, 290-301.

Quine, Williard Van Orman (1951) "Two dogmas of empiricism," reprinted in Quine, *From a Logical Point of View*, Harvard University Press, Cambridge, MA.

Quine, Willard Van Orman (1995) "Peirce's logic," in Ketner (1995) pp. 23-31.

Rescher, Nicholas (1962) "The revolt against process," *Journal of Philosophy*, vol. 59, pp. 410-417.

Sowa, John F. (1984) *Conceptual Structures: Information Processing in Mind and Machine*, Addison-Wesley, Reading, MA.

Sowa, John F. (2000) *Knowledge Representation: Logical, Philosophical, and Computational Foundations*, Brooks/Cole Publishing Co., Pacific Grove, CA.

Sowa, John F. (2003) "Laws, facts, and contexts: Foundations for multimodal reasoning," in *Knowledge Contributors*, edited by V. F. Hendricks, K. F. Jørgensen, and S. A. Pedersen, Kluwer Academic Publishers, Dordrecht, pp. 145-184.

Sowa, John F., & Arun K. Majumdar (2003) "Analogical reasoning," in A. de Moor, W. Lex, & B. Ganter, eds., *Conceptual Structures for Knowledge Creation and Communication*, LNAI 2746, Springer-Verlag, Berlin, pp. 16-36. http://www.jfsowa.com/pubs/analog.htm

Strawson, Peter F. (1959) *Individuals: An Essay in Descriptive Metaphysics*, Methuen and Co., London.

Strawson, Peter F. (1992) *Analysis and Metaphysics: An Introduction to Philosophy*, Oxford University Press, Oxford.

Tenny, Carol L., & James Pustejovsky, eds. (2000) *Events as Grammatical Objects: The Converging Perspectives of Lexical Semantics and Syntax*, CSLI Publications, Stanford, CA.

Wang, Hao (1986) *Beyond Analytic Philosophy: Doing Justice to What We Know*, MIT Press, Cambridge, MA.

Watson, John B. (1913) "Psychology as the behaviorist views it," *Psychological Review* **20**, pp. 158-177.

Whitehead, Alfred North (1937) "Analysis of Meaning," *Philosophical Review*, reprinted in A. N. Whitehead, *Essays in Science and Philosophy*, Philosophical Library, New York, pp. 122-131.

Whitehead, Alfred North, & Bertrand Russell (1910) *Principia Mathematica*, 2nd edition, Cambridge University Press, Cambridge, 1925.

Winograd, Terry (1972) *Understanding Natural Language*, Academic Press, New York.

Winograd, Terry, & Fernando Flores (1986) *Understanding Computers and Cognition*, Ablex, Norwood, NJ.

Wittgenstein, Ludwig (1921) *Tractatus Logico-Philosophicus*, Routledge & Kegan Paul, London.

Two Iconicity Notions in Peirce's Diagrammatology

Frederik Stjernfelt

Learning Lab Denmark – The Danish University of Education
stjern@lld.dk

Two different concepts of iconicity compete in Peirce's diagrammatical logic. One is articulated in his general reflections on the role of diagrams in thought, in what could be termed his diagrammatology – the other is articulated in his construction of Existential Graphs as an iconic system for logic representation. One is operational and defines iconicity in terms of which information may be derived from a given diagram or diagram system – the other has stronger demands on iconicity, adding to the operational criterion a demand for as high a degree of similarity as possible.

This paper investigates the two iconicity notions and addresses some of the issues they involve.

1 Operational Iconicity

The basic concept of iconicity in Peirce's semiotics and logic is presented in his second tricotomy of sign types, the well-known distinction between icons, indices, and symbols, respectively.[i] This tricotomy deals with the relation between the sign and its dynamic object, and the idea is that this relation may take three different forms. Icons function by means of a similarity between the sign and the object, or, as Peirce may also say, by shared characteristics between the sign and its object. Indices function by means of an actual connection between the sign and its object, either of a causal character (the footprint on the beach) or of a purposive character (deictics, pronomina or proper names in language). Symbols, finally, function by means of a habit, in mind or in nature, of connecting two otherwise unconnected entities to a sign. It should immediately be added, that the sign types of this tricotomy, just as is the case in the later Peirce's other nine tricotomies, do not correspond directly to distinct, natural kinds of signs. They rather pertain to *aspects* of signs, so that pure icons, indices, and symbols, respectively, may be conceived of as borderline cases only, while most typical, and indeed most interesting signs involve all three aspects to different degrees. It is possible, though, in many cases, to point out which of the three aspects is *basic* in a given sign or a given sign type – so as for instance diagrams being basically icons, and only secondarily (but still necessarily) having also indexical and symbolical aspects.

In this basic iconicity definition by similarity or shared characteristics, however, none of the two iconicity concepts to be discussed here, is obvious. They only appear when a further determination of similarity is attempted. The first, operational, definition appears exactly in the discussion of diagrams, and is developed by Peirce already in the 80s, even if the full articulation of it awaits Peirce's mature philosophy of the years after the turn of the century. To continue in Peirce's detailed taxonomy of

H. Schärfe, P. Hitzler, and P. Øhrstrøm (Eds.): ICCS 2006, LNAI 4068, pp. 70–86, 2006.

signs from that period, icons come in three subtypes, images, diagrams, and metaphors, respectively. Images are to be taken in a special, technical sense not corresponding to our everyday image notion: they are icons whose similarity functions by means of simple qualities only, colour, sound, shape, form, etc. Thus, images are very simple icons, functioning by one or few such qualities only. The recognition of a crescent form as a sign for the moon may serve as an example. The simplicity of images is made clear by their contrast to diagrams. Diagrams are skelettal icons, representing their object analyzed into parts among which "rational relations" hold, be they explicit or implicit. Such relations may be spatial, logical, mathematical, or any other type which may make clear the kind of relation holding between parts. So, as soon as the icon consists of parts whose relations mirror the relations between the corresponding parts of the object, and the sign is used to gain information about those parts and their relations, a diagram is at stake.[ii] In contrast to the technical notion of image, being much more narrow than the everyday use of the word, Peirce's technical notion of diagram is much more wide than the everyday diagram notion: it must include any use of, e.g. a painting, in which the relation between its parts plays a role in the interpretation – and it must include also algrabraic notations which may not, at a first glance, seem diagrammatical. Metaphors, to finish this tricotnomy, are icons functioning through the mediation of a third object, so as for instance an ancestral tree, charting family relationships in a branching diagram structure through the intermediate icon of a tree. The important notion here is the very wide sense of the notion of diagram which stems, in fact, from the operational criterion for iconicity. An icon is a sign "... from which information may be derived.", Peirce says ("Syllabus", ca. 1902, CP 2.309), and this forms the basic idea in the operational criterion: icons as the only sign type able to provide information which is why all more complex sign types must involve or lead to icons in order to convey information. Later in the same paper, Peirce adds that "An Icon, however, is strictly a possibility involving a possibility ..." (CP.2.311), and in this enigmatic formula, the first "possibility" should be read as referring to an icon being a possible sign of everything which resembles it in the respect so highlighted (only an index may make explicit which object or class of objects the sign more precisely refers to, so only the combination of icon and index holds the possibility of actually conveying information in the shape of a proposition). The second "possibility", however, refers to the fact that the similarity characteristics defined by the first possibility involve ,in themselves, possibilities which are not explicit and which may be further developed:

> "For a great distinguishing property of the icon is that by the direct observation of it other truths concerning its object can be discovered than those which suffice to determine its construction". ("That Categorical and Hypothetical Propositions are one in essence, with some connected matters," c. 1895, CP 2.279).

I have earlier argued (Stjernfelt 2000, Stjernfelt (forthcoming)) that this idea constitutes an epistemologically crucial property of the icon: it is nothing but an operational elaboration on the concept of similarity. The icon is not only the only kind of sign directly presenting some of the qualities of its object; it is also the only sign by the contemplation of which more can be learnt than lies in the directions for the construction of the sign. This definition immediately separates the icon from any psychologism: it does not matter whether sign and object for a first (or second) glance seem or are experienced as similar; the decisive test for iconicity lies in whether it is possible to manipulate or develop the sign so that new information as to its object appears. This definition is non-trivial because it avoids the circularity threat in most definitions of similarity which has so often been noted.[iii] At the same time, it connects the concept of icon intimately to that of deduction. This is because in order to discover these initially unknown pieces of information about the object involved in the icon, some deductive experiment on the icon must be performed. The prototypical icon deduction in Peirce's account is the rule-governed manipulation of a geometrical figure in order to observe a theorem - but the idea is quite general: an icon is characterized by containing implicit information about its object which in order to appear must be made explicit by some more or less complicated deductive manipulation or experiment procedure accompanied by observation. Thus, Peirce's diagrammatical logic rests on the basic idea that all knowledge, including logical knowledge, indispensably involves a moment of observation. Peirce thus writes, as early as 1885:

> "The truth, however, appears to be that all deductive reasoning, even simple syllogism, involves an element of observation; namely, deduction consists in constructing an icon or diagram the relations of whose parts shall present a complete analogy with those of the parts of the object of reasoning, of experimenting upon this image in the imagination, and of observing the result so as to discover unnoticed and hidden relations among the parts." ("On the Algebra of Logic. A Contribution to the Philosophy of Notation" (1885), CP 3.363)

This operational criterion makes obvious the breadth of the diagram category within icons. As soon as rationally related parts of an icon is distinguished, and the manipulation of such parts is undertaken, we perform a diagram manipulation, developing some of the implicit possibilities involved in the icon.

A very important use of this operational criterion of similiarity is now the appreciation of iconicity where it may not be, at a first glance, obvious. Peirce himself makes this use of the operational criterion when arguing that syllogistic logic or algebra are, in fact, instances of diagrammatical iconicity. In what I believe is Peirce's most detailed account for the diagrammatical reasoning process in general, abstracted from particular diagram systems, he thus argues this point (in "PAP" (1906), a parallel version to "Prologomena to an Apology for Pragmaticism" from the same year), Peirce (1976), p. 317-18):

> "Now necessary reasoning makes its conclusion *evident*. What is this "Evidence"? It consists in the fact that the truth of the conclusion is *perceived*, in all its generality, and in the generality of the how and the why of the truth is perceived. What sort of a Sign can communicate this Evidence? No index, surely, can it be; since it is by brute force that the Index thrusts its Object into the Field of Interpretation, the consciousness, as if disdaining gentle "evidence". No Symbol can do more than apply a "rule of thumb" resting as it does entirely on Habit (including under this term natural disposition); and a Habit is no evidence. I suppose it would be the general opinion of logicians, as it certainly was long mine, that the Syllogism is a Symbol, because of its Generality. But

there is an inaccurate analysis and confusion of thought at the bottom of that view; for so understood it would fail to furnish Evidence. It is true that ordinary Icons, - the only class of Signs that remains for necessary inference, - merely suggest the possibility of that which they represent, being percepts *minus* the insistency and percussivity of percepts. In themselves, they are mere Semes, predicating of nothing, not even so much as interrogatively. It is, therefore, a very extraordinary feature of Diagrams that they *show*, - as literally *show* as a Percept shows the Perceptual Judgment to be true, - that a consequence does follow, and more marvellous yet, that it *would* follow under all varieties of circumstances accompanying the premisses."

Here, the operational criterion is used in order to include traditional syllogistic reasoning within the field of diagrams: the structure of syllogism simply *is* a diagram, even when presented in the clothing of ordinary language. The same criterion was early used by Peirce in order to include algebra as icons, even as involving icons "par excellence" in the patterns of manipulation permitted:

"As for algebra, the very idea of the art is that it presents formulæ which can be manipulated, and that by observing the effects of such manipulation we find properties not to be otherwise discerned. In such manipulation, we are guided by previous discoveries which are embodied in general formulæ. These are patterns which we have the right to imitate in our procedure, and are the icons par excellence of algebra." ("On the Algebra of Logic. A Contribution to the Philosophy of Notation" (1885), CP 3.363)

Even if Peirce in this very paper tries to develop a notation of logic which, unlike his later entiative and existential graphs, sticks to traditional algebraic representations, he already here acknowledges that such algebraic representations must necessarily be diagrammatic, as measured on the operational criterion of iconicity. Elsewhere, the extends that criterion to include also aspects of linguistic grammar in the diagram category.

This operational criterion of iconicity thus becomes a very strong tool for a Peircean trying to chart the limits of iconicity. Unfortunately, Peirce never went into a further taxonomical exercise in order to chart the possible subtypes of diagrams – the only reference I found in this direction is a brief comment upon the diagram types of maps, algebra, and graphs, respectively.[iv] In any case, the operational criterion forms a very strong argument in a Peircean diagrammatology – yielding the means of a similarity test which is immune against psychologism and any subjective similarity impressions or confusions.

This broad iconicity and diagram criterion is not, however, without any problems. One terminological issue is that the technical, Peircean notion of diagram is now extended to such a degree that the common-sense notion of diagrams vanishes in the haze and seems to constitute only a small subset of the new, enlarged category. Another, more serious problem, is that Peirce still tends to take such diagrams as *prototypical* diagrams in many discussions, generalizing diagram notions taken from them to the whole category of diagrams. This goes, e.g., for his distinction between corollarial and theorematical reasoning, distinguishing conclusions which may be directly read off the diagram, on the one hand, and more difficult inferences requiring the introduction of new entities in the diagram. This distinction is taken from the prototypical diagram case of Euclidean geometrical diagrams where the new entities introduced are helping lines, etc. As Hintikka has argued, however, this distinction may be valid and indeed highly valuable when extrapolated to the more general category of diagrams. The most serious problem, however, in the generalization of the diagram concept, is connected to the lack of a rational sub-taxonomy of diagrams, namely: by which semiotic means should we now distinguish between, e.g. algebraical representations and topological-geometrical representations of the same content, as for instance the graphical and algebraical-arithmetical representations of the same mathematical functions? If the same amount of information may be operationally derived from such representations, they are, to the exact same degree, diagrammatical representations, and Peirce's diagram category offers no means for us to distinguish the particular properties of these different representations.

2 Optimal Iconicity

This problem seems, indeed, to lie behind Peirce's introduction of a second, moredemanding , notion of iconicity. It is well known that Peirce, in the latter half of the 90's, gave up his early attempts from the 80's at an algebra of logic (two versions of which were developed in 1880 and 1885), now preferring the development of graphical systems known as entiative and existential graphs. Especially the development of the latter was seen by Peirce himself as one of his major achievements, and they have been a central inspiration for diagrammatical or multimodal logic of our day, because they involve "iconical" representations which differ highly from algebraical or "symbolical" representation systems of formal logic, e.g. in the Peano-Russell tradition. I place "iconical" and "symbolical" in quotation marks here to emphasize that the use of such words in this context run directly counter to Peirce's operational iconicity criterion. For according to this criterion, such representation systems are indeed diagrammatical and iconical *to the exact same degree*, provided they yield similar possibilities for extracting new information about their object. If the same theorems may be inferred from such systems, they are, on the operational criterion, both of them operationally iconical. And if we take Peirce's two completed systems of "iconical" logic graphs, the Alpha and Beta systems of existential graphs, they have indeed been proved complete and consistent representations of propositional logic and first order predicate logic, respectively. So, in terms of which theorems may be derived from them, the Alpha and Beta graphs are just as iconical as propositional logic and first order predicate logic, as developed within mainstream formal logic, and vice versa. Peirce's operational iconicity criterion does, it is true, provide the strong insight that these results of mainstream formal logic are *not,* contrary to widespread belief, "symbolical" in the sense that they do not involve iconical representations. They may, of course, be termed "symbolical" understood in the sense that they employ symbols to a larger degree than Peirce's graphs (which also NB employ symbols), but this term may no longer be taken, implicitly, also to imply that they do not contain iconical representations of their object. This is, indeed, a very strong and to some extent counter-intuitive result of Peirce's operational iconicity criterion. But it immediately raises a further question: *what is then the difference between "iconical" and "symbolical" logic representations when it may no longer be expressed in terms of operational iconicity?*

Even if Peirce does not explicitly (at least where I have searched in his writings) pose the question in these terms, this issue is involved in his introduction of a second, stronger iconicity criterion. This takes place especially in his discussion of the conventions used in his Beta system equivalent to first order predicate logic. While the Alpha system required only a sheet of assertion, letters representing propositions, same location of graphs indicating conjunctions, and cuts representing negations, the Beta system adds to these entities further conventions representing quantifications, variables, and predicates. The whole machinery of these issues isintroduced by means of a very simple convention. Predicates with up to three variables (equivalent to functions with arguments in the Fregean tradition) are introduces by means of the verbal/predicative kernel of the predicate written directly on the graph with the corresponding subject slots indicated by blanks to be filled in by symbols for the subjects involved (nouns, pronouns, or proper names). In ordinary text, such blanks are indicated by underlinings such as in "_____ gives _____ to _____" involving three blanks. In the Existential Graphs, similar lines are interpreted as "lines of identity" so that any further determination of the identity of the subjects of these blanks are to be added to the ends of the lines. The very line of identity thus refers to a variable, and the line may branch in order to tie to different slots in different predicates, indicating that the individual(s) referred to by that line has those predicates. The spots at the end of such lines are, consequently, the second convention added: they refer, as indices, to the binding of the variables bearing the predicates in issue. Thus, the whole logical machinery of quantification, variables, and predicates is represented by these very simple means. If a line of identity abuts on the sheet of assertion (or on any evenly enclosed part of it, that is, by 2, 4, 6, ... cuts), then this immediately indicates the existential quantifier of "Something exists which ..." and the three dots are then filled in by the predicates to which the line of identity connects this implicit quantification. Similarly, any such line of identity ending in an unevenly enclosed cut immediately indicates a negative universal quantifier.[v]

In his development of the Beta system, Peirce lays a great emphasis on the fact that the representation of quantification and bound variables by the means of lines of identity is *more iconical* than the representation of the same issues by means of repeated identification of the same bound variables represented by symbols,[vi] so as for instance when he writes that

"A diagram ought to be as iconic as possible, that is, it should represent relations by visible relations analogous to them." ("Logical Tracts, vol. 2", 1903, CP 4.432)

In quotes such as this, it may remain ambiguous which iconicity concept is exactly at stake, but the fact that Peirce considers alternative, more or less iconic, ways of representation of the same propositions and arguments, shows an alternative iconicity concept being considered. Peirce thus considers alternative representation as substitutes for Identity Lines (here "Ligatures" as a concept for systems of Identity Lines meeting across cuts) under the headline of "Selectives":

"A Ligature crossing a Cut is to be interpreted as unchanged in meaning by erasing the part that crosses to the Cut and attaching to the two Loose Ends so produced two Instances of a Proper Name nowhere else used; such a Proper name (for which a capital letter will serve) being termed a *Selective*." ("Prolegomena to an Apology for Pragmaticism" (1906), CP 4.561)

In cases where the web of Lines of Identity in a Beta graph becomes so entangled that it is difficult to survey, some of these lines may be cut, and the identity of the now severed and scattered bits of Identity Line may be secured by the addition of identical symbolical letters to the outermost end of the remaining Identity Line bits. When reading the graph outside-in, the reader must now take note of the quantification indicated by the location of that outermost Identity Line end, remember the letter representing the Selective and identify the more innerly appearances of the same letter with the first quantification. Peirce explicitly regrets the introduction of these Selectives because they lack the iconicity of identity lying in the continuous line connecting the different predicate which this Identity Line takes:[vii]

> "[The] purpose of the System of Existential Graphs, as it is stated in the Prolegomena [533], [is] to afford a method (1) as *simple* as possible (that is to say, with as small a number of arbitrary conventions as possible), for representing propositions (2) as *iconically*, or diagrammatically and (3) as *analytically* as possible. [...] These three essential aims of the system are, every one of them, missed by Selectives." ("The Bedrock beneath Pragmaticism" (2), 1906, CP 4.561 n.1)

The substition for the Identity Line by Selectives is less iconic because it requires the symbolic convention of identifying different line segments by means of attached identical symbols. The Identity Line, on the other hand, is immediately an icon of identity because it makes use of the continuity of the line which so to speak just stretches the identity represented by the spot – and which is, at the same time, a natural iconical representation of a general concept:

> "The second aim, to make the representations as iconical as possible, is likewise missed; since Ligatures are far more iconic than Selectives. For the comparison of the above figures shows what a Selective can only serve its purpose through a special habit of interpretation that is otherwise needless in the system, and that makes the Selective a Symbol and not an Icon; while a Ligature expresses the same thing as a necessary consequence regarding each sizeable dot as an Icon of what we call an "individual object"; and it must be such an Icon if we are to regard an invisible mathematical point as an Icon of the strict individual, absolute determinate in all respects, which imagination cannot realize." (ibid.)

The Peircean Selective, of course, does exactly the same as quantification with bound variables undertake in the traditional system: the first presentation of the variable determines the quantification of it, and later occurrences of that variable in the logical expression remains under the scope of that quantifier. But it remains a second-rate, anti-iconic representation when one and the same bound variable is no longer represented by one entity only (the line of identity) but is, instead, represented by a series of different lines of identity identified only by the addition of symbolical indices, or, as in ordinary formal logic, by the series of x's or y's, identified only by their merely symbolical identity.

The reason why Peirce considers the introduction of Selectives at all is, of course, that in sufficiently complicated Beta graphs involving many variables taking many predicates, the network of Identity Lines may form a thicket hard to get a simple visual grasp of. The reason for introducing Selectives is thus heuristic and psychological, pointing to the specific competences and limitatins of a human observer; we might imagine a mind better equipped than ours which would be able to survey in one glance any complicated web of Identity Lines without having to resort to Selectives.

But the important issue here is Peirce's very motivation for preferring Identity Lines to Selectives in the first place: they are *more iconical*, because they represent in one icon entity what is also, in the object, one entity. This thus forms an additional, stronger iconicity criterion in addition to the operational iconicity criterion. One could object that Peirce was in no position to know the informational equivalence between his Beta system and what was only later named first order predicate logic – but still his argument was implicitly aimed against his own earlier algebraical attempts at logic formalization (a formalization, we should add, which through Schröder yielded a huge impact on Peano's formalization merging with Russell to result in mainstream "symbolic" formal logic). In any case, Peirce realized that the two versions of Beta graphs, with Identity Lines and with Selectives, respectively, was logically equivalent, and the latter even in some cases heuristically superior. And still he preferred the former version in as many cases as possible, thereby indicating a criterion for distinguishing more and less iconical (2) representations among iconical (1) representations being equivalent under the operational criterion. We may indicate these two different concepts of iconicity by iconicity (1), referring to the operational criterion, and iconicity (2), referring to the "more iconical", optimal type of iconicity. Peirce's arguments pro et con Identity Lines and Selectives display two different constraints on logic representations. What counts for the Selectives was heuristic, practical issues tied to the psychology of the reasoner – obviously a constraint deemed less noble by an avowed anti-psychologist like Peirce. What counts for the Identity Lines is rather an *ontological* argument: the idea that using them, Beta graphs more appropriately depict logical relations *like they really are*, thus adding to the pragmatist operational criterion of iconicity an ontologically motivated extra criterion. According to this criterion, if two icons are equivalent according to iconicity (1), still the representation which is most iconical according to iconicity (2) must be preferred – if heuristic arguments do not count against it, that is.

This implies that the addition of iconicity (2) to Peirce's iconicity doctrine is connected to his *realism*. It is well known that Peirce's realism developed over the years, such as is documented most famously by his own diamond example from the very birthplace of pragmatism, *How To Make Our Ideas Clear* (1878), to which he returns in *Issues of Pragmatism* (1905) in order to correct what he now sees as a youthful failure. In his early doctrine, he claimed that if a diamond was formed within a bed of cotton and remained there until it was consumed by fire, it would be a mere convention to call that diamond hard, because it was never put to any test. In his mature correction, Peirce says that his earlier idea was nominalist and tied to an actualist conception of being. Now, he refers to the "real possibilities" inherent in the very concept of diamond which implies that it is hard because it *would be* tested hard if subjected to the adequate testing – the hardness of the diamond is not only subject to testing but connected to other pieces of knowledge of diamonds' molecular structure, reflection abilities, heat development during burning, etc. While earlier only admitting subjective possibilities – possibilities due to the fact that we possess incomplete knowledge about the fact in issue (in this sense, it is possible that there are living beings on other planets, because we do not know it is not the case) – Peirce now admit that certain such possibilities also have a *real* character, laws of nature being the most clear expressions of such real possibilities (if I held a stone and let go, the stone would fall to the ground). Peirce's admission of such real possibilities in the latter half of the 90's considerably changes and enriches his concept of thirdness as well as his conception of the pragmatic maxim in terms of *would-be*s. Still, this realism was never really incorporated into his logic graphs.

In Max Fisch's famous charting of Peirce's almost life-long development into a still more extreme – or consequent – realism, the last step, only hinted at in some of Peirce's late writings, was the rejection of material implication – the nomal logical interpretation of the implication $p \to q$ according to which it is equivalent to *non-p or q*. Of course, the traditional uneasiness with this interpretation is that according to this interpretation, all cases of p being false automatically render $p \to q$ true, in contrast to different versions of strong implication, among those implication in everyday language where p being false rather makes the implication irrelevant than true. Most of his lifetime, Peirce was a strong defender of material implication (under the title of "Philonian", as opposed to "Diodoran" implication, the names stemming from Cicero's reference to two competing Hellenistic logicians), but Fisch is right in indicating that the mature Peirce expressed increasing doubts as to the possible nominalism inherent in material implication, admitting as early as 1898 that it does indeed seems strange that an occurrence of non-lightning should really support the implication that "If it is lightening, it will thunder."[viii]:

"For my part, I am a Philonian; but I do not think that justice has ever been done to the Diodoran side of the question. The Diodoran vaguely feels that there is something wrong about the statement that the proposition "If it is lightening, it will thunder," can be made true merely by its not lightening." ("Types of Reasoning" (1898), Peirce 1976, 169). .

One even stronger locus of such doubt appears eight years later, and interestingly it addresses the interpretation of exactly the issue of Identity Lines in Beta and Gamma graphs:

"Second, In a certain partly printed but unpublished "Syllabus of Logic," which contains the only formal or full description of Existential Graphs that I have ever undertaken to give, I laid it down, as a rule, that no graph could be partly in one area and partly in another; and this I said simply because I could attach no interpretation to a graph which should cross a cut. As soon, however, as I discovered that the verso of the sheet represents a universe of possibility, I saw clearly that such a graph was not only interpretable, but that it fills the great lacuna in all my previous developments of the logic of relatives. For although I have always recognized that a possibility may be real, that it is sheer insanity to deny the reality of the possibility of my raising my arm, even if, when the time comes, I do not raise it; and although, in all my attempts to classify relations, I have invariably recognized, as one great class of relations, the class of references, as I have called them, where one correlate is an existent, and another is a mere possibility; yet whenever I have undertaken to develop the logic of relations, I have always left these references out of account, notwithstanding their manifest importance, simply because the algebras or other forms of diagrammatization which I employed did not seem to afford me any means of representing them. I need hardly say that the moment I discovered in the verso of the sheet of Existential Graphs a representation of a universe of possibility, I perceived that a reference would be represented by a graph which should cross a cut, thus subduing a vast field of thought to the governance and control of exact logic.

Third, My previous account of Existential Graphs

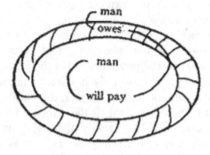

Fig. 1

was marred by a certain rule which, from the point of view from which I thought the system ought to be regarded, seemed quite out of place and inacceptable, and yet which I found myself unable to dispute. I will just illustrate this matter by an example. Suppose we wish to assert that there is a man every dollar of whose indebtedness will be paid by some man

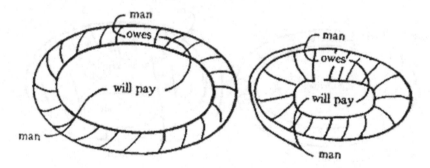

Figs. 2-3

or other, perhaps one dollar being paid by one man and another by another man, or perhaps all paid by the same man. We do not wish to say how that will be. Here will be our graph, Fig. 1. But if we wish to assert that one man will pay the whole, without saying in what relation the payer stands to the debtor, here will be our graph, Fig. 2. Now suppose we wish to add that this man who will pay all those debts is the very same man who owes them. Then we insert two graphs of teridentity and a line of identity as in Fig. 3. The difference between the graph with and without this added line is obvious, and is perfectly represented in all my systems. But here it will be observed that the graph "owes" and the graph "pays" are not only united on the left by a line outside the smallest area that contains them both, but likewise on the right, by a line inside that smallest common area. Now let us consider a case in which this inner connection is lacking. Let us assert that there is a man A and a man B, who may or may not be the same man, and if A becomes bankrupt then B will suicide. Then, if we add that A and B are the same man, by drawing a line outside the smallest common area of the graphs joined, which are here bankrupt and suicide, the strange rule to which I refer is that such outer line, because there is no connecting line within the smallest common area, is null and void, that is, it does not affect the interpretation in the least. . . . The proposition that there is a man who if he goes bankrupt will commit suicide is false only in case, taking any man you please, he will go bankrupt, and will not suicide. That is, it is falsified only if every man goes bankrupt without suiciding. But this is the same as the state of things under which the other proposition is false; namely, that every man goes broke while no man suicides. This reasoning is irrefragable as long as a mere possibility is treated as an absolute nullity. Some years ago, however, when in consequence of an invitation to deliver a course of lectures in Harvard University upon Pragmatism, I was led to revise that doctrine, in which I had already found difficulties, I soon discovered, upon a critical analysis, that it was absolutely necessary to insist upon and bring to the front, the truth that a mere possibility may be quite real. That admitted, it can no longer be granted that every conditional proposition whose antecedent does not happen to be realized is true, and the whole reasoning just given breaks down.

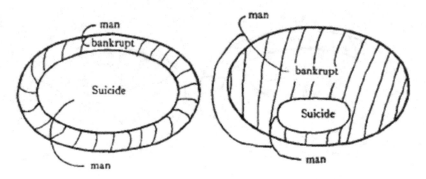

Figs. 4-5

I often think that we logicians are the most obtuse of men, and the most devoid of common sense. As soon as I saw that this strange rule, so foreign to the general idea of the System of Existential Graphs, could by no means be deduced from the other rules nor from the general idea of the system, but has to be accepted, if at all, as an arbitrary first principle -- I ought to have asked myself, and should have asked myself if I had not been afflicted with the logician's bêtise, What compels the adoption of this rule? The answer to that must have been that the interpretation requires it; and the inference of common sense from that answer would have been that the interpretation was too narrow. Yet I did not think of that until my operose method like that of a hydrographic surveyor sounding out a harbour, suddenly brought me up to the important truth that the verso of the sheet of Existential Graphs represents a universe of possibilities. This, taken in connection with other premisses, led me back to the same conclusion to which my studies of Pragmatism had already brought me, the reality of some possibilities. This is a striking proof of the superiority of the System of Existential Graphs to either of my algebras of logic. For in both of them the incongruity of this strange rule is completely hidden behind the superfluous machinery which is introduced in order to give an appearance of symmetry to logical law, and in order to facilitate the working of these algebras considered as reasoning machines. I cannot let this remark pass without protesting, however, that in the construction of no algebra was the idea of making a calculus which would turn out conclusions by a regular routine other than a very secondary purpose. . . ." ("For the National Academy of Sciences, 1906 April Meeting in Washington", CP 4.579-81)

In this long quotation, Peirce considerably revises the whole foundation of Beta and Gamma graphs. Cuts no longer represent negation, but merely possibility – they only represent negation if they enclose a further blank cut (meaning everything can be derived from the contents of the first cut, evidently making those contents false). Furthermore, material implication is given up or at least relativized: not all conditional propositions with false antecedents are true. References as relations are included as represented by graphs connecting actuality and possibility, evenly and unevenly enclosed cuts.

Finally, there is the relation between Identity Line conventions and real possibilities which Peirce admitted in his metaphysics from the later 90's onwards (cf. the diamond discussion). The "strange rule" which Peirce refers to in the quote is presented earlier that very same year and says in its brief form that "... there is some one individual of which one or other of two predicates is true is no more than to say that there either is some individual of which one is true or else there is some individual of which the other is true." ("Prolegomena to an Apology for Pragmatism", 1906, CP 4.569). Now, this rule will imply that the two graphs representing "if A becomes bankrupt, B will suicide", and "if A becomes bankrupt, A will suicide", are identical. Both are falsified if every man goes bankrupt without any man suiciding. However, the two propositions are, evidently, not identical, A and B being potentially different persons in the former proposition, not so in the latter. But the "strange rule" exactly makes of such possibilities mere "nullities". Peirce's hasty and difficult reasoning at this point must refer to the fact that the possibility of A and B being identical is not a mere subjective possibility but a real possibility, given by the possible causal link between bankruptcy and suicidal tendencies, constituting a real tendency in social life.

The fact that it is the very system of Existential Graphs which leads Peirce to these conclusions is taken to count among the chief virtues of that system. While his own algebras hid such facts behind "superfluous machinery" constructed with their (secondary) aim as reasoning machines, the Existential Graphs are not so constructed, but with the aim of displaying to the highest degree of detail and clarity every single logical step taken in reasoning. The efficiency of the algebras is thus contrasted to the logical detail of the graphs – this is an argument referring to the larger degree of iconicity (2) of the graphs, even if they may be equivalent as reasoning machines, that is, with respect to iconicity (1).

This also leads to a further reinterpretation of the iconicity inherent in Identity Lines:

> "The System of Existential Graphs recognizes but one mode of combination of ideas, that by which two indefinite propositions define, or rather partially define, each other on the recto and by which two general propositions mutually limit each other upon the verso; or, in a unitary formula, by which two indeterminate propositions mutually determine each other in a measure. I say in a measure, for it is impossible that any sign whether mental or external should be perfectly determinate. If it were possible such sign must remain absolutely unconnected with any other. It would quite obviously be such a sign of its entire universe, as Leibniz and others have described the omniscience of God to be, an intuitive representation amounting to an indecomposable feeling of the whole in all its details, from which those details would not be separable. For no reasoning, and consequently no abstraction, could connect itself with such a sign. This consideration, which is obviously correct, is a strong argument to show that what the system of existential graphs represents to be true of propositions and which must be true of them, since every proposition can be analytically expressed in existential graphs, equally holds good of concepts that are not propositional; and this argument is supported by the evident truth that no sign of a thing or kind of thing -- the ideas of signs to which concepts belong -- can arise except in a proposition; and no logical operation upon a proposition can result in anything but a proposition; so that non-propositional signs can only exist as constituents of propositions. But it is not true, as ordinarily represented, that a proposition can be built up of non-propositional signs. The truth is that concepts are nothing but indefinite problematic judgments. The concept of man necessarily involves the thought of the possible being of a man; and thus it is precisely the judgment, "There may be a man." Since no perfectly determinate proposition is possible, there is one more reform that needs to be made in the system of existential graphs. Namely, the line of identity must be totally abolished, or rather must be understood quite differently. We must hereafter understand it to be potentially the graph of teridentity by which means there always will virtually be at least one loose end in every graph. In fact, it will not be truly a graph of teridentity but a graph of indefinitely multiple identity.
>
> We here reach a point at which novel considerations about the constitution of knowledge and therefore of the constitution of nature burst in upon the mind with cataclysmal multitude and resistlessness." (op.cit., CP 4.583-84)

All Identity Lines are now to be considered implicitly polyadic – for the realist reason that the entities referred to may have other predicates in common than the ones explicitly mentioned in the graph, thus sharing real possibilities which are not referred to in the explicit graph. Peirce never consistently revised the Graphs according tothe cataclysms of ideas proposed here, but it is obvious that the revisions proposed pertain to the overall idea of iconicity (2) – the attempt at making the graphs match ontological structure to as large a degree as possible.

3   The Pragmatic Maxim and the Two Iconicity Notions

The coexistence of two competing iconicity criteria in the mature philosophy of Peirce raises further questions. What about the pragmatic maxim, Peirce's basic idea that the content of all conceptions may be exhausted by considering which practical effects those conceptions would be conceived to have under imagined circumstances? The operational iconicity criterion seems moulded after the pragmatic maxim due to the reductivist action foundation of both: anything which does not have conceived consequences, practically or theoretically, may be discarded. The investigation of possible practical consequences in the former case mirrors the investigation of possible theorems to be inferred in the latter. But this interpretation leaves iconicity (2) in a strange vacuum. If optimal iconicity remains without any practically conceivable consequences, it may be thought to belong to what may be discarded by the maxim as superfluous verbiage. For is there any conceivable practical difference between Identity Lines and Selectives in Existential Graphs? Of course there is the realist conviction that Identity Lines may refer to real generals which may be easier grasped (in some cases, at least) by Identity Lines than by Selectives? And of course there is the practical issue that in complicated cases, Selectives may facilitate an easier use of the graphs than Identity Lines. But at the same time, the amount of theorems, of new information, accessible by the two means are supposed to be exactly the same? Maybe, this difference corresponds to two different readings of the pragmatic maxim, cf. Peirce's own two readings without and with the hardness of the untested diamond, respectively. The untested diamond hardness and the realist interpretation of the pragmatic maxim seems to correspond to the addition of iconicity (2) as a criterion with its possibilities for distinguishing between more and less iconical representations in addition to the provision of new information, while the earlier, nominalist idea corresponds to the version of the maxim where it charts testable regularities and nothing more. Just like existence is no predicate, it seems like Peircean reality is no predicate neither, and the addition of reality does not add to the amount of information which may be taken out of any given predicate. But Iconicity (2) may add, in some cases, to the heuristics of working with representation systems, just like it presents the same information in a so to speak ontologically more valid form.   If that interpretation is correct, then the introduction of iconicity (2) as a criterion constitutes yet another step in Peirce's lifelong movement towards realism, as charted by Max Fisch. In that case, Iconicity (2) is tightly interwoven with the step leading from the Real Possibilities introduced in the latter half of the 90's as the central mode of Thirdness on the one hand, and to Peirce's final and most realist position in search for stronger implications than material implication in the years after 1900, on the other hand.

References

Fisch, Max (1986) "Peirce's Progress from Nominalism Towards Realism" (1967), in Max Fisch (eds. K.L.Ketner and C.J.W.Kloesel) *Peirce, Semeiotic, and Pragmatism* (1986) Bloomington: Indiana University Press, 184-200

Houser, Roberts, and Van Evra (eds.) (1997) *Studies in the Logic of Charles Sanders Peirce*, Bloomington: Indiana University Press

Peirce, C. *Collected Papers* [CP] (1998) I-VIII, (ed. Hartshorne and Weiss; Burks) London: Thoemmes Press (1931-58)

New Elements of Mathematics [NEM] (1976), (ed. C. Eisele) I-IV, The Hague: Mouton

Reasoning and the Logic of Things [RLOT] (1992), (eds. K.Ketner and H.Putnam), Camb.Mass

The Essential Peirce, vol. I. (1867-1893) [EPI], *vol. II (1893-1913)* [EPII] (eds. N. Houser and C. Kloesel) (1992; 1998), Bloomington: Indiana University Press

"Logic, Considered as Semeiotic" [LCS], constructed from manuscript L 75 by Joseph Ransdell

"Existential Graphs", version of MS 514 commented by John Sowa, http://www.jfsowa.com/peirce/ms514.htm

Pietarinen, Ahti-Veikko (forthcoming) *Signs of Logic. Peircean Themes on the Philosophy of Language, Games, and Communication*, Dordrecht: Springer

Roberts, Don (1973) *The Existential Graphs of Charles S. Peirce*, The Hague: Mouton

Shin, Sun-Joo (2002) *The Iconic Logic of Peirce's Graphs*, Camb. Mass.: MIT Press

Sowa, John (2005) Commented version of Peirce MS 514 "Existential Graphs" (1909), http://www.jfsowa.com/peirce/ms514.htmStjernfelt, Frederik (2000) "Diagrams as Centerpiece in a Peircean Epistemology", in *Transactions of the Charles S. Peirce Society*, Summer, 2000, vol. XXXVI, no. 3, p. 357-92.

(2000a) "How to Learn More. An Apology for a Strong Concept of Iconicity" in M. Skov et al. (eds.) *Iconicity*, Copenhagen: NSU Press, 21-58

(forthcoming) *Diagrammatology. An Investigation on the Borderlines of Phenomenology, Ontology, and Semiotics*, Dordrecht: Springer

Zeman, Jay (1964) *The Graphical Logic of C.S. Peirce*, http://www.clas.ufl.edu/users/jzeman/graphicallogic/index.htm

---

NOTES

[i] The tricotomy is the second out of Peirce's three major tricotomies, referring to the sign's relation to itself, to its object, and to its interpretant, respectively. In Peirce's more developed series of ten tricotomies from his later years, it is the fourth.

[ii] It is important to note that Peirce's distinctions pertain to sign *use* rather than to the specific sign vehicles, based on his dictum "A sign is only a sign *in actu* ..." ("Truth and Falsity and Error," *Dictionary of Philosophy and Psychology*, ed. J.M. Baldwin, pp. 718-20, vol. 2 (1901); CP 3.569). Thus, the very same sign token may be used in some contexts as an image – paying no attention to what can be learnt from the relation between its parts – and in other contexts as a diagram. If, for instance, we took the crescent shape, image of the moon, and performed observations on it pertaining to the relation between its parts, if we, say, measured its area in comparison to the implicit full moon area, we would treat exactly the same sign token as a diagram.

[iii] It is an interesting fact in the history of science that such attacks on the notion of similarity have come from otherwise completely opposed camps, namely the analytical tradition (.e.g. Nelson Goodman) on the one hand, and the (post-) structuralists in the continental tradition on the other (e.g. Umberto Eco). See Stjernfelt (2000a) and Stjernfelt (forthcoming).

[iv] In "On Quantity" (ca. 1895, in Peirce 1976, p. 275).

---

^v Peirce had already, in his algebras of logic and independently of Frege, invented the "symbolic" quantifier notion. Peirce's version became later, through Schröder and Peano, the standard notation of ∀ and ∃ (in Peirce's version ∏ and Σ, respectively).

^{vi} The issue of the iconicity of different aspects and conventions of Existential Graphs is far wider than the alternative between Identity Lines and Selectives which is chosen as the main case in our context because Peirce himself highlights it so thoroughly. The overall iconical motivation in the construction of the graphs is well indicated by Peirce when introducing the details of the graphs:

"I dwell on these details which from our ordinary point of view appear unspeakably trifling, — not to say idiotic, — because they go to show that this syntax is truly **diagrammatic**, that is to say that its parts are really related to one another in forms of relation analogous to those of the assertions they represent, and that consequently in studying this syntax we may be assured that we are studying the real relations of the parts of the assertions and reasonings; which is by no means the case with the syntax of speech." (MS 514, "Existential Graphs" (1909), quoted from John Sowa's commented version of that text).

Shin (2002, 53-58) lists three basic iconical features of Beta graphs, namely Identity Lines, quantifiers and scope. Quantifiers do seem to come naturally because the end of an Identity Line in an unenclosed graph is simply taken to mean "something is ...", but it deserves mention that in Peirce's earlier formalization attempt from the 90's known as Entiative Graphs, in many respects dual to Existential Graphs, the very same sign is taken to stand for the universal quantifier. Maybe it could be argued that a point in a plane does indeed more naturally mean "something" than "all". Scope seems to come natural in the endoporeutic, outside-in, reading of the graphs (which Shin is otherwise out to dismantle), because the outermost occurrence of part of an Identity Line defines the scope of the corresponding quantifier, and more innerly located quantifiers are taken to lie within the scope of the more outerly ones.

In addition to these iconicities, a basic iconicity in Existential Graphs is one of its very motivating ideas in Peirce, namely the representation of material implication by means of a "scroll", that is, two nested cuts where the premiss is placed within the outer cut but outside the inner cut, while the conclusion is placed in the inner cut. This geometrical inclusion of the conclusion within the premiss furnishes a simple iconic representation of the idea that the conclusion lies in, is inherent in, or is im-plicated by the premiss. Peirce proudly refers to this in CP 4,553 n1 (from "The Bedrock beneath Pragmaticism", 1906) while at the same time complaining about the lack of iconic representation of *modality* in the Graphs, a lack he attempts to remedy not much later, cf. below.

Another issue discussed by Shin – but not in relation to iconicity – is Peirce's distinction between logic systems as result-oriented calculi and logic systems as representations of logical thought process (a distinction she strangely thinks loses its relevance in graphical systems). Here, the former aims at quick and easy results, and a plurality of logical connectors and rules may be used to further that aim as expediently as possible. In the dissection of logical inference steps, on the other hand, as few connectors and rules as possible should be chosen, in order to be able to compare the single steps taken – a guideline explicitly followed in Peirce's graphs. In this connection, Peirce remarks that it is "... a defect of a system intended for logical study that it has two ways of expressing the same fact, or any superfluity of symbols, although it would not be a serious defect for a calculus to have two ways of expressing a fact." ("Symbolic Logic", in *Baldwin's Dictionary*, 1901/1911, CP 4.373). This requirement – which Existential Graphs do not perfectly satisfy – is obviously iconical, demanding the extinction of arbitrary, that is, non-iconical, choices between parallel representations.

Finally, Pietarinen's (forthcoming, 128-31) argument against Shin runs along these lines: her rewriting of the inference rules of Peirce's graphs gives many more rules and connectors than does Peirce's own version, and so is less analytical and iconical than his (even if maybe facilitating easier readability on some points). In his defense of the endoporeutic, outside-in, interpretation of the graphs against Shin's attacks, Pietarinen highlights a further and very basic iconical feature in them: the dialogic structure, rhythmically changing between a Graphist and a Grapheus, responsible for existentially and universally quantified propositions, respectively, and thus responsible for taking turns in a dialogue where each of them manipulates the graph according to Peirce's rules. Pietarinen of course makes this point in order to facilitate his interesting, Hintikkan interpretation of the graphs in terms of game-theoretical semantics, where the two interlocutors hold opposed atrategic aims in the conversation: the proof or disproof of the initial proposition, respectively.In our context, we may emphasize the basic iconicity inherent in this conversational structure of the graphs, motivated in the supposedly dialogical structure of thought, be it between persons or between positions in one person's thought and mind.

vii Given the equivalence between Identity Line and Selective representations, we might use this idea in reconsidering ordinary Peano-Russell-style formal logic – here, we might see the different instances of the same bound variable in a symbolic expression as invisibly connected by an erased Identity Line running in an additional line parallel to the line of the normal expression.

viii Two years earlier, not long before the introduction of Real Possibilities in January 1897, the doubt is awakening: "It may, however, be suspected that the Diodoran view has suffered from incompetent advocacy, and that if it were modified somewhat, it might prove the preferable one." ("The Regenerated Logic", 1896, CP 3.442-3). But as early as the second "On the Algebra of Logic" (1885, 3.374), Peirce states that "If, on the other hand, A [the premiss] is in no case true, throughout the range of possibility, it is a matter of indifference whether the hypothetical be understood to be true or not, since it is useless. But it will be more simple to class it among true propositions, because the cases in which the antecedent is false do not, in any other case, falsify a hypothetical." Here, Peirce observes the problem, but accepts material implication out of simplicity (and not iconicity) reasons.

Simple Conceptual Graphs and Simple Concept Graphs

J.P. Aubert[1], J.-F. Baget[2], and M. Chein[1]

[1] LIRMM
{aubert, chein}@lirmm.fr
[2] INRIA/LIRMM
baget@inrialpes.fr

Abstract. Sowa's Conceptual Graphs and Formal Concept Analysis have been combined into another knowledge representation formalism named Concept Graphs. In this paper, we compare Simple Conceptual Graphs with Simple Concept Graphs, by successively studying their different syntaxes, semantics, and entailment calculus. We show that these graphs are almost identical mathematical objects, have equivalent semantics, and similar inference mechanisms. We highlight the respective benefits of these two graph-based knowledge representation formalisms, and propose to unify them.

1 Introduction

Introduced in [19], Conceptual Graphs were extended in [20]. Since [5], the "Montpellier school of conceptual graphs" has been studying this knowledge representation formalism as a family of formal languages whose objects are graphs and where inferences are computed using graph-based operations (e.g. [3]). In the same way, [22] has proposed to combine conceptual graphs with Formal Concept Analysis (FCA). This work has been developed in [18,7,8].

In this paper, we compare these two approches and focus on the mathematical and computational viewpoints. Since we are interested in conceptual graphs and concept graphs as logics, we will successively compare the syntax (Sect. 2), semantics (Sect. 3), and calculus (Sect. 4) of these two languages.

2 Syntax

We show here that simple conceptual graphs and simple concept graphs are avatars of the notion introduced by Sowa [19]. Only simple conceptual (or concept) graphs are considered, thus the adjective *simple* is implicit hereafter.

In the first subsection, we show that, up to a well-known transformation, the objects described by bipartite graphs and directed hypergraphs have the same structure. Then we show that the *vocabulary* (or support) upon which conceptual graphs are defined and the *alphabet* used for concept graphs are identical, with some minor variants. Finally, we compare various definitions used for conceptual and concept graphs.

H. Schärfe, P. Hitzler, and P. Øhrstrøm (Eds.): ICCS 2006, LNAI 4068, pp. 87–101, 2006.

2.1   Hypergraphs and Bipartite Graphs

Let us recall a very well known bijection between hypergraphs and bipartite graphs (see [4] for relationships between graphs and hypergraphs and [10] for a more recent introduction to graph theory). Let $H = (X, \mathcal{E})$ be a hypergraph over X, that is X is a set (of vertices) and $\mathcal{E}$ a set of hyperedges i.e. non-empty subsets of X. Let $\alpha(H)$ be the bipartite graph (X, R, E) defined as follows:

- R is disjoint from X and there is a bijection f from $\mathcal{E}$ to R,
- let $c \in C$ and $r \in R$, (c, r) is in E iff $c \in f^{-1}(r)$.

It is simple to check that α is a bijection from the set of hypergraphs over X to the set of bipartite graphs with the first vertex set X.

$\alpha(H)$ is called the *incidence (bipartite) graph of H*. It is straightforward to extend the bijection α to a bijection from multi-hypergraph (i.e. $\mathcal{E}$ is no longer a set of subsets but rather a family of subsets) to bipartite multi-graphs (E is a family of edges). Let us call *ordered multi-hypergraph* a multi-hypergraph in which any hyperedge is totally ordered. Let us call *ordered bipartite multi-graph* a bipartite multi-graph in which any set of edges incident to a vertex of R is totally ordered. α can be trivially extended to the ordered objects and one gets:

Property 1. The application α from the set of ordered multi-hypergraphs over X to the set of ordered bipartite multi-graphs with first vertex set X is a bijection.

A bipartite graph is a graph, this trivial remark leads to a first important consequence (other will be discussed after introducing conceptual graphs). When one wants to graphically represent a hypergraph, a representation of its incidence bipartite graph is generally drawn (see FIG. 1). Although conceptual graphs are usually defined via bipartite graphs, the alternative hypergraph definition is sometimes used (*e.g.* [2], for more efficient algorithms).

2.2   Vocabulary and Alphabet

The structure, called *support* in [5], encoding *terms*, as well as type orderings is the core of the *canon* [20]. Here we use the name *vocabulary*, which is more standard in KR. A similar structure, named *alphabet*, is used in the concept graphs formalism [18,7]. In this subsection we compare these two structures.

Definition 1 (Vocabulary). *A vocabulary is a triple* $(T_C, T_R, \mathcal{I})$ *where:*

- $T_C, T_R, \mathcal{I}$ *are pairwise disjoint sets.*
- T_C, *the set of concept types, is partially ordered by a relation $\leq$ and has a greatest element denoted* $\top$.
- T_R, *the set of relation symbols, is partially ordered by a relation $\leq$, and is partitioned into subsets* $T_R^1, \ldots, T_R^k$ *of relation symbols of arity* $1, \ldots, k$ *respectively. The arity of a relation r is denoted* $arity(r)$. *Furthermore, any two relations with different arities are not comparable.*
- $\mathcal{I}$ *is the set of individual markers.*

Definition 2 (Alphabet). *Relationships between an alphabet* $(\mathcal{G}, \mathcal{C}, \mathcal{R})$ *and a vocabulary are as follows:*

- $\mathcal{G} = \mathcal{I}$ *the object names are the individual markers,*
- $\mathcal{C} = T_C$ *the concept names are the type of concepts,*
- $\mathcal{R} = T_R \cup \{=\}$ *the relation names include the equality symbol.*

Some definitions incorporate the generic marker $*$ in the vocabulary. This is not necessary, since it is the same for all vocabularies, so we will only introduce it in the definition of conceptual graphs. In the same way, the equality symbol is not required in the definition of an alphabet. In logics, it is usually considered separately from the relation symbols.

Both vocabulary and alphabet encode the same information, which could be represented in logics by an ordered FOL language without function symbols in which some unary predicates are distinguished.

2.3 Comparing Conceptual Graphs and Concept Graphs

Conceptual Graphs. The following definition is directly inspired from [5]. We will add in Sect. 4 conditions about coreference (they are only relevant for computational purposes).

Definition 3 (Conceptual graph). *A* conceptual graph *over a vocabulary* $\mathcal{V}$ *is a 5-tuple* $(C, R, E, l, \text{coref})$ *such that:*

- (C, R, E) *is a multi-bipartite graph,*
- *coref is an equivalence relation over* C,
- l *is a labelling function of* $C \cup R$ *such that:*
  - *for any* $x \in C$, $l(x) \in T_C \times (\mathcal{I} \cup \{*\})$,
  - *for any* $x \in R$, $l(x) \in T_R$,
  - *for any* $x \in R$, *the edges incident to* x *are labelled* $\{1, \ldots, arity(l(x))\}$

Concept Graphs. The first definition of a concept graph was proposed in [22]. We also present here the definitions from [18] and [7].

Definition 4 (Concept graph, [22]). *An* abstract concept graph *is a structure* $\mathfrak{G} = (V, F, \nu, D, \kappa, \theta)$ *for which:*

- V *and* F *are finite sets and* ν *is a mapping of* E *to* $\bigcup_{k=1}^{n} V^k$ *($n \geq 2$ s. t. (V, F, ν) can be considered as a finite directed multi-hypergraph with vertices from V and edges from F (we define $\mid e \mid = k$ if $\nu(e) = (v_1, \ldots, v_k)$),*
- D *is a finite set and* κ *a mapping of* $V \cup F$ *to* D *s. t.* $\kappa(e_1) = \kappa(e_2) \Rightarrow \mid e_1 \mid = \mid e_2 \mid$ *(the elements of D may be understood as abstract concepts),*
- θ *is an equivalence relation on* V.

Prediger [18] slightly transforms the previous definition by removing the label set D, and replacing it by the (exterior) notion of an alphabet:

Definition 5 (Concept graph, [18]). *A concept graph over the alphabet* $(\mathcal{C}, \mathcal{G}, \mathcal{R})$ *is a structure* $\mathfrak{G} = (V, F, \nu, \kappa, \rho)$, *where*

- *(V, F, ν) is a finite directed multi-hypergraph*
- *$\kappa: V \cup F \to \mathcal{C} \cup \mathcal{R}$ is a mapping such that $\kappa(V) \subseteq \mathcal{C}$ and $\kappa(F) \subseteq \mathcal{R}$, and all $e \in F$ with $\nu(e) = (v_1, \ldots, v_k)$ satisfy $\kappa(e) \in \mathcal{R}_k$,*
- *$\rho: V \to \mathfrak{P}(\mathcal{G}) \setminus \{\emptyset\}$ is a mapping.*

There are two other syntactical differences between Wille and Prediger, in Wille's definition there is an equivalence relation over V, which is not the case in Prediger, and in Prediger two labels are associated to an element of V: an element of $\mathcal{C}$ and a non-empty subset of $\mathcal{G}$. Thus, as Prediger said [18]:

> *"Apart from some little differences, the concept graphs correspond to the simple conceptual graphs as defined in [5] or [20]."*

More precisely, it is straightforward to extend the canonical bijection α from a class of ordered multi-hypergraphs to the class of their incidence graphs to an injective mapping, also called α, from the class of concept graphs over $\mathcal{V}$ to the class of conceptual graphs over $\mathcal{V}$.

Let $\mathfrak{G} = (V, F, \nu, \kappa, \theta)$ be a concept graph, $G = \alpha(\mathfrak{G}) = (C, R, E, l, coref)$ is defined as follows. Any $x \in V$ with $\rho(x) = \{g_1, \ldots, g_k\}$, $k \geq 2$, is duplicated into k nodes $x_1, \ldots, x_k$. C is the union of the $\{x \in V \text{s.t.} |\rho(x)| = 1$ and the set of duplicated nodes. If $\rho(x) = \{g\}$ then $l(x) = (\kappa(x), g)$. If $\rho(x) = \{g_1, \ldots, g_k\}$ then $l(x_i) = (\kappa(x), g_i)$.

Any $e \in F$ with $\nu(e) = (v_1, \ldots, v_k)$ is transformed into $|\rho(v_1)| \times \ldots \times |\rho(v_k)|$ relation nodes of R with label $\kappa(e)$ and whose neighborhood are the $arity(\kappa(e))$-tuples associated with $\rho(v_1) \times \ldots \times \rho(v_k)$. The equivalence $coref$ is the discrete equivalence. This coding preserves the graphs semantics (Sect. 3).

Let's consider now the third definition of concept graphs [7].

Definition 6 (Concept graphs, [7]). *A concept graph over $\mathcal{V}$ is a structure* $\mathfrak{G} = (V, F, \nu, \kappa, \rho)$, *where:*

- *V and F are pairwise disjoint, finite sets whose elements are called vertices and edges,*
- *$\nu: F \to \bigcup_{k \in N} V^k$ is a mapping (we write $| e | = k$ for $\nu(e) \in V^k$,*
- *$\kappa: V \cup F \to \mathcal{C} \cup \mathcal{R}$ is a mapping such that $\kappa(V) \subseteq \mathcal{C}$ and $\kappa(F) \subseteq \mathcal{R}$, and all $e \in F$ with $| e | = k$) satisfy $\kappa(e) \in \mathcal{R}_k$,*
- *$\rho: V \to \mathfrak{G} \cup \{*\}$ is a mapping.*

This is almost the definition of conceptual graphs (modulo α). Instead of considering the equivalence relation induced by coreference links, it keeps, as Sowa, coreference links. Considering two steps as in DEF 6 (a symmetric relation over C or V, then its reflexo-transitive closure $coref$ or θ), or directly the equivalence relation as in DEF 3 and 4 is a matter of taste.

Let $\mathfrak{G} = (V, E, \nu, \kappa, \rho)$ be a Dau's concept graph over $\mathcal{A}$. The conceptual graph $\alpha(G) = (C, R, E, l, coref)$ is defined as follows:

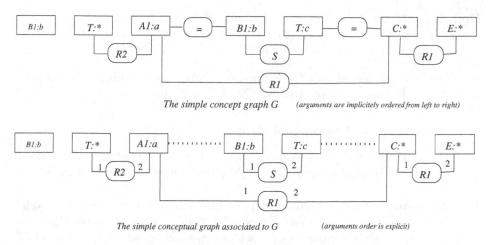

The simple concept graph G   (arguments are implicitely ordered from left to right)

The simple conceptual graph associated to G   (arguments order is explicit)

Fig. 1. Drawings of G and of $\alpha(G)$

- $C = V$, the concept nodes of $\alpha(G)$ are the vertices of G,
- R is a set in bijection with E, to each edge e of G a relation node noted $\alpha(e)$ is associated (α is a bijection and C and R are disjoint),
- if $\nu(e) = (c_1, \ldots, c_k)$, then for $i = 1, \ldots, k$, $\{r = \alpha(e), c_i\}$ is an edge of $\alpha(G)$,
- the label of a concept node c is $l(c) = (\kappa(c), \rho(c))$,
- the label of a relation node r is $l(r) = \kappa(\alpha^{-1}(r))$,
- the label of an edge $\{r = \alpha(e), c_i\}$ of $\alpha(G)$ is i,
- a class of *coref* is composed of a connected component of the graph $(V, =)$.

Let us consider the concept graph in FIG. 1 (copied from [7]). If the equality relation is replaced by coreference links, this is the drawing of $\alpha(\mathfrak{G})$. Note that, in the drawing of $\mathfrak{G}$ there are no labels on the edges, but at least for the edges incident to the oval vertices labelled R_1, R_2, S they must be added since it is not stated that these relations are symmetrical. We think that it is interesting to consider that the equality is a specific relation, and this is achieved by *coref*, which is an equivalence relation and by drawing it with specific edges (coreference links). In Figure 1 a drawing of $\alpha(G)$ is reproduced besides the drawing of G.

At this moment of our discussion, preferring concept or conceptual graphs is a matter of taste not of real mathematical differences, since they are almost the same mathematical objects. In the rest of the paper, we will now call CGs conceptual as well as concept graphs, and will consider them as the same objects.

3   Semantics

In logics, semantics are provided to define the conditions under which an assertion is true. An *interpretation* is a structure encoding a possible world. An interpretation is a *model* of a formula F if the assertion encoded by F is true in that world. The notions of interpretations and models lead to *logical consequence* (or *entailment*), whose calculus will be detailed in Sect. 4.

Definition 7. *Let $\mathcal{L}$ be a logic, and G and H be two formulas of $\mathcal{L}$. We say that G entails H (or that H is a logical consequence of G) and note $G \models_{\mathcal{L}} H$ iff every interpretation that is a model of G is also a model of H.*

We show here that models of CGs, defined in standard model-theoretic semantics [20,16] or in Formal Concept Analysis [22,18,7], are equivalent.

3.1   Semantics for Conceptual Graphs

Historically, conceptual graphs semantics have been presented by a translation Φ to FOL. This "logical semantics" is equivalent to model-theoretic semantics.

FOL Semantics of Conceptual Graphs. FOL semantics of conceptual graphs described below were introduced by Sowa [20]. A FOL language is associated to a vocabulary , and is composed of a set of constants equal to $\mathcal{I}$ and a set of predicates equal to $T_C \cup T_R$ with their arities. The order over the symbol types is represented by the following set of formulas.

Definition 8 ($\Phi(\mathcal{V})$). *Type ordering is translated by: $\forall t_1, t_2$ types of $\mathcal{V}$ of arity p such that $t_2 \leq t_1$, we obtain the formula $\forall x_1 ... x_p(t_2(x_1, ..., x_p) \rightarrow t_1(x_1, ..., x_p))$.*

Definition 9 ($\Phi(G)$). *Given any CG G, the formula $\Phi(G)$ is built as follows.*

1. *A term $term(c)$ is assigned to each concept c in the following way. If c is generic (labelled by $*$), then $term(c)$ is a variable, and if c and c' are two different generic concepts, then $term(c) \neq term(c')$. If c is labelled by the individual i, then $term(c) = i$.*
2. *An atom is assigned to each relation or concept:*
   - *the atom $t(term(c))$ is assigned to each concept c of type t;*
   - *the atom $r(term(c_1), \ldots, term(c_k))$ is assigned to each relation node x, where r is its type, k the arity of r and c_i denotes the i-th neighbor of x.*
3. *To any coreference link between two nodes c and c' is associated the formula $term(c) = term(c')$. Let $\varphi(G)$ be the conjunction of all preceding atoms.*
4. *$\Phi(G)$ is the existential closure of $\varphi(G)$.*

It is simple to check that definition 6 of a concept graph is simply a graph reformulation of such a formula.

Model Theoretic Semantics for Conceptual Graphs. It is a direct translation [16] of the model-theoretic semantics of the formulas obtained by Φ.

Definition 10 (Interpretation of terms). *The terms of a vocabulary $(T_C, T_R, \mathcal{I})$ are the elements of $T_C \cup T_R \cup \mathcal{I}$. Their interpretation is a pair (D, δ) s.t.:*

- *D is a non-empty set;*
- *δ maps each marker of $\mathcal{I}$ to an element of D, each concept type of T_C to a subset of D, and each relation type of arity k in T_R to a subset of D^k.*

Definition 11 (Model of a vocabulary). *A* model *of a vocabulary $\mathcal{V}$ is an interpretation $I = (D, \delta)$ of its terms s. t.:*

- $\forall c, c' \in T_C,\ c \le c' \Rightarrow \delta(c) \subseteq \delta(c')$
- $\forall r, r' \in T_R,\ r \le r' \Rightarrow \delta(r) \subseteq \delta(r')$

Definition 12 (Model of a graph). *Let $G = (C, R, E, l, \mathrm{coref})$ be a conceptual graph over a vocabulary $\mathcal{V}$. An interpretation (D, δ) of the terms of $\mathcal{V}$ is a model of G iff there is a mapping (an* assignment*) α from C to D s. t.:*

- *For any individual concept c with marker i: $\alpha(c) = \delta(i)$;*
- *$\forall c \in C,\ \alpha(c) \in \delta(type(c))$;*
- *$\forall r \in R$ with neighbors $(c_1, \ldots, c_k),\ (\alpha(c_1), \ldots, \alpha(c_k)) \in \delta(type(r))$;*
- *$\forall c, c' \in C,\ (c, c') \in \mathrm{coref} \Rightarrow \alpha(c) = \alpha(c')$.*

It is easy to check that the models (as usually defined in FOL) of $\Phi(\mathcal{V})$ (resp. $\Phi(G)$) are exactly the models of $\mathcal{V}$ (resp. G).

Definition 13 (Deduction for Conceptual Graphs). *Let $\mathcal{V}$ be a vocabulary, and G and H be two conceptual graphs over $\mathcal{V}$. We say that H is deducible from G and $\mathcal{V}$ and note $\mathcal{V}, G \models H$ iff every interpretation that is a model of $\mathcal{V}$ and G is also a model of H.*

3.2 Semantics for Concept Graphs

The first semantics, based upon power context families [22], for concept graphs was given by Prediger [18], we present here the slightly different version in [7].

Definition 14 (Power context family). *A* power context family *is a family $\vec{\mathbb{K}} := (\mathbb{K}_0, \mathbb{K}_1, \ldots)$ of formal contexts $\mathbb{K}_k := (G_k, M_k, I_k)$ such that $G_0 \ne \emptyset$ and for every $k : G_k \subseteq (G_0)^k$. The elements of G_0 are the objects of $\vec{\mathbb{K}}$. A pair (A,B) with $A \subseteq G_k$ and $B \subseteq M_k$ is called a concept of $\mathbb{K}_k$ if and only if $A = \{g \in G_k \mid g\, I_k\, b \text{ for all } b \in B\}$ and $B = \{m \in M_k \mid a\, I_k\, m \text{ for all } a \in A\}$. A is called the extension $ext((A,B))$ and B is called the intension $int((A,B))$ of the concept (A,B). The set of all concepts of a formal context $\mathbb{K}_k$ is denoted by $\mathfrak{B}(\mathbb{K}_k)$. The elements of $\bigcup_{k \in \mathbb{N}_0} \mathfrak{B}(\mathbb{K}_k)$ are called concepts, and the elements of $\mathfrak{R}_{\vec{\mathbb{K}}} = \bigcup_{k \in \mathbb{N}} \mathfrak{B}(\mathbb{K}_k)$ are called relation-concepts.*

The structure used to interpret concept graphs is a power context family. Below we split the definition of [7] (to differentiate interpretations and models), and use conceptual graphs notations to facilitate comparison.

Definition 15 (Contextual interpretation). *Let $\mathcal{V}$ be a vocabulary. A contextual interpretation of the terms of $\mathcal{V}$ is a pair $(\vec{\mathbb{K}}, \lambda)$, where $\vec{\mathbb{K}}$ is a power context family and λ is a mapping that maps each marker of $\mathcal{I}$ to an element of G_0, each concept type of T_C to an element of $\mathfrak{B}(\mathbb{K}_0)$ (i.e. a formal concept of $\mathbb{K}_0$), and each relation type of arity k in T_R to an element of $\mathfrak{B}(\mathbb{K}_k)$.*

Definition 16 (Contextual model of a vocabulary). *Let $\mathcal{V}$ be a vocabulary. A contextual interpretation of the terms of $\mathcal{V}$ is a contextual model of $\mathcal{V}$ iff the mapping λ is order-preserving.*

Equivalence Between Interpretations and Contextual Interpretations.
Here we present two transformations, the first $c2i$ is from contextual interpretations into interpretations, and the second $i2c$ from interpretations into contextual interpretations. These transformations preserve the models of a vocabulary.

(c2i) Let $C = (\overrightarrow{\mathbb{K}}, \lambda)$ be a contextual interpretation of $\mathcal{V}$. We obtain $c2i(C) = (G_0, \delta)$ where δ is defined by:
  - $\forall i \in \mathcal{I}$, $\delta(i) = \lambda(i)$;
  - $\forall t \in T_C$, $\delta(t) = \text{ext}(\lambda(t))$;
  - $\forall r \in T_R$, $\delta(r) = \text{ext}(\lambda(r))$.

(i2c) Let (D, δ) be an interpretation of $\mathcal{V}$. We obtain $i2c(I) = (\overrightarrow{\mathbb{K}}, \lambda)$ as follows:
  - $G_0 = D$;
  - $\forall c, c' \in T_C$, we note $c \leq_\delta c'$ iff $\delta(c) \subseteq \delta(c')$. $\mathbb{K}_0$ is then the power context over G_0 associated with the partial order $\leq_\delta$ (Dedekind-MacNeille Completion theorem, [11], pp. 48);
  - The power contexts $\mathbb{K}_i$ are constructed in the same way from the sets of relation types of arity i.

Property 2. Let $\mathcal{V}$ be a vocabulary. I is a model of $\mathcal{V} \Rightarrow i2c(i)$ is a contextual model of $\mathcal{V}$; conversely, C is a contextual model of $\mathcal{V} \Rightarrow c2i(C)$ is a model of $\mathcal{V}$.

Proof. We successively prove the two assertions of this property:

  - $t \leq t' \Rightarrow \delta(t) \subseteq \delta(t')$ (since I is a model of $\mathcal{V}$, with t and t' being concept or relation types) $\Leftrightarrow t \leq_\delta t'$ (by construction of $i2c$) $\Leftrightarrow \lambda(t) \leq \lambda(t')$ (Def. 16).
  - $t \leq t' \Leftrightarrow \lambda(t) \leq \lambda(t') \Leftrightarrow \text{ext}(\lambda(t)) \subseteq \text{ext}(\lambda(t')) \Leftrightarrow \delta(t) \subseteq \delta(t')$.

$\square$

Definition 17 (Contextual model of a graph). *Let $\mathcal{V}$ be a vocabulary and $G = (C, R, E, l, \text{coref})$ be a CG over $\mathcal{V}$. A contextual interpretation of the terms of $\mathcal{V}$ is a contextual model of G iff there is a mapping α from C into G_0 s.t.:*

  - *if c is an individual concept node having marker i, $\alpha(c) = \lambda(i)$;*
  - *$\forall c \in C$, $\alpha(c) \in \text{ext}(\lambda(type(c)))$;*
  - *$\forall r \in R$, with neighbors $(c_1, \ldots, c_k)$, $(\alpha(c_1), \ldots, \alpha(c_k)) \in \text{ext}(\lambda(type(r)))$;*
  - *$(c, c') \in \text{coref} \Rightarrow \alpha(c) = \alpha(c')$.*

It is simple to check that the following property holds:

Property 3. Let $\mathcal{V}$ be a vocabulary, and G be a CG over $\mathcal{V}$. I is a model of $G \Rightarrow i2c(I)$ is a contextual model of G; conversely, C is a contextual model of $G \Rightarrow c2i(C)$ is a model of G.

Definition 18 (Deduction for Concept Graphs). *Let $\mathcal{V}$ be a vocabulary, and G and H be two concept graphs defined over $\mathcal{V}$. We say that H is deducible from G and note $G \models_c H$ iff all contextual models of G are also contextual models of H.*

The following theorem proves the equivalence between the two semantics. Thanks to Props. 2 and 3., its proof is straightforward.

Theorem 1 (Equivalence of deductions). *Let $\mathcal{V}$ be a vocabulary, and G and H be two CGs over $\mathcal{V}$. Then $\mathcal{V}, G \models H$ iff $\mathcal{V}, G \models_c H$.*

In concept graphs, concept lattices are used to define the order relation on concepts and relations as well as their interpretations. In conceptual graphs, there is a separation between the syntax (the orders) and the semantics (set inclusions). By considering interpretations at a syntactic level, concept lattices theory provide useful tools to build a vocabulary from examples.

4   Calculus

In this section, we discuss the various calculi proposed to compute entailment in conceptual and concept graphs. In the first subsection (4.1), we compare the derivation rules used as a sound and complete calculus for conceptual graphs [15] and concept graphs [18,7]. Then (4.2) we compare their reformulation as a kind of graph homomorphism named projection [5,18], and discuss the interests of this global operation for efficiency purpose. Finally (4.3), we discuss the normality requirement for a sound and complete projection mechanism, and the various methods proposed to ensure that any graph could be put into its normal form.

4.1   Elementary Generalization/Specialization Rules

To compute conceptual graphs entailment, [20] proposed a sound set of derivation rules that transform one graph into another one. This set of derivation rules has been corrected in [15] to achieve completeness w.r.t. CGs semantics. Similar sets of rules have been proposed in [18,7] for concept graphs.

These sets of derivation rules respect the same behavior: let G be a conceptual or concept graph, and $\mathcal{R}$ be a set of derivation rules. A CG G' is *immediately derived* from G in $\mathcal{R}$ if G' is obtained by applying a rule of $\mathcal{R}$ to G. A CG H is *derived* from G in $\mathcal{R}$ if there is a sequence $G = G_0, G_1, \ldots, G_k = H$ where, for $i = 1$ to k, G_i is immediately derived from G_{i-1} in $\mathcal{R}$. We note $G \vdash_{\mathcal{R}} H$.

Rules for Conceptual Graphs. Two sets of rules have been proposed for CGs in [20]. The first set $\mathcal{S}$ of rules, *specialization rules*, transforms a CG into a more specific one, *i.e.* $G \vdash_{\mathcal{S}} H$ iff $G, \mathcal{V} \models H$. The second set $\mathcal{G}$ of generalization rules transforms a CG into a more general one *i.e.* $H \vdash_{\mathcal{G}} G$ iff $G, \mathcal{V} \models H$. We present here the sound and complete version of these rules proposed in [15].

Specialization rules

1. *Relation simplify:* If two relation nodes have the same label and the same ordered neighbours, delete one of them.
2. *Restrict:* Replace the label of any node by a more specific one.
3. *Join:* Merge two concept nodes having the same label.
4. *Disjoint sum:* Draw another CG next to the original one.
5. *Co-reference addition:* Merge two co-reference classes.
6. *Co-identical join:* Merge two concept-nodes that belong to the same co-reference class.

Generalization rules The set of generalization rules is obtained by building the inverse rules of the specialization rules presented above.

1. *Relation duplicate:* Duplicate a relation node (with the same ordered neighbors and the same label).
2. *Unrestrict:* Replace the label of a relation or a concept node by a more general one.
3. *Detach:* Split a concept node into two nodes of the same label. The union of their neighbors is the original set.
4. *Substract:* Delete a connected component.
5. *Co-reference deletion:* Split a co-reference class.
6. *Co-identical split:* Split a node into two co-referent ones. The union of their neighbors is the original set.

Rules for Concept Graphs. The following set of rules in [7] correspond to the generalization rules of [15]. They update the rules in [18] to take co-reference into account. The twelve rules in [7] are named *Erasure, Iteration:, Deiteration, Generalization, Isomorphism, Exchanging references, Merging two vertices, Splitting a vertex, ⊤-erasure, ⊤-insertion, Identify erasure* and *Identify deletion*.

Since we have proven in Sect. 3 that conceptual graphs and concept graphs have equivalent semantics, and since both generalization rules in [15] and [7] are sound and complete w.r.t. these equivalent semantics, it follows that these two sets of rules create the same graphs (up to the bijection in Sect. 2).

4.2   From Specialization Rules to Graph Homomorphism

Although with generalization/specialization rules we have a sound and complete calculus for CGs, the need for efficient algorithms led us to consider another operation: a graph homomorphism named projection [5]. We first show that the equivalence between specialization rules and graph homomorphism is a well-known characterization in graph theory. We then present two versions of projection, although the first does not require any normality condition, the second is more efficient.

Graph Theoretical Background. Let us first consider the two specialization rules in [15] that have a direct impact on the structure of the graph, *Join* and *Disjoint Sum*. Without any consideration on the labels, using these rules consists in checking whether or not we can obtain a graph G from a graph H by making a disjoint sum of H and a graph D, then by merging its vertices, *i.e.* by checking whether or not a sequence of merges on H leads to a subgraph of G. This is a well-known caracterization of the *graph homomorphism* problem, where the merges are ususally called *retracts* [13].

Basically, a graph homomorphism is a mapping from the nodes of a graph into the nodes of another one that preserves neighborhood. This standard definition is easily updated to bipartite graphs (it must also preserve the bipartition) and to labels (it must preserve some order relation on the labels). The main difficulty in extending this definition to CGs is to take coreference into account.

CGs Homomorphism. We present here an extension of the usual graph homomorphism (usually called *projection*) that takes the particular features of CGs specialization rules into account. This version does not require any normality condition, since it projects coreference classes into other coreference classes (instead of nodes into nodes). The following definition [6] translates the algorithm in [12].

Definition 19 (Coreference projection). *Let G and H be two CGs over a vocabulary $\mathcal{V}$. A coreference projection (or coref-projection) from H into G is a mapping Π from the coreference classes of H into the coreference classes of G such that:*

- *For each co-reference class C in H, let us consider the set of individual markers $I = \{i_1, \ldots, i_k\}$ labelling the nodes of C. Then I is a subset of the individual markers of $\Pi(C)$.*
- *For each relation node r in H, with neighbors $x_1, \ldots, x_k$ and label t, let us consider C_i the coreference class of x_i, for $1 \leq i \leq k$. Then there is a relation node r' in G whose label is more specific than t and whose neighbors $y_1, \ldots, y_k$ are such that $y_i \in \Pi(C_i), 1 \leq i \leq k$.*

Theorem 2 (Soundness and completeness [6]). *Let G and H be two CGs over a vocabulary $\mathcal{V}$. Then H coref-projects into G iff $G, \mathcal{V} \models H$*

Generalization/specialization rules and *coref*-projection are thus two calculi for entailment of CGs. An immediate advantage of generalization/specialization rules is that they allow us to generate all CGs that are more general/specific than the given one. However, *coref*-projection is more efficient w.r.t. computing entailmment between two CGs. Let us consider the difference between "compute a graph homomorphism from H into G" and "check if a sequence of retracts of H generates a graph isomorphic to a subgraph of G". This simplification of our problem corresponds to the core, which is the NP-complete part of our entailment problem. Deciding on the existence of an homomorphism is an NP-complete problem, and efficient algorithms can be used (*e.g.* see [15,2] for relationships with constraint networks). However, for specialization rules, even checking if a graph is isomorphic to another one is an ISO-complete problem (an intermediary class between P and NP), and this test must be done after applying each application of a rule. And even with a carefully written algorithm, there can be an exponential number of rule applications.

The Need for a Normal Form. The normal form of a CG is an artefact used to optimize *coref*-projection. It is neither a syntactic (see Sect. 2) nor a semantic (see Sect. 3) requirement of CGs. A CG is said in *normal form* if every coreference class contains a single node. If the graph G we look for a projection into is in normal form, *coref*-projection becomes the standard projection [16] (and also [18] for concept graphs), as expressed by the following definition. Since we only have to examine the edges incident to the current node, and not those incident to all the nodes belonging to the same coreference class, the calculus is more efficient.

Definition 20 (Projection). *Let G and H be two CGs over a vocabulary $\mathcal{V}$. A projection from H into G is a mapping π from the nodes of H into the nodes of G such that:*

- *For each concept node $c \in C(H)$, the type of c is more general than the type of $\pi(c)$, and if the marker of c is individual, $\pi(c)$ has the same marker.*
- *For each coreferent concept nodes c, c' in H, $\pi(c) = \pi(c')$.*
- *For each relation node r in H, with neighbors $x_1, \ldots, x_k$ and label t, there is a relation node r' in G having a more specific type and whose neighbors are $\pi(x_1), \ldots, \pi(x_k)$.*

Since projection is equivalent to *coref*-projection when G is in normal form, the following theorem is a direct consequence of the previous one:

Theorem 3 (Soundness and completeness [20,16]). *Let G and H be two CGs over a vocabulary $\mathcal{V}$, G being in normal form. Then H projects into G iff $G, \mathcal{V} \models H$.*

Note that, historically, projection was proposed in [5] without any normality condition. A counterexample was exhibited simultaneously in [16,21], leading to two corrections: the normal form presented here, and the antinormal form [21] for the query, which is less efficient for computational purposes.

4.3 CGs and Normalization

Although projection is an interesting, efficient algorithm to compute entailment of CGs, it requires putting a CG into its normal form. This is done by merging all vertices that belong to the same coreference class. However, this is not always possible (what is the resulting type of the merge of two nodes having different types?). Different solutions to this problem have been proposed:

1. *Syntactic restrictions:* The *conformity relation* (assigning a type to each individual marker), as well as explicit restrictions on co-reference are used to force all vertices belonging to the same coreference class to have the same type (*e.g.* [16]). A weakness of this solution is to impose syntactic restrictions to solve calculus problems. It is of interest from a KR standpoint: the conformity relation defined in the vocabulary is a modelling guide.
2. *Semantic modifications:* When merging concept nodes having different types, the resulting type is their greatest common subtype. This *lattice-theoretic interpretation* changes the semantics of the vocabulary, and also imposes a syntactic restriction: the order on concept types must be a lattice (see CG-list: *CG: Individual Markers refer to unique entities?* for a discussion on this topic).
3. *Syntactic extensions:* Using type conjunction (*e.g.* [2,6]) in CGs naturally solves this problem, but does not extend the expressivity of the language (we have seen in Sect. 3 that concept types and unary relations have the same interpretation, so a conjunction of concept types could be represented by multiple unary relations).

5 Conclusion

In this paper, we have compared simple conceptual graphs and simple concept graphs w.r.t. their syntax, semantics, and calculus.

Syntax. As mathematical objects, we have proven that, up to superficial differences, they are identical objects (there are simple injective mappings from one class of objects to the other one).

Semantics. Concerning the interpretation of CGs, power context families are not specific since they can define any ordered FOL structure. This shows that the (model-theoretical) semantics for conceptual graphs and for concept graphs are identical. Furthermore, power context families are not wholly used in the definition of the entailment relation, only the order relation between concepts (and relation-concepts) is used. Thus, in the development of a unified CG theory, we propose to use power context families only for the construction of vocabularies (i.e. ordered FOL languages).

Calculus. If the aim is to build software tools in order to solve actual problems, one has to go beyond the decision problem of deduction and consider algorithms for constructing solutions and thus computational efficiency. This explains why, besides the interesting visual aspect of graphs in knowledge representation, we emphasize the graph viewpoint in dealing with CGs. Graph theory is a mature mathematical theory with many mathematical and algorithmic results that can be imported into CGs, especially the homomorphism (projection) notion, which is central in many computational and combinatorial problems (from graph coloring to category representation along with constraint satisfaction problems or query inclusion problems in relational databases).

Further Works. During our work on the different CGs semantics, it appeared that conceptual graphs canonical models [14] (or isomorphic interpretations [1,2]) and concept graphs canonical models [18] (or Standard models in [7]) are similar notions. Altogether, they correspond to Herbrand models in FOL. For space requirements, this part of our work was not included in this paper, but will be developed later.

Finally, we have shown in this paper that power context families were a too expressive a structure for the reasonings involved in CGs. We intend to study if this conclusion is still valid in different extensions of conceptual and concept graphs (*e.g.* negation).

References

1. J.-F. Baget. *Représenter des connaissances et raisonner avec des hypergraphes: de la projection à la dérivation sous contraintes.* PhD thesis, Université Montpellier II, Nov. 2001.
2. J.-F. Baget. Simple Conceptual Graphs Revisited: Hypergraphs and Conjunctive Types for Efficient Projection Algorithms. In de Moor et al. [9], pages 229–242.
3. J.-F. Baget and M.-L. Mugnier. The Complexity of Rules and Constraints. *JAIR*, 16:425–465, 2002.
4. Claude Berge. *Graphes et hypergraphes.* Dunod, 1970.
5. M. Chein and M.-L. Mugnier. Conceptual Graphs: Fundamental Notions. *Revue d'Intelligence Artificielle*, 6(4):365–406, 1992.
6. M. Chein and M.-L. Mugnier. Types and Coreference in Simple Conceptual Graphs. In K.E. Wolff et al, editor, *Proc. ICCS'04*, volume 3127 of *LNAI*. Springer, 2004. to appear.
7. F. Dau. Concept graphs without negations: Standardmodels and standardgraphs. In de Moor et al. [9], pages 243–256.
8. Frithjof Dau. *The Logic System of Concept Graphs with Negations (And its Relationship to Predicate Logic)*, volume 2892 of *Lecture Notes in Artificial Intelligence.* Springer-Verlag, 2003. PhD-Thesis.
9. Aldo de Moor, Wilfried Lex, and Bernhard Ganter, editors. *Conceptual Structures for Knowledge Creation and Communication, 11th International Conference on Conceptual Structures, ICCS 2003 Dresden, Germany, July 21-25, 2003 Proceedings*, volume 2746 of *Lecture Notes in Computer Science.* Springer, 2003.
10. Reinhard Diestel. *Graph Theory*, volume 173 of *Graduate Texts in Mathematics.* Springer-Verlag, 3 edition, 2000.
11. B. Ganter and R. Wille. *Formal Concept Analysis.* Springer-Verlag, 1999.
12. O. Guinaldo and O. Haemmerlé. Kowledge Querying in the Conceptual Graph Model: the RAP Module. In Mugnier and Chein [17], pages 287–294.
13. P. Hell and J. Nesetril. *Graphs and Homomorphisms*, volume 121. Oxford University Press, 2004.
14. G. Kerdiles. *Saying it with Pictures: a logical landscape of conceptual graphs.* PhD thesis, Univ. Montpellier II / Amsterdam, Nov. 2001.
15. M.-L. Mugnier. Knowledge Representation and Reasoning based on Graph Homomorphism. In Bernhard Ganter and Guy W. Mineau, editors, *ICCS*, volume 1867 of *Lecture Notes in Computer Science*, pages 172–192. Springer, 2000.

16. M.-L. Mugnier and M. Chein. Représenter des connaissances et raisonner avec des graphes. *Revue d'Intelligence Artificielle*, 10(1):7–56, 1996. Available at http://www.lirmm.fr/~mugnier/.

17. Marie-Laure Mugnier and Michel Chein, editors. *Conceptual Structures: Theory, Tools and Applications, 6th International Conference on Conceptual Structures, ICCS '98, Montpellier, France, August 10-12, 1998, Proceedings*, volume 1453 of *Lecture Notes in Computer Science*. Springer, 1998.

18. S. Prediger. Simple concept graphs: A logic approach. In Mugnier and Chein [17], pages 225–239.

19. J. F. Sowa. Conceptual Graphs. *IBM Journal of Research and Development*, 1976.

20. J. F. Sowa. *Conceptual Structures: Information Processing in Mind and Machine*. Addison-Wesley, 1984.

21. M. Wermelinger. Conceptual Graphs and First-Order Logic. pages 323–337.

22. R. Wille. Conceptual graphs and formal context analysis. In Dickson Lukose, Harry S. Delugach, Mary Keeler, Leroy Searle, and John F. Sowa, editors, *ICCS*, volume 1257 of *Lecture Notes in Computer Science*, pages 290–303. Springer, 1997.

Rules Dependencies in Backward Chaining
of Conceptual Graphs Rules

Jean-François Baget[1] and Éric Salvat[2]

[1] INRIA Rhône-Alpes/LIRMM
jean-francois.baget@inrialpes.fr
[2] IMERIR
salvat@imerir.com

Abstract. Conceptual Graphs Rules were proposed as an extension of
Simple Conceptual Graphs (CGs) to represent knowledge of form "IF A
THEN B", where A and B are simple CGs. Optimizations of the deduction
calculus in this KR formalism include a Backward Chaining that unifies
at the same time whole subgraphs of a rule, and a Forward Chaining
that relies on compiling dependencies between rules.
In this paper, we show that the unification used in the first algorithm
is exactly the operation required to compute dependencies in the second
one. We also combine the benefits of the two approaches, by using the
graph of rules dependencies in a Backward Chaining framework.

1 Introduction

Conceptual graphs (CG) rules [13] were proposed as an extension of simple CGs
[12] to represent knowledge of form "IF A THEN B", where A and B are simple
CGs. This graph-based knowledge representation (KR) formalism (named $\mathcal{SR}$
in [3]) was further formalized in [11]. Notwithstanding the interest of graphical
representation of knowledge for an human interaction purpose, we are mainly
motivated in using the graph structure of CGs to improve sound and complete
deduction algorithms. Using graph-theoretical operations, instead of translating CGs into their equivalent formulae and use a FOL solver, the algorithms
presented in this paper explore a different optimization paradigm in KR.

Simple CGs [12] form the basic KR formalism (named $\mathcal{SG}$ in [3]) on which
CG rules are built. The semantics Φ identifies them with formulae in positive,
conjunctive, existential FOL (without function symbols) [13]. Sound and complete reasonings in $\mathcal{SG}$ (a NP-hard problem) can be computed with a kind of
graph homomorphism named *projection* [5].

Projection is also the elementary operation in Forward Chaining (FC) of
CG rules [11], a graph-based algorithm computing deduction in $\mathcal{SR}$. Since CG
Rules can be translated into FOL formulae having the form of Tuple Generating
Dependencies (TGDs) [7], $\mathcal{SR}$-DEDUCTION is semi-decidable.

A Backward Chaining (BC) framework is often used to avoid a major pitfall in
FC: applying rules that are unrelated to the query. Though CG Rules deduction
can be computed using a PROLOG-like BC algorithm, successively unifying

H. Schärfe, P. Hitzler, and P. Øhrstrøm (Eds.): ICCS 2006, LNAI 4068, pp. 102–116, 2006.

predicate after predicate in the equivalent FOL formulae, [10] proposed to rely upon the structure of the graph and unify at the same time whole subgraphs of the rule (called pieces), effectively reducing the number of backtracks [7].

To optimize FC, [4] defines neutrality: a CG Rule R_1 is neutral w.r.t. a rule R_2 if no application of R_1 on a CG can create a new application of R_2. The resulting graph of rules dependencies (GRD) allows to reduce the number of checks for rule applicability as well as the cost of these checks in FC.

In this paper, we show that the criterium used to compute dependencies in [4] and the piece unification of [10] are similar operations. In particular, piece unification generalizes computation of dependencies to rules having individual markers in their conclusion (excluded in [4]). On the other hand, we generalize piece unification to any type hierarchy (and not only lattices, as in [10]). We propose solutions to use the GRD in a BC framework.

Organization of the paper. Sect. 2 and 3 are respectively devoted to simple CGs (the $\mathcal{SG}$ language) and CG rules ($\mathcal{SR}$). We present the syntax, the semantics (via the translation Φ to FOL), and a sound and complete calculus (projection in the first case, basic FC in the latter) of both languages. The first enhancement of $\mathcal{SR}$-DEDUCTION, the BC based upon piece unification [11,10], is presented in Sect. 4. The graph of rules dependencies (GRD) [4], its use in FC, and its relationships with piece unification, are presented in Sect. 5. Finally, in Sect. 6, we show how to efficiently use the GRD in a BC framework.

2 Simple Conceptual Graphs

We recall fundamental results on simple CGs (without coreference links) [12,13]. Sect. 2.1 presents their syntax, and Sect. 2.2, their semantics [13]. We use these formulas to define simple CGs deduction ($\mathcal{SG}$-DEDUCTION in [3]). In Sect. 2.3, we use projection [5] as a calculus for $\mathcal{SG}$-DEDUCTION.

2.1 Syntax

Definition 1 (Vocabulary). *A vocabulary is a tuple $(T_C, (T_R^1, \ldots, T_R^N), \mathcal{I}, \kappa)$ where $T_C, T_R^1, \ldots, T_R^N$ are pairwise disjoint partially ordered sets (partial orders are denoted by $\leq$), $\mathcal{I}$ is a set, and $\kappa : \mathcal{I} \to T_C$ is a mapping. Elements of T_C are called* concept types, *elements of T_R^i* relation types *of arity i, elements of $\mathcal{I}$* individual markers, *and κ is the* conformity relation.

Definition 2 (Simple CGs). *A simple CG over a vocabulary $\mathcal{V}$ is a tuple $G = (E, R, \epsilon, \gamma)$ where E and R are two disjoint sets, respectively of* entities *and* relations. *The mapping ϵ labels each entity of E by a pair of $T_C \times (\mathcal{I} \cup \{*\})$ (its* type *and* marker*). An entity whose marker is $*$ is called* generic, *otherwise it is an* individual. *For each individual $x \in E$, $\text{type}(x) = \kappa(\text{marker}(x))$. The mapping ϵ also labels each relation of R by a relation type (its* type*). We call* degree *of a relation the arity of its type. The mapping* γ *maps each relation of degree* k *to a* k*-tuple of* E^k. *If* $\gamma(r) = (x_1, \ldots, x_k)$ *we denote by* $\gamma_i(r) = x_i$ *the* i*th argument of* r. *If* x *and* y *are two arguments of* r, x *and* y *are* neighbours.

Simple CGs can be seen both as *bipartite multigraphs*, as in [5] ($\gamma_i(r) = e$ means that there is an edge labelled i between the concept node e and the relation node r); or as *directed multiple hypergraphs*, as in [2] ($\gamma(r) = (x_1, \ldots, x_k)$ is a directed hyperarc whose ends are the concept nodes $x1, \ldots, x_k$).

Whatever the structure used to encode them, they share the same drawing. An entity e with $\epsilon(e) = (t, m)$ is represented by a rectangle enclosing the string "t: m". A relation r typed t is represented by an oval enclosing the string "t". If $\gamma(r) = (x_1, \ldots, x_k)$, then for $1 \leq i \leq k$, we draw a line between the oval representing r and the rectangle representing x_i, and write the number i next to it.

2.2   Semantics

Simple CGs semantics are often expressed via a translation Φ to first-order logics [13], and deduction is defined by the logical consequence of associated formulas. This translation Φ is explicited in [13,8].

- The interpretation $\Phi(\mathcal{V})$ of a vocabulary $\mathcal{V}$ is a FOL formula translating the order on concept and relation types; *i.e.* a conjunction of formulae $\forall x_1 \ldots \forall x_k (t(x_1, \ldots, x_k) \rightarrow t'(x_1, \ldots, x_k))$ where t is a type (concept or relation) more specific than t'.
- The interpretation $\Phi(G)$ of a simple CG G is the existential closure of a conjunction of atoms interpreting concepts and relations between them.

Definition 3 ($\mathcal{SG}$-Deduction). *Let* G *and* H *be two simple CGs over a vocabulary* $\mathcal{V}$. *We say that* G *entails* H *in* $\mathcal{V}$ *(and note* $G \models_{\mathcal{V}} H$*) iff* $\Phi(H)$ *is a logical consequence of* $\Phi(G)$ *and* $\Phi(\mathcal{V})$.

2.3   Calculus

Definition 4 (Projection). *Let* G *and* H *be two simple CGs over a vocabulary* $\mathcal{V}$, *with* $G = (E_G, R_G, \epsilon_G, \kappa_G)$ *and* $H = (E_H, R_H, \epsilon_H, \kappa_H)$. *A projection from* H *into* G *(according to* $\mathcal{V}$*) is a mapping* $\pi : E_H \rightarrow E_G$ *such that:*

- *For each entity* $e \in E_H$, *type*($\pi(e)$) $\leq$ *type*(e). *If, moreover,* e *is an individual, then* marker($\pi(e)$) = marker(e).
- *For each relation* $r \in R_H$, *with* $\gamma_H(r) = (x_1, \ldots, x_p)$, *there exists a relation* $r' \in R_G$ *such that type*(r') $\leq$ *type*(r) *and* $\gamma_G(r') = (\pi(x_1), \ldots, \pi(x_k))$.

As a generalization of GRAPH HOMOMORPHISM, PROJECTION is NP-complete.

Normal form of a simple CG. A simple CG G over $\mathcal{V}$ is said *normal* if all its individuals have distinct markers. A simple CG G is put into its *normal form* nf(G) by successively joining all pairs of individuals having the same marker. We note $join(e_1, e_2)$ the individual resulting from a join: it has same marker and same type (thanks to the conformity relation) as e_1 and e_2. Putting a simple CG into its normal form is linear in the size of the graph.

Theorem 1 (Soundness and completeness [9]). *Let G and H be two simple CGs over a vocabulary $\mathcal{V}$. Then $G \models_{\mathcal{V}} H$ if and only if there is a projection from H into nf(G), the normal form of G, according to $\mathcal{V}$.*

3   Conceptual Graphs Rules

CG rules have been introduced in [13] as an extension of simple CGs allowing to represent knowledge of form "IF H THEN C", where H and C are simple CGs. As for simple CGs, we first present their syntax (Sect. 3.1) and semantics. As a sound and complete calculus for $\mathcal{SR}$-DEDUCTION, we present Forward Chaining (FC) [11], based upon projection of simple CGs.

3.1   Syntax

Definition 5 (CG rules). *A conceptual graph rule (or CG rule) over a vocabulary $\mathcal{V}$ is a triple $\mathbb{R} = (\lambda, H, C)$ where $H = (E_H, R_H, \epsilon_H, \gamma_H)$ and $C = (E_C, R_C, \epsilon_C, \gamma_C)$ are two simple CGs over $\mathcal{V}$, and λ is a bijection between a distinguished subset of generic entities of E_H (called connecting entities of H) and a subset of generic entities of E_C (called connecting entities of C), s.t. $\lambda(e) = e' \rightarrow \text{type}(e) = \text{type}(e')$. The simple CG H is called the hypothesis of $\mathbb{R}$, and C its conclusion. They are respectively denoted by $\text{Hyp}(\mathbb{R})$ and $\text{Conc}(\mathbb{R})$.*

This definition of CG rules clearly relates to a pair of λ-abstractions [11].

The usual way to represent such a rule is by drawing two boxes next to each other. The box to the left is the hypothesis box, and the box to the right the conclusion box. Draw between these boxes an implication symbol $\Rightarrow$. Draw the simple CG H (as done in Sect. 2.1) in the hypothesis box and the simple CG G in the conclusion box. Finally, for each pair $(e, \lambda(e))$ of connecting entities, draw a dashed line (a coreference link) between the rectangle representing e and the rectangle representing $\lambda(e)$.

3.2   Semantics

Interpretation of a CG Ruleset. Let $\mathcal{R}$ be a *CG ruleset* (a set of CG rules) over $\mathcal{V}$. Its interpretation $\Phi(\mathcal{R})$ is the conjunction of the FOL formulas $\Phi(\mathbb{R})$ of form $\forall x_1 \ldots \forall x_p(\Phi(\mathbb{R}) \rightarrow (\exists y_1 \ldots \exists y_q \Phi(\mathbb{R})))$ interpreting its CG rules [11].

Definition 6 ($\mathcal{SR}$-Deduction). *Let G and H be two simple CGs over $\mathcal{V}$, and $\mathcal{R}$ be a CG ruleset. We say that $G, \mathcal{R}$ entails H in $\mathcal{V}$ (and note $G, \mathcal{R} \models_{\mathcal{V}} H$) iff $\Phi(H)$ is a logical consequence of $\Phi(G)$, $\Phi(\mathcal{R})$ and $\Phi(\mathcal{V})$.*

3.3   Calculus

Application of a CG rule. Let $\mathbb{R} = (\lambda, H, C)$ be a CG rule and $G = (E, R, \epsilon, \gamma)$ be a simple CG over $\mathcal{V}$. The CG rule $\mathbb{R}$ is said *applicable* to G iff there is a projection π from $Hyp(\mathbb{R})$ into $nf(G)$. In that case, the *application of* $\mathbb{R}$ *on* G *following* π produces a simple CG $G' = \alpha(G, \mathbb{R}, \pi)$ built as follows. We define the *disjoint union* of two graphs G_1, G_2 as the graph whose drawing is the juxtaposition of those of G_1 and G_2. We build the disjoint union of a copy of $nf(G)$ and of a copy of $Conc(\mathbb{R})$. Then, for each pair $(e, \lambda(e))$ of connecting entities in $\mathbb{R}$, we join the entity x in the copy of $nf(G)$ obtained from $\pi(e)$ and the entity y in the copy of $Conc(\mathbb{R})$ obtained from $\lambda(e)$. Since $\epsilon(e) = \epsilon(\lambda(e))$, the label of x (*i.e.* the label of $\pi(e)$) is a specialization of the label of y, and $\epsilon(x)$ is used as the label of $join(x, y)$.

Deriving a simple CG with CG rules. Let $\mathcal{R}$ be a CG ruleset and G, G' be two simple CGs over a vocabulary $\mathcal{V}$. We say that G' is *immediately derived from* G *in* $\mathcal{R}$ (and note $G \overset{\mathcal{R}}{\mapsto} G'$) iff there is a rule $\mathbb{R} \in \mathcal{R}$ and a projection π from $Hyp(\mathbb{R})$ into G such that $G' = \alpha(G, \mathbb{R}, \pi)$. We say that G' is *derived from* G *in* $\mathcal{R}$ (and note $G \overset{\mathcal{R}}{\leadsto} G'$) iff there is a sequence $G = G_0, G_1, \ldots, G_n = G'$ of simple CGs over $\mathcal{V}$ such that, for $1 \leq i \leq n$, $G_{i-1} \overset{\mathcal{R}}{\mapsto} G_i$.

Theorem 2 (Soundness and completeness [11]). *Let $\mathcal{R}$ be a CG ruleset, and G and H be two simple CGs over a vocabulary $\mathcal{V}$. Then $G, \mathcal{R} \models_{\mathcal{V}} H$ if and only if there is a simple CG G' such that $G \overset{\mathcal{R}}{\leadsto} G'$ and H projects into $nf(G')$.*

Forward Chaining of CG rules. The Forward Chaining (FC) algorithm [11] immediately follows from theorem 2 and the property of confluence (Prop. 1).

Property 1 (Confluence). Let $\mathcal{R}$ be a CG ruleset, and G and H be two simple CGs over a vocabulary $\mathcal{V}$. Let us suppose that $G, \mathcal{R} \models_{\mathcal{V}} H$. Then for every simple CG G' such that $G \overset{\mathcal{R}}{\leadsto} G'$, the entailment $G', \mathcal{R} \models_{\mathcal{V}} H$ holds.

Any algorithm exploring all rule applications (Th. 2), *e.g.* using a breadth-first method, in any order (Prop. 1), will lead to a simple CG entailing the query H, if it exists. Such an algorithm, named FC, is proposed here (Alg. 1).

Algorithm 1. Forward Chaining

Data: A vocabulary $\mathcal{V}$, a CG ruleset $\mathcal{R}$, two simple CGs G and H over $\mathcal{V}$.
Result: YES iff $G, \mathcal{R} \models_{\mathcal{V}} H$ (infinite calculus otherwise).
ProjList $\leftarrow \emptyset$;
while TRUE **do**
    for $\mathbb{R} \in \mathcal{R}$ **do**
        for $\pi \in Projections(Hyp(\mathbb{R}), G)$ **do**
            ProjList $\leftarrow$ ProjList $\cup \{(\mathbb{R}, \pi)\}$;
    for $(\mathbb{R}, \pi) \in ProjList$ **do**
        $G \leftarrow \alpha(G, \mathbb{R}, \pi)$;
        if $Projects?(H, G)$ **then return** YES ;

Decidability. Since FOL formulae associated with CG rules have the same form as TGDs [7], $\mathcal{SR}$-DEDUCTION is semi-decidable (a sound and complete algorithm can compute in finite time whenever the answer is YES, but cannot always halt otherwise). Some decidable subclasses of the problem are proposed in [3]: let us suppose that, after the nth execution of the **while** loop in Alg. 1, the simple CG G obtained is equivalent to G as it was at the beginning of this loop. In that case, the algorithm could safely stop and answer NO. A CG ruleset ensured to have this behavior is called a *finite expansion set*. Examples of finite expansion sets are *disconnected* CG rules (having no connecting entities) or *range restricted* CG rules (having no generic entity in the conclusion). Note that the union of two finite expansion rulesets is not necessarily a finite expansion ruleset.

4   Piece Unification and Backward Chaining

FC generates explicitly knowledge implicitly encoded in CG rules. By opposition, a Backward Chaining (BC) algorithm starts with the query H and rewrites it using *unification*. The interest of *piece unification* [11,10] w.r.t. a PROLOG-like unification, is that it unifies at the same time a whole subgraph, instead of a simple predicate. Sect. 4.1 present preliminary definitions and Sect. 4.2 piece unification. A BC algorithm using piece unification is presented in Sect. 4.3.

4.1   Preliminary Definitions

Definition 7 (Cut points, pieces). *Let $\mathbb{R} = (\lambda, H, C)$ be a CG rule over $\mathcal{V}$. A cut point of C is either a connecting entity (Def. 5) or an individual of C. A cut point of H is either a connecting entity of H or an individual of H whose marker also appears in C. A piece P of C is a subgraph of C whose entities are a maximal subset of those of C s.t. two entities e_1 and e_2 of C belong to P if there is a path $e_1, x_1, \ldots, x_k, e_2$ where the x_i are not cut points of C.*

Conjunctive CGs. When a CG rule $\mathbb{R}$ is applied to a simple CG G, the entities of $\alpha(G, \mathbb{R}, \pi)$ obtained from a join between a connecting entity of $Conc(\mathbb{R})$ and an entity of G may have a more specific label than the former entities (Sect. 3.3). So to compute unification, we have to find which cut points of $Conc(\mathbb{R})$ have a common specialization with entities of the query. In [11,10], such common specialization of two entities e_1 and e_2 was typed by the *greatest lower-bound (glb)* of $type(e_1)$ and $type(e_2)$. The existence of the glb was ensured by using a lattice as partial order on concept types. We generalize the previous approach by considering, as in [2,6], conjunctive types.

A *conjunctive CG* is defined as a simple CG, but the type of an entity can be the conjunction of types of T_C. The interpretation of an entity e with $\epsilon(e) = (t_1 \sqcap \ldots \sqcap t_p, m)$ is the conjunction $\phi(e) = t_1(f(e)) \wedge \ldots \wedge t_p(f(e))$. The partial order on T_C is extended to the partial order $\leq_\sqcap$ on conjunctive types: $t_1 \sqcap \ldots \sqcap t_p \leq_\sqcap t'_1 \sqcap \ldots \sqcap t'_q$ iff $\forall t'_i, \exists t_j$ with $t_j \leq t'_i$. We define the *join* operation between two entities e_1 and e_2 having different (conjunctive) types: the type of $e = join(e_1, e_2)$ is the conjunction of the types of e_1 and e_2. If both e_1 and e_2 are individuals with same marker m, or generic entities with $m = *$, the marker of e is also m. If e_1 has individual marker m and e_2 is generic, the marker of e is m. The label $\epsilon(e)$ defined here is the *common specialization* of $\epsilon(e_1)$ and $\epsilon(e_2)$. The projection algorithm is the same as in Sect. 2.3, but relies on $\leq_\sqcap$ to compare conjunctive types. Normalization relies on the above-mentioned join operation. Up to these two differences, the soundness and completeness result (Th. 1) remains the same.

Compatible partitions. A set of entities E is *join compatible* iff there is a concept type of T_C more specific than all types in E and there is at most one individual marker in E. Let G be a simple or conjunctive CG and E be a join compatible subset $\{e_1, \ldots, e_p\}$ of entities of G. The *join of G according to E* is the conjunctive CG obtained by joining e_1 and e_2 into e, then by joining G according to $\{e, e_3, \ldots, e_p\}$, until this subset contains a single entity e: we note $e = join(E)$. Let S and S' be two disjoint sets of entities. Let $P = (P_1, \ldots, P_n)$ and $P' = (P'_1, \ldots, P'_n)$ be two ordered partitions, resp. of S and S' (a partition of X is a set of pairwise disjoint sets whose union equals X). P and P' are *compatible partitions* of S and S' iff $P_i \cup P'_i$ is a join compatible set, for $1 \leq i \leq n$.

Definition 8 (Specialization according to a compatible partition). *Let G and G' be two simple or conjunctive CGs over $\mathcal{V}$. Let E and E' be respective subsets of entities of G and G'. Let $P = (P_1, \ldots, P_n)$ and $P' = (P'_1, \ldots, P'_n)$ be two compatible partitions of E and E'. The specialization of G according to (P, P') is the conjunctive CG $sp(G, (P, P'))$ built from G by building the join of G according to P_i, for $1 \leq i \leq n$, then by replacing the label of each $join(P_i)$ with its common specialization $join(P'_i)$.*

The *join of G and G' according to compatible partitions P and P'* is the conjunctive CG obtained by making the disjoint union of $sp(G, (P, P'))$ and of $sp(G', (P, P'))$, then by joining each $join(P_i)$ with $join(P'_i)$.

4.2   Piece Unification

Definition 9 (Piece unification). *Let Q be a simple (or conjunctive) CG (the query) and $\mathbb{R} = (\lambda, H, C)$ be a CG rule over $\mathcal{V}$. Q and $\mathbb{R}$ are said unifiable iff there is a piece unification between Q and $\mathbb{R}$, i.e. a triple $\mu = (P^C, P^Q, \Pi)$ where:*

- P^C *and* P^Q *are two compatible partitions, resp. of a subset of cut points of C and a of subset of entities of Q that will be considered as cut points of Q;*

- Π is a projection from a non-empty set of pieces of $\mu(Q) = sp(Q, (P^C, P^Q))$ (cut points of $\mu(Q)$ are entities resulting from the join of cut points of Q) into $\mu(\mathbb{R}) = sp(C, (P^C, P^Q))$ such that $\Pi(join(P_i^Q)) = join(P_i^C)$.

Rewriting of a query. An unification μ between a query Q and a CG rule $\mathbb{R}$ determines a rewriting of Q (that can become a conjunctive CG). Simply put, we remove from the new query the conclusion of $\mathbb{R}$ and add its hypothesis.

More precisely, let Q be a simple (or conjunctive) CG, $\mathbb{R} = (\lambda, H, C)$ be a CG rule, and $\mu = (P^C, P^Q, \Pi)$ be a piece unification between Q and $\mathbb{R}$. We call unification result of μ on Q and note $\beta(Q, \mathbb{R}, \mu)$ the conjunctive CG built as follows:

1. Let S^C and S^Q be the sub-partitions of P^C and P^Q formed respectively from the codomain and the domain of Π;

2. Let S^H be a partition of the subset of cut points of H that correspond to the partition S^C of cut points of C (if e is an entity of a partition S_i^C of S^C, the entities $g_1, \ldots, g_q$ of H that correspond to e, i.e. either $q = 1$ and $\lambda(g_1) = e$ or $g_1, \ldots, g_q$ and e have the same individual marker, belong to the partition S_i^H);

3. Build the conjunctive CGs $Q' = sp(Q, (S^H, S^Q))$ and $H' = sp(H, (S^H, S^Q))$;

4. Let P be a piece of Q whose entities are in the domain of Π. We remove from Q' all relations of P and all entities of P that are not cut points of Q';

5. We finally join Q' and H' according to (S^H, S^Q).

Definition 10 (Resolution). Let H be a simple CGs, and $\mathcal{R}$ be a CG ruleset (that includes the facts CG G as a rule with an empty hypothesis) over $\mathcal{V}$. We call resolution of H in $\mathcal{R}$ a sequence $H = H_1, H_2, \ldots, H_{p+1}$ of conjunctive CGs such that, for $1 \leq i \leq p$, there is a piece unification μ between H_i and a rule $\mathbb{R} \in \mathcal{R}$, $H_{i+1} = \beta(H_i, \mathbb{R}, \mu)$ and H_{p+1} is the empty CG.

Theorem 3 (Soundness and completeness [11]). Let G and H be two simple CGs, and $\mathcal{R}$ be a CG ruleset over $\mathcal{V}$. Then $G, \mathcal{R} \models_{\mathcal{V}} H$ if and only if there is a resolution of H in $\mathcal{R} \cup \{\mathbb{G}\}$ ($\mathbb{G} = (\lambda, \emptyset, G)$ is a CG rule equivalent to G).

Proof. [11,10] proves that if $H = H_1, H_2, \ldots, H_{p+1} = \emptyset$ is a resolution of H in $\mathcal{R}$ using successively the rules $\mathbb{R}_{i_1}, \ldots \mathbb{R}_{i_p} = \mathbb{G}$, then there is a FC derivation sequence $G = G_1, \ldots, G_p$ that successively applies the rules $\mathbb{R}_{i_1}, \ldots \mathbb{R}_{i_{p-1}}$ in reverse order, and such that H projects into G_p. Conversely, from a FC derivation, we can extract a subsequence that corresponds to a resolution using the same rules in reverse order. The theorem is a consequence of this correspondences between FC and BC.

4.3 Backward Chaining

Algorithm 2. Backward Chaining

Data: A vocabulary $\mathcal{V}$, a CG ruleset $\mathcal{R}$, two simple CGs G and H over $\mathcal{V}$.
Result: If YES, then $G, \mathcal{R} \models_{\mathcal{V}} H$, if NO, then $G, \mathcal{R} \not\models_{\mathcal{V}} H$ (no halting ensured).
UnifList ← NewFilo() ;
for $\mathbb{R} \in \mathcal{R} \cup \{\mathbb{G}\}$ **do**

    for $\mu \in Unifications(\mathbb{R}, H)$ **do**

        UnifList ← AddFilo(UnifList, $(\mu, \mathbb{R}, H)$) ;

while $UnifList \neq \emptyset$ **do**

    $(\mu, \mathbb{R}, H)$ ← FiloRemove(UnifList) ;
    H' ← Rewrite$(\mu, \mathbb{R}, H)$;
    if $H' = \emptyset$ **then return** YES ;
    for $\mathbb{R}' \in \mathcal{R}$ **do**

        for $\mu' \in Unifications(\mathbb{R}', H')$ **do**

            UnifList ← AddFilo(UnifList, $(\mu', \mathbb{R}', H')$) ;

return NO ;

Comparing FC and BC. It is well known (*e.g.* [1]) in Logic Programing that, from BC or FC, no algorithm is always better. The main differences are that 1) FC enriches the facts until they contain an answer to the query while BC rewrites the query until all its components have been proven; 2) FC derivation is a confluent mechanism, while BC rewritings depends upon the order of these rewritings, and thus requires a backtrack; and 3) FC enumerates all solutions to the query by applying rules breadth-first, while BC *usually* (as in Alg. 2) tries to find them quicker by rewriting the query depth-first (eventually missing solutions). A breadth-first version of BC, that misses no solution, can be implemented by replacing the Filo structure of *UnifList* in Alg. 2 by a Fifo. Completeness is then achieved at the expense of efficiency. [7] compares piece unification with the standard PROLOG that unifies one predicate at a time. Though piece unification leads to fewer backtracks in query rewriting, it does not always translate to the overall efficiency of the algorithm, since these backtracks are hidden in unifications. Optimization and compilation of unifications in the graph of rules dependencies (Sect. 6) can be solutions to this problem.

5 Rules Dependencies in Forward Chaining

The notions of neutrality/dependency between CG rules were introduced in [4] to enhance the basic FC (Alg. 1). The basic idea is expressed as follows: suppose that the conclusion of $\mathbb{R}_1$ contains no entity or relation that is a specialization of an entity or a relation in the hypothesis of $\mathbb{R}_2$. Then an application of $\mathbb{R}_1$ on a given simple CG does not create any new application of $\mathbb{R}_2$. This is a simple case of neutrality between rules. A general definition is provided in Sect. 5.1. We present in Sect. 5.2 a characterization of dependency (the inverse notion of neutrality), based upon piece unification, that generalizes the characterization of [4]. Finally, in Sect. 5.3, we enhance FC by encoding all dependencies of a CG ruleset (in the graph of rules dependencies [4]).

5.1 Neutrality and Dependency

Though the definition of neutrality and dependency expressed below seems strictly identical to [4], it is indeed more general. A component of this definition is rule application (Sect. 3.3). In this paper, the graph on which the rule is applied is put into normal form, and not in [4]. As a consequence, the algorithm was not complete for CG rulesets containing rules having individuals in the conclusion. Since our definition of derivation takes into account the need to put a simple CG into its normal form after each application of a rule, the following definition of neutrality/dependency is more adapted to $\mathcal{SR}$-DEDUCTION.

Definition 11 (Neutrality, Dependency). *Let $\mathbb{R}_1$ and $\mathbb{R}_2$ be two CG rules over a vocabulary $\mathcal{V}$. We say that $\mathbb{R}_1$ is neutral w.r.t. $\mathbb{R}_2$ iff, for every simple CG G over $\mathcal{V}$, for every projection π of $Hyp(\mathbb{R}_1)$ into G, the set of all projections of $Hyp(\mathbb{R}_2)$ into $\alpha(G, \mathbb{R}_1, \pi)$ and the set of all projections of $Hyp(\mathbb{R}_2)$ into G are equal. If $\mathbb{R}_1$ is not neutral w.r.t. $\mathbb{R}_2$, we say that $\mathbb{R}_2$ depends upon $\mathbb{R}_1$.*

5.2 Piece Unification and Dependency

Since we have changed the definition of derivation used in [4] the characterization of dependency must take that change into account. We prove here that this updated characterization corresponds to the piece unification of [11,10], for CG rules that are not trivially useless. A CG rule $\mathbb{R}$ is said *trivially useless* if, for every simple CG G, for every projection π of $Hyp(\mathbb{R})$ on G, $G = \alpha(G, \mathbb{R}, \pi)$. We can remove in linear time all trivially useless rules from a CG ruleset.

Theorem 4. *Let $\mathbb{R}_1$ and $\mathbb{R}_2$ be two CG rules over a vocabulary $\mathcal{V}$, where $\mathbb{R}_1$ is not trivially useless. Then $\mathbb{R}_2$ depends upon $\mathbb{R}_1$ if and only if $Hyp(\mathbb{R}_2)$ and $\mathbb{R}_1$ are unifiable (see Def. 9).*

Composition of unification and projection (noted $\odot$). Let G and H be a simple CG, and $\mathbb{R}$ be a CG rule over $\mathcal{V}$. Let $\mu = (P^C, P^Q, \Pi)$ be a unification between H and $\mathbb{R}$. Let π be a projection from $Hyp(\mathbb{R})$ into G. We say that μ and π are *composable* iff for each compatible partition P_i^H whose join belongs to the domain of Π, the entities of $Hyp(\mathbb{R})$ associated (by λ^{-1} or by sharing the same individual marker) with the compatible partition P_i^C of $Conc(\mathbb{R})$ are all mapped by π into the same entity noted $f(P_i^H)$. If μ and π are composable, then we note $\mu \odot \pi : H \to \alpha(G, \mathbb{R}, \pi)$ the partial mapping defined as follows: if e is a cut point of P_i^H in the domain of Π, then $\mu \odot \pi(e) = f(P_i^H)$, otherwise, if e is an entity in the domain of Π that is not a cut point, $\mu \odot \pi(e)$ is the entity of $\alpha(G, \mathbb{R}, \pi)$ that corresponds to $\Pi(e)$ in $Conc(\mathbb{R})$. It is immediate to check that $\mu \odot \pi$ is a partial projection from H into $\alpha(G, \mathbb{R}, \pi)$.

Proof. Let us successively prove both directions of the equivalence:

($\Leftarrow$) Suppose that $Hyp(\mathbb{R}_2)$ and $\mathbb{R}_1$ are unifiable, and note μ such an unification. Let us consider the conjunctive CG $G = \beta(Hyp(\mathbb{R}_2), \mathbb{R}_1, \mu)$. We transform it into a simple CG by replacing all its conjunctive types by one of their specializations in T_C (it exists, by definition of compatible partitions, Sect. 4.1). There exists a projection π from $Hyp(\mathbb{R}_1)$ into G: if e has been joined in G, $\pi(e)$ is this join, and $\pi(e) = e$ otherwise. This mapping π is a projection. It is immediate to check that μ and π are composable (see above). Then $\mu \odot \pi$ is a partial projection from $Hyp(\mathbb{R}_2)$ into $G' = \alpha(G, \mathbb{R}_1, \pi)$ that uses an entity or relation of G' that is not in G (or $\mathbb{R}_1$ would have been trivially useless). Since BC is sound and complete, $\mu \odot \pi$ can be extended to a projection π' of $Hyp(\mathbb{R}_2)$ into G', and π' is not a projection from $Hyp(\mathbb{R}_2)$ into G. Then $\mathbb{R}_2$ depends upon $\mathbb{R}_1$.

($\Rightarrow$) Suppose that $H = Hyp(\mathbb{R}_2)$ and $\mathbb{R}_1$ are not unifiable. Let us consider a simple CG G, and a projection π from $H = Hyp(\mathbb{R}_1)$ into G. If there is a projection from H) into $\alpha(G, \mathbb{R}_1, \pi)$ that is not a projection of H into G, it means that there is a solution to the query H that requires the application of $\mathbb{R}_1$. Since H and $\mathbb{R}_1$ are not unifiable, such a solution could not be found by BC, which is absurd. $\qquad \square$

5.3   Graph of Rules Dependencies in Forward Chaining

In this section, we present an enhancement of FC (Alg. 1) that relies upon the graph of rules dependencies (GRD) [4].

Building the Graph of Rules Dependencies. Let $\mathcal{R}$ be a CG ruleset over $\mathcal{V}$. We call *graph of rules dependencies* (GRD) of $\mathcal{R}$, and note $GRD_{\mathcal{V}}(\mathcal{R})$ the (binary) directed graph whose nodes are the rules of $\mathcal{R}$, and where two nodes $\mathbb{R}_1$ and $\mathbb{R}_2$ are linked by an arc $(\mathbb{R}_1, \mathbb{R}_2)$ iff $\mathbb{R}_2$ depends upon $\mathbb{R}_1$. In that case, the arc $(\mathbb{R}_1, \mathbb{R}_2)$ is labelled by the set of all unifications between $Hyp(\mathbb{R}_2)$ and $\mathbb{R}_1$. By considering the simple CG G encoding the facts as a CG rule with empty hypothesis and the simple CG H encoding the query as a CG rule with empty conclusion, we can integrate them in the GRD, obtaining the graph $GRD_{\mathcal{V}}(\mathcal{R}, G, H)$. Finally, we point out that if a rule $\mathbb{R}$ is not on a path from G to H, then no application of $\mathbb{R}$ is required when solving $\mathcal{SR}$-DEDUCTION [4]. The graph $SGRD_{\mathcal{V}}(\mathcal{R})$ obtained by removing all nodes that are not on a path from G to H, called the *simplified GRD*, is used to restrain the number of unnecessary rules applications.

The problem $\mathcal{SR}$-DEPENDENCY (deciding if a CG rule $\mathbb{R}_2$ depends upon a CG rule $\mathbb{R}_1$) is NP-complete (since a unification is a polynomial certificate, and when

$\mathbb{R}_1$ is disconnected, a unification is exactly a projection). Building the GRD is thus a costly operation, that requires $|\mathcal{R}|^2$ calls to a NP-hard operation.

Using the Graph of Rules Dependencies in Forward Chaining. The GRD (or its simplified version) can be used to enhance FC (Alg. 1) as follows. Let us consider a *step* of FC (an execution of the main **while** loop). The *PartialProjList* contains all partial projections from the hypothesis of the CG rules in $\mathcal{R}$ into G. If one of these partial projections can be extended to a full projection π of the hypothesis of a rule $\mathbb{R}$, then $\mathbb{R}$ is applicable and the only rules that will be applicable on $\alpha(G, \mathbb{R}, \pi)$ (apart from those already in *PartialProjList*) are the successors of $\mathbb{R}$ in the GRD. Moreover, the operator $\odot$ is used to efficiently generate partial projections of the hypothesis of these rules.

Algorithm 3. Forward Chaining using Rules Dependencies

Data: A vocabulary $\mathcal{V}$, a CG ruleset $\mathcal{R}$, two simple CGs G and H over $\mathcal{V}$.
Result: YES iff $G, \mathcal{R} \models_{\mathcal{V}} H$ (infinite calculus otherwise).
$D \leftarrow$ SimplifiedRulesDependenciesGraph($\mathcal{R}, G, H$) ;
PartialProjList $\leftarrow$ NewFifo() ;
for $\mathbb{R} \neq H \in$ *Successors(D, G)* **do**
    for $\mu \in$ *Unifications(D, G, $\mathbb{R}$)* **do**
        PartialProjList $\leftarrow$ AddFifo(PartialProjList, $(\mathbb{R}, \mu)$) ;

while TRUE **do**
    $(\mathbb{R}, \pi) \leftarrow$ FifoRemove(PartialProjList) ;
    for $\pi' \in$ *ExtendPartialtoFullProjections(Hyp($\mathbb{R}$), G, π)* **do**
        $G \leftarrow \alpha(G, \mathbb{R}, \pi')$;
        if *Projects?(H, G)* **then return** YES ;
        for $\mathbb{R}' \neq H \in$ *Successors(D, $\mathbb{R}$)* **do**
            for $\mu \in$ *Unifications(D, $\mathbb{R}, \mathbb{R}'$)* **do**
                if *Composable(μ, π')* **then**
                    PartialProjList $\leftarrow$ AddFifo(PartialProjList, $\mathbb{R}', \mu \odot \pi'$) ;

Evaluating the algorithm. With respect to the standard FC, FC with rules dependencies (FCRD, Alg. 3) relies on three different optimizations:

1. using the simplified GRD allow to ignore some CG rules during derivation;
2. though FC, at each step, checks applicability of all rules in $\mathcal{R}$, FCRD only checks the successors of the rules applied at the previous step;
3. the operator $\odot$, by combining projections and unifications into a partial projection, reduces the search space when checking applicability of a rule.

Though generating the GRD is a lengthy operation, it can be done once and for all for a knowledge base $(G, \mathcal{R})$, leaving only to compute the $|\mathcal{R}|$ unifications of the query Q at run time. Moreover, even if the KB is used only once, the cost of the operations required to compute the GRD is included in the two first steps (the main **while** loop) of the basic FC algorithm.

Finally, the GRD has been used in [4] to obtain new decidability result. If the GRD (or the simplified GRD) has no circuit, then $\mathcal{SR}$-DEDUCTION is decidable. Moreover, if all strongly connected components of the GRD (or simplified GRD) are finite expansion sets (see Sect. 3.3), then $\mathcal{SR}$-DEDUCTION is decidable.

6   Rules Dependencies in Backward Chaining

The identification of dependencies and unifications (Th. 4) naturally leads to the following question: how to efficiently use the GRD in a Backward Chaining framework ? We consider the three interests of the simplified GRD in a FC framework, at the end of Sect. 5.3, and show how they translate to a BC framework (Sect. 6.1). In Sect. 6.2, we provide an update of BC (Alg. 2) that relies on the simplified GRD. Further works on that algorithm are discussed in Sect. 6.3.

6.1   Reducing the Number of Searches for Unification

The simplified GRD can be used as in Forward Chaining to remove rules that are not involved in reasonings: if there is no derivation sequence from G into a solution of H that involves the rule $\mathbb{R}$, then the correspondence between FC and BC proves that no rewriting of H into $\emptyset$ involves that same CG rule $\mathbb{R}$. We should note that, if there is a path from $\mathbb{R}$ to H, but no path from G to $\mathbb{R}$ in the GRD, simplifying the GRD removes this rule though the standard Backward Chaining may try to use it in a rewriting sequence.

The second optimization brought by the GRD to Forward Chaining consists in reducing the number of checks for applicability of a rule. To translate that feature to Backward Chaining, we must ask if, after unifying a query with a rule and rewriting this query w.r.t. this unification, we need to compute the unifications of this new query with all the rules in the CG ruleset $\mathcal{R}$. By giving a negative answer to this question, Th. 5 shows that the GRD can be used during BC for added efficiency.

Theorem 5. *Let H be a simple CG, and $\mathcal{R}$ be a CG ruleset over a vocabulary $\mathcal{V}$. Let μ be an unification between H and $\mathbb{R} \in \mathcal{R}$. Let $H' = \alpha(H, \mathbb{R}, \mu)$ be the rewriting of H according to μ. The following property holds: if $\mathbb{R}'$ and H' are unifiable then $\mathbb{R}'$ is a predecessor of H or $\mathbb{R}$ in $GRD(\mathcal{R}, G, H)$.*

Proof. Suppose $\mathbb{R}'$ and H' are unifiable, by a unification μ'. We note $H'' = \beta(H', \mathbb{R}', \mu')$. Let us consider the simple CG G' that specializes the conjunctive CG H'', built in the same way as in the proof of Th. 4. Since G' proves H'', the correspondence between FC and BC implies that there exists a derivation sequence $G', G'' = \alpha(G', \mathbb{R}', \pi_1), G''' = \alpha(G'', \mathbb{R}, \pi_2)$ such that H projects into G'''. Since FC with rules dependencies is complete, it means that either H depends upon $\mathbb{R}'$, or that $\mathbb{R}$ depends upon $\mathbb{R}'$.   □

6.2   Backward Chaining with Rules Dependencies

The following algorithm uses the graph of rules dependencies in a Backward Chaining framework to include the two optimizations discussed in Sect. 6.1.

Algorithm 4. Backward Chaining using Rules Dependencies

Data: A vocabulary $\mathcal{V}$, a CG ruleset $\mathcal{R}$, two simple CGs G and H over $\mathcal{V}$.
Result: If Backward Chaining halts on YES, then $G, \mathcal{R} \models_{\mathcal{V}} H$, if it halts on NO, then $G, \mathcal{R} \not\models_{\mathcal{V}} H$ (but it can run infinitely).

$D \leftarrow$ SimplifiedRulesDependenciesGraph$(\mathcal{R}, G, H)$;
UnifList $\leftarrow$ NewFilo() ;
for $\mathbb{R} \in$ *Predecessors(D, H)* **do**
    for $\mu \in$ *Unifications(D, $\mathbb{R}$, H)* **do**
        UnifList $\leftarrow$ AddFilo(UnifList, $(\mu, \mathbb{R}, H)$) ;

while *UnifList* $\neq \emptyset$ **do**
    $(\mu, \mathbb{R}, H) \leftarrow$ FiloRemove(UnifList) ;
    $H' \leftarrow$ Rewrite$(\mu, \mathbb{R}, H)$;
    if $H' = \emptyset$ **then return** YES ;
    for $\mathbb{R}' \in$ *Predecessors($\mathbb{R}$)* **do**
        for $\mu' \in$ *ComputeNewUnifications($\mathbb{R}', H'$)* **do**
            UnifList $\leftarrow$ AddFilo(UnifList, $(\mu', \mathbb{R}', H')$) ;

return NO ;

6.3   Further Work: Combining Unifications

Finally, we point out that we have not used in this BC framework the third optimization of FC brought by the GRD. In FC, the composition operator $\odot$ between the current projection and unifications is used to reduce the size of projections that have to be computed during the following execution of the main **while** loop. A similar operator, composing unifications into a partial unification, would be required to achieve the same optimization result in BC.

7   Conclusion

In this paper, we have unified two optimization schemes used for computing deduction with conceptual graphs rules [13,11] ($\mathcal{SR}$-DEDUCTION), namely piece unification in Backward Chaining [11,10], and the graph of rules dependencies in Forward Chaining [4]. Our main contributions are listed below:

1. **Unification of syntax:** [11,10] defines simple CGs as bipartite multigraphs and CG rules as pairs of λ-abstractions, while [4] defines them as directed hypergraphs and colored CGs. We have unified these different syntaxes.
2. **Generalization of piece unification:** the definition of piece unification in [11,10] does no longer rely on concept types being ordered by a lattice.
3. **Generalization of dependencies:** the definition of dependencies in [4] is restricted to CG rules having no individual in the conclusion. This restriction is dropped here.

4. **Identification of piece unification and dependencies:** Up to the generalizations above, we prove that piece unification and neutrality (the inverse of dependency) are equivalent (Th. 4 in Sect. 5.2).
5. **Use of the graph of rules dependencies in a Backward Chaining framework:** we show how the optimizations allowed by the GRD of [4] in a FC framework are adapted to the BC framework of [11,10] (Th. 5 in Sect. 6).

Though the GRD already increases efficiency in both FC and BC, we are now considering the following problems as research perspectives:

1. **Traversals of the GRD:** FC and BC rely respectively on a breadth and depth-first traversal of the GRD. Different types of traversals can be tested.
2. **Rewriting of a CG ruleset:** Some transformations of rules preserve their semantics (*e.g.* a rule with k pieces is equivalent to k rules with one piece). What transformations can give a more efficient FC or BC?
3. **Finding a composition operator for unifications:** (Sect. 6.3)

References

1. S. Abiteboul, R. Hull, and V. Vianu. *Foundations of Databases.* Addison-Wesley, 1995.
2. J.-F. Baget. Simple Conceptual Graphs Revisited: Hypergraphs and Conjunctive Types for Efficient Projection Algorithms. In *Proc. of ICCS'03*, volume 2746 of *LNAI.* Springer, 2003.
3. J.-F. Baget and M.-L. Mugnier. The Complexity of Rules and Constraints. *JAIR*, 16:425–465, 2002.
4. Jean-François Baget. Improving the forward chaining algorithm for conceptual graphs rules. In *Proc. of KR2004)*, pages 407–414. AAAI Press, 2004.
5. M. Chein and M.-L. Mugnier. Conceptual Graphs: Fundamental Notions. *Revue d'Intelligence Artificielle*, 6(4):365–406, 1992.
6. M. Chein and M.-L. Mugnier. Types and Coreference in Simple Conceptual Graphs. In *Proc. ICCS'04*, volume 3127 of *LNAI.* Springer, 2004.
7. S. Coulondre and E. Salvat. Piece Resolution: Towards Larger Perspectives. In *Proc. of ICCS'98*, volume 1453 of *LNAI*, pages 179–193. Springer, 1998.
8. M.-L. Mugnier. Knowledge Representation and Reasoning based on Graph Homomorphism. In *Proc. ICCS'00*, volume 1867 of *LNAI*, pages 172–192. Springer, 2000.
9. M.-L. Mugnier and M. Chein. Représenter des connaissances et raisonner avec des graphes. *Revue d'Intelligence Artificielle*, 10(1):7–56, 1996.
10. E. Salvat. Theorem proving using graph operations in the conceptual graphs formalism. In *Proc. of ECAI'98*, pages 356–360, 1998.
11. E. Salvat and M.-L. Mugnier. Sound and Complete Forward and Backward Chainings of Graph Rules. In *Proc. of ICCS'96*, volume 1115 of *LNAI*, pages 248–262. Springer, 1996.
12. J. F. Sowa. Conceptual Graphs. *IBM Journal of Research and Development*, 1976.
13. J. F. Sowa. *Conceptual Structures: Information Processing in Mind and Machine.* Addison-Wesley, 1984.

Thresholds and Shifted Attributes in Formal Concept Analysis of Data with Fuzzy Attributes*

Radim Bělohlávek, Jan Outrata and Vilém Vychodil

Department of Computer Science, Palacky University, Olomouc
Tomkova 40, CZ-779 00 Olomouc, Czech Republic
{radim.belohlavek, jan.outrata, vilem.vychodil}@upol.cz

Abstract. We focus on two approaches to formal concept analysis (FCA) of data with fuzzy attributes recently proposed in the literature, namely, on the approach via hedges and the approach via thresholds. Both of the approaches present parameterized ways to FCA of data with fuzzy attributes. Our paper shows basic relationships between the two of the approaches. Furthermore, we show that the approaches can be combined in a natural way, i.e. we present an approach in which one deals with both thresholds and hedges. We argue that while the approach via thresholds is intuitively appealing, it can be considered a special case of the approach via hedges. An important role in this analysis is played by so-called shifts of fuzzy attributes which appeared earlier in the study of factorization of fuzzy concept lattices. In addition to fuzzy concept lattices, we consider the idea of thresholds for the treatment of attribute implications from tables with fuzzy attributes and prove basic results concerning validity and non-redundant bases.

1 Introduction and Motivation

Recently, there have been proposed several approaches to formal concept analysis (FCA) of data with fuzzy attributes, i.e. attributes which apply to objects to various degrees taken from a scale L of degrees. In particular, parameterized approaches are of interest where the parameters control the number of the extracted formal concepts. In this paper, we deal with two of these approaches, namely the approach via hedges and the approach via thresholds. Hedges were proposed as parameters for formal concept analysis of data with fuzzy attributes in [10], see also [8, 11]. For particular choices of hedges, one obtains the original approach by Pollandt and Bělohlávek [3, 23] and one-sided fuzzy approach, see [9, 22, 14]. The idea of thresholds in formal concept analysis of data with fuzzy attributes is the following. In a fuzzy setting, given a collection A of objects, the collection $A^\uparrow$ of all attributes shared by all objects from A is in general a fuzzy

* Supported by grant No. 1ET101370417 of GA AV ČR, by grant No. 201/05/0079 of the Czech Science Foundation, and by institutional support, research plan MSM 6198959214.

H. Schärfe, P. Hitzler, and P. Øhrstrøm (Eds.): ICCS 2006, LNAI 4068, pp. 117–130, 2006.

set, i.e. attributes y belong to $A^\uparrow$ in various degrees $A^\uparrow(y) \in L$. It is then intuitively appealing to pick a threshold δ and to consider a set ${}^\delta A^\uparrow = \{y \mid A^\uparrow(y) \geq \delta\}$ of all attributes which belong to $A^\uparrow$ in a degree greater than or equal to δ. With $\delta = 1$, this approach was proposed independently in [22, 14]. In [15], this was extended to arbitrary δ. However, the extent- and intent-forming operators defined in [15] do not form a Galois connection. This shortcoming was recognized and removed in [16] where the authors proposed new operators based on the idea of thresholds for general δ.

In our paper, we take a closer look at [16]. We show that while conceptually natural and appealing, the approach via thresholds, as proposed in [16], can be seen as a particular case of the approach via hedges. In particular, given a data with fuzzy attributes, the fuzzy concept lattices induced by the operators of [16] are isomorphic (and in fact, almost the same) to fuzzy concept lattices with hedges induced from a data containing so-called shifts of the given fuzzy attributes. This observation suggests a combination of the approaches via hedges and via thresholds which we also explore. It is interesting to note that shifts of fuzzy attributes play an important role for an efficient computation in a factorization by similarity of a fuzzy concept lattice, see [2, 7]. In addition to that, we apply the idea of thresholds to attribute implications from data with fuzzy attributes and extend some of our previous results, see e.g. [6, 12].

2   Fuzzy Concept Lattices with Hedges and Thresholds

2.1   Preliminaries from Fuzzy Logic

We first briefly recall the necessary notions from fuzzy sets and fuzzy logic (we refer to [3, 20] for further details). As a structure of truth degrees, we use an arbitrary complete residuated lattice $\mathbf{L} = \langle L, \wedge, \vee, \otimes, \rightarrow, 0, 1 \rangle$, i.e. $\langle L, \wedge, \vee, 0, 1 \rangle$ is a complete lattice with 0 and 1 being the least and greatest element of L, respectively (for instance, L is $[0, 1]$, a finite chain, etc.); $\langle L, \otimes, 1 \rangle$ is a commutative monoid (i.e. $\otimes$ is commutative, associative, and $a \otimes 1 = 1 \otimes a = a$ for each $a \in L$); and $\otimes$ and $\rightarrow$ satisfy so-called adjointness property, i.e. $a \otimes b \leq c$ iff $a \leq b \rightarrow c$ for each $a, b, c \in L$. Elements a of L are called truth degrees (usually, $L \subseteq [0, 1]$). $\otimes$ and $\rightarrow$ are (truth functions of) "fuzzy conjunction" and "fuzzy implication". Note that in [16], the authors do not require commutativity of $\otimes$ (but this plays no role in our note). Note that complete residuated lattices are basic structures of truth degrees used in fuzzy logic, see [18, 20]. Residuated lattices cover many structures used in applications.

For a complete residuated lattice $\mathbf{L}$, a (truth-stressing) hedge is a unary function $*$ satisfying (i) $1^* = 1$, (ii) $a^* \leq a$, (iii) $(a \rightarrow b)^* \leq a^* \rightarrow b^*$, (iv) $a^{**} = a^*$, for all $a, b \in L$. A hedge $*$ is a (truth function of) logical connective "very true" [21]. The largest hedge (by pointwise ordering) is identity, the least hedge is globalization which is defined by $a^* = 1$ for $a = 1$ and $a^* = 0$ for $a < 1$.

For $L = \{0, 1\}$, there exists exactly one complete residuated lattice $\mathbf{L}$ (the two-element Boolean algebra) and exactly one hedge (the identity on $\{0, 1\}$).

By $\mathbf{L}^U$ or L^U we denote the set of all fuzzy sets (**L**-sets) in universe U, i.e. $L^U = \{A \mid A$ is a mapping of U to $L\}$, $A(u)$ being interpreted as a degree to which u belongs to A; by 2^U we denote the set of all ordinary subsets of U, and by abuse of notation we sometimes identify ordinary subsets of U with crisp fuzzy sets from L^U, i.e. with those $A \in L^U$ for which $A(u) = 0$ or $A(u) = 1$ for each $u \in U$. For $A \in L^U$ and $a \in L$, a set ${}^aA = \{u \in U \mid A(u) \geq a\}$ is called an a-cut of A; a fuzzy set $a \to A$ in U defined by $(a \to A)(u) = a \to A(u)$ is called an a-shift of A. Given $A, B \in \mathbf{L}^U$, we define a subsethood degree

$$S(A, B) = \bigwedge_{u \in U} (A(u) \to B(u)),$$

which generalizes the classical subsethood relation $\subseteq$. $S(A, B)$ represents a degree to which A is a subset of B. In particular, we write $A \subseteq B$ iff $S(A, B) = 1$ (A is fully contained in B). As a consequence, $A \subseteq B$ iff $A(u) \leq B(u)$ for each $u \in U$.

2.2   Fuzzy Concept Lattices with Hedges

A formal fuzzy context can be identified with a triplet $\langle X, Y, I \rangle$ where X is a non-empty set of objects, Y is a non-empty set of attributes, and I is a fuzzy relation between X and Y, i.e. $I : X \times Y \to L$. For $x \in X$ and $y \in Y$, a degree $I(x, y) \in L$ is interpreted as a degree to which object x has attribute y. A formal fuzzy context $\langle X, Y, I \rangle$ can be seen as a data table with fuzzy attributes with rows and columns corresponding to objects and attributes, and table entries filled with truth degrees $I(x, y)$. For $L = \{0, 1\}$, formal fuzzy contexts can be identified in an obvious way with ordinary formal contexts.

Let ^{*X} and ^{*Y} be hedges. For fuzzy sets $A \in L^X$ and $B \in L^Y$, consider fuzzy sets $A^\uparrow \in L^Y$ and $B^\downarrow \in L^X$ (denoted also $A^{\uparrow I}$ and $B^{\downarrow I}$ to make I explicit) defined by

$$A^\uparrow(y) = \bigwedge_{x \in X} (A^{*X}(x) \to I(x, y)), \tag{1}$$

$$B^\downarrow(x) = \bigwedge_{y \in Y} (B^{*Y}(y) \to I(x, y)). \tag{2}$$

Using basic rules of predicate fuzzy logic, $A^\uparrow$ is a fuzzy set of all attributes common to all objects (for which it is very true that they are) from A, and $B^\downarrow$ is a fuzzy set of all objects sharing all attributes (for which it is very true that they are) from B. The set

$$\mathcal{B}(X^{*X}, Y^{*Y}, I) = \{\langle A, B \rangle \mid A^\uparrow = B, \ B^\downarrow = A\}$$

of all fixpoints of $\langle \uparrow, \downarrow \rangle$ is called a fuzzy concept lattice of $\langle X, Y, I \rangle$; elements $\langle A, B \rangle \in \mathcal{B}(X^{*X}, Y^{*Y}, I)$ will be called formal concepts of $\langle X, Y, I \rangle$; A and B are called the extent and intent of $\langle A, B \rangle$, respectively. Under a partial order $\leq$ defined on $\mathcal{B}(X^{*X}, Y^{*Y}, I)$ by

$$\langle A_1, B_1 \rangle \leq \langle A_2, B_2 \rangle \quad \text{iff} \quad A_1 \subseteq A_2,$$

$\mathcal{B}(X^{*X}, Y^{*Y}, I)$ happens to be a complete lattice and we refer to [10] for results describing the structure of $\mathcal{B}(X^{*X}, Y^{*Y}, I)$. Note that $\mathcal{B}(X^{*X}, Y^{*Y}, I)$ is the basic structure used for formal concept analysis of the data table represented by $\langle X, Y, I \rangle$.

Remark 1. Operators $^\uparrow$ and $^\downarrow$ were introduced in [8, 10] as a parameterization of operators $A^\Uparrow(y) = \bigwedge_{x \in X}(A(x) \to I(x,y))$ and $B^\Downarrow(x) = \bigwedge_{y \in Y}(B(y) \to I(x,y))$ which were studied before, see [1, 4, 23]. Clearly, if both ^{*X} are ^{*Y} are identities on L, $^\uparrow$ and $^\downarrow$ coincide with $^\Uparrow$ and $^\Downarrow$, respectively. If ^{*X} or ^{*Y} is the identity on L, we omit ^{*X} or ^{*Y} in $\mathcal{B}(X^{*X}, Y^{*Y}, I)$, e.g. we write just $\mathcal{B}(X^{*X}, Y, I)$ if $^{*Y} = \mathrm{id}_L$.

2.3   Fuzzy Concept Lattices Defined by Thresholds

In addition to the pair of operators $^\Uparrow : L^X \to L^Y$ and $^\Downarrow : L^Y \to L^X$, the authors in [16] define pairs of operators (we keep the notation of [16]) $^* : 2^X \to 2^Y$ and $^* : 2^Y \to 2^X$, $^\square : 2^X \to L^Y$ and $^\square : L^Y \to 2^X$, and $^\diamond : L^X \to 2^Y$ and $^\diamond : 2^Y \to L^X$, as follows. Let δ be an arbitrary truth degree from L (δ plays a role of a threshold). For $A \in L^X$, $C \in 2^X$, $B \in L^Y$, $D \in 2^Y$ define $C^* \in 2^Y$ and $D^* \in 2^X$ by

$$C^* = \{y \in Y \mid \bigwedge_{x \in X}(C(x) \to I(x,y)) \geq \delta\}, \tag{3}$$

$$D^* = \{x \in X \mid \bigwedge_{y \in Y}(D(y) \to I(x,y)) \geq \delta\}; \tag{4}$$

$C^\square \in L^Y$ and $B^\square \in 2^X$ by

$$C^\square(y) = \delta \to \bigwedge_{x \in C} I(x,y), \tag{5}$$

$$B^\square = \{x \in X \mid \bigwedge_{y \in Y}(B(y) \to I(x,y)) \geq \delta\}; \tag{6}$$

and $A^\diamond \in 2^Y$ and $D^\diamond \in L^X$ by

$$A^\diamond = \{y \in Y \mid \bigwedge_{x \in X}(A(x) \to I(x,y)) \geq \delta\}, \tag{7}$$

$$D^\diamond(x) = \delta \to \bigwedge_{y \in D} I(x,y), \tag{8}$$

for each $x \in X$, $y \in Y$.

Denote now the corresponding set of fixpoints of these pairs of operators by

$$\mathcal{B}(X_\star, Y_\star, I) = \{\langle A, B \rangle \in 2^X \times 2^Y \mid A^* = B, B^* = A\},$$
$$\mathcal{B}(X_\square, Y_\square, I) = \{\langle A, B \rangle \in 2^X \times L^Y \mid A^\square = B, B^\square = A\},$$
$$\mathcal{B}(X_\diamond, Y_\diamond, I) = \{\langle A, B \rangle \in L^X \times 2^Y \mid A^\diamond = B, B^\diamond = A\},$$
$$\mathcal{B}(X_\Uparrow, Y_\Downarrow, I) = \{\langle A, B \rangle \in L^X \times L^Y \mid A^\Uparrow = B, B^\Downarrow = A\} \quad (= \mathcal{B}(X, Y, I)).$$

2.4   Fuzzy Concept Lattices with Hedges and Thresholds

We now introduce a new pair of operators induced by a formal fuzzy context $\langle X, Y, I \rangle$. For $\delta, \varepsilon \in L$, fuzzy sets $A \in L^X$ and $B \in L^Y$, consider fuzzy sets $A^{\uparrow_{I,\delta}} \in L^Y$ and $B^{\downarrow_{I,\varepsilon}} \in L^X$ defined by

$$A^{\uparrow_{I,\delta}}(y) = \delta \to \bigwedge_{x \in X}(A^{*X}(x) \to I(x,y)), \tag{9}$$

$$B^{\downarrow_{I,\varepsilon}}(x) = \varepsilon \to \bigwedge_{y \in Y}(B^{*Y}(y) \to I(x,y)). \tag{10}$$

We will often write just $A^\uparrow$ and $B^\downarrow$ if I, δ, and ε are obvious, particularly if $\delta = \varepsilon$.

Remark 2. Note that, due to the properties of $\rightarrow$, we have that $A^{\uparrow I,\delta}(y) = 1$ iff

$$\delta \leq \bigwedge\nolimits_{x \in X}(A^{*X}(x) \rightarrow I(x,y)),$$

i.e. iff the degree to which y is shared by all objects from A is at least δ. In general, $A^{\uparrow I,\delta}(y)$ can be thought of as a truth degree of *the degree to which y is shared by all objects from A is at least δ.* We will show that this general approach involving the idea of thresholds subsumes the proposals of [16] as special cases. Moreover, unlike formulas (5) and (6), and (7) and (8), formulas for operators $\uparrow I,\delta$ and $\downarrow I,\delta$ are symmetric.

The set

$$\mathcal{B}(X_\delta^{*X}, Y_\varepsilon^{*Y}, I) = \{\langle A, B \rangle \mid A^\uparrow = B,\ B^\downarrow = A\}$$

of all fixpoints of $\langle \uparrow, \downarrow \rangle$ is called a fuzzy concept lattice of $\langle X, Y, I \rangle$; elements $\langle A, B \rangle \in \mathcal{B}(X_\delta^{*X}, Y_\varepsilon^{*Y}, I)$ will be called formal concepts of $\langle X, Y, I \rangle$; A and B are called the extent and intent of $\langle A, B \rangle$, respectively.

Remark 3. Since $1 \rightarrow a = a$ for each $a \in L$, we have $A^{\uparrow I,1} = A^{\uparrow I}$ and $B^{\downarrow I,1} = B^{\downarrow I}$ and, therefore, $\mathcal{B}(X_1^{*X}, Y_1^{*Y}, I) = \mathcal{B}(X^{*X}, Y^{*Y}, I)$.

Basic Relationships to Earlier Approaches. The following theorem shows that from a mathematical point of view, $\mathcal{B}(X_\delta^{*X}, Y_\delta^{*Y}, I)$ is, in fact, a fuzzy concept lattice with hedges (i.e. without thresholds) induced by a δ-shift $\delta \rightarrow I$ of I.

Theorem 1. *For any $\delta \in L$, $\uparrow I,\delta$ coincides with $\uparrow \delta \rightarrow I$, and $\downarrow I,\delta$ coincides with $\downarrow \delta \rightarrow I$. Therefore, $\mathcal{B}(X_\delta^{*X}, Y_\delta^{*Y}, I) = \mathcal{B}(X^{*X}, Y^{*Y}, \delta \rightarrow I)$.*

Proof. Using $a \rightarrow (b \rightarrow c) = b \rightarrow (a \rightarrow c)$ and $a \rightarrow (\bigwedge_{j \in J} b_j) = \bigwedge_{j \in J}(a \rightarrow b_j)$ we get

$$
\begin{aligned}
A^{\uparrow I,\delta}(y) &= \delta \rightarrow \bigwedge\nolimits_{x \in X}(A^{*X}(x) \rightarrow I(x,y)) = \\
&= \bigwedge\nolimits_{x \in X}(\delta \rightarrow (A^{*X}(x) \rightarrow I(x,y))) = \\
&= \bigwedge\nolimits_{x \in X}(A^{*X}(x) \rightarrow (\delta \rightarrow I(x,y))) = A^{\uparrow \delta \rightarrow I}(y).
\end{aligned}
$$

One can proceed analogously to show that $\downarrow I,\delta$ coincides with $\downarrow \delta \rightarrow I$. Then the equality $\mathcal{B}(X_\delta^{*X}, Y_\delta^{*Y}, I) = \mathcal{B}(X^{*X}, Y^{*Y}, \delta \rightarrow I)$ follows immediately.

Remark 4. (1) Using [10], Theorem 1 yields that $\mathcal{B}(X_\delta^{*X}, Y_\delta^{*Y}, I)$ is a complete lattice; we show a main theorem for $\mathcal{B}(X_\delta^{*X}, Y_\delta^{*Y}, I)$ below.

(2) In addition to $A^{\uparrow I,\delta}(y) = A^{\uparrow \delta \rightarrow I}$ we also have $A^{\uparrow I,\delta}(y) = (\delta \otimes A^{*X})^{\Uparrow I}$; similarly for $B^{\downarrow I,\delta}$.

Remark 5. Note that shifted fuzzy contexts $\langle X, Y, a \rightarrow I \rangle$ play an important role in fast factorization of a fuzzy concept lattice $\mathcal{B}(X, Y, I)$ by a similarity given by a parameter a, see [2, 7]. Briefly, $\mathcal{B}(X, Y, a \rightarrow I)$ is isomorphic to a factor lattice $\mathcal{B}(X, Y, I)/^a\approx$ where $^a\approx$ is an a-cut of a fuzzy equivalence relation $\approx$ defined on $\mathcal{B}(X, Y, I)$ as in [2]. An investigation of the role of $a \rightarrow I$ in factorization of fuzzy concept lattices involving hedges is an important topic which will be a subject of a forthcoming paper.

The next theorem and Remark 6 show that the fuzzy concept lattices defined in [16] are isomorphic, and in fact identical, to fuzzy concept lattices defined by (9) and (10) with appropriate choices of ^{*X} and ^{*Y}.

Theorem 2. *Let $\mathcal{B}(X_\star, Y_\star, I)$, $\mathcal{B}(X_\square, Y_\square, I)$, and $\mathcal{B}(X_\lozenge, Y_\lozenge, I)$ denote the concept lattices defined in Section 2.3 using a parameter δ.*

(1) $\mathcal{B}(X_\star, Y_\star, I)$ *is isomorphic to $\mathcal{B}(X_\delta^{*X}, Y_\delta^{*Y}, I)$, and due to Theorem 1 also to $\mathcal{B}(X^{*X}, Y^{*Y}, \delta \to I)$, where both ^{*X} and ^{*Y} are globalizations on L.*

(2) $\mathcal{B}(X_\square, Y_\square, I)$ *is isomorphic to $\mathcal{B}(X_\delta^{*X}, Y_\delta^{*Y}, I)$, and due to Theorem 1 also to $\mathcal{B}(X^{*X}, Y^{*Y}, \delta \to I)$, where ^{*X} is globalization and ^{*Y} is the identity on L.*

(3) $\mathcal{B}(X_\lozenge, Y_\lozenge, I)$ *is isomorphic to $\mathcal{B}(X_\delta^{*X}, Y_\delta^{*Y}, I)$, and due to Theorem 1 also to $\mathcal{B}(X^{*X}, Y^{*Y}, \delta \to I)$, where ^{*X} is the identity and ^{*Y} is globalization on L.*

Proof. We prove only (2); the proofs for (1) and (3) are similar. First, we show that for $\langle C, D \rangle \in \mathcal{B}(X_\delta^{*X}, Y_\delta^{*Y}, I)$ we have $\langle {}^1C, D \rangle \in \mathcal{B}(X_\square, Y_\square, I)$. Indeed, for ^{*X} being globalization we have $^1C = C^{*X}$ and thus

$$({}^1C)^\square = \delta \to \bigwedge\nolimits_{x \in {}^1C} I(x,y) = \delta \to \bigwedge\nolimits_{x \in X}(({}^1C)(x) \to I(x,y)) =$$
$$= \delta \to \bigwedge\nolimits_{x \in X}(C^{*X}(x) \to I(x,y)) = C^{\uparrow_{I,\delta}},$$

and

$$D^\square = \{x \in X \mid \bigwedge\nolimits_{y \in Y}(D(y) \to I(x,y)) \geq \delta\} =$$
$$= \{x \in X \mid \delta \to \bigwedge\nolimits_{y \in Y}(D(y) \to I(x,y)) = 1\} =$$
$$= \{x \in X \mid D^{\downarrow_{I,\delta}}(x) = 1\} = {}^1(D^{\downarrow_{I,\delta}}) = {}^1C.$$

Clearly, $\langle C, D \rangle \mapsto \langle {}^1C, D \rangle$ defines an injective mapping of $\mathcal{B}(X_\delta^{*X}, Y_\delta^{*Y}, I)$ to $\mathcal{B}(X_\square, Y_\square, I)$. This mapping is also surjective. Namely, for $\langle A, B \rangle \in \mathcal{B}(X_\square, Y_\square, I)$ we have $\langle A^{\uparrow_{I,\delta}\downarrow_{I,\delta}}, B \rangle \in \mathcal{B}(X_\delta^{*X}, Y_\delta^{*Y}, I)$ and $A = {}^1(A^{\uparrow_{I,\delta}\downarrow_{I,\delta}})$. Indeed, since $A = A^{*X}$, [8], $^{\uparrow_{I,\delta}} = {}^{\uparrow_{\delta \to I}}$, and $^{\downarrow_{I,\delta}} = {}^{\downarrow_{\delta \to I}}$ give $A^{\uparrow_{I,\delta}\downarrow_{I,\delta}\uparrow_{I,\delta}} = A^{\uparrow_{I,\delta}} = A^\square = B$. Furthermore, $B^{\downarrow_{I,\delta}} = A^{\uparrow_{I,\delta}\downarrow_{I,\delta}}$. This shows $\langle A^{\uparrow_{I,\delta}\downarrow_{I,\delta}}, B \rangle \in \mathcal{B}(X_\delta^{*X}, Y_\delta^{*Y}, I)$. Observing

$$B^\square = {}^\delta(B^{\downarrow_I}) = {}^1(B^{\downarrow_{\delta \to I}}) = {}^1(B^{\downarrow_{I,\delta}}) = {}^1(A^{\uparrow_{I,\delta}\downarrow_{I,\delta}})$$

finishes the proof.

Remark 6. (1) As one can see from the proof of Theorem 2, an isomorphism exists such that the corresponding elements $\langle A, B \rangle \in \mathcal{B}(X_\square, Y_\square, I)$ and $\langle C, D \rangle \in \mathcal{B}(X_\delta^{*X}, Y_\delta^{*Y}, I)$ are almost the same, namely, $\langle A, B \rangle = \langle {}^1C, D \rangle$. A similar fact pertains to (1) and (3) of Theorem 2 as well.

(2) Alternatively, Theorem 2 can be proved using results from [11]. Consider e.g. $\mathcal{B}(X_\square, Y_\square, I)$: It can be shown that $\mathcal{B}(X_\square, Y_\square, I)$ coincides with "one-sided fuzzy concept lattice" of $\langle X, Y, \delta \to I \rangle$ (in the sense of [22]); therefore, by [11], $\mathcal{B}(X_\square, Y_\square, I)$ is isomorphic to a fuzzy concept lattice with hedges where ^{*X} is globalization and ^{*Y} is identity, i.e. to $\mathcal{B}(X^{*X}, Y, \delta \to I)$.

From (9) and (10) one easily obtains the following assertion.

Corollary 1. $\mathcal{B}(X_*, Y_*, I)$ *coincides with an ordinary concept lattice* $\mathcal{B}(X, Y, {}^\delta I)$ *where* ${}^\delta I = \{\langle x, y \rangle \mid I(x, y) \geq \delta\}$ *is the δ-cut of I.*

Remark 7. The foregoing results show that $\mathcal{B}(X_\square, Y_\square, I)$ and $\mathcal{B}(X_\delta^{*_X}, Y_\delta^{*_Y}, I)$ are isomorphic (with appropriate $*_X$ and $*_Y$). Moreover, $\mathcal{B}(X_\square, Y_\square, I)$ is almost identical to $\mathcal{B}(X_\delta^{*_X}, Y_\delta^{*_Y}, I)$, but they are not equal. Alternatively, one can proceed so as to define our operators by

$$A^{\uparrow_{I,\delta}}(y) = \left(\delta \to \bigwedge_{x \in X}(A(x) \to I(x, y))\right)^{*_Y}, \tag{11}$$

$$B^{\downarrow_{I,\varepsilon}}(x) = \left(\varepsilon \to \bigwedge_{y \in Y}(B(y) \to I(x, y))\right)^{*_X}. \tag{12}$$

Then, we even have $\mathcal{B}(X_\square, Y_\square, I) = \mathcal{B}(X_\delta^{*_X}, Y_\delta^{*_Y}, I)$ (with the same choices of $*_X$ and $*_Y$). We still prefer (9) and (10) to (11) and (12) for reasons we omit here due to lack of space.

Main Theorem of Fuzzy Concept Lattices Defined by Thresholds and Hedges. Due to Theorem 1 and Theorem 2, we can obtain main theorems for fuzzy concept lattices defined by thresholds. Omitting the proof due to lack of space, we only give here a version for the general case of $\mathcal{B}(X_\delta^{*_X}, Y_\delta^{*_Y}, I)$ for the sake of illustration.

Theorem 3. (1) $\mathcal{B}(X_\delta^{*_X}, Y_\delta^{*_Y}, I)$ *is under $\leq$ a complete lattice where the infima and suprema are given by*

$$\bigwedge_{j \in J} \langle A_j, B_j \rangle = \langle (\bigcap_{j \in J} A_j)^{\uparrow_{I,\delta} \downarrow_{I,\delta}}, (\bigcup_{j \in J} B_j^{*_Y})^{\downarrow_{I,\delta} \uparrow_{I,\delta}} \rangle, \tag{13}$$

$$\bigvee_{j \in J} \langle A_j, B_j \rangle = \langle (\bigcup_{j \in J} A_j^{*_X})^{\uparrow_{I,\delta} \downarrow_{I,\delta}}, (\bigcap_{j \in J} B_j)^{\downarrow_{I,\delta} \uparrow_{I,\delta}} \rangle. \tag{14}$$

(2) *Moreover, an arbitrary complete lattice* $\mathbf{K} = \langle K, \leq \rangle$ *is isomorphic to* $\mathcal{B}(X_\delta^{*_X}, Y_\delta^{*_Y}, I)$ *iff there are mappings* $\gamma : X \times \mathrm{fix}(*_X) \to K, \mu : Y \times \mathrm{fix}(*_Y) \to K$ *such that*

(i) $\gamma(X \times \mathrm{fix}(*_X))$ *is $\bigvee$-dense in K, $\mu(Y \times \mathrm{fix}(*_Y))$ is $\bigwedge$-dense in K;*
(ii) $\gamma(x, a) \leq \mu(y, b)$ *iff* $a \otimes b \otimes \delta \leq I(x, y)$,

with $\mathrm{fix}(*) = \{a \mid a^* = a\}$ *denoting the set of all fixpoints of $*$.*

3   Attribute Implications from Shifted Fuzzy Attributes

Let Y be a finite set of attributes (each $y \in Y$ is called an attribute). A fuzzy attribute implication (over Y) is an expression $A \Rightarrow B$, where $A, B \in \mathbf{L}^Y$ are fuzzy sets of attributes. In [6, 12, 13] we showed that (i) fuzzy attribute implications can be interpreted in data tables with fuzzy attributes (i.e., in formal fuzzy contexts); (ii) truth (validity) of fuzzy attribute implications (FAIs) in data tables with fuzzy attributes can be described as truth of implications in fuzzy concept intents; (iii) FAIs which are true in a data table with fuzzy attributes can be fully characterized by a so-called non-redundant basis of that table and the basis itself can be computed with polynomial time delay; (iv) semantic entailment

from collections of fuzzy attribute implications can be characterized syntactically by an Armstrong-like set of deduction rules (two versions of completeness:
characterization of FAIs which are fully entailed and characterization of degrees
of entailment).

In this section we show that using the idea of thresholds one can generalize
the notion of a truth of an attribute implication to a notion of δ-truth, where δ
is a truth degree acting as a threshold degree. We show results answering basic
questions arising with the notion of a δ-truth.

For an **L**-set $M \in \mathbf{L}^Y$ of attributes and a truth degree $\delta \in L$, define a *degree*
$||A \Rightarrow B||_M^\delta \in L$ *to which* $A \Rightarrow B$ *is δ-true in M* by

$$||A \Rightarrow B||_M^\delta = (\delta \to S(A, M))^{*X} \to (\delta \to S(B, M)). \tag{15}$$

Since $S(B, M)$ can be interpreted as "a degree to which M has each attribute
from B", $\delta \to S(B, M)$ expresses a truth degree of proposition "a degree to which
M has each attribute from B is at least δ". Thus, one can see that $||A \Rightarrow B||_M^\delta$
is interpreted as a degree to which it is true that "if it is very true that M has
all attributes from A at least to degree δ, then M has all attributes from B at
least to degree δ". Hence, δ acts as a threshold for antecedent and consequent
of $A \Rightarrow B$ which influences the truth of $A \Rightarrow B$ in M. The notion of truth
$||\cdots||_M$ being used in [6, 12, 13] is now but a particular case for $\delta = 1$, i.e.
$||A \Rightarrow B||_M = ||A \Rightarrow B||_M^1$. For $\delta = 0$, which is the other borderline case,
$||A \Rightarrow B||_M^0 = 1$ for each $A, B, M \in \mathbf{L}^Y$.

Theorem 4. *For each $A, B, M \in \mathbf{L}^Y$ and $\delta \in L$,*

$$||A \Rightarrow B||_M^\delta = ||A \Rightarrow B||_{\delta \to M}^1 = ||\delta \otimes A \Rightarrow \delta \otimes B||_M^1 = \delta \to ||\delta \otimes A \Rightarrow B||_M^1. \tag{16}$$

Proof. Using $a \to (b \to c) = b \to (a \to c)$, $a \to \bigwedge_i b_i = \bigwedge_i (a \to b_i)$, and
$1 \to a = a$, see [3], one can conclude $\delta \to S(C, M) = S(C, \delta \to M) = 1 \to$
$S(C, \delta \to M)$. Thus, $||A \Rightarrow B||_M^\delta = ||A \Rightarrow B||_{\delta \to M}^1$. The second equality follows
by using $a \to (b \to c) = (a \otimes b) \to c$. The last one is also clear.

For technical reasons we introduce the following convention. For a set $\mathcal{M} \subseteq \mathbf{L}^Y$
(i.e. $\mathcal{M}$ is an ordinary set of **L**-sets) we define a degree $||A \Rightarrow B||_{\mathcal{M}}^\delta \in L$ to which
$A \Rightarrow B$ is δ-true in $\mathcal{M}$ by $||A \Rightarrow B||_{\mathcal{M}}^\delta = \bigwedge_{M \in \mathcal{M}} ||A \Rightarrow B||_M^\delta$. Obviously,

$$||A \Rightarrow B||_{\mathcal{M}}^\delta = \bigwedge_{M \in \mathcal{M}} ||A \Rightarrow B||_M^\delta = \bigwedge_{M \in \mathcal{M}} ||A \Rightarrow B||_{\delta \to M}^1 = ||A \Rightarrow B||_{\delta \to \mathcal{M}}^1,$$

where $\delta \to \mathcal{M} = \{\delta \to M \mid M \in \mathcal{M}\}$. For $\langle X, Y, I \rangle$, let $I_x \in \mathbf{L}^Y$ $(x \in X)$ be
an **L**-set of attributes such that, for each $y \in Y$, $I_x(y) = I(x, y)$. Described
verbally, I_x is the **L**-set of all attributes of object $x \in X$ in $\langle X, Y, I \rangle$. Now, a
degree $||A \Rightarrow B||_{\langle X, Y, I \rangle}^\delta \in L$ *to which $A \Rightarrow B$ is δ-true in (each row of) $\langle X, Y, I \rangle$*
is defined by

$$||A \Rightarrow B||_{\langle X, Y, I \rangle}^\delta = ||A \Rightarrow B||_{\mathcal{M}}^\delta, \text{ where } \mathcal{M} = \{I_x \mid x \in X\}. \tag{17}$$

Using previous observations, we get the following

Corollary 2. *Let $\langle X, Y, I \rangle$ be a data table with fuzzy attributes, $\delta \in L$. Then*

$$||A \Rightarrow B||^{\delta}_{\langle X,Y,I \rangle} = ||A \Rightarrow B||^{1}_{\langle X,Y,\delta \rightarrow I \rangle}. \tag{18}$$

The following assertion generalizes a well-known characterization of a degree of truth of an attribute implication. It also shows that the notion of a δ-truth is well-connected to the formulas for $\uparrow_{I,\delta}$ and $\downarrow_{I,\delta}$.

Theorem 5. *Let $\langle X, Y, I \rangle$ be a data table with fuzzy attributes, $\delta \in L$. Then*

$$||A \Rightarrow B||^{\delta}_{\langle X,Y,I \rangle} = S(B, A^{\uparrow_{I,\delta}\downarrow_{I,\delta}}).$$

Proof. Using [12], we have $||A \Rightarrow B||^{1}_{\langle X,Y,J \rangle} = S(B, A^{\uparrow_J \downarrow_J})$ for any fuzzy relation J between X and Y. Therefore, by Theorem 1 and Corollary 2,

$$||A \Rightarrow B||^{\delta}_{\langle X,Y,I \rangle} = ||A \Rightarrow B||^{1}_{\langle X,Y,\delta \rightarrow I \rangle} = S(B, A^{\uparrow_{\delta \rightarrow I}\downarrow_{\delta \rightarrow I}}) = S(B, A^{\uparrow_{I,\delta}\downarrow_{I,\delta}}).$$

Using the concept of δ-truth, we can define appropriate notions of a model and a semantic entailment from collections of FAIs. Let T be a set of FAIs, $\delta \in L$. $M \in \mathbf{L}^Y$ is called a δ-*model* of T if $||A \Rightarrow B||^{\delta}_{M} = 1$ for each $A \Rightarrow B \in T$. The set of all δ-models of T will be denoted by $\text{Mod}^{\delta}(T)$, i.e.

$$\text{Mod}^{\delta}(T) = \{M \in \mathbf{L}^Y \mid \text{for each } A \Rightarrow B \in T \colon ||A \Rightarrow B||^{\delta}_{M} = 1\}. \tag{19}$$

In our terminology, models used in [6, 12, 13] are the 1-models. Using the notion of a δ-model, we define a degree of semantic δ-entailment from T. A degree $||A \Rightarrow B||^{\delta}_{T} \in L$ to which $A \Rightarrow B$ is *semantically δ-entailed from* T is defined by

$$||A \Rightarrow B||^{\delta}_{T} = ||A \Rightarrow B||^{\delta}_{\text{Mod}^{\delta}(T)}. \tag{20}$$

Again, semantic 1-entailment coincides with the semantic entailment as it was introduced in [6, 12, 13]. The following assertion shows relationship between various degrees of δ-entailment.

Theorem 6. *Let $A, B \in \mathbf{L}^Y$, $\delta \in L$, T be a set of FAIs. Then*

(i) $\text{Mod}^{\delta}(T) = \{M \in \mathbf{L}^Y \mid \delta \rightarrow M \in \text{Mod}^1(T)\}$,
(ii) $||A \Rightarrow B||^{1}_{T} \leq ||A \Rightarrow B||^{\delta}_{T} \leq ||A \Rightarrow B||^{0}_{T}$,
(iii) $||A \Rightarrow B||^{1}_{T} = \bigwedge_{\delta \in L} ||A \Rightarrow B||^{\delta}_{T}$.

Proof. (i): By definition and using (16), $\text{Mod}^{\delta}(T) = \{M \in \mathbf{L}^Y \mid \text{for each } A \Rightarrow B \in T \colon ||A \Rightarrow B||^{\delta}_{M} = 1\} = \{M \in \mathbf{L}^Y \mid \text{for each } A \Rightarrow B \in T \colon ||A \Rightarrow B||^{1}_{\delta \rightarrow M} = 1\} = \{M \in \mathbf{L}^Y \mid \delta \rightarrow M \in \text{Mod}^1(T)\}$.

(ii): Taking into account (i), we get $||A \Rightarrow B||^{1}_{T} = \bigwedge_{M \in \text{Mod}^1(T)} ||A \Rightarrow B||^{1}_{M} \leq \bigwedge_{\delta \rightarrow M \in \text{Mod}^1(T)} ||A \Rightarrow B||^{1}_{\delta \rightarrow M} = \bigwedge_{M \in \text{Mod}^{\delta}(T)} ||A \Rightarrow B||^{\delta}_{M} = ||A \Rightarrow B||^{\delta}_{T}$. The rest is true because $0 \rightarrow S(B, M) = 1$ for all $B, M \in \mathbf{L}^Y$.

(iii): The "$\leq$"-part follows from (ii); the "$\geq$"-part is trivial since $1 \in L$.

Remark 8. In some cases we even have

$$||A \Rightarrow B||_T^1 = ||A \Rightarrow B||_T^\delta$$

for $\delta > 0$. Inspecting the proof of Theorem 6, one can see that this is, for instance, the case when each $M \in \mathbf{L}^Y$ is of the form $M = \delta \rightarrow N$ for some $N \in \mathbf{L}^Y$. This condition is satisfied for a product structure on $[0,1]$, i.e. when $a \otimes b = a \cdot b$. Then, $M = \delta \rightarrow (\delta \otimes M)$ as one can verify.

The following assertion shows that if $*_X$ is a globalization, then the degrees of semantic δ-entailment can be expressed as degrees of semantic 1-entailment.

Theorem 7. *Let $*_X$ be globalization. For each set T of fuzzy attribute implications and $\delta \in L$ there is a set $T' \supseteq T$ of fuzzy attribute implications such that, for each $A \Rightarrow B$,*

$$||A \Rightarrow B||_T^\delta = ||A \Rightarrow B||_{T'}^1. \tag{21}$$

Proof. Take any T and $\delta \in L$. Since $||A \Rightarrow B||_T^\delta = \bigwedge_{M \in \mathrm{Mod}^\delta(T)} ||A \Rightarrow B||_M^\delta = \bigwedge_{\delta \rightarrow M \in \mathrm{Mod}^1(T)} ||A \Rightarrow B||_{\delta \rightarrow M}^1$, it suffices to find $T' \supseteq T$ so that $\mathrm{Mod}^1(T') = \mathrm{Mod}^1(T) \cap \{\delta \rightarrow M \mid M \in \mathbf{L}^Y\}$. From [6, 12, 13] we have that $\mathrm{Mod}^1(T)$ is a closure system, i.e., an intersection of arbitrary 1-models of T is again a 1-model of T. In addition, $\bigcap_{i \in I}(\delta \rightarrow M_i) = \delta \rightarrow \bigcap_{i \in I} M_i$ is true for each $\{M_i \in \mathrm{Mod}^1(T) \mid i \in I\}$ from which we get that $\mathcal{M}_\delta = \mathrm{Mod}^1(T) \cap \{\delta \rightarrow M \mid M \in \mathbf{L}^Y\}$ is closed under arbitrary intersections. Thus, for each $M \in \mathbf{L}^Y$ let $cl_\delta(M) \in \mathbf{L}^Y$ denote the least fuzzy set of attributes (w.r.t. "$\subseteq$") which belongs to $\mathcal{M}_\delta$. Moreover, put $T' = T \cup \{M \Rightarrow cl_\delta(M) \mid M \in \mathbf{L}^Y\}$. Clearly, $\mathrm{Mod}^1(T') \subseteq \mathcal{M}_\delta$ because $T \subseteq T'$, and for each $M \in \mathrm{Mod}^1(T')$ there is $N \in \mathbf{L}^Y$ such that $M = \delta \rightarrow N$ (the existence of N follows from the fact that M is a 1-model of $\{M \Rightarrow cl_\delta(M) \mid M \in \mathbf{L}^Y\}$, i.e., it belongs to $\{\delta \rightarrow M \mid M \in \mathbf{L}^Y\}$). The "$\supseteq$"-part is true because if $M \notin \mathrm{Mod}^1(T')$, then either $M \notin \mathrm{Mod}^1(T)$ or there is $N \in \mathbf{L}^Y$ such that $||N \Rightarrow cl_\delta(N)||_M^1 \neq 1$ from which we further obtain $N \subseteq M$ and $cl_\delta(N) \nsubseteq M$ yielding $M \notin \{\delta \rightarrow M \mid M \in \mathbf{L}^Y\}$. In either case, assuming $M \notin \mathrm{Mod}^1(T')$, we get $M \notin \mathcal{M}_\delta$. Finally, $||A \Rightarrow B||_T^\delta = \bigwedge_{\delta \rightarrow M \in \mathrm{Mod}^1(T)} ||A \Rightarrow B||_{\delta \rightarrow M}^1 = \bigwedge_{M \in \mathrm{Mod}^1(T')} ||A \Rightarrow B||_M^1 = ||A \Rightarrow B||_{T'}^1$.

We now turn our attention to particular sets of FAIs which describe δ-truth of attribute implications in a given data table via semantic entailment. Let $\langle X, Y, I \rangle$ be a data table with fuzzy attributes, $\delta \in L$ be a truth degree. A set T of FAIs is called *δ-complete in* $\langle X, Y, I \rangle$ if, for each $A \Rightarrow B$, $||A \Rightarrow B||_T^1 = ||A \Rightarrow B||_{\langle X, Y, I \rangle}^\delta$. If T is δ-complete and no proper subset of T is δ-complete, then T is called a *non-redundant δ-basis* of $\langle X, Y, I \rangle$. The following assertion gives a criterion of δ-completeness.

Theorem 8. *Let $\langle X, Y, I \rangle$ be a data table with fuzzy attributes, $\delta \in L$, $*_Y$ be identity. Then T is δ-complete in $\langle X, Y, I \rangle$ iff $\mathrm{Mod}^1(T) = \mathrm{Int}(X_\delta^{*_X}, Y_\delta^{*_Y}, I)$.*

Proof. By definition, we get that T is δ-complete in $\langle X, Y, I \rangle$ iff, for each $A \Rightarrow B$, $\|A \Rightarrow B\|^1_T = \|A \Rightarrow B\|^\delta_{\langle X,Y,I\rangle}$, which is true iff $\|A \Rightarrow B\|^1_T = \|A \Rightarrow B\|^1_{\langle X,Y,\delta\to I\rangle}$, i.e., iff T is 1-complete in $\langle X, Y, \delta \to I \rangle$. The latter is true, by results on 1-completeness [6, 12], if and only if $\mathrm{Mod}^1(T) = \mathrm{Int}(X^{*X}, Y^{*Y}, \delta \to I)$. By Theorem 1, $\mathrm{Int}(X^{*X}, Y^{*Y}, \delta \to I) = \mathrm{Int}(X^{*X}_\delta, Y^{*Y}_\delta, I)$, finishing the proof.

	size		distance	
	small (s)	large (l)	far (f)	near (n)
Mercury	1	0	0	1
Venus	0.75	0	0	1
Earth	0.75	0	0	0.75
Mars	1	0	0.5	0.75
Jupiter	0	1	0.75	0.5
Saturn	0	1	0.75	0.5
Uranus	0.25	0.5	1	0.25
Neptune	0.25	0.5	1	0
Pluto	1	0	1	0

Fig. 1. Data table with fuzzy attributes and fuzzy concept lattice

Remark 9. (1) Theorem 8 says that a set T of FAIs which is δ-complete in a given data table with fuzzy attributes not only describes truth of all FAIs in the table, but also fully determines the corresponding concept lattice (intents of $\mathcal{B}(X^{*X}_\delta, Y^{*Y}_\delta, I)$ are exactly the models of T). More importantly, the claim was proven due to existing results on FAIs and due to a reduction of the problem of δ-completeness to the problem of 1-completeness.

(2) Previous results [6, 12] allow us to determine a non-redundant basis of a data table with fuzzy attributes. The procedure is the following. Given $\langle X, Y, I \rangle$ and $\delta \in L$, first determine $\langle X, Y, \delta \to I \rangle$, then find a non-redundant basis T of $\langle X, Y, \delta \to I \rangle$ (in the sense of [6, 12]) which is, in consequence, a non-redundant δ-basis of $\langle X, Y, I \rangle$. Note that the well-known Guigues-Duquenne basis [17, 19] is a particular case of the above-described basis for $\mathbf{L} = 2$ and $\delta = 1$.

4 Illustrative Example

Take a finite Łukasiewicz chain $\mathbf{L}$ with $L = \{0, 0.25, 0.5, 0.75, 1\}$ as a structure of truth degrees. Consider an input data table $\langle X, Y, I \rangle$ depicted in Fig 1 (left) which describes properties of planets of our solar system. The set X of object consists of objects "Mercury", "Venus", ..., set Y contains four attributes: size of the planet (small / large), distance from the sun (far / near). Let $*_X$ be globalization and $*_Y$ be identity. Fuzzy concept lattice $\mathcal{B}(X^{*X}, Y^{*Y}, I)$ (i.e., $\mathcal{B}(X^{*X}_1, Y^{*Y}_1, I)$) is depicted in Fig.1 (right). A non-redundant (minimal) basis (i.e., 1-basis) of $\langle X, Y, I \rangle$ consists of the following fuzzy attribute implications.

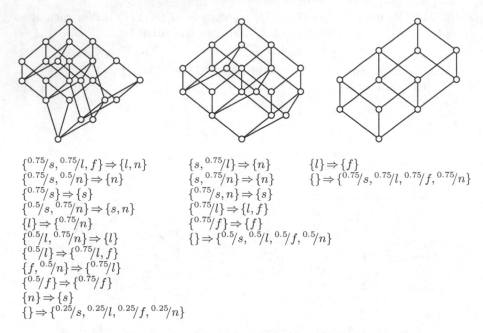

$$\{^{0.75}/s, ^{0.75}/l, f\} \Rightarrow \{l, n\}$$
$$\{^{0.75}/s, ^{0.5}/n\} \Rightarrow \{n\}$$
$$\{^{0.75}/s\} \Rightarrow \{s\}$$
$$\{^{0.5}/s, ^{0.75}/n\} \Rightarrow \{s, n\}$$
$$\{l\} \Rightarrow \{^{0.75}/n\}$$
$$\{^{0.5}/l, ^{0.75}/n\} \Rightarrow \{l\}$$
$$\{^{0.5}/l\} \Rightarrow \{^{0.75}/l, f\}$$
$$\{f, ^{0.5}/n\} \Rightarrow \{^{0.75}/l\}$$
$$\{^{0.5}/f\} \Rightarrow \{^{0.75}/f\}$$
$$\{n\} \Rightarrow \{s\}$$
$$\{\} \Rightarrow \{^{0.25}/s, ^{0.25}/l, ^{0.25}/f, ^{0.25}/n\}$$

$$\{s, ^{0.75}/l\} \Rightarrow \{n\}$$
$$\{s, ^{0.75}/n\} \Rightarrow \{n\}$$
$$\{^{0.75}/s, n\} \Rightarrow \{s\}$$
$$\{^{0.75}/l\} \Rightarrow \{l, f\}$$
$$\{^{0.75}/f\} \Rightarrow \{f\}$$
$$\{\} \Rightarrow \{^{0.5}/s, ^{0.5}/l, ^{0.5}/f, ^{0.5}/n\}$$

$$\{l\} \Rightarrow \{f\}$$
$$\{\} \Rightarrow \{^{0.75}/s, ^{0.75}/l, ^{0.75}/f, ^{0.75}/n\}$$

Fig. 2. Fuzzy concept lattices and corresponding non-redundant bases

$$\{s, ^{0.5}/l, f\} \Rightarrow \{l, n\}$$
$$\{s, ^{0.5}/f, n\} \Rightarrow \{l, f\}$$
$$\{^{0.75}/s, ^{0.5}/f\} \Rightarrow \{s\}$$
$$\{^{0.75}/s, ^{0.25}/n\} \Rightarrow \{^{0.75}/n\}$$
$$\{^{0.5}/s\} \Rightarrow \{^{0.75}/s\}$$
$$\{^{0.25}/s, ^{0.75}/f\} \Rightarrow \{f\}$$
$$\{^{0.25}/s, ^{0.5}/n\} \Rightarrow \{^{0.75}/s, ^{0.75}/n\}$$

$$\{^{0.75}/l\} \Rightarrow \{l, ^{0.5}/n\}$$
$$\{^{0.25}/l, ^{0.5}/n\} \Rightarrow \{l\}$$
$$\{^{0.25}/l\} \Rightarrow \{^{0.5}/l, ^{0.75}/f\}$$
$$\{f\} \Rightarrow \{^{0.25}/s\}$$
$$\{^{0.75}/f, ^{0.25}/n\} \Rightarrow \{^{0.5}/l\}$$
$$\{^{0.25}/f\} \Rightarrow \{^{0.5}/f\}$$
$$\{^{0.75}/n\} \Rightarrow \{^{0.75}/s\}$$

Models of the basis are exactly the intents of $\mathcal{B}(X^{*_X}, Y^{*_Y}, I)$, see [6, 12]. We now show how the fuzzy concept lattice and its minimal basis change when we consider thresholds $\delta \in L$. Recall that if $\delta = 1$, then $\mathcal{B}(X_1^{*_X}, Y_1^{*_Y}, I) = \mathcal{B}(X^{*_X}, Y^{*_Y}, I)$, and a 1-basis of $\langle X, Y, I \rangle$ is the previous set of FAIs. For $\delta = 0$ the concept lattice is trivial (one-element) and the basis consists of a single fuzzy attribute implication $\{\} \Rightarrow \{s, l, f, n\}$. Fig. 2 (left) depicts fuzzy concept lattice $\mathcal{B}(X_{0.75}^{*_X}, Y_{0.75}^{*_Y}, I)$ and its non-redundant basis (below the lattice); Fig. 2 (middle) depicts fuzzy concept lattice $\mathcal{B}(X_{0.5}^{*_X}, Y_{0.5}^{*_Y}, I)$ and the corresponding basis. Finally, Fig. 2 (right) depicts $\mathcal{B}(X_{0.25}^{*_X}, Y_{0.25}^{*_Y}, I)$ and its basis.

5   Conclusions

We showed that the extent- and intent-forming operators from [16], based on the idea of thresholds, form, in fact, a particular case of Galois connections with hedges. Furthermore, we showed that the formulas for Galois connections with hedges can be extended using the idea of thresholds and that this extension still reduces to the original formulas. This enables us to reduce the problems of Galois connections with hedges and thresholds and their concept lattices to problems of Galois connections with hedges and their concept lattices. Nevertheless, the concept of Galois connections with hedges and thresholds is intuitively appealing, the thresholds being parameters which influence the size of the resulting concept lattices. In addition to that, we introduced thresholds to the definition of truth of fuzzy attribute implication and proved some results concerning reduction to the case without thresholds and some further results.

Further research will deal with the following problems:

- the role of shifted attributes in FCA of data with fuzzy attributes,
- analysis of the relationship between δ_1 and δ_2, and the corresponding structures $\mathcal{B}\left(X_{\delta_1}^{*_X}, Y_{\delta_1}^{*_Y}, I\right)$ and $\mathcal{B}\left(X_{\delta_2}^{*_X}, Y_{\delta_2}^{*_Y}, I\right)$,
- further investigation of thresholds in fuzzy attribute implications.

References

1. Bělohlávek R.: Fuzzy Galois connections. *Math. Logic Quarterly* **45**, 4 (1999), 497–504.
2. Bělohlávek R.: Similarity relations in concept lattices. *J. Logic and Computation* Vol. **10** No. **6**(2000), 823–845.
3. Bělohlávek R.: *Fuzzy Relational Systems: Foundations and Principles*. Kluwer, Academic/Plenum Publishers, New York, 2002.
4. Bělohlávek R.: Concept lattices and order in fuzzy logic. *Ann. Pure Appl. Logic* **128**(2004), 277–298.
5. Bělohlávek R.: A note on variable precision concept lattices. Draft, 2006.
6. Bělohlávek R., Chlupová M., Vychodil V.: Implications from data with fuzzy attributes. In: AISTA 2004 in cooperation with IEEE Computer Society Proceedings, 15–18 November 2004, Kirchberg - Luxembourg, 5 pp.
7. Bělohlávek R., Dvořák J., Outrata J.: Direct factorization in formal concept analysis by factorization of input data. *Proc. 5th Int. Conf. on Recent Advances in Soft Computing, RASC 2004*. Nottingham, United Kingdom, 16–18 December, 2004, pp. 578–583.
8. Bělohlávek R., Funioková T., Vychodil V.: Galois connections with hedges. In: Yingming Liu, Guoqing Chen, Mingsheng Ying (Eds.): *Fuzzy Logic, Soft Computing & Computational Intelligence: Eleventh International Fuzzy Systems Association World Congress* (Vol. II), 2005, pp. 1250–1255. Tsinghua University Press and Springer, ISBN 7–302–11377–7.
9. Bělohlávek R., Sklenář V., Zacpal J.: Crisply Generated Fuzzy Concepts. In: B. Ganter and R. Godin (Eds.): ICFCA 2005, *LNCS* **3403**, pp. 268–283, Springer-Verlag, Berlin/Heidelberg, 2005.
10. Bělohlávek R., Vychodil V.: Reducing the size of fuzzy concept lattices by hedges. In: FUZZ-IEEE 2005, The IEEE International Conference on Fuzzy Systems, May 22–25, 2005, Reno (Nevada, USA), pp. 663–668.
11. Bělohlávek R., Vychodil V.: What is a fuzzy concept lattice? In: Proc. CLA 2005, 3rd Int. Conference on Concept Lattices and Their Applications, September 7–9, 2005, Olomouc, Czech Republic, pp. 34–45, URL: http://ceur-ws.org/Vol-162/.

12. Bělohlávek R., Vychodil V.: Fuzzy attribute logic: attribute implications, their validity, entailment, and non-redundant basis. In: Yingming Liu, Guoqing Chen, Mingsheng Ying (Eds.): *Fuzzy Logic, Soft Computing & Computational Intelligence: Eleventh International Fuzzy Systems Association World Congress* (Vol. I), 2005, pp. 622–627. Tsinghua University Press and Springer, ISBN 7–302–11377–7.
13. Bělohlávek R., Vychodil V.: Axiomatizations of fuzzy attribute logic. In: Prasad B. (Ed.): *IICAI 2005, Proceedings of the 2nd Indian International Conference on Artificial Intelligence*, pp. 2178–2193, IICAI 2005, ISBN 0–9727412–1–6.
14. Ben Yahia S., Jaoua A.: Discovering knowledge from fuzzy concept lattice. In: Kandel A., Last M., Bunke H.: *Data Mining and Computational Intelligence*, pp. 167–190, Physica-Verlag, 2001.
15. Elloumi S. *et al.*: A multi-level conceptual data reduction approach based in the Łukasiewicz implication. *Inf. Sci.* **163**(4)(2004), 253–264.
16. Fan S. Q., Zhang W. X.: Variable threshold concept lattice. *Inf. Sci.* (submitted).
17. Ganter B., Wille R.: *Formal Concept Analysis. Mathematical Foundations.* Springer-Verlag, Berlin, 1999.
18. Goguen J. A.: The logic of inexact concepts. *Synthese* **18**(1968-9), 325–373.
19. Guigues J.-L., Duquenne V.: Familles minimales d'implications informatives resultant d'un tableau de données binaires. *Math. Sci. Humaines* **95**(1986), 5–18.
20. Hájek P.: *Metamathematics of Fuzzy Logic.* Kluwer, Dordrecht, 1998.
21. Hájek P.: On very true. *Fuzzy Sets and Systems* **124**(2001), 329–333.
22. Krajči S.: Cluster based efficient generation of fuzzy concepts. *Neural Network World* **5**(2003), 521–530.
23. Pollandt S.: *Fuzzy Begriffe.* Springer-Verlag, Berlin/Heidelberg, 1997.

Formal Concept Analysis with Constraints by Closure Operators*

Radim Bělohlávek and Vilém Vychodil

Department of Computer Science, Palacky University, Olomouc
Tomkova 40, CZ-779 00 Olomouc, Czech Republic
Phone: +420 585 634 700, Fax: +420 585 411 643
{radim.belohlavek, vilem.vychodil}@upol.cz

Abstract. The paper presents a general method of imposing constraints in formal concept analysis of tabular data describing objects and their attributes. The constraints represent a user-defined requirements which are supplied along with the input data table. The main effect is to filter-out outputs of the analysis (conceptual clusters and if-then rules) which are not compatible with the constraint, in a computationally efficient way (polynomial time delay algorithm without the need to compute all outputs). Our approach covers several examples studied before, e.g. extraction of closed frequent itemsets in generation of non-redundant association rules. We present motivations, foundations, and examples.

1 Introduction and Motivation

Formal concept analysis (FCA) is a method of data analysis and visualization which deals with input data in the form of a table describing objects (rows), their attributes (columns), and their relationship (table entries ×'s and blanks indicate whether or not object has attribute) [4, 6]. Basic outputs of FCA are the following: First, a collection of maximal rectangles of the table which are full of ×'s. These rectangles are interpreted as concept-clusters (so-called formal concepts), can be hierarchically ordered and form a so-called concept lattice. Second, a (non-redundant) set of if-then rules describing attribute dependencies (so-called attribute implications). FCA proved to be useful in several fields either as a direct method of data analysis, see e.g. [4], the references therein and also [5, 11], or as a preprocessing method, see e.g. [12]. In the basic setting, it is assumed that no further information is supplied at the input except for the data table. However, it is often the case that there is an additional information available in the form of a constraint (requirement) specified by a user. In such a case, one is not interested in all the outputs (maximal full rectangles or if-then rules) but only in those which satisfy the constraint. The other outputs may be left out as non-interesting. This way, the number of outputs is reduced by

---

* Supported by grant No. 1ET101370417 of GA AV ČR, by grant No. 201/05/0079 of the Czech Science Foundation, and by institutional support, research plan MSM 6198959214.

H. Schärfe, P. Hitzler, and P. Øhrstrøm (Eds.): ICCS 2006, LNAI 4068, pp. 131–143, 2006.

focusing on the "interesting ones". Needless to say, the general idea of constraints is not new. A reader can find examples of using constraints in data mining in [3].

In this paper, we develop a method of constraints in FCA which are expressed by means of closure operators. The constraints can be used both for constraining maximal rectangles and if-then rules. Our approach is theoretically and computationally tractable and covers several interesting forms of constraints. For instance, one can set the closure operator in such a way that the maximal full rectangles satisfying the constraint correspond exactly to closed frequent itemsets [10], used e.g. in generating non-redundant association rules [12], see also Section 4. As another example, one can set the closure operator in such a way that at the output one gets exactly the formal concepts respecting a given hierarchy of attributes (a user tells some attributes are more important than others), see [1].

In Section 2, we present preliminaries from FCA. Section 3 presents our approach, theoretical foundations, and algorithms. In Section 4, we present several examples of constraints by closure operators and demonstrating examples. Section 5 is a summary and an outline of future research.

2   Preliminaries

In what follows, we summarize basic notions of FCA. An object-attribute data table describing which objects have which attributes can be identified with a triplet $\langle X, Y, I \rangle$ where X is a non-empty set (of objects), Y is a non-empty set (of attributes), and $I \subseteq X \times Y$ is an (object-attribute) relation. Objects and attributes correspond to table rows and columns, respectively, and $\langle x, y \rangle \in I$ indicates that object x has attribute y (table entry corresponding to row x and column y contains $\times$; if $\langle x, y \rangle \notin I$ the table entry contains blank symbol). In the terminology of FCA, a triplet $\langle X, Y, I \rangle$ is called a formal context. For each $A \subseteq X$ and $B \subseteq Y$ denote by $A^\uparrow$ a subset of Y and by $B^\downarrow$ a subset of X defined by

$$A^\uparrow = \{y \in Y \mid \text{for each } x \in A : \langle x, y \rangle \in I\},$$
$$B^\downarrow = \{x \in X \mid \text{for each } y \in B : \langle x, y \rangle \in I\}.$$

That is, $A^\uparrow$ is the set of all attributes from Y shared by all objects from A (and similarly for $B^\downarrow$). A formal concept in $\langle X, Y, I \rangle$ is a pair $\langle A, B \rangle$ of $A \subseteq X$ and $B \subseteq Y$ satisfying $A^\uparrow = B$ and $B^\downarrow = A$. That is, a formal concept consists of a set A (so-called extent) of objects which fall under the concept and a set B (so-called intent) of attributes which fall under the concept such that A is the set of all objects sharing all attributes from B and, conversely, B is the collection of all attributes from Y shared by all objects from A. Alternatively, formal concepts can be defined as maximal rectangles of $\langle X, Y, I \rangle$ which are full of $\times$'s: For $A \subseteq X$ and $B \subseteq Y$, $\langle A, B \rangle$ is a formal concept in $\langle X, Y, I \rangle$ iff $A \times B \subseteq I$ and there is no $A' \supset A$ or $B' \supset B$ such that $A' \times B \subseteq I$ or $A \times B' \subseteq I$.

A set $\mathcal{B}(X, Y, I) = \{\langle A, B \rangle \mid A^\uparrow = B, B^\downarrow = A\}$ of all formal concepts in data $\langle X, Y, I \rangle$ can be equipped with a partial order $\leq$ (modeling the subconcept-superconcept hierarchy, e.g. dog $\leq$ mammal) defined by

$$\langle A_1, B_1 \rangle \leq \langle A_2, B_2 \rangle \text{ iff } A_1 \subseteq A_2 \text{ (iff } B_2 \subseteq B_1). \tag{1}$$

Note that $^\uparrow$ and $^\downarrow$ form a so-called Galois connection [6] and that $\mathcal{B}(X,Y,I)$ is in fact a set of all fixed points of $^\uparrow$ and $^\downarrow$. Under $\leq$, $\mathcal{B}(X,Y,I)$ happens to be a complete lattice, called a concept lattice of $\langle X, Y, I \rangle$, the basic structure of which is described by the so-called main theorem of concept lattices [6]:

Theorem 1. (1) *The set $\mathcal{B}(X,Y,I)$ is under $\leq$ a complete lattice where the infima and suprema are given by*

$$\bigwedge\nolimits_{j \in J} \langle A_j, B_j \rangle = \langle \bigcap\nolimits_{j \in J} A_j, (\bigcup\nolimits_{j \in J} B_j)^{\downarrow \uparrow} \rangle,$$
$$\bigvee\nolimits_{j \in J} \langle A_j, B_j \rangle = \langle (\bigcup\nolimits_{j \in J} A_j)^{\uparrow \downarrow}, \bigcap\nolimits_{j \in J} B_j \rangle.$$

(2) *Moreover, an arbitrary complete lattice $\mathbf{V} = \langle V, \leq \rangle$ is isomorphic to $\mathcal{B}(X,Y,I)$ iff there are mappings $\gamma : X \to V$, $\mu : Y \to V$ such that*

(i) *$\gamma(X)$ is $\bigvee$-dense in V, $\mu(Y)$ is $\bigwedge$-dense in V;*
(ii) *$\gamma(x) \leq \mu(y)$ iff $\langle x, y \rangle \in I$.*

For a detailed information on formal concept analysis we refer to [4,6] where a reader can find theoretical foundations, methods and algorithms, and applications in various areas.

Recall that a closure operator in a set Y is a mapping $C : 2^Y \to 2^Y$ satisfying

$$B \subseteq C(B),$$
$$B_1 \subseteq B_2 \text{ implies } C(B_1) \subseteq C(B_2),$$
$$C(C(B)) = C(B)$$

for any $B, B_1, B_2 \in 2^Y$, see e.g. [6].

3   Constraints by Closure Operators

Selecting "interesting" formal concepts from $\mathcal{B}(X,Y,I)$ needs to be accompanied by a criterion of what is interesting. Such a criterion can be seen as a constraint and depends on particular data and application. Therefore, the constraint should be supplied by a user along with the input data $\langle X, Y, I \rangle$. One way to specify "interesting concepts" is to focus on concepts whose sets of attributes are "interesting". This seems to be natural because "interesting concepts" are determined by "interesting attributes/properties of objects". Thus, for a formal context $\langle X, Y, I \rangle$, the user may specify a subset $Y' \subseteq 2^Y$ such that $B \in Y'$ iff the user considers B to be an interesting set of attributes. A formal concept $\langle A, B \rangle \in \mathcal{B}(X,Y,I)$ can be then seen as "interesting" if $B \in Y'$. In this section we develop this idea provided that the selected sets of attributes which are taken as "interesting" form a closure system on Y.

3.1   Interesting Formal Concepts (Maximal Full Rectangles)

We start by formalizing interesting sets of attributes using closure operators.

Definition 1. *Let Y be a set of attributes, $C: 2^Y \to 2^Y$ be a closure operator on Y. A set $B \subseteq Y$ of attributes is called a C-interesting set of attributes (shortly, a set of C-attributes) if $B = C(B)$.*

Throughout the paper, Y denotes a set of attributes and $C: 2^Y \to 2^Y$ denotes a closure operator on Y. Described verbally, Definition 1 says that C-interesting sets of attributes are exactly the fixed points of the closure operator C. Thus, given any set $B \subseteq Y$ of attributes, $C(B)$ can be seen as *the least set of C-interesting attributes containing B.*

Remark 1. (1) Representing interesting sets of attributes by closure operators has technical as well as epistemic reasons. Specifying particular $C: 2^Y \to 2^Y$, we prescribe a particular meaning of "being interesting". Given a set $B \subseteq Y$ of attributes, either we have $B = C(B)$, i.e. B is C-interesting, or $B \subset C(B)$ which can be read: "B is not C-interesting, but additional attributes $C(B) - B$ would make B interesting". Thus, C can be seen as an operator describing which attributes must be added to a set of attributes to make it interesting.

(2) A definition of C depends on particular application. In our approach, we assume that C is any closure operator, covering thus all possible choices of C. On the other hand, in real applications, it is necessary to have a collection of easy-to-understand definitions of such closure operators. In Section 4 we give several examples to define C which are intuitively clear for an inexperienced user.

Definition 2. *Let $\langle X, Y, I \rangle$ be a formal context, $C : 2^Y \to 2^Y$ be a closure operator on Y. We put*

$$\mathcal{B}_C(X, Y, I) = \{\langle A, B \rangle \in 2^X \times 2^Y \mid A^\uparrow = B, \, B^\downarrow = A, \, B = C(B)\}, \tag{2}$$
$$\text{Ext}_C(X, Y, I) = \{A \subseteq X \mid \text{there is } B \subseteq Y \text{ such that } \langle A, B \rangle \in \mathcal{B}_C(X, Y, I)\}, \tag{3}$$
$$\text{Int}_C(X, Y, I) = \{B \subseteq Y \mid \text{there is } A \subseteq X \text{ such that } \langle A, B \rangle \in \mathcal{B}_C(X, Y, I)\}. \tag{4}$$

Each $\langle A, B \rangle \in \mathcal{B}_C(X, Y, I)$ is called a C-interesting concept (C-concept); $A \in \text{Ext}_C(X, Y, I)$ is called a C-interesting extent (C-extent); $B \in \text{Int}_C(X, Y, I)$ is called a C-interesting intent (C-intent).

Remark 2. (1) According to Definition 2, $\langle A, B \rangle$ is a C-concept iff $\langle A, B \rangle$ is a concept (in the ordinary sense) such that B is a set of C-attributes. Therefore, C-concepts $\langle A, B \rangle$ can be seen as maximal rectangles in the input data table which are full of $\times$'s, see Section 2, with B being closed under C. Notice that two boundary cases of closure operators on Y are (i) $C(B) = B$ ($B \in 2^Y$), (ii) $C(B) = Y$ ($B \in 2^Y$). For C defined by (i), the notion of a C-concept coincides with that of a concept. In this case, $\mathcal{B}_C(X, Y, I)$ equals $\mathcal{B}(X, Y, I)$. In case of (ii), $\mathcal{B}_C(X, Y, I)$ is a one-element set (not interesting).

(2) Observe that B is a C-intent iff $B = B^{\downarrow\uparrow} = C(B)$. Denoting the set of all fixed points of C by $\text{fix}(C)$, we have $\text{Int}_C(X, Y, I) = \text{Int}(X, Y, I) \cap \text{fix}(C)$.

The following assertion characterizes the structure of C-concepts:

Theorem 2. *Let $\langle X, Y, I \rangle$ be a formal context, $C \colon 2^Y \to 2^Y$ be a closure operator. Then $\mathcal{B}_C(X, Y, I)$ equipped with $\leq$ defined by (1) is a complete lattice which is a $\bigvee$-sublattice of $\mathcal{B}(X, Y, I)$.*

Proof. In order to show that $\mathcal{B}_C(X, Y, I)$ equipped with $\leq$ is a complete lattice, it suffices to check that Int_C is closed under arbitrary infima. Take an indexed system $\{ B_i \in \mathrm{Int}_C(X, Y, I) \mid i \in I \}$ of C-intents. Since $B_i \in \mathrm{Int}(X, Y, I)$, Theorem 1 gives that $\bigcap_{i \in I} B_i \in \mathrm{Int}(X, Y, I)$. Now, it remains to prove that $B = \bigcap_{i \in I} B_i$ is a set of C-attributes. Since each B_i is a set of C-attributes and C is a closure operator, we get $B = \bigcap_{i \in I} B_i = \bigcap_{i \in I} C(B_i) = C(\bigcap_{i \in I} C(B_i)) = C(\bigcap_{i \in I} B_i) = C(B)$. Hence, $B = \bigcap_{i \in I} B_i$ is a set of C-attributes. Altogether, $B \in \mathrm{Int}_C(X, Y, I)$. To see that $\mathcal{B}_C(X, Y, I)$ is a $\bigvee$-sublattice of $\mathcal{B}(X, Y, I)$ observe that $\mathrm{Int}(X, Y, I)$ and $\mathrm{Int}_C(X, Y, I)$ agree on arbitrary intersections and then apply Theorem 1. $\qquad\square$

Remark 3. For each context $\langle X, Y, I \rangle$, $Y \in \mathrm{Int}(X, Y, I)$ and $C(Y) = Y$ because C is extensive. Therefore, $Y \in \mathrm{Int}_C(X, Y, I)$, i.e. the set of all attributes determines the least C-concept of $\mathcal{B}_C(X, Y, I)$, see (1). This might seem strange at first sight because the least C-concept of $\mathcal{B}_C(X, Y, I)$ which is also the least concept of $\mathcal{B}(X, Y, I)$ is rather not interesting—it is basically a concept of objects having all attributes. It might be tempting to "remove this concept from $\mathcal{B}_C(X, Y, I)$", however, this would dissolve important structural properties of $\mathcal{B}_C(X, Y, I)$. For instance, after the removal, $\mathcal{B}_C(X, Y, I)$ would not be a lattice in general.

We now focus on the computational aspects of generating all C-concepts. The naive way to compute $\mathcal{B}_C(X, Y, I)$ is to find $\mathcal{B}(X, Y, I)$ first and then go through all of its concepts and filter out the C-concepts. This method is not efficient because in general, $\mathcal{B}_C(X, Y, I)$ can be considerably smaller than $\mathcal{B}(X, Y, I)$. In the sequel we show that $\mathcal{B}_C(X, Y, I)$ can be directly computed using Ganter's NextClosure [6] algorithm without the need to compute $\mathcal{B}(X, Y, I)$.

In order to use the NextClosure [6] algorithm, we need to combine together two closure operators: $^{\downarrow\uparrow}$ (operator induced by the Galois connection given by a formal context $\langle X, Y, I \rangle$) and C (operator specifying interesting sets of attributes). For any $B \subseteq Y$ define sets B_i ($i \in \mathbb{N}_0$) and $\mathcal{C}(B)$ of attributes as follows:

$$B_i = \begin{cases} B & \text{if } i = 0, \\ C(B_{i-1}{}^{\downarrow\uparrow}) & \text{if } i \geq 1. \end{cases} \tag{5}$$

$$\mathcal{C}(B) = \bigcup_{i=1}^{\infty} B_i. \tag{6}$$

Theorem 3. *Let Y be a finite set of attributes, $\langle X, Y, I \rangle$ be a formal context, $C \colon 2^Y \to 2^Y$ be a closure operator on Y, $\mathcal{C}$ be defined by (6). Then $\mathcal{C} \colon 2^Y \to 2^Y$ is a closure operator such that $B = \mathcal{C}(B)$ iff $B \in \mathrm{Int}_C(X, Y, I)$.*

Proof. Since both $^{\downarrow\uparrow}$ and C are closure operators, $B_0 \subseteq B_1 \subseteq \cdots$, and $B_i \subseteq \mathcal{C}(B)$ for each $i \in \mathbb{N}_0$. Extensivity and monotony of $^{\downarrow\uparrow}$ and C yield extensivity and monotony of $\mathcal{C}$. To check idempotency of $\mathcal{C}$, we show $C((\mathcal{C}(B))^{\downarrow\uparrow}) \subseteq \mathcal{C}(B)$ for

Input: $\langle X, Y, I \rangle$
Output: $\text{Int}_C(X, Y, I)$
    $B := \emptyset,\ \text{Int}_C(X, Y, I) := \emptyset$
    while $B \neq Y$:
        $B := B^+$
        add B *to* $\text{Int}_C(X, Y, I)$

Input: $\langle X, Y, I \rangle$
Output: $\mathcal{P}_C$ (C-pseudo-intents of $\langle X, Y, I \rangle$)
    $B := \emptyset,\ \mathcal{P}_C := \emptyset$
    if $B \neq \mathcal{C}(B)$: *add* B *to* $\mathcal{P}_C$
    while $B \neq Y$:
        $T := \{P \Rightarrow \mathcal{C}(P) \mid P \in \mathcal{P}_C\}$
        $B := B_T^+$
        if $B \neq \mathcal{C}(B)$: *add* B *to* $\mathcal{P}_C$

Fig. 1. Algorithms for computing C-intents (left) and C-pseudo-intents (right); B^+ denotes the lectically smallest fixed point of $\mathcal{C}$ which is a successor of B; B_T^+ denotes the lectically smallest fixed point of cl_T which is a successor of B

each $B \subseteq Y$. For each $y \in \mathcal{C}(B)$ denote by i_y an index $i_y \in \mathbb{N}$ such that $y \in B_{i_y}$, where B_{i_y} is defined by (5). We have $\mathcal{C}(B) = \bigcup_{y \in \mathcal{C}(B)} B_{i_y}$. Since Y is finite, $\mathcal{C}(B)$ is finite, i.e. for an index $i = \max\{i_y \mid y \in \mathcal{C}(B)\}$, we have $\mathcal{C}(B) = B_i$, where B_i is defined by (5). Therefore, $C((\mathcal{C}(B))^{\downarrow\uparrow}) = C(B_i^{\downarrow\uparrow}) = B_{i+1} \subseteq \mathcal{C}(B)$, i.e. $\mathcal{C}$ is idempotent. Altogether, $\mathcal{C}$ is a closure operator. We now prove that $B = \mathcal{C}(B)$ iff $B \in \text{Int}_C(X, Y, I)$.

"$\Rightarrow$": Let $B = \mathcal{C}(B)$. Using the above idea, $\mathcal{C}(B) = B_i$ for some index $i \in I$. Therefore, $B = B_i = C(B_{i-1}^{\downarrow\uparrow})$ for some $i \in I$ which proves that B is a set of C-attributes. Moreover, $B^{\downarrow\uparrow} = B_i^{\downarrow\uparrow} \subseteq C(B_i^{\downarrow\uparrow}) = B_{i+1} \subseteq \mathcal{C}(B) = B$, i.e. $B \in \text{Int}(X, Y, I)$. Putting it together, $B \in \text{Int}_C(X, Y, I)$.

"$\Leftarrow$": Let $B \in \text{Int}_C(X, Y, I)$. By definition, $B = C(B)$ and $B = B^{\downarrow\uparrow}$. Thus, for each $i \in \mathbb{N}$, $B_i = B$, yielding $B = \mathcal{C}(B)$. $\qquad\square$

Theorem 3 gives a way to compute C-interesting intents and thus the complete lattice of C-concepts in case of finite Y: we can use Ganter's NextClosure [6] algorithm for computing fixed points of closure operators because the C-interesting intents are exactly the fixed points of $\mathcal{C}$. The algorithm is depicted in Fig. 1 (left).

Remark 4. Notice that NextClosure, being used in Fig.1 to compute the fixed points of $\mathcal{C}$, works with polynomial time delay provided that $C(B)$ ($B \subseteq Y$) can be computed with a polynomial time complexity. Indeed, for each $B \subseteq Y$, $B^{\downarrow\uparrow}$ can be computed with a polynomial time delay (well-known fact). Since Y is finite, there is an index $i \leq |Y|$ such that $\mathcal{C}(B) = B_i$, where B_i is defined by (5). Thus, if $C(B)$ can be computed in a polynomial time, NextClosure can use $\mathcal{C}$ with a polynomial time delay (the same complexity as if NextClosure were using $^{\downarrow\uparrow}$). In practical applications, the computation of $\mathcal{C}(B)$ is usually more time consuming than the computation of $B^{\downarrow\uparrow}$. Still, the number of C-interesting concepts is usually much smaller than the number of all concepts, thus, NextClosure with $\mathcal{C}$ is in most situations considerably faster than NextClosure with $^{\downarrow\uparrow}$.

3.2   Bases of Interesting Attribute Implications

In this section we show that each lattice of C-concepts can be alternatively described by particular sets of "interesting" implications between attributes. We present a way to compute minimal sets of such implications. We suppose that Y is finite. Recall basic notions of attribute implications and their validity [6, 7]: an *attribute implication* (*over attributes Y*) is an expression $A \Rightarrow B$, where $A, B \in 2^Y$ are sets of attributes. An attribute implication $A \Rightarrow B$ is true in $M \subseteq Y$, written $M \models A \Rightarrow B$, if $A \subseteq M$ implies $B \subseteq M$. Given a set T of attribute implications, $M \subseteq Y$ is called a *model* of T if, for each $A \Rightarrow B \in T$, $M \models A \Rightarrow B$. The system of all models of T is denoted by $\mathrm{Mod}(T)$.

If we focus only on "interesting models" of sets of attribute implications (or sets of "interesting attribute implications"), we naturally come to the following notions of a C-implication and a C-model:

Definition 3. *Let Y be a set of attributes, $C \colon 2^Y \to 2^Y$ be a closure operator, T be a set of attribute implications in Y. An attribute implication $A \Rightarrow B$ in Y is called a C-implication if A and B are sets of C-attributes. $M \subseteq Y$ is called a C-model of T if M is a set of C-attributes and $M \in \mathrm{Mod}(T)$. Denote by $\mathrm{Mod}_C(T)$ the system of all C-models of T.*

Using the notion of a C-model, we define sets of attribute implications which are C-complete in a given formal context:

Definition 4. *Let $\langle X, Y, I \rangle$ be a formal context, $C \colon 2^Y \to 2^Y$ be a closure operator. A set T of attribute implications is called C-complete in $\langle X, Y, I \rangle$ if*

$$\mathrm{Mod}_C(T) = \mathrm{Int}_C(X, Y, I). \tag{7}$$

A set T of C-implications is called a C-basis of $\langle X, Y, I \rangle$ if T is C-complete in $\langle X, Y, I \rangle$ and no proper subset of T is C-complete in $\langle X, Y, I \rangle$.

Remark 5. (1) Described verbally, a set T of attribute implications is C-complete if the C-models of T are exactly the C-interesting intents. From this point of view, a C-complete set of attribute implications fully describes the lattice of C-concepts using the notion of a C-model. A C-basis is a set of C-implications T (i.e., implications of the form "set of C-attributes A implies a set of C-attributes B") fully describing C-concepts so that one cannot remove any C-implication from T without losing C-completeness. Hence, C-bases are the least C-complete sets of C-implications.

(2) In general, a C-complete set T of attribute implications (C-implications) has models which are not C-models. Also note that if C is given by $C(B) = B$ ($B \in 2^Y$), then the notions of a C-model and a C-completeness coincide with that of a model and a completeness [6].

We now show a way to find particular C-bases. For that purpose, we introduce the following generalized notion of a pseudo-intent:

Definition 5. *Let $\langle X, Y, I \rangle$ be a formal context, $C \colon 2^Y \to 2^Y$ be a closure operator, $\mathcal{C}$ be defined by (6). A set P of C-attributes is called a C-pseudo-intent of $\langle X, Y, I \rangle$ if $P \subset \mathcal{C}(P)$ and, for each C-pseudo-intent Q of $\langle X, Y, I \rangle$ such that $Q \subset P$, we have $\mathcal{C}(Q) \subseteq P$.*

If C is the identity mapping, the notion of a C-pseudo-intent coincides with the notion of a pseudo-intent, see [6, 7]. All C-pseudo-intents determine a C-basis of a given formal context:

Theorem 4. *Let $\langle X, Y, I \rangle$ be a formal context, $C : 2^Y \to 2^Y$ be a closure operator, C be defined by (6). Then*

$$T = \{P \Rightarrow C(P) \mid P \text{ is a } C\text{-pseudo-intent of } \langle X, Y, I \rangle\} \tag{8}$$

is a C-basis of $\langle X, Y, I \rangle$.

Proof. We first check that T given by (8) is C-complete, i.e. we check equality (7) by showing both inclusions.

"$\subseteq$": Let $M \in \mathrm{Mod}_C(T)$. Thus, M is a set of C-attributes. By contradiction, let $M \neq C(M)$, i.e. $M \subset C(M)$ because C is extensive. Now, for each C-pseudo-intent Q, we have $M \models Q \Rightarrow C(Q)$ because M is a model of T. Therefore, for each C-pseudo-intent Q, if $Q \subset P$ then $C(Q) \subseteq P$, i.e. M is a C-pseudo-indent by Definition 5. On the other hand, $M \not\models M \Rightarrow C(M)$ because $C(M) \not\subseteq M$, a contradiction to $M \in \mathrm{Mod}_C(T)$.

"$\supseteq$": Let $M \in \mathrm{Int}_C(X, Y, I)$. Then $M = C(M)$. For each C-pseudo-intent P, if $P \subseteq M$ then $C(P) \subseteq C(M) = M$, i.e. $M \models P \Rightarrow C(P)$.

T is a C-basis: T is obviously a set of C-implications; for each C-pseudo-intent P, $P \models Q \Rightarrow C(Q)$ where $Q \neq P$ is any C-pseudo-intent. Thus, P is a C-model of $T_P = T - \{P \Rightarrow C(P)\}$ which gives $\mathrm{Mod}_C(T_P) \supseteq \mathrm{Int}_C(X, Y, I)$, i.e. T_P is not C-complete. $\qquad\square$

Due to Theorem 4, in order to get a C-basis of $\langle X, Y, I \rangle$, it suffices to compute all C-pseudo-intents. We now turn our attention to the computation of C-pseudo-intents. Given a set T of attribute implications define sets B^{T_i}, $cl_T(B)$ ($i \in \mathbb{N}_0$):

$$B^{T_i} = \begin{cases} B & \text{if } i = 0, \\ C\left(B^{T_{i-1}} \cup \bigcup\{D \mid A \Rightarrow D \in T \text{ and } A \subset B^{T_{i-1}}\}\right) & \text{if } i \geq 1, \end{cases} \tag{9}$$

$$cl_T(B) = \bigcup_{i=0}^{\infty} B^{T_i}. \tag{10}$$

Operator $cl_T : 2^Y \to 2^Y$ has the following property:

Theorem 5. *Let $\langle X, Y, I \rangle$ be a formal context, T be defined by (8), $\mathcal{P}_C$ be the system of all C-pseudo-intents of $\langle X, Y, I \rangle$. Then cl_T defined by (10) is a closure operator such that $\{cl_T(B) \mid B \subseteq Y\} = \mathcal{P}_C \cup \mathrm{Int}_C(X, Y, I)$.*

Proof. cl_T is a closure operator (apply arguments from the proof of Theorem 3). We check that $\{cl_T(B) \mid B \subseteq Y\} = \mathcal{P}_C \cup \mathrm{Int}_C(X, Y, I)$.

"$\subseteq$": Let $B = cl_T(B)$. If $B \notin \mathrm{Int}_C(X, Y, I)$, it suffices to check that B is a C-pseudo-intent. Since Y is finite, $B = cl_T(B) = B^{T_{i_0}}$ for some $i_0 \in \mathbb{N}$. That is, B is of the form $C(\cdots)$, yielding that B is a set of C-attributes. Moreover, for each C-pseudo-intent Q, if $Q \subset B$ then $C(Q) \subseteq B$ because $B = cl_T(B) = B^{T_{i_0}} = B^{T_{i_0+1}}$. Therefore, B is a C-pseudo-intent.

"$\supseteq$": Clearly, for each C-intent B, $B^{T_i} = B$ ($i \in \mathbb{N}$), i.e. B is a fixed point of cl_T. The same is true if B is a C-pseudo-intent. $\qquad\square$

		a	b	c	d	e	f	g	h	i
leech	1	×	×					×		
bream	2	×	×					×	×	
frog	3	×	×	×				×	×	
dog	4	×		×				×	×	×
spike-weed	5	×	×		×		×			
reed	6	×	×	×	×		×			
bean	7	×		×	×	×				
maize	8	×		×	×		×			

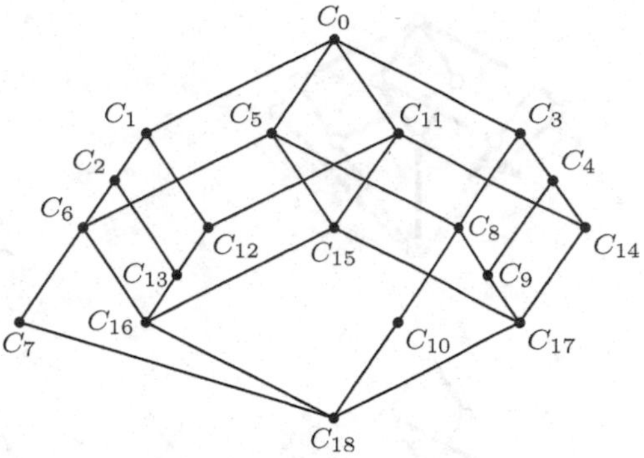

Fig. 2. Context (left) and concept lattice (right); the attributes are: a: needs water to live, b: lives in water, c: lives on land, d: needs chlorophyll to produce food, e: two seed leaves, f: one seed leaf, g: can move around, h: has limbs, i: suckles its offspring

Theorem 5 says that the set of all C-pseudo-intents and all C-intents is the set of all fixed points of cl_T. This provides us with a way to determine a C-basis: we can use the NextClosure [6] algorithm to compute all fixed points of cl_T and then $\{P \mid P = cl_T(P) \text{ and } P \neq C(P)\}$ is the system of all C-pseudo-intents, i.e.

$$T = \{P \Rightarrow C(P) \mid P = cl_T(P) \text{ and } P \neq C(P)\}$$

is a C-basis due to Theorem 4. The algorithm is depicted in Fig. 1 (right).

4   Examples

Consider an illustrative formal context [6] $\langle X, Y, I \rangle$ given by Fig. 2 (left). The set X of objects contains objects $1, 2, \ldots$ denoting organisms "leech", "bream", $\ldots$; the set Y contains attributes $a, b, \ldots$ denoting certain properties of organisms, see the comment under Fig. 2. The concept lattice $\mathcal{B}(X, Y, I)$ corresponding with $\langle X, Y, I \rangle$ has 19 concepts, here denoted by $C_0, \ldots, C_{18}$:

$C_0 = \langle \{1, 2, 3, 4, 5, 6, 7, 8\}, \{a\} \rangle$, $C_1 = \langle \{1, 2, 3, 4\}, \{a, g\} \rangle$,
$C_2 = \langle \{2, 3, 4\}, \{a, g, h\} \rangle$, $C_3 = \langle \{5, 6, 7, 8\}, \{a, d\} \rangle$,
$C_4 = \langle \{5, 6, 8\}, \{a, d, f\} \rangle$, $C_5 = \langle \{3, 4, 6, 7, 8\}, \{a, c\} \rangle$,
$C_6 = \langle \{3, 4\}, \{a, c, g, h\} \rangle$, $C_7 = \langle \{4\}, \{a, c, g, h, i\} \rangle$,
$C_8 = \langle \{6, 7, 8\}, \{a, c, d\} \rangle$, $C_9 = \langle \{6, 8\}, \{a, c, d, f\} \rangle$,
$C_{10} = \langle \{7\}, \{a, c, d, e\} \rangle$, $C_{11} = \langle \{1, 2, 3, 5, 6\}, \{a, b\} \rangle$,
$C_{12} = \langle \{1, 2, 3\}, \{a, b, g\} \rangle$, $C_{13} = \langle \{2, 3\}, \{a, b, g, h\} \rangle$,
$C_{14} = \langle \{5, 6\}, \{a, b, d, f\} \rangle$, $C_{15} = \langle \{3, 6\}, \{a, b, c\} \rangle$,
$C_{16} = \langle \{3\}, \{a, b, c, g, h\} \rangle$, $C_{17} = \langle \{6\}, \{a, b, c, d, f\} \rangle$,
$C_{18} = \langle \{\}, \{a, b, c, d, e, f, g, h, i\} \rangle$.

Fig. 2 (right) depicts the concept lattice $\mathcal{B}(X, Y, I)$ [6].

(a) Define C so that B is C-interesting iff $B = Y$ or $|B^{\downarrow}| \geq s$ where s is a non-negative integer. It is easy to see that C-interesting sets form a closure

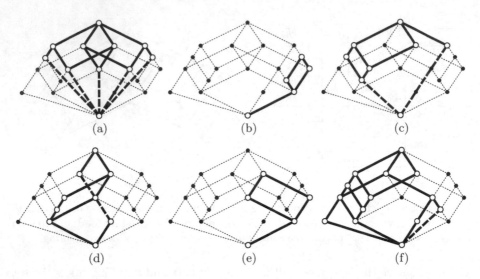

Fig. 3. Concept lattice constrained by various closure operators

system. $|B^{\downarrow}| \geq s$ means that the number of objects sharing all attributes from B exceeds a user-defined parameter s called *support* in association rules [13]. Condition $B = Y$ is a technical one to ensure that C-interesting sets form a closure system. The corresponding closure operator is defined by

$$C(B) = \begin{cases} B & \text{if } |B^{\downarrow}| \geq s, \\ Y & \text{otherwise.} \end{cases}$$

Then, the set $\text{Int}_C(X, Y, I) - \{Y\}$ of all C-interesting intents without Y coincides with the set of closed frequent itemsets defined by Zaki [12] in order to get non-redundant association rules.

(b) Define C by

$$C(B) = \begin{cases} B & \text{if } |B| \leq n, \\ Y & \text{otherwise.} \end{cases}$$

That is, $B \subseteq Y$ is C-interesting iff B contains at most n attributes or $B = Y$. That is, C can be used to determine intents with at most n attributes. Fig. 3 (a) depicts the situation for $n = 3$: "•" denote concepts of $\mathcal{B}(X, Y, I)$ which are not present in $\mathcal{B}_C(X, Y, I)$; "∘" denote C-concepts; dotted lines denote edges of the original concept lattice which are not present in $\mathcal{B}_C(X, Y, I)$; bold solid lines denote edges which are presented in both $\mathcal{B}(X, Y, I)$ and $\mathcal{B}_C(X, Y, I)$; bold dashed lines denote new edges which are in $\mathcal{B}_C(X, Y, I)$ but are not in the original concept lattice.

(c) For any $Z \subseteq Y$, C defined by $C(B) = B \cup Z$ is a closure operator. This closure operator determines intents containing Z. Notice that the boundary cases mentioned in Remark 2 (1) are given by choices $Z = \emptyset$ and $Z = Y$, respectively. For instance, $Z = \{d, f\}$ determines a constraint on "organisms with one seed leaf that need chlorophyll to produce food", see Fig. 3 (b).

(d) For any $Z \subseteq Y$, we can define C so that B is C-interesting iff B does not contain any attribute from Z (or $B = Y$) by putting

$$C(B) = \begin{cases} B & \text{if } B \cap Z = \emptyset, \\ Y & \text{otherwise.} \end{cases}$$

Fig. 3 (c) contains a lattice for $Z = \{c, e, f\}$.

(e) A general method for defining C is the following. Consider a binary relation R on Y, $\langle y_1, y_2 \rangle \in R$ meaning that if y_1 is an attribute of a concept then y_2 should also be an attribute of that concept. Now, put

$$B_i = \begin{cases} B & \text{if } i = 0, \\ B_{i-1} \cup \{y' \mid \text{there is } y \in B_{i-1} : \langle y, y' \rangle \in R\} & \text{if } i \geq 1, \end{cases}$$
$$C(B) = \bigcup_{i=0}^{\infty} B_i.$$

Since Y is finite, we have $C(B) = B_{i_0}$ for some $i_0 \in \mathbb{N}$. B is C-interesting iff all dependencies given by R are satisfied. In more detail, B is C-interesting iff IF $\langle y_1, y_2 \rangle \in R$ and $y_1 \in B$ THEN $y_2 \in B$. Fig. 3 (d) depicts the resulting structure for $R = \{\langle g, b \rangle, \langle d, e \rangle\}$.

(f) A particular case of (e) is a constraint given by an equivalence relation (i.e., R is reflexive, symmetric, and transitive), see also [2]. In this case, $C(B) = \bigcup \{[y]_R \mid y \in B\}$, where $[y]_R$ denotes the class of R containing y. Fig. 3 (e) contains a structure determined by an equivalence R induced by a partition $\{\{a, b\}, \{c\}, \{d\}, \{e\}, \{f\}, \{g, h, i\}\}$.

(g) Let T be a set of attribute implications. The system of all models of T is a closure system [6], the corresponding closure operator C can be described as follows [6]:

$$B_i = \begin{cases} B & \text{if } i = 0, \\ B_{i-1} \cup \bigcup \{D \mid A \Rightarrow D \in T \text{ and } A \subseteq B_{i-1}\} & \text{if } i \geq 1, \end{cases}$$
$$C(B) = \bigcup_{i=0}^{\infty} B_i.$$

$C(B)$ is the least model of T containing B. Hence, B is C-interesting iff B is a model of attribute implications from T. For $T = \{\{b, c\} \Rightarrow \{h\}, \{d\} \Rightarrow \{c\}\}$, the resulting structure in depicted in Fig. 3 (f). Notice that this type of definition of a closure operator is, in fact, the most general one, because each closure operator on a finite set of attributes can be completely described by a set of attribute implications (i.e., the fixed points of C are exactly the models of some set of attribute implications).

Consider a closure operator C such that $C(B) = B$ ($B \in 2^Y$). Let T be the C-basis given by (8). Since C is an identical operator, T is a basis of the concept lattice $\mathcal{B}(X, Y, I)$. In this particular case, T is the following:

$$T = \{\{a, b, c, g, h, i\} \Rightarrow Y, \{a, b, d\} \Rightarrow \{a, b, d, f\}, \{a, c, d, e, f\} \Rightarrow Y,$$
$$\{a, c, g\} \Rightarrow \{a, c, g, h\}, \{a, d, g\} \Rightarrow Y, \{a, e\} \Rightarrow \{a, c, d, e\}, \{a, f\} \Rightarrow \{a, d, f\},$$
$$\{a, h\} \Rightarrow \{a, g, h\}, \{a, i\} \Rightarrow \{a, c, g, h, i\}, \{\} \Rightarrow \{a\}\}.$$

If we define C as in Example (b), T defined by (8) is a C-basis of the constrained lattice of C-concepts depicted in Fig. 3 (a):

$$T = \{\{a, b, d\} \Rightarrow Y, \{a, c, g\} \Rightarrow Y, \{a, d, g\} \Rightarrow Y, \{a, e\} \Rightarrow Y, \{a, f\} \Rightarrow \{a, d, f\},$$
$$\{a, h\} \Rightarrow \{a, g, h\}, \{a, i\} \Rightarrow Y, \{\} \Rightarrow \{a\}\}.$$

Observe that since C determines concepts with at most three attributes, each implication in the latter T has at most three attributes on both sides of "$\Rightarrow$" or the right-hand side of the implication consists of the whole set of attributes Y.

5   Further Issues

For limited scope, we did not present the following topics some of which will appear in a full paper or are subject of future research:

– Interactive specification of constraining closure operators. An expert might not be able to explicitly describe a constraining closure operator. However, he/she is usually able to tell which formal concepts from the whole $\mathcal{B}(X, Y, I)$ are interesting. If $\mathcal{I}$ is a subset of $\mathcal{B}(X, Y, I)$ identified as (examples of) interesting formal concepts, an important problem is to describe a possibly largest closure operator C such that each $\langle A, B \rangle \in \mathcal{I}$ is C-interesting. Namely, putting $C_1 \leq C_2$ iff for each $B \in 2^Y$ we have $C_1(B) \subseteq C_2(B)$ for closure operators C_1 and C_2, we have $C_1 \leq C_2$ iff $\mathrm{fix}(C_2) \subseteq \mathrm{fix}(C_1)$ where $\mathrm{fix}(C_i)$ is a set of all fixed points of C_i. Therefore, since we require $B \in \mathrm{fix}(C)$ for each $\langle A, B \rangle \in \mathcal{I}$, larger C means a better approximation of $\mathcal{I}$ by $\mathcal{B}_C(X, Y, I)$. The problem is to find a tractable description of C. For instance, if C is supposed to be given by an equivalence relation R, see Section 4 (e), then given $\mathcal{I}$, the largest closure operator C we look for is the one induced by a relation $R = R_{\mathcal{I}}$ where

$$\langle y_1, y_2 \rangle \in R_{\mathcal{I}} \quad \text{iff} \quad \text{for each } \langle A, B \rangle \in \mathcal{I} : y_1 \in B \text{ iff } y_2 \in B.$$

  Then, one can present $\mathcal{B}_C(X, Y, I)$ to the expert who might then revise the selection of $\mathcal{I}$, etc., to finally arrive at a satisfactory closure operator C.
– Entailment of constraints. Intuitively, a constraint $\mathcal{C}_1$ (semantically) entails a constraint $\mathcal{C}_2$ iff each $B \subseteq Y$ satisfying $\mathcal{C}_1$ satisfies $\mathcal{C}_2$ as well. A study of entailment is important for obtaining small descriptions of constraining closure operators.
– More detailed results and more efficient algorithms for particular closure operators can be obtained (we omit details).

References

1. Bělohlávek R., Sklenář V., Zacpal J.: Formal concept analysis with hierarchically ordered attributes. *Int. J. General Systems* **33**(4)(2004), 283–294.
2. Bělohlávek R., Sklenář V.: Formal concept analysis constrained by attribute-dependency formulas. In: B. Ganter and R. Godin (Eds.): ICFCA 2005, *Lect. Notes Comp. Sci.* **3403**, pp. 176–191, Springer-Verlag, Berlin/Heidelberg, 2005.

3. Boulicaut J.-F., Jeudy B.: Constraint-based data mining. In: Maimon O., Rokach L. (Eds.): *The Data Mining and Knowledge Discovery Handbook*, Springer, 2005. pp. 399–416.
4. Carpineto C., Romano G.: *Concept Data Analysis. Theory and Applications*. J. Wiley, 2004.
5. Dekel U., Gill Y.: Visualizing class interfaces with formal concept analysis. In *OOPSLA'03*, pages 288–289, Anaheim, CA, October 2003.
6. Ganter B., Wille R.: *Formal Concept Analysis. Mathematical Foundations*. Springer, Berlin, 1999.
7. Guigues J.-L., Duquenne V.: Familles minimales d'implications informatives resultant d'un tableau de données binaires. *Math. Sci. Humaines* **95**(1986), 5–18.
8. Maier D.: *The Theory of Relational Databases*. Computer Science Press, Rockville, 1983.
9. Norris E. M.: An algorithm for computing the maximal rectangles of a binary relation. *Journal of ACM* 21:356–266, 1974.
10. Pasquier N., Bastide Y., Taouil R., Lakhal L.: Efficient Mining of Association Rules Using Closed Itemset Lattices. *Information Systems* **24**(1)(1999), 25–46.
11. Snelting G., Tip F.: Understanding class hierarchies using concept analysis. *ACM Trans. Program. Lang. Syst.* 22(3):540–582, May 2000.
12. Zaki M. J.: Mining non-redundant association rules. *Data Mining and Knowledge Discovery* **9**(2004), 223–248.
13. Zhang C., Zhang S.: *Association Rule Mining. Models and Algorithms*. Springer, Berlin, 2002.

Mining a New Fault-Tolerant Pattern Type as an Alternative to Formal Concept Discovery

Jérémy Besson[1,2], Céline Robardet[3],
and Jean-François Boulicaut[1]

[1] INSA Lyon, LIRIS CNRS UMR 5205, F-69621 Villeurbanne cedex, France
[2] UMR INRA/INSERM 1235, F-69372 Lyon cedex 08, France
[3] INSA Lyon, PRISMa, F-69621 Villeurbanne cedex, France
`Firstname.Name@insa-lyon.fr`

Abstract. Formal concept analysis has been proved to be useful to support knowledge discovery from boolean matrices. In many applications, such 0/1 data have to be computed from experimental data and it is common to miss some one values. Therefore, we extend formal concepts towards fault-tolerance. We define the DR-bi-set pattern domain by allowing some zero values to be inside the pattern. Crucial properties of formal concepts are preserved (number of zero values bounded on objects and attributes, maximality and availability of functions which "connect" the set components). DR-bi-sets are defined by constraints which are actively used by our correct and complete algorithm. Experimentation on both synthetic and real data validates the added-value of the DR-bi-sets.

1 Introduction

Many application domains can lead to possibly huge boolean matrices whose rows denote objects and columns denote attributes. Mining such 0/1 data has been studied extensively and quite popular data mining techniques have been designed for set pattern extraction (e.g., frequent sets or association rules which capture some regularities among the one values within the data). We are interested in bi-set mining, i.e., the computation of sets of objects and sets of attributes which are somehow "associated". An interesting case concerns Conceptual Knowledge Discovery [8,9,10,11,6]. It is based on the formal concepts contained in the data, i.e., the maximal bi-sets of one values [17]. Examples of formal concepts in $\mathbf{r}_1$ (Table 1) are $(\{o_1, o_2, o_3, o_4\}, \{a_1, a_2\})$ and $(\{o_4\}, \{a_1, a_2, a_3, a_4\})$. Formal concept discovery is related to the popular frequent (closed) set computation. Efficient algorithms can nowadays compute complete collections of constrained formal concepts (see, e.g., [15,2]).

In this paper, we address one fundamental limitation of Knowledge Discovery processes based on formal concepts. Within such local patterns, the strength of the association of the two set components is often too strong in real-life data. Indeed, errors of measurement and boolean encoding techniques may lead to erroneous zero values which will give rise to a combinatorial explosion of the number of formal concepts. Assume that K_1 represents a real phenomenon but

H. Schärfe, P. Hitzler, and P. Øhrstrøm (Eds.): ICCS 2006, LNAI 4068, pp. 144–157, 2006.

Table 1. A formal context K_1 (left), K_2 with 17% of noise (right)

	a_1	a_2	a_3	a_4
o_1	1	1	0	0
o_2	1	1	0	0
o_3	1	1	0	0
o_4	1	1	1	1
o_5	0	0	1	1
o_6	0	0	1	1

	a_1	a_2	a_3	a_4
o_1	1	1	0	0
o_2	1	0	1	0
o_3	1	1	0	1
o_4	1	1	1	1
o_5	0	0	1	0
o_6	0	0	1	1

that data collection and preprocessing lead to the data K_2. The number of formal concepts in K_2 is approximately twice larger than in K_1. Based on our expertise in real-life data mining, it is now clear that the extraction of formal concepts, their post-processing and their interpretation is not that relevant in noisy data which encode measured and/or computed boolean relationships. Our hypothesis is that mining formal concepts with some zero values might be useful and should be considered as a valuable alternative to formal concept discovery. For example, the bi-set ($\{o_1, o_2, o_3, o_4\}, \{a_1, a_2\}$) appears to be relevant in K_2: its objects and attributes are strongly associated (only one zero value) and the outside objects and attributes contain more zero values.

Therefore, we propose to extend formal concepts towards such fault-tolerant patterns by specifying a new type of bi-sets, the so-called DR-bi-sets. The main challenge is to preserve important properties of formal concepts which have been proved useful during pattern interpretation:

- The numbers of zero values are bounded on objects and attributes.
- These bi-sets are maximal on both dimensions.
- It does not exist an outside pattern object (resp. attribute) which is identical to an inside pattern object (resp. attribute). It increases pattern relevancy.
- There exist two functions, one which associates to a set of objects (resp. attributes) a unique set of attributes (resp. objects). Such functions ensure that every DR-bi-set captures a relevant association between the two set components. As such it provides powerful characterization mechanisms.

Section 2 discusses related work. Section 3 is a formalization of our new pattern domain. It is shown that DR-bi-sets are a fairly natural extensions of formal concepts. Section 4 sketches our correct and complete algorithm which computes every DR-bi-set. Section 5 provides experimental results on both synthetic and real data. Section 6 concludes.

2   Related Work

Looking for fault-tolerant pattern has been already studied. To the best of our knowledge, most of the related work has concerned mono-dimensional patterns and/or the use of heuristic techniques. In [18], the frequent set mining task is extended towards fault-tolerance. A level-wise algorithm is proposed but their fault-tolerant property is not anti-monotonic while this is needed to achieve tractability. Therefore, [18] provides a greedy algorithm leading to an incomplete computation. [14] revisits this work and it looks for an anti-monotonic constraint such that a level-wise algorithm can provide every set whose density of one values is greater than δ in at least σ situations. Anti-monotonicity is obtained by enforcing that every subset of extracted sets satisfies the constraint as well. The extension of such dense sets to dense bi-sets is difficult: the connection which associates objects to properties and vice-versa is not decreasing while this is an appreciated property of formal concepts. Instead of using a relative density definition, [12] considers an absolute threshold to define fault-tolerant frequent patterns: given a threshold δ, a set of attributes P, such that $\sharp P > \delta$, holds in an object X iff $\sharp(X \cap P) \geq \sharp P - \delta$ where $\sharp X$ denotes the size of X. To ensure that the support is significant for each attribute, they use a minimum support threshold per attribute beside the classical minimum support. Thus, each object of an extracted pattern contains less than δ zero values and each attribute contains more one values than the given minimum support for each attribute. This definition is not symmetrical on the object and attribute dimension, and the more the support increases, the less the patterns are relevant. In [7], the authors are interested in geometrical tiles (i.e., dense bi-sets which involve contiguous elements given orders on both dimensions). Their local optimization algorithm is not deterministic and thus can not guarantee the global quality of the extracted patterns. Furthermore, the hypothesis on built-in orders can not be accepted on many data.

Some fault-tolerant extensions of formal concepts have been recently proposed as well. In [1], available formal concepts are merged while checking for a bounded number of exceptions on both dimensions. The proposed technique is however incomplete, and the mapping between set components of the extracted bi-sets is not guaranteed. The proposal in [13] concerns an extension which can be computed efficiently but none of the appreciated properties are available. This research is also related to condensed representations of concept lattices or dense bi-sets. [16] introduces a "zooming" approach on concept lattices. The so-called α-Galois lattices exploit a partition on the objects to reduce the collection of the extracted bi-sets: a situation s is associated to a set G if $\alpha\%$ of the objects which have the same class value than s are associated to elements from G and if s is associated to G as well. Our context is different since we want to preserve the duality between objects and attributes as far as possible.

3   Formalization

Let G and M be sets, called the set of objects and attributes respectively. Let I be a relation $I \subseteq G \times M$ between objects and attributes: for $g \in G, m \in M, (g, m) \in I$ holds iff the object g has the attribute m. The triple $K = (G, M, I)$ is called a (formal) context.

A bi-set (X, Y) is a couple of sets from $2^G \times 2^M$. Some specific types of bi-sets have been extensively studied. This is the case of formal concepts which can be defined thanks to Galois connection [17]:

Definition 1. *Given $X \subseteq G$ and $Y \subseteq M$, the Galois connection on K is the couple of functions (ϕ, ψ) s.t. $\psi(X) = \{m \in M \mid \forall g \in X, (g,m) \in I\}$ and $\phi(Y) = \{g \in G \mid \forall m \in Y, (g,m) \in I\}$. A bi-set (X,Y) is a formal concept with extent X and intent Y iff $X = \phi(Y)$ and $Y = \psi(X)$.*

We now give a new way to define formal concepts which will be generalised to DR-bi-sets.

Definition 2. *Let us denote by $\mathcal{Z}_o(x, Y)$ the number of zero values of an object x on the attributes in Y: $\mathcal{Z}_o(x, Y) = \sharp\{y \in Y \mid (x,y) \notin I\}$. Similarly $\mathcal{Z}_a(y, X) = \sharp\{x \in X \mid (x,y) \notin I\}$ denotes the number of zero values of an attribute y on the objects in X.*

Formal concepts can now be characterized by the following lemma:

Lemma 1. *A bi-set (X,Y) is a formal concept of the context K iff:*
$$\forall x \in X, \; \mathcal{Z}_o(x, Y) = 0 \text{ or similarly, } \forall y \in Y, \; \mathcal{Z}_a(y, X) = 0 \qquad (1)$$
$$(\forall x \in G \setminus X, \; \mathcal{Z}_o(x, Y) \geq 1) \text{ and } (\forall y \in M \setminus Y, \; \mathcal{Z}_a(y, X) \geq 1) \qquad (2)$$

It introduces constraints which can be used to compute formal concepts [2]. Interestingly, these constraints ensure the maximality (w.r.t. set inclusion) of the bi-sets which satisfy them. It is well-known that constraint monotonicity properties are extremely important for a clever exploration of the associated search space. These properties are related to a specialization relation. Let us consider an unusual specialization relation for building concept lattices.

Definition 3. *Our specialization relation $\preceq$ on bi-sets is defined as follows: $(X_1, Y_1) \preceq (X_2, Y_2)$ iff $X_1 \subseteq X_2$ and $Y_1 \subseteq Y_2$. A constraint $\mathcal{C}$ is said anti-monotonic w.r.t. $\preceq$ iff $\forall D, E \in 2^G \times 2^M$ s.t. $D \preceq E$, $\mathcal{C}(E) \Rightarrow \mathcal{C}(D)$. Dually, $\mathcal{C}$ is said monotonic w.r.t. $\preceq$ iff $\mathcal{C}(D) \Rightarrow \mathcal{C}(E)$. Notice that $\mathcal{C}(D)$ denotes that the constraint $\mathcal{C}$ is satisfied by the bi-set D.*

For instance, we might use a minimal size constraint $\mathcal{C}_{ms}(\sigma_1, \sigma_2, (X,Y)) \equiv \sharp X \geq \sigma_1 \wedge \sharp Y \geq \sigma_2$. Such a constraint is monotonic w.r.t. $\preceq$.

3.1 Dense Bi-sets

We want to compute bi-sets with a strong association between the two sets and such that its number of zero values can be controlled. We can decide to bound the number of zero values per object/attribute or on the whole bi-set (strong density vs. weak density). We can also look at relative or absolute density, i.e., to take into account the density w.r.t. the size of the whole bi-set or not. If we use the weak density, we can obtain bi-sets containing objects or attributes with only zero values. In this case, these objects (resp. attributes) are never associated to the bi-set attributes (resp. objects). We decided to use an absolute strong density constraint that enforces an upper bound for the number of zero values per object and per attribute. Using strong density enables to get the important monotonicity property.

Definition 4. *Given $(X,Y) \in 2^G \times 2^M$ and a positive integer value α, (X,Y) is said dense iff it satisfies the anti-monotonic constraint $\mathcal{C}_d(\alpha, (X,Y)) \equiv (\forall x \in X, \mathcal{Z}_o(x,Y) \leq \alpha)$ and $(\forall y \in Y, \mathcal{Z}_a(y,X) \leq \alpha)$.*

3.2   Relevant Bi-sets

We want to extract bi-sets (X,Y) such that the objects of X (resp. the attributes of Y) have a larger density of one values on the attributes from Y (resp. on the objects from X) than on the other attributes, i.e., $M \setminus Y$ (resp. objects, i.e., $G \setminus X$). It leads to the formalisation of a relevancy constraint where the parameter δ is used to enforce the difference of zero values inside and outside the bi-set.

Definition 5. *Given $(X,Y) \in 2^G \times 2^M$, and a positive integer value δ, (X,Y) is said relevant iff it satisfies the following constraint:*

$$\mathcal{C}_r(\delta, (X,Y)) \equiv (\forall g \in G \setminus X, \forall x \in X, \mathcal{Z}_o(g,Y) \geq \mathcal{Z}_o(x,Y) + \delta)$$
$$and \ (\forall m \in M \setminus Y, \forall y \in Y, \mathcal{Z}_a(m,X) \geq \mathcal{Z}_a(y,X) + \delta)$$

3.3   DR-Bi-sets

The bi-sets which satisfy both $\mathcal{C}_d$ and $\mathcal{C}_r$ constraints are a new type of fault-tolerant patterns. Dense and relevant bi-sets are indeed a generalisation of formal concepts (bi-sets with $\alpha = 0$ and $\delta = 1$). $\mathcal{C}_d$ is a straightforward generalisation of the first equation in Lemma 1. $\mathcal{C}_r$ generalizes the second equation in Lemma 1 by enforcing that all outside elements of the bi-set contain at least δ zero values in addition to the one of every inside element. Parameter α controls the density of the bi-sets whereas the parameter δ enforces a significant difference with the outside elements. $\mathcal{C}_d$ is anti-monotonic w.r.t. $\preceq$ (see Definition 3) and can give rise to efficient pruning. $\mathcal{C}_r$ is neither monotonic nor anti-monotonic but we explain in Section 4 how to exploit this constraint efficiently. Fig. 1 shows the collection of bi-sets in K_3 which satisfy $\mathcal{C}_d \wedge \mathcal{C}_r$ when $\alpha = 5$ and $\delta = 1$ ordered w.r.t. $\preceq$. Each level indicates the maximal number of zero values per object and per attribute. For instance, if $\alpha = 1$, a sub-collection containing five bi-sets is

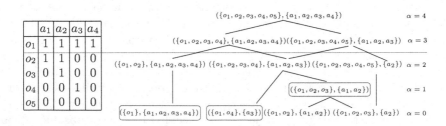

Fig. 1. A formal context K_3 and the bi-sets satisfying $\mathcal{C}_d \wedge \mathcal{C}_r$ with $\alpha = 5$ and $\delta = 1$

extracted, four of them being formal concepts ($\alpha = 0$). Density and relevancy constraints do not ensure maximality which is a desired property. For instance, in Fig. 1, if B denotes $(\{o_1, o_2, o_3\}, \{a_1, a_2\})$, we have $(\{o_1, o_2\}, \{a_1, a_2\}) \preceq B$ and $(\{o_1, o_2, o_3\}, \{a_2\}) \preceq B$. As a result, to increase bi-set relevancy, we finally consider the so-called DR-bi-sets which are the maximal dense and relevant bi-sets.

Definition 6. *Let* $(X, Y) \in 2^G \times 2^M$ *be a dense and relevant bi-set (i.e., satisfying* $\mathcal{C}_d \wedge \mathcal{C}_r$*).* (X, Y) *is called a DR-bi-set iff it is maximal w.r.t.* $\preceq$*, i.e. it does not exist* $(X', Y') \in 2^G \times 2^M$ *s.t.* (X', Y') *satisfies* $\mathcal{C}_d \wedge \mathcal{C}_r$ *and* $(X, Y) \preceq (X', Y')$.

This collection is denoted $DR_{\alpha\delta}$. For example, DR_{11} on K_3 contains the three circled bi-sets of Fig. 1. It is important to notice that different threshold values might be considered on objects/attributes (say α/α' for the density constraint and δ/δ' for the relevancy constraint).

3.4 Properties

Let us first emphasize that the DR-bi-set size increases with parameter α.

Property 1. Given $0 \leq \alpha_1 \leq \alpha$, $\forall (X_1, Y_1) \in DR_{\alpha_1\delta}$, $\exists (X, Y) \in DR_{\alpha\delta}$ such that $(X_1, Y_1) \preceq (X, Y)$.

Proof. $\forall (X, Y)$ satisfying $\mathcal{C}_d(\alpha_1, (X, Y)) \wedge \mathcal{C}_r(\delta, (X, Y))$ then (X, Y) satisfies $\mathcal{C}_d(\alpha, (X, Y)) \wedge \mathcal{C}_r(\delta, (X, Y))$. $DR_{\alpha\delta}$ contains (X, Y) or a bi-set (X', Y') s. t. $(X, Y) \preceq (X', Y')$. □

The larger α is, the more the size of each extracted bi-set from $DR_{\alpha\delta}$ increases while extracted associations with smaller α value are preserved. In practice, an important reduction on the size of the extracted collections is observed when the parameters are well chosen (see Section 5). As a result, a zooming effect is obtained when α is varying. Parameter δ enables to select more relevant patterns. For example, when $\delta = 2$ and $\alpha \leq 1$ the collection in K_3 is reduced to the DR-bi-set $(\{o_1\}, \{a_1, a_2, a_3, a_4\})$.

The following property ensures that DR-bi-sets are actually a generalisation of formal concepts, i.e., they are related by two functions.

Property 2. For $\delta > 0$, there exists two functions called ψ_{DR} and ϕ_{DR} such that $\psi_{DR} : 2^G \rightarrow 2^M$ and $\phi_{DR} : 2^M \rightarrow 2^G$ such that (X, Y) is a DR-bi-set iff $X = \phi_{DR}(Y)$ and $Y = \psi_{DR}(X)$.

Proof. Let $(S_1, S_2), (S_1, S_3) \in DR_{\alpha\delta}$ such that $S_2 \neq S_3$. Let $Max\mathcal{Z}_a(X, Y) \equiv \max_{m \in X} \mathcal{Z}_a(m, Y)$ and $Min\mathcal{Z}_a(X, Y) \equiv \min_{m \in X} \mathcal{Z}_a(m, Y)$

As $DR_{\alpha\delta}$ contains maximal bi-sets, $S_2 \not\subseteq S_3$ and $S_3 \not\subseteq S_2$. We have

$$Max\mathcal{Z}_a(S_1, S_3) \leq Min\mathcal{Z}_a(S_1, M \setminus S_3) - \delta \ (\mathcal{C}_r \text{ constraint})$$
$$\leq Min\mathcal{Z}_a(S_1, S_2 \setminus S_3) - \delta \text{ (set inclusion)}$$
$$< Min\mathcal{Z}_a(S_1, S_2 \setminus S_3) \ (\delta > 0) \leq Max\mathcal{Z}_a(S_1, S_2 \setminus S_3)$$
$$\leq Max\mathcal{Z}_a(S_1, S_2)$$

Then, we have $MaxZ_a(S_1, S_3) < MaxZ_a(S_1, S_2)$ and similarly we can derive $MaxZ_a(S_1, S_2) < MaxZ_a(S_1, S_3)$ which leads to a contradiction.

Thus, we have a function between 2^G and 2^M. The existence of a function between 2^M and 2^G can be proved in a similar way. $\square$

These functions are extremely useful to support pattern interpretation: to a set of objects X corresponds at most one set of attributes. Typically, they were missing in previous approaches for fault-tolerance extensions of formal concepts [1,12]. Unfortunately, we do not have an explicit definition of these functions. This remains an open problem.

4   A Complete Algorithm

The whole collection of bi-sets ordered by $\preceq$ forms a lattice whose bottom is $(\perp_G, \perp_M) = (\emptyset, \emptyset)$ and top is $(\top_G, \top_M) = (G, M)$. Let us note by $\mathcal{B}$ the set of sublattices[1] of $((\emptyset, \emptyset), (G, M))$, $\mathcal{B} = \{((X_1, Y_1), (X_2, Y_2))$ s.t. $X_1, X_2 \in 2^G, Y_1, Y_2 \in 2^M$ and $X_1 \subseteq X_2$, $Y_1 \subseteq Y_2\}$, where the first (resp. the second) bi-set is the bottom (resp. the top) element. The algorithm DR-MINER explores some of the sublattices of $\mathcal{B}$ built by means of three mechanisms: enumeration, pruning and propagation.

Table 2. DR-MINER pseudo-code

$K = (G, M, I)$ is a formal context, $\mathcal{C}$ a conjunction of monotonic and anti-monotonic constraints on $2^G \times 2^M$ and α, δ are positive integer values.
DR-Miner
          Generate$((\emptyset, \emptyset), (G, M))$
End DR-Miner

Generate$(\mathcal{L})$
          Let $\mathcal{L} = ((\perp_G, \perp_M), (\top_G, \top_M))$
          $\mathcal{L} \leftarrow$ **Prop**$(\mathcal{L})$
          If **Prune**$(\mathcal{L})$ then
                    If $(\perp_G, \perp_M) \neq (\top_G, \top_M)$ then
                              $(\mathcal{L}_1, \mathcal{L}_2) \leftarrow$ **Enum**$(\mathcal{L}, \textbf{Choose}(\mathcal{L}))$
                              Generate$(\mathcal{L}_1)$
                              Generate$(\mathcal{L}_2)$
                    Else Store $(\perp_G, \perp_M)$
                    End if
          End if
End Generate

---

[1] X is a sublattice of Y if Y is a lattice, X is a subset of Y and X is a lattice with the same join and meet operations as Y.

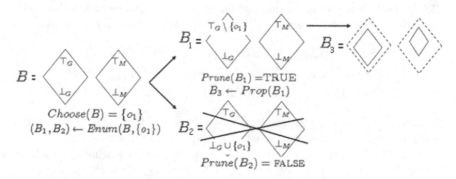

Fig. 2. Example of DR-MINER execution

DR-MINER starts with the complete lattice $((\emptyset, \emptyset), (G, M))$ and then recursively propagates the constraints using *Prop* function, check the consistency of the obtained sublattice with *Prune* function and then generates two new sublattices thanks to *Enum* function (see Table 2). The Figure 2 shows an example of DR-MINER execution.

- **Enumeration:** Let $Enum : \mathcal{B} \times G \cup M \rightarrow \mathcal{B}^2$ such that

$$Enum(((\bot_G, \bot_M), (\top_G, \top_M)), e)$$
$$= \begin{cases} (((\bot_G \cup \{e\}, \bot_M), (\top_G, \top_M)), ((\bot_G, \bot_M), (\top_G \setminus \{e\}, \top_M))) \text{ if } e \in G \\ (((\bot_G, \bot_M \cup \{e\}), (\top_G, \top_M)), ((\bot_G, \bot_M), (\top_G, \top_M \setminus \{e\}))) \text{ if } e \in M \end{cases}$$

  where $e \in \top_G \setminus \bot_G$ or $e \in \top_M \setminus \bot_M$. *Enum* generates two new sublattices which are a partition of its input parameter.
  Let $Choose : \mathcal{B} \rightarrow G \cup M$ be a function which returns (one of) the element $e \in \top_G \setminus \bot_G \cup \top_M \setminus \bot_M$ containing the largest number of zero values on $\top_M$ if $e \in G$ or on $\top_G$ if $e \in M$. It is an heuristic which tends to increase the efficiency of propagation mechanisms by reducing the search space as soon as possible.
- **Pruning:** We prune a sublattice if we are sure that none of its bi-sets satisfies the constraint. Let $Prune_{\mathcal{C}}^m : \mathcal{B} \rightarrow \{\text{TRUE,FALSE}\}$ be a function which returns TRUE iff the monotonic constraint $\mathcal{C}$ (w.r.t. $\preceq$) is satisfied by the top of the sublattice: $Prune_{\mathcal{C}}^m((\bot_G, \bot_M), (\top_G, \top_M)) \equiv \mathcal{C}(\top_G, \top_M)$

  Let $Prune_{\mathcal{C}}^{am} : \mathcal{B} \rightarrow \{\text{TRUE,FALSE}\}$ be a function which returns TRUE iff the anti-monotonic constraint $\mathcal{C}$ (w.r.t $\preceq$) is satisfied by the bottom of the sublattice: $Prune_{\mathcal{C}}^{am}((\bot_G, \bot_M), (\top_G, \top_M)) \equiv \mathcal{C}(\bot_G, \bot_M)$

  $\mathcal{C}_d$ is anti-monotonic and thus it can be used as $Prune_{\mathcal{C}_d}^{am}$. Nevertheless, $\mathcal{C}_r$ is neither monotonic nor anti-monotonic. The $\mathcal{C}_r$ constraint is adapted to ensure that the elements which do not belong to the sublattice might contain more zero values on the top (the elements that can be included in the bi-sets) than the inside ones do on the bottom (the elements that belong to each bi-set). Let $Prune_{C_r} : \mathcal{B} \to \{\text{TRUE},\text{FALSE}\}$ be a function such that

$$Prune_{C_r}((\bot_G, \bot_M), (\top_G, \top_M)) \equiv$$
$$\forall s \in G \setminus \top_G, \forall t \in \bot_G, \mathcal{Z}_o(s, \top_M) \geq \mathcal{Z}_o(t, \bot_M) + \delta \text{ and}$$
$$\forall s \in M \setminus \top_M, \forall t \in \bot_M, \mathcal{Z}_a(s, \top_G) \geq \mathcal{Z}_a(t, \bot_G) + \delta$$

If $Prune_{C_1}^m(\mathcal{L})$ (resp. $Prune_{C_2}^{am}(\mathcal{L})$ and $Prune_{C_r}(\mathcal{L})$) is FALSE, then any bi-set contained in $\mathcal{L}$ does not satisfy C_1 (resp. C_2 and C_r).

In DR-MINER, we use $Prune : \mathcal{B} \to \{\text{TRUE},\text{FALSE}\}$ which is such that $Prune(\mathcal{L}) \equiv Prune_{C_1}^m(\mathcal{L}) \wedge Prune_{C_2}^{am}(\mathcal{L}) \wedge Prune_{C_r}(\mathcal{L}) \wedge Prune_{C_d}^m(\mathcal{L})$

- **Propagation:** C_d and C_r can be used to reduce the size of the sublattices by moving objects of $\top_G \setminus \bot_G$ into $\bot_G$ or outside $\top_G$. The fonctions $Prop_{in} : \mathcal{B} \to \mathcal{B}$ and $Prop_{out} : \mathcal{B} \to \mathcal{B}$ are used to do it as follow:

$$Prop_{in}((\bot_G, \bot_M), (\top_G, \top_M)) = \{((\bot_G', \bot_M'), (\top_G, \top_M)) \in \mathcal{B} \mid$$
$$\bot_G' = \bot_G \cup \{x \in \top_G \setminus \bot_G \mid \exists t \in \bot_G, \mathcal{Z}_o(x, \top_M) < \mathcal{Z}_o(t, \bot_M) + \delta\}$$
$$\bot_M' = \bot_M \cup \{x \in \top_M \setminus \bot_M \mid \exists t \in \bot_M, \mathcal{Z}_a(x, \top_G) < \mathcal{Z}_a(t, \bot_G) + \delta\}\}$$

$$Prop_{out}((\bot_G, \bot_M), (\top_G, \top_M)) = \{((\bot_G, \bot_M), (\top_G', \top_M')) \in \mathcal{B} \mid$$
$$\top_G' = \top_G \setminus \{x \in \top_G \setminus \bot_G \mid \mathcal{Z}_o(x, \bot_M) > \alpha\}$$
$$\top_M' = \top_M \setminus \{x \in \top_M \setminus \bot_M \mid \mathcal{Z}_a(x, \bot_G) > \alpha\}\}$$

$Prop : \mathcal{B} \to \mathcal{B}$ is defined as $Prop(\mathcal{L}) = Prop_{in}(Prop_{out}(\mathcal{L}))$. It is recursively applied as long as its result changes.

To prove the correctness and completeness of DR-MINER, a sublattice $\mathcal{L} = ((\bot_G, \bot_M), (\top_G, \top_M))$ is called a leaf when it contains only one bi-set i.e., $(\bot_G, \bot_M) = (\top_G, \top_M)$. DR-bi-sets are these maximal bi-sets. To extract only maximal dense and relevant ones, we have adapted the DUAL-MINER strategy for pushing maximality constraints [4].

DR-Miner correctness: Every bi-set (X, Y) belonging to leaf $\mathcal{L}$ satisfies $C_d \wedge C_r$ according to $Prune_{C_d}^{am}$ and $Prune_{C_r}$.

DR-Miner completeness: Let $T_1 = ((\bot_G^1, \bot_M^1), (\top_G^1, \top_G^1))$ and $T_2 = ((\bot_G^2, \bot_M^2), (\top_G^2, \top_G^2))$. Let $\sqsubseteq$ be a partial order on $\mathcal{B}$ defined as $T_1 \sqsubseteq T_2$ iff $(\bot_G^2, \bot_M^2) \preceq (\bot_G^1, \bot_M^1)$ and $(\top_G^1, \top_G^1) \preceq (\top_G^2, \top_G^2)$ (see Definition 3). $\sqsubseteq$ is the partial order used to generate the sublattices.

We show that for each bi-set (X, Y) satisfying $C_d \wedge C_r$, it exists a leaf $\mathcal{L} = ((X, Y), (X, Y))$ which is generated by the algorithm.

Property 3. If $\mathcal{F}$ is a sublattice such that $\mathcal{L} \sqsubseteq \mathcal{F}$ then among the two sublattices obtained by the enumeration of $\mathcal{F}$ ($Enum(\mathcal{F}, Choose(\mathcal{F}))$) one and only one is a super-set of $\mathcal{L}$ w.r.t. $\sqsubseteq$. This property is conserved by function $Prop$.

Proof. Let $\mathcal{F} = ((\perp_G, \perp_M), (\top_G, \top_M)) \in \mathcal{B}$ such that $\mathcal{L} \sqsubseteq \mathcal{F}$. Assume that the enumeration is done on objects (it is similar on attributes) and that the two sublattices generated by the enumeration of $o \in \top_G \setminus \perp_G$ are $\mathcal{L}_1$ and $\mathcal{L}_2$. If $o \in X$ then $\mathcal{L} \sqsubseteq \mathcal{L}_1$ and $\mathcal{L} \not\sqsubseteq \mathcal{L}_2$, otherwise $\mathcal{L} \sqsubseteq \mathcal{L}_2$ and $\mathcal{L} \not\sqsubseteq \mathcal{L}_1$.

Let us now show that constraint propagation (function *Prop*) on any sub-lattice $\mathcal{F} = ((\perp_G, \perp_M), (\top_G, \top_M))$ such that $\mathcal{L} \sqsubseteq \mathcal{F}$ preserves this order. More precisely, no element of X is removed of $\top_G$ due to $Prop_{out}$ (Case 1) and no element of $G \setminus X$ is moved to $\perp_G$ due to $Prop_{in}$ (Case 2).

- Case 1: (X, Y) satisfies $\mathcal{C}_r$ then $\forall p \in \top_G \setminus \perp_G$ s.t. $p \in G \setminus X$ and $\forall t \in \perp_G$, we have $\mathcal{Z}_o(p, Y) \geq \mathcal{Z}_o(t, Y) + \delta$. But $\perp_M \subseteq Y \subseteq \top_M$, and thus $\mathcal{Z}_o(p, \top_M) \geq \mathcal{Z}_o(p, Y) \geq \mathcal{Z}_o(t, Y) + \delta \geq \mathcal{Z}_o(t, \perp_M) + \delta$. Consequently, $\mathcal{Z}_o(p, \top_M) < \mathcal{Z}_o(t, \perp_M) + \delta$ is false. Consequently, p is not moved to $\perp_G$.
- Case 2: (X, Y) satisfies $\mathcal{C}_d$ then $\forall p \in \top_G \setminus \perp_G$ s.t. $p \in X$, we have $\mathcal{Z}_o(p, Y) \leq \alpha$. But $\perp_M \subseteq Y$, and thus $\mathcal{Z}_o(p, \perp_M) \leq \mathcal{Z}_o(p, Y) \leq \alpha$. Consequently, p is not removed from $\top_G$. $\square$

Since DR-MINER starts with $((\emptyset, \emptyset), (G, M))$ which is a super-set of $\mathcal{L}$, given that $\mathcal{B}$ is finite and that recursively it exists always a sublattice which is an super-set of $\mathcal{L}$ w.r.t. $\sqsubseteq$ even after the propagation has been applied, then we can affirm that every bi-set satisfying $\mathcal{C}_d \wedge \mathcal{C}_r$ is extracted by DR-MINER.

5 Experimentation

5.1 Robustness on Synthetic Data

Let us first illustrate the added-value of DR-bi-set mining in synthetic data. Our goal is to show that the extraction of these patterns in noisy data sets enables to find some originally built-in formal concepts blurred by some random noise. Our raw synthetic data is a matrix 30×15 in which three disjoint formal concepts of size 10×5 hold. Then, we introduced a uniform random noise on the whole matrix and 5 different data sets have been produced for each level of noise, i.e., from 1% to 30% (each zero or one value has a probability of X% to be changed).

To compare the extracted collections with the three original built-in formal concepts, we used a measure which tests the presence of a subset of the original pattern collection in the extracted ones. This measure σ associates to each pattern of one collection $\mathcal{C}_1$ the closest pattern of the other one $\mathcal{C}_2$ (and reciprocally). It is based on a distance measure taking into account their shared area:

$$\sigma(\mathcal{C}_1, \mathcal{C}_2) = \frac{\rho(\mathcal{C}_1, \mathcal{C}_2) + \rho(\mathcal{C}_2, \mathcal{C}_1)}{2}$$

$$\rho(\mathcal{C}_1, \mathcal{C}_2) = \frac{1}{\sharp \mathcal{C}_1} \sum_{(X_i, Y_i) \in \mathcal{C}_1} \max_{(X_j, Y_j) \in \mathcal{C}_2} \frac{\sharp(X_i \cap X_j) * \sharp(Y_i \cap Y_j)}{\sharp(X_i \cup X_j) * \sharp(Y_i \cup Y_j)}$$

when $\rho(\mathcal{C}_1, \mathcal{C}_2) = 1$, each pattern of $\mathcal{C}_1$ has an identical instance in $\mathcal{C}_2$, and when $\sigma = 1$, the two collections are identical. High values of σ mean that (a) we can

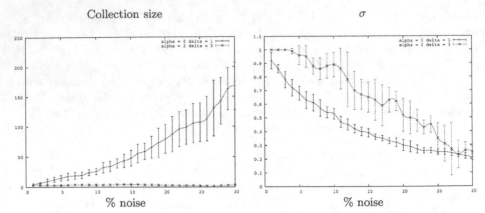

Fig. 3. Mean and standard deviation of the number of bi-sets (5 trials) (left) and of σ (right) w.r.t. the percentage of noise

find all the formal concepts of the reference collection within the noised matrix, and (b) the collection extracted from noised matrices does not contain many bi-sets that are too different from the reference ones.

Figure 3 presents the mean and the standard deviation of the number of extracted bi-sets (left) and the mean and standard deviation of σ (right) for each level of noise. Two collections are represented: one for $\alpha = 0$ and $\delta = 1$ (i.e., the case of formal concepts), and the second one for $\alpha = 2$ and $\delta = 3$. On both collections, a minimal size constraint is added which enforces that each pattern contains at least 3 elements on each dimension (i.e., satisfying $\mathcal{C}_{ms}(3,3)$). It avoids the computation of the smallest bi-sets which can indeed be due to noise.

We can observe that when the noise level increases, the number of extracted formal concepts (i.e., $\alpha = 0$ and $\delta = 1$) increases drastically, whereas σ decreases drastically as well. For $\alpha = 2$ and $\delta = 3$, we observe an important reduction of the number of extracted DR-bi-sets and an important increase of the DR-bi-set quality: for 10 % of noise the collection is similar to the built-in formal concept collection. These graphics emphasize the difference between the use of formal concepts and DR-bi-sets in noisy data: the first one constitutes a large collection (tens to hundreds of patterns) of poorly relevant patterns, whereas the second one is clearly closer to the three built-in patterns. Indeed, we get between 2 and 4 patterns with higher σ values. When the level of noise is very high (say over 20%), the DR-bi-sets are not relevant any more. Indeed, with such level of noise, the data turns to be random.

5.2 Impact of Parameters α and δ

To study the influence of the α parameter, we performed several mining tasks on the UCI data set Internet Advertisements which is large on both dimensions (matrix 3 279 × 1 555) [3].

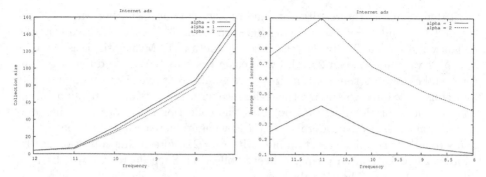

Fig. 4. Number of extracted DR-bi-sets (left) and average increase of bi-set size w.r.t. formal concepts (right) for several frequency thresholds on both dimensions (‰), with $\delta = 1$ and $\alpha \in 0..2$

We have extracted DR-bi-set collections with a minimal size constraint on both dimensions varying between 12‰ and 7‰, where $\delta = 1$ and α varying between 0 and 2. Figure 4 (left) shows the size of DR-bi-set collections. In this data set, the collection sizes decrease with α. Figure 4 (right) shows the average number of added objects and attributes of each formal concept. More formally, if C_0 denotes the collection of formal concepts and if C_α denotes the collection of DR-bi-sets obtained with parameter α, the measure is computed as follow:

$$\frac{1}{\sharp C_0} \sum_{(X_0,Y_0)\in C_0} \max_{(X_\alpha,Y_\alpha)\in \mathcal{A}(X_0,Y_0)} \sharp(X_\alpha \setminus X_0) * \sharp(Y_\alpha \setminus Y_0)$$

where $\mathcal{A}(X_0,Y_0) = \{(X,Y) \in C_\alpha \, such \, that (X_0,Y_0) \preceq (X,Y)\}$ and $\preceq$ is the order of Definition 3. As proved in Property 1, the average sizes of the extracted bi-sets increase with α. But we can observe that this increase is quite important: for example, for $\alpha = 2$ and $frequency = 11$, one element has been added to each formal concept in average.

To study the influence of the δ parameter, we have also performed experiments on the UCI data set Mushroom (matrix 8 124 × 128) [3] and on the real world medical data set Meningitis [5]. Meningitis data have been gathered from children hospitalized for acute meningitis. The pre-processed Boolean data set is composed of 329 patients described by 60 Boolean properties.

A straightforward approach to avoid some irrelevant patterns and to reduce the pattern collection size is to use size constraints on bi-set components. For these experiments, we use the constraint $C_{ms}(500, 10)$ on Mushroom and $C_{ms}(10, 5)$ on Meningitis. Using D-MINER [2], we have computed the collection of such large enough formal concepts and we got more than 1 000 formal concepts on Mushroom and more than 300 000 formal concepts on Meningitis (see Table 3). We used different values of δ on G (denoted δ) and on M (denoted δ').

Table 3 gathers the results obtained on the two data sets. For Mushroom, α is fixed to 0 and $\delta = \delta'$ are varying between 2 and 6. We can observe that the collection sizes drastically decrease with δ and δ'. On Meningitis, α is set to 1 and δ' is varying between 2 and 6 whereas δ is set to 1. We use different values for δ and δ' because the pattern sizes were greater on the object set components and thus we wanted to enforce the difference with the outside elements on these components. For this data set, not only the collection sizes, but also the computational times are considerably reduced when δ' increases. Notice that $\delta' = 1$ leads to an intractable extraction but, with $\delta' = 2$, the resulting collection is 80% smaller than the related formal concept collection. Such decreases are observed when considering higher δ' values.

Table 3. DR-bi-set collection sizes and extraction time when δ' is varying from 1 to 6 on Mushroom and Meningitis

Mushroom ($\mathcal{C}_{ms}(500, 10)$, $\alpha = 0$)						
$\delta = \delta'$	Concepts	1	2	3	4	5	6
size	1 102	1 102	11	6	2	1	0
time	1.6s	10s	4s	4s	3s	2s	2s
Meningitis ($\mathcal{C}_{ms}(10, 5)$, $\alpha = 1$, $\delta = 1$)						
δ'	Concepts	1	2	3	4	5	6
size	354 366	-	75 376	22 882	8 810	4 164	2 021
time	5s	-	693s	327s	181s	109s	70s

6   Conclusion

We have considered the challenging problem of computing fault-tolerant bi-sets. Formal concepts fail to emphasize relevant associations when the data is intrinsically noisy. We have formalized a new task, maximal dense and relevant bi-set mining, within the constraint-based data mining framework. We propose a complete algorithm DR-MINER which computes every DR-bi-set by pushing these constraints during an enumeration process. Density refers to the bounded number of zero values and relevancy refers to the specificities of the elements involved in the extracted bi-sets when considering the whole data set. We experimentally validated the added-value of this approach on both synthetic and real data. Fixing the various parameters might appear difficult (it is often driven by tractability issues) but this is balanced by the valuable counterpart of completeness: the user knows exactly which properties are satisfied by the extracted collections.

Acknowledgements. This research is partially funded by ACI Masse de Données Bingo (CNRS STIC MD 46) and the EU contract IQ FP6-516169 (FET arm of the IST programme). We thank Ruggero G. Pensa for his contribution to the experimental validation and Jean-Marc Petit for his comments.

References

1. J. Besson, C. Robardet, and J.-F. Boulicaut. Mining formal concepts with a bounded number of exceptions from transactional data. In *Post-Workshop KDID'04*, volume 3377 of *LNCS*, pages 33–45. Springer, 2005.
2. J. Besson, C. Robardet, J.-F. Boulicaut, and S. Rome. Constraint-based bi-set mining for biologically relevant pattern discovery in microarray data. *IDA journal*, 9(1):59–82, 2005.
3. C. Blake and C. Merz. UCI repository of machine learning databases, 1998.
4. C. Bucila, J. E. Gehrke, D. Kifer, and W. White. Dualminer: A dual-pruning algorithm for itemsets with constraints. In *ACM SIGKDD*, pages 42–51, 2002.
5. P. François, C. Robert, B. Cremilleux, C. Bucharles, and J. Demongeot. Variables processing in expert system building: application to the aetiological diagnosis of infantile meningitis. *Med. Inf.*, 15(2):115–124, 1990.
6. B. Ganter, G. Stumme, and R. Wille, editors. *Formal Concept Analysis, Foundations and Applications*, volume 3626 of *LNCS*. springer, 2005.
7. A. Gionis, H. Mannila, and J. K. Seppänen. Geometric and combinatorial tiles in 0-1 data. In *PKDD'04*, volume 3202 of *LNAI*, pages 173–184. Springer, 2004.
8. A. Guenoche and I. V. Mechelen. Galois approach to the induction of concepts. *Categories and concepts : Theorical views and inductive data analysis*, pages 287–308, 1993.
9. J. Hereth, G. Stumme, R. Wille, and U. Wille. Conceptual knowledge discovery and data analysis. In *ICCS'00*, pages 421–437, 2000.
10. S. O. Kuznetsov and S. A. Obiedkov. Comparing performance of algorithms for generating concept lattices. *JETAI*, 14 (2-3):189–216, 2002.
11. E. M. Nguifo, V. Duquenne, and M. Liquiere. Concept lattice-based knowledge discovery in databases. *JETAI*, 14((2-3)):75–79, 2002.
12. J. Pei, A. K. H. Tung, and J. Han. Fault-tolerant frequent pattern mining: Problems and challenges. In *DMKD*. Workshop, 2001.
13. R. G. Pensa and J.-F. Boulicaut. Towards fault-tolerant formal concept analysis. In *AI*IA'05*, volume 3673 of *LNAI*, pages 212–223. Springer-Verlag, 2005.
14. J. K. Seppänen and H. Mannila. Dense itemsets. In *ACM SIGKDD'04*, pages 683–688, 2004.
15. G. Stumme, R. Taouil, Y. Bastide, N. Pasqier, and L. Lakhal. Computing iceberg concept lattices with TITANIC. *DKE*, 42:189–222, 2002.
16. V. Ventos, H. Soldano, and T. Lamadon. Alpha galois lattices. In *ICDM IEEE*, pages 555–558, 2004.
17. R. Wille. Restructuring lattice theory: an approach based on hierarchies of concepts. In I. Rival, editor, *Ordered sets*, pages 445–470. Reidel, 1982.
18. C. Yang, U. Fayyad, and P. S. Bradley. Efficient discovery of error-tolerant frequent itemsets in high dimensions. In *ACM SIGKDD*, pages 194–203. ACM Press, 2001.

The MIEL++ Architecture
When RDB, CGs and XML Meet for the Sake
of Risk Assessment in Food Products

Patrice Buche[1], Juliette Dibie-Barthélemy[1],
Ollivier Haemmerlé[2], and Rallou Thomopoulos[3]

[1]Unité INRA Mét@risk, 16 rue Claude Bernard, F-75231 Paris Cedex 05
[2]GRIMM-ISYCOM, Université de Toulouse le Mirail, Département de
Mathématiques-Informatique, 5 allées Antonio Machado, F-31058 Toulouse Cedex,
[3]INRA - UMR IATE - bat. 31, 2 Place Viala, F-34060 Montpellier Cedex 1
{Patrice.Buche,Juliette.Dibie}@inapg.fr,
Ollivier.Haemmerle@univ-tlse2.fr,rallou@ensam.inra.fr

Abstract. This article presents a data warehouse used for risk assessment in food products. The experimental data stored in this warehouse are heterogeneous, they may be imprecise; the data warehouse itself is incomplete by nature. The MIEL++ system – which is partially commercialized – is composed of three databases which are queried simultaneously, and which are expressed in three different data models: the relational model, the Conceptual Graph model and XML. Those models have been extended in order to allow the representation of fuzzy values. In the MIEL++ language, used to query the data warehouse, the end-users can express preferences in their queries by means of fuzzy sets. Fuzzy pattern matching techniques are used in order to compare preferences and imprecise values.

Preamble

In ICCS 2000, we presented an article that summarized our project to build a tool that aimed at preventing microbiological risk in food products [1]. That led us to work on the integration of a Relational Database and a Conceptual Graph database, and on an extension of the CG model allowing the representation of fuzzy values in the concept vertices [2]. Our work which took place in an important project called Sym'Previus[1] has significantly evolved since 2000. Indeed, it was the basis for a new French project, called e.dot, which aimed at building thematic data warehouses automatically fed from data extracted from the Web. We think that it could be interesting to present to the Conceptual Structure community the 2006 version of the MIEL++ system which is the result of 5-years work. That system, which involves Relational Databases, XML data and, of course, Conceptual Graphs, is now partially commercialized, and it is still being developed.

---

[1] This project is backed by the French Ministries of Agriculture and Research.

H. Schärfe, P. Hitzler, and P. Øhrstrøm (Eds.): ICCS 2006, LNAI 4068, pp. 158–171, 2006.

Note that some parts of this work have been already published in more details, particularly in the Journal of Intelligent Information Systems and in IEEE Transactions on Fuzzy Systems [3, 4, 5]. Our goal in this article is to provide a synthetic overview of the system, which has never been presented globally in an international event.

1 Introduction

Since 1999, we have been working with industrial[2] and academic[3] partners on several projects which concern knowledge representation and data integration in the field of predictive microbiology. In the Sym'Previus [6] and e.dot [7] projects, we worked on the building of a data warehouse composed of data concerning the behaviour of pathogenic germs in food products. Those data are designed to be used in a tool dedicated to researchers in microbiology or to industrials. Our goal is to help them in a decision support approach in order to prevent food products from contamination.

The information we have to store in our data warehouse presents several specificities. It is *weakly-structured* because information comes from heterogeneous sources (scientific literature, industrial partners, Web sites...) and is still rapidly evolving since predictive microbiology is a research field. It is *imprecise* because of the complexity of the underlying biological processes, and because of the internal imprecision of the measurement tools. The data warehouse is *incomplete* by nature since the number of experiments is potentially infinite: it will never contain information about all the possible food products and all the possible pathogenic germs in any possible experimental conditions.

Those three characteristics are taken into account in the following ways. The weak structure of the data led us to build a data warehouse composed of three bases: a Relational Database which contains the stable part of the information, a Conceptual Graph base which contains the weakly-structured part of the information and an XML base filled with data semi-automatically extracted from the Web. The imprecision of the data is represented by means of possibility distributions expressed by fuzzy sets, in each of the three bases. Finally, the incompleteness is partially solved by allowing the end-users to express large queries with expression of preferences in the selection criteria; we also propose a mechanism of generalization of the queries. The knowledge of the application domain is represented by means of the *MIEL++ ontology* which was built by experts of the domain during the Sym'Previus project.

The three bases are queried in a transparent way by means of the user interface of the MIEL++ system. The MIEL++ system is a kind of mediated architecture [8] between three different databases; each piece of information is stored in the most suited base.

This article aims at presenting the data warehouse as a whole. In section 2 we make an overall presentation of the MIEL++ architecture. In the next sections,

---

[2] Danone, Pernod-Ricard...

[3] INRIA, LRI, Institut Pasteur...

we present the two most innovative subsystems among the three which compose our data warehouse: the CG subsystem in section 3 and the XML subsystem in section 4. More detailed explanations about the Relational Database subsystem (called RDB subsystem in the following) can be found in [5].

2   The MIEL++ Architecture

The MIEL++ architecture is composed of three distinct databases – which are called subsystems in the following – which have been added successively during the development of the projects we are involved in. The first one is a Relational Database which contains the stable part of the information which can fit a given relational schema. Since the evolution of a database schema is an expensive operation, we proposed to add a second database dedicated to the less structured part of the data. We chose to use the Conceptual Graph model. Finally, we have a third database composed of XML data. That XML database contains data semi-automatically extracted from the Web. Fig. 1 presents an overview of the architecture of the MIEL++ system.

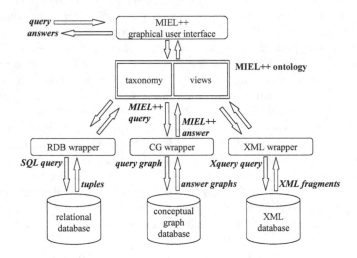

Fig. 1. A general overview of the MIEL++ architecture

When a query is asked to the MIEL++ system, that query is asked through a single graphical user interface, which is based on the MIEL++ ontology. The query is translated by each subsystem's wrapper into a query expressed in the query language of the subsystem (an SQL query in the RDB subsystem, a Conceptual Graph in the CG subsystem and a XQuery query in the XML subsystem). Finally, the global answer to the query is the union of the local results of the three subsystems. Note that, for the moment, the MIEL++ ontology is partially duplicated in each subsystem, as we will see for example in section 3.

2.1 The Data in MIEL++

The MIEL++ ontology. The MIEL++ ontology is notably composed of:

1. a *taxonomy of terms*, composed of the set of attributes which can be queried on by the end-user, and their corresponding definition domains. Each attribute has a definition domain which can be: (1) numeric, (2) "flat" symbolic (unordered constants such as a set of authors) or (3) hierarchized symbolic (constants partially ordered by the "kind-of" relation). Fig. 2 is a part of the taxonomy composed of the attribute Substrate and its hierarchized symbolic definition domain. The taxonomy contains for instance the food products, the pathogenic germs, etc.
2. a *relational schema*, which corresponds to the schema of the Relational Database of the MIEL++ system. That schema is composed of a set of signatures of the possible relations between the terms of the taxonomy. For example, the relation *FoodProductPH* is used to link a food product and its pH value.
3. a *set of views*, which consists of pre-written queries, which are given to help the end-users express their queries.

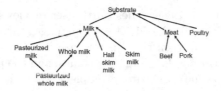

Fig. 2. A part of the taxonomy corresponding to the attribute Substrate

The fuzzy values in MIEL++. As we mentioned in the introduction of this article, the MIEL++ data can be imprecise, due to the complexity of the biological processes as well as the intrinsic imprecision of the measurement tools. We decided to allow the representation of such imprecise values by means of possibility distributions, expressed by means of fuzzy sets. Thus we proposed a representation of fuzzy values in our three databases. We use the representation of fuzzy sets proposed in [9, 10].

Definition 1. A **fuzzy set** f on a definition domain $Dom(f)$ is defined by a membership function μ_f from $Dom(f)$ to $[0,1]$ that associates the degree to which x belongs to f with each element x of $Dom(f)$.

2.2 The Queries in MIEL++

In the MIEL++ system, the query processing is done through the MIEL++ query language. We do not introduce extensively the MIEL++ query language in this article. The reader who wants a formal description of the MIEL++ query language can refer to [3]. In this article, we present the query language through the graphical user interface which is dedicated to end-users who are non-computer scientists.

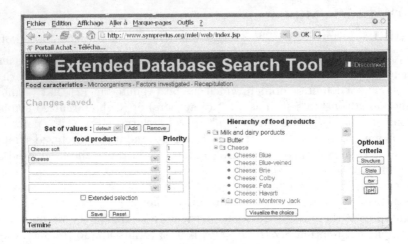

Fig. 3. The GUI permits to select an ordered list composed of *soft cheese* and *cheese* in the hierarchy of food products

This MIEL++ query language relies on the set of views and the taxonomy of the MIEL++ ontology. In the MIEL++ language, the end-users select the view they are interested in, then they instantiate it by specifying the selection attributes and their corresponding searched values, and the projection attributes of the query. As the three databases are incomplete (since the number of potential experiments is very large), we propose a mechanism of query enlargement by means of expression of preferences – represented by fuzzy values – in the values of the searched attributes.

The first screenshot (see Fig. 3) presents the first step of the expression of a query. The end-users choose in the taxonomy (here the hierarchy of food products) an ordered list of food names which represent their preferences for the attribute *FoodProduct*. In this example, the end-user expresses that he/she is first interested in *soft cheese*, but if there is no information about it in the databases, he/she accepts to enlarge to all kind of *cheese* with a lower preference degree.

The second screenshot (see Fig. 4) presents the second step of the expression of a query. Here, the end-user expresses his/her preferences for the attribute *pH* defined on a numeric domain. In this example, the end-user is first interested by pH values in the interval $[6, 7]$, but he/she accepts to enlarge the querying till the interval $[4, 8]$ with decreasing degrees of preference.

The third screenshot (see Fig. 5) presents the answers returned by the MIEL++ system. The resulting tuples are presented to the end-user ordered by their adequation degree δ which is presented in the following. In the screenshot, the two first answers fully match the preferences of the end-user ($\delta = 1$) for the *Food-Product* (*soft cheese*) and *pH* ($[6, 7]$) attributes. The next three answers ($\delta = .9$) correspond to the second choice expressed for the attribute *FoodProduct* attribute (*cheese*). The other ones ($\delta < .9$) also correspond to a kind of *cheese*, but with a pH value which goes away from the interval $[6, 7]$. It can be noticed that the *pH*

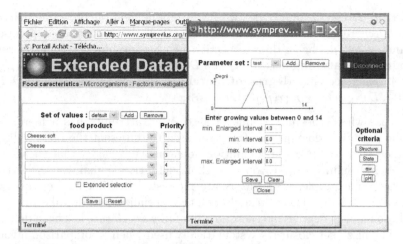

Fig. 4. The GUI permits to define a trapezoidal fuzzy set which represents the end-user's preferences for the numerical attribute *pH*

	B11			f_x	Indus20-i			
	A	B	C	H	K	M	O	P
1	pertinence	source	food product	microorganism	factor	type of answer	ph_min	ph_max
2	1.0	Little, 1994	Cheese: soft with bloomy rind	Bacillus cereus	Storage temperature	Graphs	6.8	6.8
3	1.0	Little, 1994	Cheese: soft with bloomy rind	Bacillus cereus	Storage temperature	Graphs	6.8	6.8
4	0.9	Murphy 96	Cheese	Listeria monocytogenes	Storage temperature	Generation time	6.4	6.4
5	0.9	Murphy 96	Cheese	Listeria monocytogenes	Storage temperature	Growth rate	6.4	6.4
6	0.9	Murphy 96	Cheese	Listeria monocytogenes	Storage temperature	Lag time	6.4	6.4
7	0.9	Indus20-b	Processed cheese	Escherichia coli	Storage temperature	Kinetics	5.9	5.9
8	0.9	Indus20-b	Processed cheese	Escherichia coli	Storage temperature	Kinetics	5.9	5.9
9	0.89	Indus20-d	Processed cheese	Listeria innocua	Temperature	Kinetics	5.8	5.8
10	0.89	Indus20-d	Processed cheese	Staphylococcus aureus	Temperature	Kinetics	5.8	5.8
11	0.85	Indus20-i	Processed cheese	Bacillus cereus	Storage temperature	Kinetics	5.65	5.75
12	0.85	Indus20-h	Processed cheese	Hafnia alvei	Storage temperature	Kinetics	5.7	5.7
13	0.85	Indus20-h	Processed cheese	Staphylococcus aureus	Storage temperature	Kinetics	5.7	5.7
14	0.85	Indus20-h	Processed cheese	Salmonella	Storage temperature	Kinetics	5.7	5.7
15	0.85	Indus20-h	Processed cheese	Listeria innocua	Storage temperature	Kinetics	5.7	5.7
16	0.85	Indus20-h	Processed cheese	Escherichia coli	Storage temperature	Kinetics	5.7	5.7
17	0.81	Indus20-i	Processed cheese	Bacillus cereus	Storage temperature	Kinetics	5.6	5.65
18	0.81	Indus20-i	Processed cheese	Bacillus cereus	Storage temperature	Kinetics	5.6	5.65
19	0.62	Indus20-e	Processed cheese	Staphylococcus aureus	Storage temperature	Kinetics	5.24	5.24
20	0.62	Indus20-e	Processed cheese	Salmonella	Storage temperature	Kinetics	5.24	5.24
21	0.62	Indus20-e	Processed cheese	Listeria innocua	Storage temperature	Kinetics	5.24	5.24
22	0.62	Indus20-e	Processed cheese	Hafnia alvei	Storage temperature	Kinetics	5.24	5.24
23	0.62	Indus20-e	Processed cheese	Escherichia coli	Storage temperature	Kinetics	5.24	5.24

Fig. 5. The answers provided by the MIEL++ querying system ordered by the adequation degree δ which is stored in the first column *pertinence*

value retrieved in the answer is considered by the MIEL++ system as an imprecise datum, presented to the end-user in two columns *pH_min* and *pH_max*.

In order to quantify the adequation of an imprecise datum D to a fuzzy selection criterion Q, both being represented by a fuzzy set, two degrees are classically used: (i) the possibility degree [10] and (ii) the necessity degree [11].

Definition 2. Let Q and D be two fuzzy sets defined on the same definition domain Dom, representing respectively a selection criterion and an imprecise datum, and μ_Q and μ_D being their respective membership functions. The **possibility degree of matching** between Q and D is $\Pi(Q, D) = sup_{x \in Dom}(min(\mu_Q(x), \mu_D(x)))$. The **necessity degree of matching** between Q and D is $N(Q, D) = 1 - \Pi(\overline{Q}, D) = inf_{x \in X} max(\mu_Q(x), 1 - \mu_D(x))$.

In the case where the fuzzy value of a selection attribute has a hierarchized symbolic definition domain, the fuzzy set used to represent the fuzzy value can be defined on a subset of this definition domain. We consider that such a fuzzy set implicitly defines degrees on the whole definition domain of the selection attribute. For example, if end-users are interested in Milk in their query, we assume that they are also interested in all the specializations of Milk. In order to take those implicit degrees into account, the *fuzzy set closure* has been defined in [12, 5]. The fuzzy set closure is systematically used when a comparison involving two fuzzy sets defined on a hierarchical definition domain is considered.

3   The Conceptual Graphs in MIEL++

3.1   The Schema of the Conceptual Graph Database

The flexibility of the Conceptual Graph model [13, 14] played an important part in the choice of that knowledge representation model in the MIEL++ system: we can build pieces of information which have different shapes by adding or removing graph vertices easily, contrary to a RDB schema.

We now summarize how the terminological knowledge is built in the MIEL++ Conceptual Graph subsystem (called CG subsystem in the following).

The **concept type set** is used to represent the main part of the MIEL++ taxonomy, since it is a partially ordered set, designed to contain the concepts of a given application. It is built as follows. A concept type t_a is associated with each attribute a of the taxonomy. If a is a hierarchized attribute, then a concept type t_{v_i} is associated with each element v_i of the definition domain of a. The t_a's and t_{v_i}'s are inserted into the concept type set, w.r.t. the partial order of that definition domain.

The hierarchized structure of the concept type set allows us to store the attribute names and the values belonging to hierarchized definition domains into the same set. For example, Fig. 6 represents a part of the concept type set of the MIEL++ Conceptual Graph database. The attribute *Substrate* and its hierarchized definition domain presented in Fig. 2 appear as a partial subgraph of that concept type set.

The **set of individual markers** is used to store the definition domain of each attribute a that has a *flat symbolic* or a *numerical* definition domain. More precisely, all the values of the definition domains of the *flat symbolic* attributes as well as the values of $\mathbb{R}$ are inserted into the set of individual markers [12].

We do not detail the **set of relation types** since it does not play an important part in our Conceptual Graph database, the semantics being mainly contained in the concept vertices.

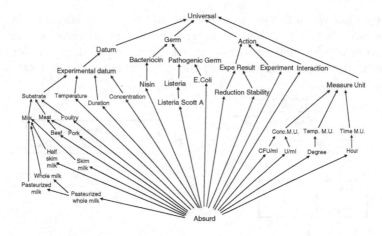

Fig. 6. A part of the concept type set of the MIEL++ CG subsystem

In order to allow a homogeneous expressivity between the three subsystems of MIEL++, we proposed an extension of the Conceptual Graph model to the representation of fuzzy values presented in [2]. A fuzzy set can appear in two ways in a concept vertex: (i) as a *fuzzy type* when the definition domain of the fuzzy set is hierarchized. A fuzzy type is a fuzzy set defined on a subset of the concept type set; (ii) as a *fuzzy marker* when the definition domain of the fuzzy set is *"flat symbolic"* or *numerical*. A fuzzy marker is a fuzzy set defined on a subset of the set of individual markers.

The Conceptual Graph database is composed of a set of Conceptual Graphs, each of them representing an elementary datum. For example, Fig. 7 is a part of a Conceptual Graph extracted from the MIEL++ CG subsystem.

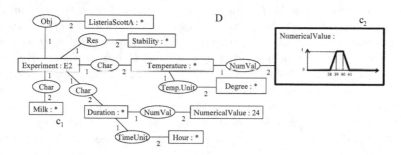

Fig. 7. An example of Conceptual Graph extracted from the MIEL++ CG subsystem. The concept vertex framed in bold is a concept with a fuzzy marker.

3.2 Query Processing in the CG Subsystem

The views. The CG subsystem uses a set of *view graphs* which allow us to define views on the Conceptual Graph database. A view graph is a pre-defined "empty"

query which has to be instantiated in order to become an actual query graph. When a query is asked in the CG subsystem, the view graph corresponding to the considered view is specialized by instantiating concept vertices in order to take into account the selection attributes of the query. The result is a *query graph*.

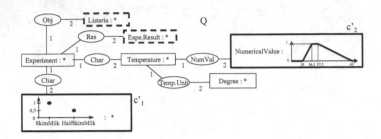

Fig. 8. An example of a query graph. The selection attributes are framed in bold, the projection attributes are dashed. One of the selection criteria is expressed by a concept with a numerical fuzzy marker.

The query processing. In the CG subsystem of the MIEL++ system, the query processing consists in searching for Conceptual Graphs which contain a more precise information than the information contained in the query (we search for specializations of the query graph) or, at least, for Conceptual Graphs which contain "approximate" answers. In order to find such Conceptual Graphs, we propose to use the δ-projection operation which is a flexible mapping operation between two Conceptual Graphs. The δ-projection is adapted from the classic projection operation, by taking into account the possibility and the necessity degrees of matching (see Definition 2).

Definition 3. *A δ-projection Π from a Conceptual Graph G into a Conceptual Graph G' is a triple (f, g, δ), f (resp. g) being a mapping from the relation (resp. concept) vertices of G into the relation (resp. concept) vertices of G' such that: (i) the edges and their labels are preserved; (ii) the labels of the relation vertices can be specialized; (iii) each concept vertex c_i of G has an image $g(c_i)$ of G' which satisfies it with the degrees π_i et n_i. The adequation degree between G and G' denoted δ is computed as the average of the minimum of the possibility degrees of adequation between the concept vertices of G and G' and the minimum of the necessity degrees of adequation between the concept vertices of G and G': $\delta = \frac{min(\pi_i) + min(n_i)}{2}$, with $1 \le i \le nb$ (nb being the number of concept vertices in G).*

The query processing in the CG subsystem consists in selecting the view graph, building the query graph, and δ-projecting that query graph into all the Conceptual Graphs of the database. Every time a δ-projection into a fact graph A_G is found, the Conceptual Graph A_G is considered an answer graph. A tuple with the adequation degree δ is built using this answer graph by extracting the values of the projection attributes.

Example 1. *If the query Q of Fig. 8 is asked on a Conceptual Graph database containing graph D of Fig. 7, the resulting tuple is: ($'ListeriaScottA'$, $'Stability'$, $\delta = 0,38$). Q can be δ-projected into D with the adequation degree $\delta = 0,38$: the vertices of Q of which the image in D is a specialization have degrees π et n equal to 1. c_1 satisfies c_1' with $\pi_1 = 1$ and $n_1 = 0$, c_2 satisfies c_2' with $\pi_2 = 0,77$ and $n_2 = 0,46$, then $\delta = \frac{min(1,\ 0,77)+min(0,\ 0,46)}{2} = 0,38$.*

4   Data Extracted from the Web

4.1   The XML Base

The XML base has been built in the MIEL++ system in order to store information retrieved from the Web. More precisely, we focus on tables included in scientific papers, which contain experimental data. The step of collecting data from the Web is achieved by a semi-automatic process called AQWEB, which is based on the MIEL++ ontology. Fig. 9 presents the AQWEB process.

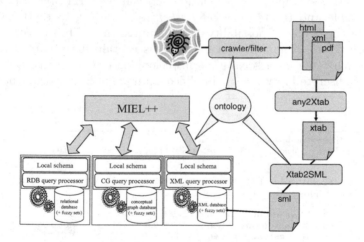

Fig. 9. The AQWEB semi-automatic process

Step 1 (crawler/filter in Fig. 9) consists in acquiring, by means of a search engine, documents on the Web which correspond to our application domain (that search is oriented by a combination of key words belonging to the MIEL++ taxonomy). We restrict our search to pdf and html documents which contain data tables. Step 2 (any2Xtab) consists in translating those data tables into XML documents, following the XTab dtd which allows one to represent a table classically by means of rows and cells. Step 3 (Xtab2SML) transforms those XTab documents into SML[4] documents, by a semantization process based on the MIEL++ taxonomy.

---

[4] For Semantic Markup Language.

That transformation allows one to enrich semantically the data tables by means of terms extracted from the MIEL++ taxonomy. Then it becomes possible to query the SML data through the MIEL++ query language.

SML process [15] achieves three kinds of semantic enrichment: (i) it associates terms of a Web table with their corresponding terms in the MIEL++ taxonomy (for example, the term *Stewed exotic fruit* of a Web table is associated with the term *Sweet fresh fruit* belonging to the taxonomy), (ii) when enough terms are identified in a given column of a Web table, it becomes possible to identify the "title" of the column; (iii) it instantiates semantic relations of the ontology which appear in the Web table schema (for example, the relation *FoodProductPH* is instantiated in a Web table that contains a column composed of food product names and another column with pH values). Some heuristics and disambiguisation mechanisms are used during this process (see [15] for a more detailed presentation of SML).

Moreover, in [16], we propose a fuzzy semantic tagging of the terms of a Web table: each association between a term of a Web table and a term belonging to the taxonomy is weighted by a possibility degree depending on their syntactic closeness (for example, the association between the term *Stewed exotic fruit* of a Web table and the term *Sweet fresh fruit* belonging to the taxonomy is weighted by the degree of possibility 0.33 computed thanks to the words belonging to both terms). The SML documents thus contain fuzzy data: for a given term of a Web table, its associated terms belonging to the taxonomy are represented by a discrete fuzzy set. A simplified example of SML document is given in Fig. 10.

```
<table> <title><table-title> </table-title>
<title-col> Item </title-col> <title-col> pH value </title-col>... </title>
<content>
...
<relLine> <FoodProductPH>
<Product><originalVal>Red onion</originalVal>
<ontoVal> <DFS>
<ValF> <item> Tree onion </item> <MD> 0.69 </MD> </ValF>
<ValF> <item> Welsh onion </item> <MD> 0.69 </MD> </ValF>
<ValF> <item> Red cabbage </item> <MD> 0.69 </MD> </ValF>
</DFS></ontoVal>
</Product>
<ph> <originalVal>5.2</originalVal> <ontoVal/></ph>
</FoodProductPH> </relLine>
</content> </origine> </table>
```

Fig. 10. *Simplified representation in SML of a Web data table*

We do not detail the query process of SML documents in this article. It has been presented in [4]. The main idea is that the query asked through the MIEL++ user interface is translated into a XQuery query by the wrapper of the XML subsystem. A mechanism allowing one to represent fuzzy queries in XQuery and fuzzy values in SML data has been proposed. The adequation of a fuzzy SML data to a fuzzy XQuery query is very close to that of the CG subsystem, which has been presented in section 3.

5 Implementation and Experimentation

The MIEL++ system has been implemented. It conforms to the J2EE standard (HTML client and servlet/JSP server). The RDB and XML subsystems have been developed in Java. The CG subsystem has been developed in C++ using the CoG-ITaNT platform [17]. At the moment, the Relational Database contains about 10.000 data. The MIEL++ RDB subsystem is used by our industrial partners of the Sym'Previus project. The Conceptual Graph database contains about 200 Conceptual Graphs manually built by analyzing the relevant sentences of scientific publications which do not fit the RDB subsystem schema. Each CG of the base is composed of about 70 vertices. The XML base contains about 200 scientific documents retrieved from the Web. Both CG and XML subsystems are currently under testing in collaboration with our partners of the Sym'Previus project.

RDB schema updating is a very rare operation (one update performed in 5 years) because it requires a huge work which can be performed only by a computer scientist. It requires firstly a schema updating and a data migration using SQL and secondly an updating of the data uploading tool and the MIEL++ querying system written in Java language. On the contrary, adding new weakly-structured data in the CG subsystem is a very less technical operation which can be performed by the database administrator. Data graphs and query graphs uploading is performed using the GUI available in the CoGITaNT platform. New concept types or new relation types only need to be registered in a text file to be available in the CG subsystem.

The RDB is regularly updated with data provided by the industrial partners, the projects financed by the French government and the scientific publications in the main journals of the domain. Data extracted from the Web thanks to AQWEB have been judged very pertinent by our partners because they come from other types of documents. They are also mainly composed of scientific information, but often this information is already synthetic, integrating a lot of results. Therefore, it contains an added value provided by experts which consists in the selection task and the treatment needed to compile the data. It is the type of information which can be found in lectures, thesis, reports published by national and international public organisations and state of the art reports realised by big research projects. In the evaluation process realised by our partners, 152 pertinent Web tables have been retrieved.

6 Perspectives

Even if the MIEL++ system is partially commercialized, it is currently being developed in several directions. Concerning each subsystem, several enhancements are being done. The RDB subsystem is stable, but its content is constantly evolving, with a lot of new pieces of information stored. The taking into account at query time of the fuzzy semantic tagging used in the XML data will be enhanced. As mentioned in section 4, in the current version of SML, fuzzy data stored in SML documents represent the mapping between terms found in Web tables and their corresponding terms in the MIEL++ taxonomy. As we accept partial instanciations of the semantic relations of the ontology in a Web table, we will also introduce fuzziness in the representation of semantic relation instanciation. Concerning the CG subsystem, we will work on an interface which will facilitate the insertion of new CGs in the database, for example by means of pre-written patterns used to build pieces of information.

At ontology level, we need to allow the use of non-taxonomic relations (for example the composition relation which will be very useful in the context of food industry).

We also have to work on a closer/tighter integration of our three subsystems. The first step will consist in uniformising our ontology, which is partially duplicated in each subsystem. Two ways are considered: (i) using the CG model in order to represent the whole ontology, then interfacing the RDB subsystem and the XML subsystem with an ontology server based on CoGITaNT; (ii) using an ontology server based on OWL-DL.

The second step will consist in integrating the subsystems by combining their partial answers in order to build global answers with pieces of information coming from different subsystems. At the moment, the global answer is only a juxtaposition/union of partial answers, but a single answer tuple comes exclusively from one subsystem.

We think about adding rules in our ontology, in order to allow a kind of fuzzy inferences in the three subsystems. Those rules could be represented in the CG formalism [18], extended to the representation of fuzzy conclusions.

Finally, we think about adapting our MIEL++ architecture to other application domains. This will be possible in the framework of a new important French project, which reunites 15 industrial and academic partners: the WebContent project, which will consist in building a platform dedicated to the integration of Semantic Web techniques.

References

1. P. Buche and O. Haemmerlé. Towards a unified querying system of both structured and semi-structured imprecise data using fuzzy views. In *Proceedings of the 8th International Conference on Conceptual Structures, Lecture Notes in Artificial Intelligence #1867*, pages 207–220, Darmstadt, Germany, August 2000. Springer.
2. R. Thomopoulos, P. Buche, and O. Haemmerlé. Representation of weakly structured imprecise data for fuzzy querying. *Fuzzy Sets and Systems*, 140-1:111–128, 2003.
3. O. Haemmerlé, P. Buche, and R. Thomopoulos. The MIEL system: uniform interrogation of structured and weakly structured imprecise data. *Journal of Intelligent Information Systems (to appear)*, 2006.
4. P. Buche, J. Dibie-Barthélemy, O. Haemmerlé, and G. Hignette. Fuzzy semantic tagging and flexible querying of xml documents extracted from the web. *Journal of Intelligent Information Systems (to appear)*, 2006.
5. P. Buche, C. Dervin, O. Haemmerlé, and R. Thomopoulos. Fuzzy querying on incomplete, imprecise and heterogeneously structured data in the relational model using ontologies and rules. *IEEE Transactions on Fuzzy Systems*, 3(13):373–383, 2005.

6. Sym'Previus. The sym'previus project. Web site, 2006. http://www.symprevius.org.
7. e.dot. The e.dot project. Web site, 2005. http://www.inria.fr/edot.
8. G. Wiederhold. Mediation in information systems. *ACM Computing Surveys*, 27(2):265–267, june 1995.
9. L.A. Zadeh. Fuzzy sets. *Information and Control*, 8:338–353, 1965.
10. L.A. Zadeh. Fuzzy sets as a basis for a theory of possibility. *Fuzzy Sets and Systems*, 1:3–28, 1978.
11. D. Dubois and H. Prade. *Possibility Theory - An Approach to Computerized Processing of Uncertainty.* Plenum Press, New York, 1988.
12. R. Thomopoulos, P. Buche, and O. Haemmerlé. Different kinds of comparisons between fuzzy conceptual graphs. In *Proceedings of the 11th International Conference on Conceptual Structures, ICCS'2003, Lecture Notes in Artificial Intelligence #2746*, pages 54–68, Dresden, Germany, July 2003. Springer.
13. J.F. Sowa. *Conceptual structures - Information processing in Mind and Machine.* Addison-Welsey, 1984.
14. M.L. Mugnier and M. Chein. Représenter des connaissances et raisonner avec des graphes. *Revue d'Intelligence Artificielle*, 10(1):7–56, 1996.
15. H. Gagliardi, O. Haemmerlé, N. Pernelle, and F. Saïs. A semantic enrichment of data tables applied to food risk assessment. In *Proceedings of the 8th International Conference on Discovery Science, DS'05, LNCS #3735*, pages 374–376, Singapore, october 2005. Springer.
16. P. Buche, J. Dibie-Barthélemy, O. Haemmerlé, and M. Houhou. Towards flexible querying of xml imprecise data in a data warehouse opened on the web. In *Proceedings of the 6th International Conference On Flexible Query Answering Systems (FQAS'04), Lecture Notes in AI #3055*, pages 28–40, Lyon, France, June 2004. Springer.
17. D. Genest. Cogitant v-5.1 - manuel de référence. Web site, 2003. http://cogitant.sourceforge.net.
18. E. Salvat and M.L. Mugnier. Sound and complete forward and backward chainings of graph rules. In *Proceedings of the 4th International Conference on Conceptual Structures, ICCS'96, Lecture Notes in Artificial Intelligence 1115, Springer-Verlag*, pages 248–262, Sydney, Australia, August 1996.

Some Notes on Proofs with Alpha Graphs

Frithjof Dau

Technische Universität Dresden, Dresden, Germany
dau@math.tu-dresden.de

Abstract. It is well-known that Peirce's Alpha graphs correspond to propositional logic (PL). Nonetheless, Peirce's calculus for Alpha graphs differs to a large extent to the common calculi for PL. In this paper, some aspects of Peirce's calculus are exploited. First of all, it is shown that the erasure-rule of Peirce's calculus, which is the only rule which does not enjoy the finite choice property, is admissible. Then it is shown that this calculus is faster than the common cut-free calculi for propositional logic by providing formal derivations with polynomial lengths of Statman's formulas. Finally a natural generalization of Peirce's calculus (including the erasure-rule) is provided such that we can find proofs linear in the number of propositional variables used in the formular, depending on the number of propositional variables in the formula.

1 Introduction

At the dawn of modern logic, Peirce invented his system of Existential Graphs (EGs), starting in 1896 and working extensively on it until he passed away in 1914 (see for example [Pei35, Pei92, PS00]). Peirce's EGs are divided into three parts which built upon each other, namely Alpha, Beta, and Gamma. Alpha corresponds to propositional logic (PL), Beta correspond to first order logic, and Gamma, which was never completed, encompasses aspects of higher order logic, modal logic and other features. Although not mathematically formalized, his Alpha and Beta EGs are one of the very early elaborations of mathematical logic.[1] But at the end of the 19th century, symbolic notations had already had taken the vast precedence in the development of formal logic, and EGs did not succeed against symbolic logic.

Several authors investigated Peirce's EGs from different perspectives, some of them aiming to elaborate a (more or less) mathematical theory of them (see for example [Zem64, Rob73, Bur91, Shi02a, Dau06]). Some works focus particularly on Alpha graphs (like [Ham95, Shi02b]) or more particular on finding proofs within Alpha ([Liu05]). But there are only few people who try to implement Peirce's calculus for automated theorem proving (see [HK05, vH03]), and one has to say that in the automated theorem proving community, Peirce's calculus is not acknowledged at all. This paper aims to exploit some aspects of Peirce's calculus which may be helpful for automated theorem proving with this calculus.

---

[1] Although Peirce did not provide any mathematical definitions for EGs, a mathematical elaboration of EGs can be obtained from a closer scrutiny of his works. See [Dau06].

H. Schärfe, P. Hitzler, and P. Øhrstrøm (Eds.): ICCS 2006, LNAI 4068, pp. 172–188, 2006.

Peirce's calculus for Alpha graphs differs to a large extent to the common calculi for propositional logic (PL). In usual calculi for PL, the transformation rules are defined along the inductive construction of formulas. That is, each transformation rule modifies formulas only on their top-level of their construction trees. In other words: We have *shallow* inferences. In contrast to that, Peirce's calculus allows to transform *arbitrary deep subformulas* of given formulas, i.e. to carry out *deep* inferences. To the best of my knowledge, there is only one proof-system which employs deep inferences as well, namely the *calculus of structures* of Gulielmi (see [Brü03]). But deep inference systems, particularly Peirce's rules for EGs, have some interesting properties which are of interest for automated theorem proving, as it is argued by Gulielmi for his calculus of structures. Some of these properties of Peirce's rules for Alpha graphs are investigated in this paper.

The organization of the paper is as follows: In Sec. 2, the basic notations, including syntax, semantics, and the calculus for Alpha graphs are introduced. Due to space limitations, we will use the linear notation for Alpha graphs. In Sec. 3, some basic theorems for Alpha graphs are provided. In Sec. 4, it is shown how Alpha graphs can be converted to normalforms, and, in order to obtain an analytic calculus, it is proven that the erasure-rule of the calculus can be removed. In Sec. 5 it is proven that the calculus is faster than the common calculi for propositional logic by showing that Statman's formulas can be proven in polynomial time. In Sec. 6, a version of the calculus with generalized rules is introduced, and it is shown that with this calculus, the number of steps of a proof for a formula f depends linearly from the number of propositional variables which occur in f. Finally in Sec. 7, the paper concludes with a discussion of the results.

2   Basic Notations for Alpha Graphs

In this paper, we will use the *linear* notion for Peirce's Alpha graphs. More precisely: Alpha graphs are introduced as formulas of propositional logic, equipped with an equivalence relation which encompasses the syntactical properties of Alpha graphs, mainly the commutativity and associativity of the juxtaposition of graphs, which corresponds on the semantical side to the commutativity and associativity of conjunction.[2]

The formulas of propositional logic, thus Alpha graphs as well, are built over a set $\mathcal{P} := \{P_1, P_2, P_3, \ldots\}$ of propositional variables and a symbol $\top \notin \mathcal{P}$ for truth, and we use the logical junctors $\neg$ and $\wedge$. Now each P_i for $i \in \mathbb{N}$ and $\top$ are formulas, if f is a formula, then $\neg f$ is a formula, and if f_1, f_2 are formulas, then $(f_1 \wedge f_2)$ is a formula. We will omit brackets if it is convenient. As usual,

---

[2] A similar approach is common in mathematical logic as well. For example, sequents in a sequent calculus are usually defined as *multisets* of formulas, thus we already have on the syntactical side encompassed commutativity and associativity of conjunction. Similarly, sometimes formulas are considered only modulo an equivalence relation. The equivalence classes are called *structures*. See for example [Brü03].

the formulas P_i and $\neg P_i$ with $i \in \mathbb{N}$ are called LITERALS. We will use the letters A, B to denote propositional variables as well, and the letters f, g, h, k, l to denote formulas.

In Peirce's calculus for EGs, the transformation rules allow to modify arbitrary subgraphs in arbitrary contexts. This idea will be carried over to the symbolic notion of propositional logic. First of all, when we speak in this paper about subformulas, we mean subformula *occurrences*. For example, for the formula $P_1 \wedge P_1$, as P_1 appears twice in this formula, we will say that it has *two* subformulas P_1. Square brackets are used to denote contexts. For example, with $f[g]$ we denote a formula f with a subformula g. A subformula g of f is EVENLY ENCLOSED resp. is PLACED IN A POSITIVE CONTEXT if it is a subformula of an even number of subformulas $\neg h$ of f. Otherwise it is said to be ODDLY ENCLOSED resp. to be PLACED IN A NEGATIVE CONTEXT. This will be denoted by $f[g]^+$ resp. $f[g]^-$. This notation can be nested. For example, with $f[P_2 \wedge g[h]]$, it is expressed that g is a formula with a subformula h, and f is a formula with the subformula $P_2 \wedge g \ (= P_2 \wedge g[h])$.

In Peirce's graphs, conjunction can only be expressed up to commutativity and associativity. Moreover, empty negations are allowed: For this reason, we had to add the symbol $\top$ to our alphabet. In the following, formulas are considered only up to the following equivalence relation $\sim$:

Commutativity:	$(f \wedge g)$	$\sim (g \wedge f)$
Associativity:	$((f \wedge g) \wedge h)$	$\sim (f \wedge (g \wedge h))$
Truthelement:	$(f \wedge \top)$	$\sim f$
Congruence:	$f[g]$	$\sim f[h]$ if $g \sim h$

Each class of formulas corresponds to a Peircean Alpha graph, thus this definition of propositional logic can be understood as a formalization of Peirce's Alpha system.

Now we are prepared to introduce the calculus. It consists of the following six rules (where f, g, h, i denote arbitrary formulas).

Erasure:	$f[g \wedge h]^+ \vdash f[g]^+$
Insertion:	$f[g]^- \vdash f[g \wedge h]^-$
Iteration:	$f[g \wedge h[i]] \vdash f[g \wedge h[g \wedge i]]$
Deiteration:	$f[g \wedge h[g \wedge i]] \vdash f[g \wedge h[i]]$
Double Cut i):	$f[\neg\neg g] \vdash f[g]$
Double Cut ii):	$f[g] \vdash f[\neg\neg g]$

Let f, g be two graphs. Then g CAN BE DERIVED FROM f (which is written $f \vdash g$), if there is a finite sequence $(f_1, f_2, \ldots, f_n)$ with $f = f_1$ and $g = f_n$ such that each f_{i+1} is derived from f_i by applying one of the rules of the calculus. The sequence is called A PROOF or DERIVATION FOR $f \vdash g$ (OF LENGTH $n-1$). Two graphs f, g with $f \vdash g$ and $g \vdash f$ are said to be PROVABLY EQUIVALENT.

If F is a set of graphs, we write $F \vdash f$ if there are $f_1, \ldots, f_i \in F$ with $f_1 \wedge \ldots \wedge f_i \vdash f$. With $f \vdash^n g$ we mean that g can be derived from f in (at most) n steps. For $\top \vdash f$, we write more simply $\vdash f$ resp. $\vdash^n f$. This set of rules is (strongly) sound and complete, as it is shown in [Dau04]. We use the usual abbreviation, i.e., $f \vee g$ is a (mere syntactical) abbreviation for $\neg(\neg f \wedge \neg g)$, $f \rightarrow g$ abbreviates $\neg(f \wedge \neg g)$, and $f \leftrightarrow g$ abbreviates $(f \rightarrow g) \wedge (f \rightarrow g)$, that is $\neg(f \wedge \neg g) \wedge \neg(g \wedge \neg f)$.

The semantics are now defined in the usual way. A VALUATION or MODEL is a mapping $val : \mathcal{P} \cup \{\top\} \mapsto \{\text{ff}, \text{tt}\}$ with $val(\top) = \text{tt}$. Let $val : \mathcal{P} \mapsto \{\text{ff}, \text{tt}\}$ be a valuation. We set $val \models P_i :\Leftrightarrow val(P_i) = \text{tt}$, $val \models (f \wedge g) :\Leftrightarrow val(f) = \text{tt} = val(g)$, and $val \models \neg f :\Leftrightarrow val(f) = \text{ff}$. For $val \models f$, we say that f HOLDS IN val. If we have two formulas f, g such that $val \models g$ for each valuation val with $val \models f$, we write $f \models g$, and we say that f ENTAILS g. Finally, a formula f is called SATISFIABLE, iff there exists a valuation val with $val \models f$, it is called VALID or A TAUTOLOGY, iff $val \models f$ for each valuation val, and it is called CONTRADICTORY, iff $val \not\models f$ for each valuation val.

3   Some Simple Theorems

In [Pei35] Peirce provided 16 useful transformation rules for EGs which he derived from his calculus. These rules are logical metalemmata in the sense that they show some *schemata* for proofs with EGs, i.e., they are derived 'macro'-rules. In this section we provide the formal Alpha graph versions for two of these transformation rules. We start with a (weakened) version of the first transformation rule of Peirce.

Lemma 1 (Reversion Theorem). *Let f and g be two formulas. Then we have:*

$$f \vdash^n g \quad \Rightarrow \quad \neg g \vdash^n \neg f \quad and \quad \neg g \vdash^n \neg f \quad \Rightarrow \quad f \vdash^{n+2} g$$

Proof: Let $(h_1, h_2, \ldots, h_n)$ with $h_1 = f$ and $g = h_n$ be a proof for $f \vdash g$. Then, due to the symmetry of the calculus, $(\neg h_n, \neg h_{n-1}, \ldots, \neg h_1)$ is a proof for $\neg g \vdash \neg f$. Analogously, from $\neg g \vdash^n \neg f$ we conclude $\neg\neg f \vdash^n \neg\neg g$. An additional application of the double cut rule at the beginning and the end of the proof yields $f \vdash^{n+2} g$.   □

Let g be a subformula of f. With $f[h/g]$ we denote the graph where g is substituted by h. If g is a subgraph in a positive context, we will more explicitly write $f[h/g]^+$, and analogously $f[h/g]^-$ for negative contexts.

All rules in the calculus which are applied in a context only depend on whether the context is positive or negative. In particular if a proof for $f \vdash g$ is given, this proof can be carried out in arbitrary positive contexts. Together with the previous lemma, this yields the following lemma. It can also be found in [Sow97] (from where we adopted the name of the theorem).

Lemma 2 (Cut-And-Paste-Theorem I). *Let $g \vdash^n h$ for formulas g, h. Then:*

$$f \vdash^n f[h/g]^+ \quad and \quad f \vdash^{n+2} f[g/h]^-$$

Particularly, tautologies can be inserted into arbitrary contexts of arbitrary formulas.

With $f[\![h/g]\!]$ we denote the formula we obtain from f by substituting *every* subformula (i.e., every occurence of the subformula) g by h.

Lemma 3 (Cut-And-Paste-Theorem II). *Let g be a formula with $\vdash^n g$, let P_i be a propositional variable and f be another formula. Then we have $\vdash^n f[\![g/P_i]\!]$.*

Proof: Let $(h_0, h_2, \ldots, h_n)$ with $h_n = f$ be a proof for f. Then it is easy to see that $(h_0[\![g/P_i]\!], h_2[\![g/P_i]\!], \ldots, h_n[\![g/P_i]\!])$ is a proof for $f[\![g/P_i]\!]$. $\square$

The next two lemmata are two other metalemmata which ease the handling of proofs (they will be needed in Sec. 6). To ease the readability of the proofs, we have sometimes underlined the subformulas which will be used in the next step (for example, by deiterating them).

Lemma 4 (Proof by Distinction). *Let f, g be formulas. Then we have*

$$(g \to f) \land (\neg g \to f) \;\; \vdash^7 f$$

Proof: $(g \to f) \land (\neg g \to f) \;=\; \underline{\neg(g \land \neg f)} \land \neg(\neg g \land \neg f)$

$\overset{\text{it.}}{\vdash} \;\; \underline{\neg(g \land \neg f)} \land \neg(\neg(g \land \neg(g \land \neg f)) \land \neg f)$

$\overset{\text{era.}}{\vdash} \;\; \neg(\neg(g \land \neg(g \land \underline{\neg f})) \land \neg f)$

$\overset{\text{deit.}}{\vdash} \;\; \neg(\neg(g \land \underline{\neg g}) \land \neg f)$

$\overset{\text{deit.}}{\vdash} \;\; \neg(\neg(\underline{g} \land \neg \top) \land \neg f)$

$\overset{\text{era.}}{\vdash} \;\; \neg(\underline{\neg \neg} \top \land \neg f)$

$\overset{\text{dc.}}{\vdash} \;\; \neg(\top \land \neg f)$

$\sim \;\; \underline{\neg \neg} f$

$\overset{\text{dc.}}{\vdash} \;\; f$ $\square$

Lemma 5. *Let f, g be formulas. Then we have* $(f \leftrightarrow g) \leftrightarrow g \;\; \vdash^{14} f$.

Proof: We provide a formal derivation of $(f \leftrightarrow g) \leftrightarrow g \;\; \vdash f$. The last step is done with Lem. 4. As we had 7 derivational steps so far, we have a total of 14 steps.

$(f \leftrightarrow g) \leftrightarrow g \;=\; (\neg(f \land \neg g) \land \neg(g \land \neg f)) \leftrightarrow g$

$=\; \neg(\neg(\neg(f \land \neg g) \land \neg(g \land \neg f) \land \neg g) \land \neg(g \land \neg(\neg(f \land \neg g) \land \neg(\underline{g} \land \neg f))))$

$\overset{\text{deit.}}{\vdash} \;\; \neg(\neg(\neg(f \land \neg g) \land \neg(g \land \neg f) \land \neg g) \land \neg(g \land \neg(\neg(f \land \neg g) \land \underline{\neg \neg} f)))$

$\overset{\text{dc.}}{\vdash} \;\; \neg(\neg(\neg(f \land \neg g) \land \neg(g \land \neg f) \land \neg g) \land \neg(g \land \neg(\neg(\underline{f} \land \neg g) \land f)))$

$\overset{\text{deit.}}{\vdash} \;\; \neg(\neg(\neg(f \land \neg g) \land \neg(g \land \neg f) \land \neg g) \land \neg(g \land \neg(\underline{\neg \neg} g \land f)))$

$$\overset{\text{dc.}}{\vdash}\ \neg(\neg(f \wedge \neg g) \wedge \neg(g \wedge \neg f) \wedge \neg g) \wedge \neg(g \wedge \neg(\underline{g} \wedge f))$$

$$\overset{\text{deit.}}{\vdash}\ \neg(\neg(f \wedge \neg g) \wedge \underline{\neg(g \wedge \neg f)} \wedge \neg g) \wedge \neg(g \wedge \neg f)$$

$$\overset{\text{deit.}}{\vdash}\ \neg(\neg(f \wedge \underline{\neg g}) \wedge \neg g) \wedge \neg(g \wedge \neg f)$$

$$\overset{\text{deit.}}{\vdash}\ \neg(\neg f \wedge \neg g) \wedge \neg(g \wedge \neg f)$$

$$\sim\ \neg(g \wedge \neg f) \wedge \neg(\neg g \wedge \neg f)$$

$$=\ (g \rightarrow f) \wedge (\neg g \rightarrow f)$$

$$\vdash^7 f \qquad\qquad\qquad\qquad\qquad\qquad\qquad\qquad\qquad\qquad\qquad\square$$

4 Normalforms and Admissibility of Erasure

In automatic theorem proving, for tracking back a proof, a desirable feature of the calculus is the so-called *subformula property* which states that all formulas in a derivation are subformulas of the endformula. The essence of the subformula property is the fact that given a conclusion, every inference rule yields a *finite* set of possible premises. Let us call this property *finite choice property* (see for example [Brü03]). It is easy to see that in Peirce's calculus, only the erasure-rule does not satisfy the finite choice property. In this section, it is shown that the erasure-rule is admissible, i.e. the remaining calculus is still complete.

The restricted version of the calculus, where the erasure-rule is removed, is denoted by $\vdash_{-e}$. Due to symmetry reasons, we will consider a calculus $\vdash_{-i}$, that is $\vdash$ without the insertion-rule, as well. In this section, it will be firstly shown how formulas can be converted to normalforms with $\vdash$ and $\vdash_{-e}$, and then how proofs with $\vdash_{-e}$ can be found in an effective way.

Lemma 6 (Reducing Transformation I). *The formulas* $\neg(f \wedge \neg(g \wedge h))$ *and* $\neg(f \wedge \neg g) \wedge \neg(f \wedge \neg h)$ *are provably equivalent in* $\vdash_{-i}$. *More precisely, we have*

$$\neg(f \wedge \neg(g \wedge h))\ \vdash^3_{-i}\ \neg(f \wedge \neg g) \wedge \neg(f \wedge \neg h)\ \vdash^4_{-i}\ \neg(f \wedge \neg(g \wedge h)) \qquad (1)$$

Proof: $\qquad \neg(f \wedge \neg(g \wedge h)) \overset{\text{it.}}{\vdash}_{-i} \neg(f \wedge \neg(g \wedge \underline{h})) \wedge \neg(f \wedge \neg(g \wedge h))$

$\qquad\qquad\qquad \overset{\text{era.}}{\vdash}_{-i} \neg(f \wedge \neg g) \wedge \neg(f \wedge \neg(\underline{g} \wedge h))$

$\qquad\qquad\qquad \overset{\text{era.}}{\vdash}_{-i} \neg(f \wedge \neg g) \wedge \underline{\neg(f \wedge \neg h)} \qquad\qquad\qquad (*)$

$\qquad\qquad\qquad \overset{\text{it.}}{\vdash}_{-i} \neg(f \wedge \neg(g \wedge \neg(f \wedge \neg h))) \wedge \underline{\neg(f \wedge \neg h)}$

$\qquad\qquad\qquad \overset{\text{era.}}{\vdash}_{-i} \neg(f \wedge \neg(g \wedge \neg(\underline{f} \wedge \neg h)))$

$\qquad\qquad\qquad \overset{\text{deit.}}{\vdash}_{-i} \neg(f \wedge \neg(g \wedge \underline{\neg\neg} h))$

$\qquad\qquad\qquad \overset{\text{dc.}}{\vdash}_{-i} \neg(f \wedge \neg(g \wedge h))$

The proof until $(*)$ shows the first part of the lemma, the remaining proof shows the second part. $\qquad\qquad\qquad\qquad\qquad\qquad\qquad\qquad\qquad\qquad\square$

The proof of this lemma shows even more. It is carried out on the sheet of assertion, thus, due to the Cut-And-Paste-Theorem I (Lem. 2), in can be carried out in positive contexts. Moreover, its inverse direction can be carried out in arbitrary negative contexts, where the rules iteration and deiteration as well as the rules erasure and insertion are mutually exchanged. Thus we immediately obtain the following corollary.

Corollary 1 (Reducing Transformation II).

$$F[\neg(f \wedge \neg(g \wedge h)]^- \vdash^4_{-e} F[\neg(f \wedge \neg g) \wedge \neg(f \wedge \neg h)]^- \vdash^3_{-e} F[\neg(f \wedge \neg(g \wedge h)]^- (2)$$
$$F[\neg(f \wedge \neg(g \wedge h)]^+ \vdash^3_{-i} F[\neg(f \wedge \neg g) \wedge \neg(f \wedge \neg h)]^+ \vdash^4_{-i} F[\neg(f \wedge \neg(g \wedge h)]^+ (3)$$

With these results, it is possible to reduce the depth of a formula and to transform it into its conjunctive normalform. Before we do so, some technical notations have to be introduced. If g is a strict subformula of f (i.e., g is a subformula of f and $g \neq f$), we write $g < f$ resp. $f > g$. A sequence $f = f_0, f_1, f_2, \ldots, f_n$ is called a NEST OF CONTEXTS OF f, if

1. $f_i = \neg f_i'$ for each $i \geq 1$ (i.e., each f_{i+1} begins with a negation sign '$\neg$'),
2. $f_i > f_{i+1}$ for each $i \geq 0$, and
3. For each $0 \leq i \leq n-1$, there is no formula $\neg g$ with $f_i > \neg g > f_{i+1}$.

The number n is called the DEPTH of the nest. A formula f is said to have depth n if n is the maximal depth of all nests of f. Such a formula is said to be NORMALIZED TO DEPTH n, if moreover for each nest $f = f_0, f_1, f_2, \ldots, f_n$, there exists a propositional variable P_i, $i \in \mathbb{N}$, with $f_n = \neg P_i$. Consider for example the following formulas:

$$f := \neg(P_1 \wedge \neg P_2 \wedge \neg P_3) \wedge \neg P_4 \qquad \text{and} \qquad g := \neg(P_1 \wedge \neg(P_2 \wedge P_3)) \wedge \neg P_4$$

Both f and g have depth 2, but only f is normalized to depth 2. A formula f which is normalized to depth 2 is a conjunction of formulas $\neg(g_1 \wedge \ldots \wedge g_n)$, where each g_i is a literal. Thus f can be understood to be in CNF (conjunctive normal form), expressed by means of $\neg$ and $\wedge$ only. As $\vdash$ is sound and complete, it is not surprising that each formula can be transformed into its CNF. This is not possible if we restrict ourselves to $\vdash_{-e}$, but even then, it is possible to normalize each formula to depth 3.

Lemma 7 (Normalform).

1. Using $\vdash_{-e}$, each formula can effectively be transformed into a provably equivalent formula which is normalized to depth 3.
2. Using $\vdash$, each formula can effectively be transformed into a provably equivalent formula which is normalized to depth 2.

Proof: We first prove 1. Let f be an arbitrary formula, assume that f is not normalized to depth 3. Then there exists a nest $f, \neg f_1, \neg f_2, \neg f_3$ where f_3 is not a propositional variable, i.e., f_3 is either of the form $\neg g_3$, or it is the conjunction of at least two nontrivial formulas, i.e., $f_3 = g_3 \wedge g_3'$, with $g_3, g_3' \neq \top$. ·

In the first case, we have more explicitely

$$f = g_0 \wedge \neg f_1 = g_0 \wedge \neg (g_1 \wedge \neg f_2) = g_0 \wedge \neg (g_1 \wedge \neg (g_2 \wedge \neg f_3)) = g_0 \wedge \neg (g_1 \wedge \neg (g_2 \wedge \neg \neg g_3))$$

Obviously, we can apply the double cut rule i) and obtain

$$f \vdash g_0 \wedge \neg (g_1 \wedge \neg (g_2 \wedge g_3)) \vdash f$$

In the latter case, we have $f = g_0 \wedge \neg (g_1 \wedge \neg (g_2 \wedge \neg (g_3 \wedge g_3')))$. Now Eqn. (2) yields

$$f \vdash g_0 \wedge \neg (g_1 \wedge \neg ((g_2 \wedge \neg g_3)) \wedge (g_2 \wedge \neg g_3'))) \vdash f$$

These transformations are carried out until we reach a formula which is normalized to depth 3. Thus 1) is proven.

A formula which is normalized to depth 3 cannot be further reduced with Eqn. (2), but Eqn. (3) can still be applied in the outermost context. Thus an analogous argument shows that with the double cut rule or Eqn. (3), each formula can be transformed into a syntactically equivalent formula normalized to depth 2.   □

Example:

$$\neg (P_1 \wedge \neg (P_2 \wedge \neg (P_3 \wedge \neg (P_4 \wedge \neg (P_5 \wedge \neg (P_6 \wedge P_7))))))$$

$$\overset{\text{Cor. 1}}{\vdash_{-e}} \neg (P_1 \wedge \neg (P_2 \wedge \neg (P_3 \wedge \neg (P_4 \wedge \neg P_5) \wedge \neg (P_4 \wedge \neg (\neg (P_6 \wedge P_7))))))$$

$$\overset{\text{dc.}}{\vdash_{-e}} \neg (P_1 \wedge \neg (P_2 \wedge \neg (P_3 \wedge \neg (P_4 \wedge \neg P_5) \wedge \neg (P_4 \wedge P_6 \wedge P_7))))$$

$$\overset{2 \times \text{Cor. 1}}{\vdash_{-e}} \neg (P_1 \wedge \neg (P_2 \wedge \neg P_3) \wedge \neg (P_2 \wedge \neg (\neg (P_4 \wedge \neg P_5))) \wedge \neg (P_2 \wedge \neg (\neg (P_4 \wedge P_6 \wedge P_7))))$$

$$\overset{2 \times \text{dc.}}{\vdash_{-e}} \neg (P_1 \wedge \neg (P_2 \wedge \neg P_3) \wedge \neg (P_2 \wedge P_4 \wedge \neg P_5) \wedge \neg (P_2 \wedge P_4 \wedge P_6 \wedge P_7))$$

In the following, we will show that each tautology can be derived with $\vdash_{-e}$. A well-known method to check the validity of a formula f is to check whether $\neg f$ is contradictory with the method of resolution. The basic idea of resolution is as follows: If k, l are formulas and if A is a propositional variable which does neither occur in k nor in l, then $(A \vee k) \wedge (\neg A \vee l)$ is satisfiable if and only if $k \vee l$ is satisfiable. Now, in order to check whether $\neg f$ is contradictory, subformulas of the form $(A \vee k) \wedge (\neg A \vee l)$ are successively replaced by $k \vee l$ until a formula is reached from which it can be easily decided whether it is satisfiable.

For the $\neg, \wedge$-formalization of propositional logic, this basic transformation can be reformulated as follows: Let k, l formulas, let A be a propositional variable which does neither occur in k nor in l. Then $\neg (A \wedge k) \wedge \neg (\neg A \wedge l)$ is satisfiable if and only if $\neg (k \wedge l)$ is satisfiable. The next lemma shows that the inverse direction of the transformation of resolution can be derived in negative contexts with $\vdash_{-e}$.

Lemma 8 (Inverse Resolution). *Let A be a propositional variable, let k, l be formulas where A does not occur. Then we have:*

$$f[\neg (k \wedge l)]^- \vdash_{-e} f[\neg (A \wedge k) \wedge \neg (\neg A \wedge l)]^-$$

Moreover, $\neg(k \wedge l)$ is satisfiable if and only if $\neg(A \wedge k) \wedge (\neg A \wedge l)$ is satisfiable.

Proof:
$$f[\neg(k \wedge l)]^- \overset{\text{ins.}}{\vdash_{-e}} f[\neg(A \wedge k) \wedge \neg(k \wedge l)]^-$$

$$\overset{\text{dc.}}{\vdash_{-e}} f[\neg(A \wedge k) \wedge \neg(\neg\neg k \wedge l)]^-$$

$$\overset{\text{ins.}}{\vdash_{-e}} f[\neg(A \wedge k) \wedge \neg(\neg(A \wedge \neg k) \wedge l)]^-$$

$$\overset{\text{it.}}{\vdash_{-e}} f[\neg(A \wedge k) \wedge \neg(\neg(A \wedge \neg(A \wedge k)) \wedge l)]^-$$

$$\overset{\text{deit.}}{\vdash_{-e}} f[\neg(A \wedge k) \wedge \neg(\neg A \wedge l)]^- \qquad \square$$

Now we are prepared to show that the erasure-rule is admissible.

Theorem 1 (Erasure is Admissible). *If f is a tautology, we have $\vdash_{-e} f$.*

Proof: Due to Lem. 7, we can assume that f is normalized to depth 3, and f cannot be a literal. For $f = g_1 \wedge g_2$, $\vdash_{-e} g_1$ and $\vdash_{-e} g_2$ yield $\vdash_{-e} f$. Thus without loss of generality, we can assume that $f = \neg g$ for a formula g. Obviously, g is normalized to depth 2, and g is contradictory (which is equivalent to f being tautologous).

Now we can resolve g to a formula h which is not resolvable (i.e., h does not contain any subformula of the form $\neg(A \wedge k) \wedge \neg(\neg A \wedge l)$, that is, the rule of resolution cannot be applied). Then g is satisfiable if and only if h is satisfiable. Next, as g is normalized to depth 2, h is normalized to depth 2, too. Moreover, as the inverse direction of the resolution is derivable in $\vdash_{-e}$ due to Lem. 8, we have $\neg h \vdash_{-e} \neg g$. Thus it is sufficient to show that $\neg h$ is derivable with $\vdash_{-e}$.

As h is not resolvable, no propositional variable appears in different subformulas $\neg h_1$, $\neg h_2$ of h one time in a positive and one time in a negative context. Moreover, due to the iteration-rule, we can assume that each propositional variable $A \in \mathcal{P}$ occurs at most once in each subformula $\neg h'$ of h. Now we can assign the following truth-values to all $P_i \in \mathcal{P}$: We set $val(P_i) := \mathbf{ff}$, if P_i occurs in a negative context of h, and we set $val(P_i) := \mathbf{tt}$ otherwise. It is easy to see that if h is not of the form $\neg\top \wedge k$, then $val \models h$. Thus h has the form $\neg\top \wedge k$. Then $\top \overset{\text{dc.}}{\vdash_{-e}} \neg\neg\top \overset{\text{ins.}}{\vdash_{-e}} \neg(\neg\top \wedge k) \; (= \neg h)$ is a derivation of $\neg h$ in $\vdash_{-e}$, thus we are done. $\qquad \square$

Due to $f \models g \;\Leftrightarrow\; \models f \to g$, we can check $f \models g$ with $\vdash_{-e}$ as well. But in general, we do not have $f \models g \;\Rightarrow\; f \vdash_{-e} g$, as the simple example $P_1 \wedge P_2 \models P_1$ shows.

5 An Exponential Speed Up

The most prominent rule in sequent-calculi is the cut-rule, a generalized version of the modus ponens: $\dfrac{\Gamma_1 \vdash \Delta_1, A \qquad A, \Gamma_2 \vdash \Delta_2}{\Gamma_1, \Gamma_2 \vdash \Delta_1, \Delta_2}$. Due to the 'erasing of A', this rule does not satisfy the finite choice property. Gentzen's famous cut-elimination-theorem states that the cut-rule is admissible: Every proof using the cut-rule can be converted into another proof without the cut-rule (proofs that do not use the cut-rule are called *analytic*). But by doing so, the size of the proof generally grows exponentially. In particular, there are classes of tautologies such that their proofs in sequent-calculi including the cut-rule grow polynomially with their size, whilst in cut-free sequent-calculi, their proofs grow exponentially. In this section, such a class will be investigated.

In [Sta78], R. Statman studied a class of polynomial-size formulas and investigated their proof-lengths in sequent calculi. First we present the formulas constructed by Statman. Let A_i, B_i with $i \geq 1$ propositional variables. We set:

$$f_i := \bigwedge_{k=1}^{i}(A_k \vee B_k)$$

$$g_1 := A_1 \qquad\qquad\qquad\qquad\qquad\qquad \text{induction start}$$

$$h_1 := B_1 \qquad\qquad\qquad\qquad\qquad\qquad \text{induction start}$$

$$g_{i+1} := f_i \rightarrow A_{i+1} = \bigwedge_{k=1}^{i}(A_k \vee B_k) \rightarrow A_{i+1} \qquad \text{induction step}$$

$$h_{i+1} := f_i \rightarrow B_{i+1} = \bigwedge_{k=1}^{i}(A_k \vee B_k) \rightarrow B_{i+1} \qquad \text{induction step}$$

$$k_n := ((g_1 \vee h_1) \wedge (g_2 \vee h_2) \wedge \ldots \wedge (g_n \vee h_n)) \rightarrow (A_n \vee B_n)$$

For example, we have

$$k_2 = [(A_1 \vee B_1) \wedge (((A_1 \vee B_1) \rightarrow A_2) \vee ((A_1 \vee B_1) \rightarrow B_2))] \rightarrow (A_1 \vee B_1)$$

It is straightforward to see that the formulas k_n are tautologies. R. Statman has proven that in cut-free sequent-calculi, the lengths of the proofs for k_n grow exponentially, whereas in sequent-calculi including the the cut-rule, it is possible to find proofs of polynomial length. Gulielmi has proven that k_n can be derived within his cut-free deep inference system, In contrast to usual sequent-calculi, in polynomial time. We provide an analogous result for so-to-speak *analytic* calculus $\vdash_e$. So in this respect, the strong rules of $\vdash_e$, yield an exponentially speed-up in the length of proofs, compared to a analytic sequent-calculus.

Theorem 2 (Statman's formulas can be proven with $\vdash_e$ in polynomial time). *For Statman's formula f_n there exists a formal proof of length $n(n+1)$.*

Proof: We provide a formal derivation of k_n. To ease the readability and to save space, we abbreviate $(A_i \vee B_i)$, i.e., $\neg(\neg A_i \wedge \neg B_i)$, by AB_i.

$$\top \qquad \vdash \qquad \neg\neg\top$$

insertion
$$\vdash \qquad \neg(AB_1 \wedge AB_2 \wedge \ldots \wedge AB_{n-2} \wedge AB_{n-1} \wedge AB_n \wedge \neg\top)$$

it. AB_n
$$\vdash \qquad \neg(AB_1 \wedge AB_2 \wedge \ldots \wedge AB_{n-2} \wedge AB_{n-1} \wedge AB_n \wedge \neg AB_n)$$

$$= \qquad \neg(AB_1 \wedge AB_2 \wedge \ldots \wedge AB_{n-2} \wedge AB_{n-1} \wedge \neg(\neg A_n \wedge \neg B_n) \wedge \neg AB_n)$$

2 ×it. of AB_{n-1}
$$\vdash \qquad \neg(AB_1 \wedge AB_2 \wedge \ldots \wedge AB_{n-2} \wedge AB_{n-1}$$
$$\wedge \neg(AB_{n-1} \wedge \neg A_n \wedge AB_{n-1} \wedge \neg B_n) \wedge \neg AB_n)$$

$$= \qquad \neg(AB_1 \wedge AB_2 \wedge \ldots \wedge AB_{n-2} \wedge \neg(\neg A_{n-1} \wedge \neg B_{n-1})$$
$$\wedge \neg(AB_{n-1} \wedge \neg A_n \wedge AB_{n-1} \wedge \neg B_n) \wedge \neg AB_n)$$

$$\overset{4 \times \text{it. of } AB_{n-2}}{\vdash} \quad \neg(AB_1 \wedge AB_2 \wedge \ldots \wedge AB_{n-2} \wedge \neg(AB_{n-2} \wedge \neg A_{n-1} \wedge AB_{n-2} \wedge \neg B_{n-1})$$
$$\wedge \neg(AB_{n-2} \wedge AB_{n-1} \wedge \neg A_n \wedge AB_{n-2} \wedge AB_{n-1} \wedge \neg B_n) \wedge \neg AB_n)$$

$$\vdots$$

$$\overset{2(n-1) \times \text{it. of } AB_1}{\vdash} \quad \neg(AB_1$$
$$\wedge \neg(AB_1 \wedge \neg A_2 \wedge AB_1 \wedge \neg B_2)$$
$$\wedge \neg(AB_1 \wedge AB_2 \wedge \neg A_2 \wedge AB_1 \wedge AB_2 \wedge \neg B_2)$$

$$\vdots$$

$$\wedge \neg(AB_1 \wedge \ldots \wedge AB_{n-1} \wedge \neg A_n \wedge AB_1 \wedge \ldots \wedge AB_{n-1} \wedge \neg B_n)$$
$$\wedge \neg AB_n)$$

$$\overset{2(n-1) \times \text{dc.}}{\vdash} \quad \neg(AB_1$$
$$\wedge \neg(\neg\neg(AB_1 \wedge \neg A_2) \wedge \neg\neg(AB_1 \wedge \neg B_2))$$
$$\wedge \neg(\neg\neg(AB_1 \wedge AB_2 \wedge \neg A_2) \wedge \neg\neg(AB_1 \wedge AB_2 \wedge \neg B_2))$$

$$\vdots$$

$$\wedge \neg(\neg\neg(AB_1 \wedge \ldots \wedge AB_{n-1} \wedge \neg A_n) \wedge \neg\neg(AB_1 \wedge \ldots \wedge AB_{n-1} \wedge \neg B_n))$$
$$\wedge \neg AB_n)$$

$$= \quad \neg(AB_1$$
$$\wedge((AB_1 \to A_2) \vee (AB_1 \to B_2))$$
$$\wedge((AB_1 \wedge AB_2 \to A_2) \vee (AB_1 \wedge AB_2 \to B_2))$$

$$\vdots$$

$$\wedge((AB_1 \wedge \ldots \wedge AB_{n-1} \to A_n) \vee (AB_1 \wedge \ldots \wedge AB_{n-1} \to B_n))$$
$$\wedge \neg AB_n)$$

$$= \quad k_n$$

So we need $1 + 1 + 2(1 + 2 + \ldots + (n-1)) + 2(n-1) = 2(1 + \ldots + n) = n(n+1)$ steps to derive f_n.   □

6   Proofs of Linear Length

In [BZ93], Baaz and Zach show that adding the scheme of equivalence (Eq), i.e.,

$$(f \leftrightarrow g) \to (h[f] \leftrightarrow h[g]) \tag{Eq}$$

to an arbitrary hilbert-style calculus H for propositional logic allows to find proofs of linear length, depending on the number of propositional variables in the formula. More precisely, if T_n is the set of all tautologies in up to n propositional variables, they show that there exists a linear function ϕ such that for all n and all $A \in T_n$ it satisfies H+EQ $\vdash^{\phi(n)} A$.

In this section, by adapting the proof of [BZ93] for our system, it will be shown that we can find a similar approximation for prooflengths. In contrast to [BZ93], it is not needed to add new rules or axioms to our calculus. Instead, it suffices to generalize the rules iteration, deiteration and double cut in a natural manner.

Recall the definition of the iteration rule: $f[g \wedge h[i]] \vdash f[g \wedge h[g \wedge i]]$. If f is a formula with a subformula g, then each subformula i such that f has the form $f[g \wedge h[i]]$ is said to be RECEIVABLE FOR THE ITERATION OF g. We now generalize the rules of the calculus. This calculus will be denoted by $\Vdash$.

gen. Iteration: If $f[g]$ is a formula, then it is allowed to add to each context i which is receivable for the iteration of g an arbitrary number of copies of g.

gen. Deiteration: Inverse direction of deiteration.

gen. Double Cut i): An arbitrary number of double negations may be removed from a formula.

gen. Double Cut ii): An arbitrary number of double negations may be added to a formula.

Some simple examples shall illustrate the rules. Consider the following proof, where in each step, the outermost subformula $A \wedge \neg B$ is iterated (one time into the outermost context, two times into the context of $D \wedge F$). In this derivation, the iterated copies of the subformula are underlined.

$$A \wedge \neg B \wedge C \wedge \neg (D \wedge F) \overset{\text{it.}}{\vdash} A \wedge \neg B \wedge \underline{A \wedge \neg B} \wedge C \wedge \neg (D \wedge F)$$

$$\overset{\text{it.}}{\vdash} A \wedge \neg B \wedge A \wedge \neg B \wedge C \wedge \neg (\underline{A \wedge \neg B} \wedge D \wedge F)$$

$$\overset{\text{it.}}{\vdash} A \wedge \neg B \wedge A \wedge \neg B \wedge C \wedge \neg (A \wedge \neg B \wedge \underline{A \wedge \neg B} \wedge D \wedge F)$$

This derivation is now consolidated to *one* application of the generalized iteration rule. But a 'nested' application of the iteration-rule is not considered as generalized iteration rule, i.e., although we have

$$A \wedge \neg B \wedge C \wedge \neg (D \wedge F) \overset{\text{it.}}{\vdash} A \wedge \neg B \wedge \underline{A \wedge \neg B} \wedge C \wedge \neg (D \wedge F)$$

$$\overset{\text{it.}}{\vdash} A \wedge \neg B \wedge A \wedge \neg (\underline{A \wedge \neg B} \wedge B) \wedge C \wedge \neg (D \wedge F)$$

the last formula is *not* obtained from the first formula with a single application of the application of the generalized iteration rule, as in the second step, the subformula $A \wedge \neg B$ is iterated into a context which was not created until the first step, i.e., into a context which does not exist in the starting formula.

The generalized double cut rule is easier to understand.

$$A \wedge \neg B \wedge C \wedge \neg (D \wedge F) \overset{\text{gen. dc.}}{\Vdash} A \wedge \neg B \wedge \underline{\neg\neg}(C \wedge \underline{\neg\neg\neg}(\underline{\neg\neg}D \wedge F))$$

We can now prove that with $\Vdash$ we can find derivations of tautologies whose length depend linearly from the number of the propositional variables in the tautology.

Theorem 3 (Proofs of linear length in the generalized calculus). *If f is a tautology with n different propositional variables, we have $\vdash^{24+14n} f$.*

Proof: The proof is done by induction over n.

So, for the induction start, let f be a tautology without propositional variables. For $f \not\sim \top$ and $f \not\sim \neg\top$, f contains $\neg\neg\top$ or $\neg\top \wedge \neg\top$ as subformula. We can successively replace subformulas $\neg\neg\top$ by $\top$ (with the double cut rule) and subformulas $\neg\top \wedge \neg\top$ by $\neg\top$ (by deiterating one occurrence of $\neg\top$). As both rules are equivalence rules, it is easy to see that f is a tautology if and only if this procedure eventually yields $\top$.

This idea is captured by the ongoing proof, which is based on Yukami's trick ([Yuk84]). In the formal derivation of f we have to construct, the manifold replacements of $\neg\neg\top$ by $\top$ of the double cut rule will be performed in one step by an application of the generalized double cut rule. But the manifold replacements of $\neg\top \wedge \neg\top$ by $\neg\top$ cannot be analogously be captured by one application of the generalized deiteration rule, as in the different applications of the deiteration rule take place in different contexts (i.e., different occurrences of $\neg\top$ are used for deiterating other occurrences of $\neg\top$). To overcome with this problem, instead of replacing $\neg\top \wedge \neg\top$ directly by $\neg\top$, we first replace each occurrence $\neg\top \wedge \neg\top$ by $\neg\neg(\neg\top \wedge \neg\top)$ with the generalized double cut rule. Then all occurrences of $\neg(\neg\top \wedge \neg\top)$ are replaced by $\top$ with the generalized deiteration rule, using a subformula $\neg(\neg\top \wedge \neg\top)$ in the uppermost context.

In order to construct the formal derivation, we first define a mapping $\Delta(f)$, which formalizes the three different modifications of formulas as follows:

1. If f contains a double negation $\neg\neg\top$ as subformula, then $\Delta(f)$ is obtained from f by removing the double negation, i.e.: For $f[\neg\neg\top]$ we set

$$\Delta(f[\neg\neg\top]) := f[\top] \quad .$$

2. If f contains $(\neg\top \wedge \neg\top)$ as subformula, then $\Delta(f)$ is obtained from f by replacing this subformula by $\neg\neg(\neg\top \wedge \neg\top)$, i.e.: For $f[\neg\top \wedge \neg\top]$ we set

$$\Delta(f[\neg\top \wedge \neg\top]) := f[\neg\neg(\neg\top \wedge \neg\top)] \quad .$$

3. If f contains $\neg(\neg\top \wedge \neg\top)$ as subformula, then $\Delta(f)$ is obtained from f by removing this subformula, i.e.: For $f[\neg(\neg\top \wedge \neg\top)]$ we set

$$\Delta(f[\neg(\neg\top \wedge \neg\top)]) := f[\top] \quad .$$

Due to the discussion at the beginning of this proof, we know that f is a tautology if and only if there is an n such that $\Delta^n(f) = \top$.

Now let f be a tautology and $n \in \mathbb{N}$ with $\Delta^n(f) = \top$. Let

$$f_d^{-1} := \Delta f \leftrightarrow (\Delta^2 f \leftrightarrow (\Delta^3 f \leftrightarrow \ldots (\Delta^{n-1} f \leftrightarrow \top) \ldots)) \quad ,$$

$$f_d := f \leftrightarrow (\Delta f \leftrightarrow (\Delta^2 f \leftrightarrow \ldots (\Delta^{n-1} f \leftrightarrow \top) \ldots))$$

$$= f \leftrightarrow (f_d^{-1}) \qquad \text{and}$$

$$f_d^\Delta := \Delta f \leftrightarrow (\Delta^2 f \leftrightarrow (\Delta^3 f \leftrightarrow \ldots (\Delta^n f \leftrightarrow \top) \ldots)$$
$$= \Delta f \leftrightarrow (\Delta^2 f \leftrightarrow (\Delta^3 f \leftrightarrow \ldots (\top \leftrightarrow \top) \ldots)$$

Now we can derive f from $\top$. We start with the construction of $\neg(\neg\top \wedge \neg\top)$, and we derive $f_d \leftrightarrow f_d$ as well.

$$
\begin{array}{ll}
\top \quad \overset{\text{gen. dc}}{\Vdash} & \neg\neg\top \wedge \neg\neg\top \\[4pt]
\overset{\text{it.}}{\Vdash} & \neg(\neg\top \wedge \neg\top) \wedge \neg\neg\top \\[4pt]
\overset{\text{ins.}}{\Vdash} & \neg(\neg\top \wedge \neg\top) \wedge \neg(f_d \wedge \neg\top) \\[4pt]
\overset{\text{it.}}{\Vdash} & \neg(\neg\top \wedge \neg\top) \wedge \neg(f_d \wedge \neg f_d) \\[4pt]
\overset{\text{it.}}{\Vdash} & \neg(\neg\top \wedge \neg\top) \wedge \neg(f_d \wedge \neg f_d) \wedge \neg(f_d \wedge \neg f_d) \\[4pt]
= & \neg(\neg\top \wedge \neg\top) \wedge (f_d \leftrightarrow f_d) \\[4pt]
= & \neg(\neg\top \wedge \neg\top) \wedge ((f \leftrightarrow (f_d^{-1})) \leftrightarrow f_d) \\[4pt]
\Vdash^3 & \neg(\neg\top \wedge \neg\top) \wedge ((f \leftrightarrow (f_d^{-1})) \leftrightarrow f_d^\Delta)
\end{array}
$$

The last step reflects the discussion at the beginning of the proof. It is carried out each with one application of:

1. the generalized double cut insertion rule
2. the generalized double cut erasure rule
3. the generalized deiteration rule

The formulas f_d^{-1} and f_d^Δ differ only in the innermost formula, which is $\top \leftrightarrow \top$ for f_d^Δ and $\top$ for f_d^{-1}. We have

$$\top \leftrightarrow \top \ = \ \neg(\top \wedge \neg\top) \wedge \neg(\top \wedge \neg\top) \ \sim \ \neg\neg\top \wedge \neg\neg\top$$

Thus the most inner formula $\top \leftrightarrow \top$ of f_d^Δ can be replaced with the generalized double cut rule by $\top$. That is, we get:

$$\neg(\neg\top \wedge \neg\top) \wedge ((f \leftrightarrow (f_d^{-1})) \leftrightarrow f_d^\Delta) \overset{\text{gen. dc.}}{\Vdash} \neg(\neg\top \wedge \neg\top) \wedge ((f \leftrightarrow f_d^{-1}) \leftrightarrow f_d^{-1})$$
$$\overset{\text{era}}{\Vdash} (f \leftrightarrow f_d^{-1}) \leftrightarrow f_d^{-1}$$

According to Lem. 5, we can derive f from this formula within 14 steps. As we needed 10 steps so far, we see that f can be derived with a total number of 24 steps from $\top$. This finishes the induction start.

Assume now we have shown that the Lemma holds for formulas with at most n propositional variables. Now let f be a tautology with $n + 1$ propositional variables, let A be one of these variables. As we have

$$\models f \ \Leftrightarrow \ \models f[\top/A] \wedge f[\neg\top/A] \ ,$$

there exists a formal derivation of $f[\top/A] \wedge f[\neg\top/A]$ with length $24+14n$. From this formula, we proceed as follows:

$$f[\top/A] \wedge f[\neg\top/A]$$

$\overset{\text{dc.}}{\Vdash} \quad \neg\neg\top \wedge f[\top/A] \wedge f[\neg\top/A]$

$\overset{\text{ins.}}{\Vdash} \quad \neg(\underline{A} \wedge \neg\top) \wedge f[\top/A] \wedge f[\neg\top/A]$

$\overset{\text{it. of } A}{\Vdash} \quad \neg(\neg A \wedge A) \wedge f[\top/A] \wedge f[\neg\top/A]$

$\overset{\text{dc.}}{\Vdash} \quad \neg(\neg A \wedge \neg\neg A) \wedge \underline{f[\top/A]} \wedge f[\neg\top/A]$

$\overset{\text{it. of } f[\top/A]}{\Vdash} \quad \neg(\neg(A \wedge f[\top/A]) \wedge \neg\neg A) \wedge f[\top/A] \wedge \underline{f[\neg\top/A]}$

$\overset{\text{it. of } f[\neg\top/A]}{\Vdash} \quad \neg(\neg(A \wedge f[\top/A]) \wedge \neg(\neg A \wedge f[\neg\top/A])) \wedge \underline{f[\top/A] \wedge f[\neg\top/A]}$

$\overset{\text{era.}}{\Vdash} \quad \neg(\neg(A \wedge f[\top/A]) \wedge \neg(\neg A \wedge f[\neg\top/A]))$

$\overset{\text{gen. it. of } A}{\Vdash} \quad \neg(\neg(A \wedge f[A/A]) \wedge \neg(\neg A \wedge f[\neg\top/A]))$

$\overset{\text{gen. it. of } \neg A}{\Vdash} \quad \neg(\neg(A \wedge f[A/A]) \wedge \neg(\neg A \wedge f[\neg\neg A/A]))$

$\overset{\text{gen. dc.}}{\Vdash} \quad \neg(\neg(A \wedge f[A/A]) \wedge \neg(\neg A \wedge f[A/A]))$

$\overset{=}{} \quad \neg(\neg(A \wedge f) \wedge \neg(\neg A \wedge f))$

$\overset{\text{era.}}{\Vdash} \quad \neg(\neg f \wedge \neg(\neg A \wedge f))$

$\overset{\text{era.}}{\Vdash} \quad \neg(\neg f \wedge \neg f)$

$\overset{\text{deit.}}{\Vdash} \quad \neg\neg f$

$\overset{\text{dc.}}{\Vdash} \quad f$

As we needed 14 further steps, we obtain $\top \Vdash^{24+14(n+1)} f$, thus we are done. $\square$

7   Further Research

This paper is a first step to the proof-theoretic foundations of Peirce's calculus for Alpha graphs. The calculus has powerful rules, and it has to be investigated whether the results of this paper can be improved. Firstly, it is natural to ask whether the deiteration rule is admissible as well. Kocura uses in [HK05] a system consisting of the rules insertion, iteration, and double cut, but a proof whether this system is complete is still missing. Secondly, one might ask whether the results of the last section hold for the non-generalized calculus as well. I strongly suspect that this is not the case. Consider the formula $f := \neg\neg\top \wedge \ldots \wedge \neg\neg\top$ consisting of 2^n subformulas $\neg\neg\top$. Then f can can be derived with $\vdash$ within $n+1$ steps as follows: First insert a double cut, then in each step, iterate the whole formula derived so far. It is likely that this is the optimal derivation of f, but so far, I did not succeed in proving that.

Besides these two questions, the results of the paper show that Peirce's calculus may be of interest for automated theorem proving, thus it should be investigated further from a proof-theoretic point of view.

References

[Brü03]   Kai Brünnler. *Deep Inference and Symmetry in Classical Proofs*. PhD thesis, Technische Universität Dresden, 2003.

[Bur91]   Robert W. Burch. *A Peircean Reduction Thesis: The Foundation of Topological Logic*. Texas Tech. University Press, Texas, Lubbock, 1991.

[BZ93]   Matthias Baaz and Richard Zach. Short proofs of tautologies using the schema of equivalence. In Egon Börger, Yuri Gurevich, and Karl Meinke, editors, *CSL*, volume 832 of *Lecture Notes in Computer Science*, pages 33–35. Springer, Berlin – Heidelberg – New York, 1993.

[Dau02]   Frithjof Dau. An embedding of existential graphs into concept graphs with negations. In Uta Priss, Dan Corbett, and Galia Angelova, editors, *ICCS*, volume 2393 of *LNAI*, pages 326–340, Borovets, Bulgaria, July, 15–19, 2002. Springer, Berlin – Heidelberg – New York.

[Dau04]   Frithjof Dau. Types and tokens for logic with diagrams: A mathematical approach. In Karl Erich Wolff, Heather D. Pfeiffer, and Harry S. Delugach, editors, *Conceptual Structures at Work: 12th International Conference on Conceptual Structures*, volume 3127 of *Lecture Notes in Computer Science*, pages 62–93. Springer, Berlin – Heidelberg – New York, 2004.

[Dau06]   Frithjof Dau. Mathematical logic with diagrams, based on the existential graphs of peirce. Habilitation thesis. To be published. Available at: http://www.dr-dau.net, 2006.

[DMS05]   Frithjof Dau, Marie-Laure Mugnier, and Gerd Stumme, editors. *Common Semantics for Sharing Knowledge: Contributions to ICCS 2005*, Kassel, Germany, July, 2005. Kassel University Press.

[Ham95]   Eric M. Hammer. *Logic and Visual Information*. CSLI Publications, Stanford, California, 1995.

[HB35]   Weiss Hartshorne and Burks, editors. *Collected Papers of Charles Sanders Peirce*, Cambridge, Massachusetts, 1931–1935. Harvard University Press.

[HK05]   David P. Hodgins and Pavel Kocura. Propositional theorem prover for peirce-logik. In Dau et al. [DMS05], pages 203–204.

[Liu05]   Xin-Wen Liu. An axiomatic system for peirce's alpha graphs. In Dau et al. [DMS05], pages 122–131.

[Pap83]   Helmut Pape. *Charles S. Peirce: Phänomen und Logik der Zeichen*. Suhrkamp Verlag Wissenschaft, Frankfurt am Main, Germany, 1983. German translation of Peirce's Syllabus of Certain Topics of Logic.

[Pei35]   Charles Sanders Peirce. *MS 478: Existential Graphs*. Harvard University Press, 1931–1935. Partly published in of [HB35] (4.394-417). Complete german translation in [Pap83].

[Pei92]   Charles Sanders Peirce. Reasoning and the logic of things. In K. L. Kremer and H. Putnam, editors, *The Cambridge Conferences Lectures of 1898*. Harvard Univ. Press, Cambridge, 1992.

[PS00]   Charles Sanders Peirce and John F. Sowa. Existential Graphs: MS 514 by
           Charles Sanders Peirce with commentary by John Sowa, 1908, 2000. Avail-
           able at: http://www.jfsowa.com/peirce/ms514.htm.

[Rob73]   Don D. Roberts. *The Existential Graphs of Charles S. Peirce.* Mouton, The
           Hague, Paris, 1973.

[Rob92]   Don D. Roberts. The existential graphs. *Computers Math. Appl..*, 23
           (6–9):639–63, 1992.

[Sch60]   Kurt Schütte. *Beweistheorie.* Springer, Berlin – Heidelberg – New York,
           1960.

[Shi02a]   Sun-Joo Shin. *The Iconic Logic of Peirce's Graphs.* Bradford Book, Mas-
           sachusetts, 2002.

[Shi02b]   Sun-Joo Shin. Multiple readings in peirce's alpha graphs. In Michael Ander-
           son, Bernd Meyer, and Patrick Olivier, editors, *Diagrammatic Representation
           and Reasoning.* Springer, Berlin – Heidelberg – New York, 2002.

[Sow84]   John F. Sowa. *Conceptual structures: information processing in mind and
           machine.* Addison-Wesley, Reading, Mass., 1984.

[Sow97]   John F. Sowa. Logic: Graphical and algebraic. manuscript, Croton-on-
           Hudson, 1997.

[Sta78]   Richard Statman. Bounds for proof-search and speed-up in predicate calcu-
           lus. *Annals of Mathematical Logic*, 15:225–287, 1978.

[vH03]   Bram van Heuveln. Existential graphs. Presentations and Applications at:
           http://www.rpi.edu/ heuveb/research/EG/eg.html, 2003.

[Yuk84]   Tsuyoshi Yukami. Some results on speed-up. *Ann. Japan Assoc. Philos. Sci.*,
           6:195–205, 1984.

[Zem64]   Jay J Zeman. *The Graphical Logic of C. S. Peirce.* PhD thesis, University
           of Chicago, 1964. Available at: http://www.clas.ufl.edu/users/jzeman/.

DOGMA-MESS: A Meaning Evolution Support System for Interorganizational Ontology Engineering

Aldo de Moor, Pieter De Leenheer, and Robert Meersman*

VUB STARLab
Semantics Technology and Applications Research Laboratory
Vrije Universiteit Brussel
Pleinlaan 2 B-1050 Brussels, Belgium
{ademoor, pdeleenh, meersman}@vub.ac.be

Abstract. In this paper, we explore the process of interorganizational ontology engineering. Scalable ontology engineering is hard to do in interorganizational settings where there are many pre-existing organizational ontologies and rapidly changing collaborative requirements. A complex socio-technical process of ontology alignment and meaning negotiation is therefore required. In particular, we are interested in how to increase the efficiency and relevance of this process using context dependencies between ontological elements. We describe the DOGMA-MESS methodology and system for scalable, community-grounded ontology engineering. We illustrate this methodology with examples taken from a case of interorganizational competency ontology evolution in the vocational training domain.

1 Introduction

In collaborative communities, people sharing goals and interests work together for a prolonged period of time. For collaboration to be successful, conceptual common ground needs to be developed. Ontologies are instrumental in this process by providing formal specifications of shared semantics. Such formal semantics are a solid basis for the development of useful collaborative services and systems. However, scalable ontology engineering is hard to do in interorganizational settings where there are many pre-existing organizational ontologies and ill-defined, rapidly evolving collaborative requirements. A complex socio-technical process of ontology alignment and meaning negotiation is therefore required. Much valuable work has been done in the Semantic Web community on the formal aspects of ontology representation and reasoning. However, the socio-technical aspects of the ontology engineering process in complex and dynamic realistic settings are still little understood. A viable methodology requires not building a single, monolithic domain ontology by a knowledge engineer, but supporting domain experts in gradually building a sequence of increasingly complex versions

* The research described in this paper was partially sponsored by EU Leonardo da Vinci CO-DRIVE project B/04/B/F/PP-144.339 and the DIP EU-FP6 507483 project. The authors wish to thank Stijn Christiaens and Ruben Verlinden for their aid in the development of the methodology and system; Luk Vervenne, Roy Ackema, and Hans Wentink for their testing of the prototype system in the CODRIVE project; and Ulrik Petersen for his fast and useful updates of the Prolog+CG tool.

H. Schärfe, P. Hitzler, and P. Øhrstrøm (Eds.): ICCS 2006, LNAI 4068, pp. 189–202, 2006.
© Springer-Verlag Berlin Heidelberg 2006

of interrelated ontologies over time. Contexts are necessary to formalize and reason about the structure, interdependencies and versioning of these ontologies, thus keeping their complexity manageable. In Sect. 2, we describe our view on interorganizational ontology engineering. Sect. 3 introduces the DOGMA-MESS methodology for scalable community-grounded ontology engineering. In Sect. 4, we describe the approach to organizational ontology alignment taken in DOGMA-MESS, focusing on the relevance of organizational definitions. We end the paper with discussion and conclusions.

2   Interorganizational Ontology Engineering

Many definitions of ontologies exist. The classical definition is that an ontology is an explicit specification of a conceptualization [12]. Other definitions, such as that an ontology is a shared and common understanding of a domain that can be communicated across people and application systems [7], stress more the community and application side of ontologies. However, the problem is not in what ontologies are, but how they *become* common formal specifications of a domain useful for building computerized services. Many open issues remain with respect to how these ontologies are to be *efficiently* engineered in communities of practice. This is all the more true in inter-organizational ontology building, where there are multiple, existing organizational ontologies that need to be aligned. In such settings, common domain ontologies need to be developed that adequately capture *relevant* interorganizational commonalities and differences in meaning. Such multiple, continuously shifting sources of meaning make knowledge sharing very difficult [8]. This is all the more true since in interorganizational settings, organizational ontologies cannot easily be merged, as they represent strong individual interests and entrenched work practices of the various participants. This means that such value-laden ontologies can only be defined in a careful and gradual process of meaning negotiation [5]. This we define as community-grounded processes for reaching the *appropriate* amount of consensus on *relevant* conceptual definitions.

Promising related work on reaching consensus on ontologies in a distributed environment has focused on architectures for consensual knowledge bases (e.g. [9]) and the cooperative construction of domain ontologies (e.g. [1]). Still, although these approaches work out basic principles for cooperative ontology engineering, they do not provide community-grounded methodologies addressing the issues of relevance and efficiency of definition processes. In interorganizational settings, however, these quality issues are of the greatest importance for ontology engineering processes to scale and be useful in daily practice. The basic question therefore is: how to develop a scalable approach to interorganizational ontology engineering? A crucial formal issue underlying such an approach is that multiple types of context dependencies need to be handled between ontological elements. Whereas much work in knowledge engineering looks at formal properties of contexts and their dependencies, in this paper we focus on how such formal approaches to handling context dependencies can be *applied* in interorganizational ontology engineering processes to increase relevance and efficiency of engineering processes. Our aim is not to be exhaustive, but to show that a systematic analysis of such context dependencies and their use in interorganizational ontology engineering processes can help optimize this very complex socio-technical process. We focus on one very important type: specialization dependencies, which play a major role in fostering both the efficiency and relevance of interorganizational ontology engineering processes.

2.1 A Model of Interorganizational Ontology Engineering

We now present a generic model for understanding interorganizational knowledge engineering.

In the model, we make the following assumptions:

- An interorganizational ontology needs to be modeled not by external knowledge engineers, but by domain experts themselves. Only they have the tacit knowledge about the domain and can sufficiently assess the real impact of the conceptualizations and derived collaborative services on their organization.
- The common interest only partially overlaps with the individual organizational interests. This means that the goal is not to produce a single common ontology, but to support organizations in interpreting common conceptualizations in their own terms, and feeding back these results. A continuous alignment of common and organizational ontologies is therefore required.
- An interorganizational ontology cannot be produced in one session, but needs to evolve over time. Due to its complexity, different versions are needed.
- Starting point for each version should be the current insight about the common interest, i.c common conceptual definitions relevant for the collaborative services for which the interorganizational ontology is going to be used.
- The end result of each version should be a careful balance of this *proposal* for a common ontology with the various *individual interpretations* represented in the organizational ontologies.

Fig. 1 shows how an interorganizational ontology (IOO) consists of various, related sub-ontologies. The engineering process starts with the creation of an upper common ontology (UCO), which contains the conceptualizations and semantic constraints that are common to and accepted by a domain. Each participating organization specializes

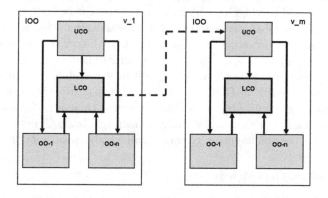

Fig. 1. A Model of Interorganizational Ontology Engineering

this ontology into its own Organizational Ontology (OO), thus resulting in a local interpretation of the commonly accepted knowledge. In the Lower Common Ontology (LCO), a new proposal for the next version of the IOO is produced, aligning relevant material from the UCO and various OOs. The part of the LCO that is accepted by the community then forms the legitimate UCO for the next version of the IOO.

Ontology engineering involves a number of knowledge engineering processes. Many, partially overlapping, classifications of these processes have been developed so far. Our intention with the model is not to add to these processes themselves, but to *position* them, indicating how they can be used in the bigger picture of interorganizational ontology engineering. Of course, many mappings of these processes are conceivable. In this paper, we connect only an initial, coarse-grained mapping of standard ontology engineering processes to the model. In future research, we will refine both the model and mappings of associated engineering processes.

This conceptual model of the interorganizational ontology engineering process is sufficiently specific to derive and organize practical methodological guidelines, yet generic enough to represent and compare many different approaches and techniques from an application point of view. This will help identify gaps in theory and methodologies, providing a conceptual lens to focus scattered research on a very confusing topic. In the next section, we show how this model underlies the development of STAR-Lab's own DOGMA-MESS methodology.

3   DOGMA-MESS

The DOGMA (Designing Ontology-Grounded Methods and Applications) approach to ontology engineering, developed at VUB STARLab, aims to satisfy real-world needs by developing a useful and scalable ontology engineering approach [17]. Its philosophy is based on a double articulation: an ontology consists of an ontology base of lexons, which holds (multiple) intuitive conceptualizations of a domain, and a layer of reified ontological commitments. These essentially are views and constraints that within a given context allow an application to commit to the selected lexons. Contexts group commitments, allowing ontological patterns to be represented and compared at various levels of granularity [2]. In this way, scalable ontological solutions for eliciting and applying complex and overlapping collaboration patterns can be built.

A fundamental DOGMA characteristic is its grounding in the linguistic representation of knowledge. This is exemplified most clearly in the linguistic nature of the lexons, with terms and role strings chosen from a given (natural) language, and that constitute the basis for all interfaces to the ontology. Linguistic "grounding" is achieved through *elicitation contexts*, which in DOGMA are just mappings from identifiers to source documents such as generalized glosses, often in natural language. As this paper however is focusing on the process architecture of interorganizational ontology building, the detailed aspects of this linguistic grounding fall mostly outside of our scope.

3.1   Outline of the Methodology

The efficiency and relevance of eliciting and applying ontological knowledge is at the heart of the DOGMA methodology. However, still undeveloped was the layer in which

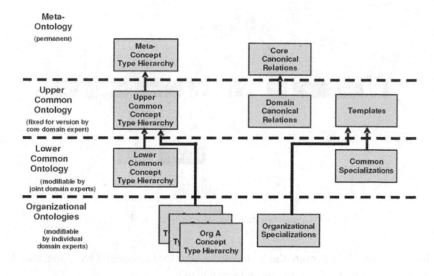

Fig. 2. Interorganizational Ontology Engineering in DOGMA-MESS

the DOGMA ontology engineering processes are grounded in communities of use. This is the purpose of the DOGMA-MESS methodology.

Based on our model of interorganizational ontology engineering, the approach adopted in DOGMA-MESS is characterized in Fig. 2. Arrows in the diagram indicate specialization dependencies between ontologies. Each version of the IOO construction consists of three stages: (1) creation of the templates; (2) definition of the organizational specializations (divergence of definitions); and (3) definition of the common specializations (convergence of definitions). After that, the relevant knowledge to be retained is moved to the first stage of the next cycle (still under design). Some important properties of the intra-version processes are:

– A (permanent) *Meta-Ontology* is the same for all applications of DOGMA-MESS and only contains stable concept types like 'Actor', 'Object', 'Process', and 'Quality'. Three main types of Actors are defined: *Core Domain Experts* represent the common interest, *Domain Experts* represent the various organizational interests, and *Knowledge Engineers* help the other experts define and analyze the various ontologies. The Meta-Ontology also contains a set of core canonical relations, similar to the ones described in [16], such as the 'Agent', 'Object', and 'Result'-relations.
– Each domain has its own *Upper Common Ontology*, and is maintained by the core domain expert. It first of all contains a specialization of the concept type hierarchy of the Meta-Ontology. This *Upper Common Concept Type Hierarchy* organizes the (evolving) concept types common to the domain. Domain canonical relations specialize core canonical relations in terms of the domain. For instance, whereas 'Agent' is a core canonical relation, in a particular domain this may be translated into 'Person'. The most important type of construct in the UCO are the *Templates*. A template describes a common knowledge definition most relevant to the common interest. At the beginning of each new version, the core domain expert defines

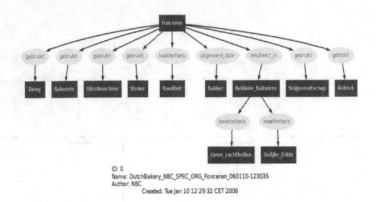

Fig. 3. An organizational specialization of a task template

templates that best capture the focus interests of that moment. Over time, templates should become more numerous and specialized.

- Templates are specialized into *Organizational Specializations* by the domain experts representing the various organizations. To this purpose, domain experts can add concept types that are particular to their organization. These *Organizational Concept Type Hierarchies* themselves need to be a specialization of the Upper Common Concept Type Hierarchy. Fig. 3 gives an example from the CODRIVE project (see below) of a definition by one of the partners in the Dutch Bakery case having specialized the task template for the task 'Panning' (='Fonceren' in Dutch).
- The most important layer for meaning negotiation is the *Lower Common Ontology*. This is where the 'specification agenda' as represented by the UCO and the, often widely differing, organizational interpretations need to be aligned and the most relevant conceptualizations for the next version need to be selected. This process is far from trivial. In the current implementation of DOGMA-MESS, there is only a very simple rule: all (selected) definitions need to be full specializations of the templates, hence they are called Common Specializations. Likewise, the Lower Common Concept Type Hierarchy needs to be a specialization of the Upper Common Concept Type Hierarchy. This, however, is overly simplified. In the meaning negotiation process, new definitions may be created that are not (complete) specializations, but represent a new category of template for the next version of the IOO, for example. This is where many of the DOGMA existing ontology analysis processes, for example based on lexon and commitment comparison, as well as ORM constraint analysis may play a role. At any rate, our framework allows for such methodology evolution to be clearly described. In Sect. 4.3, we give an example of one of the LCO processes currently being developed: organizational ontology alignment.

3.2   System Implementation

The system supporting the DOGMA-MESS methodology is being implemented as a web server that can be accessed by any web browser, thus ensuring maximum accessibility and ease-of-use (Fig. 4).

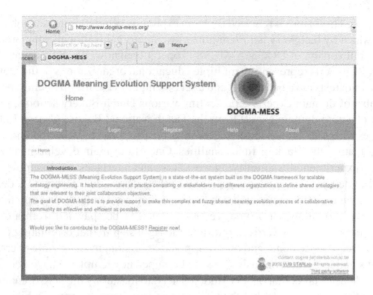

Fig. 4. The DOGMA-MESS system

The core of the server is a Java server that interacts with the DOGMA Studio server, which in turn implements the standard DOGMA ontology engineering and analysis engine. Special converters translate standard DOGMA representations to and from a simple native DOGMA-CG format (amongst other things limited to simple referents). CharGer is one of the tools allowing core domain experts to create templates in CharGer-XML, which can also be converted into DOGMA-CG. Concept type hierarchies can be imported as indented text files, easily to be edited with simple ASCII-editors. Type hierarchies and templates, like organizational specializations, can also be edited through the DOGMA-MESS web interface. This interface, for example, allows concept and relation types to be defined and graph concepts and relations to be added, changed, or removed. Prolog+CG has been embedded in DOGMA-MESS as the conceptual graph inference engine. The main operation currently being used is the projection operation for checking whether organizational specializations conform to their templates. To visualize stored and inferred graphs, AT&T's GraphViz engine is being used.

3.3   Initial User Experiences

The explicit mission of DOGMA-MESS is to increase the efficiency and relevance of the interorganizational ontology engineering process. A continuous process of testbed-like development of methodology and system is therefore essential. Its socio-technical fine-tuning is being done in the CODRIVE project. This project aims to develop a methodology for shared competency ontology definition between organizations representing the educational sector, industry, and public employment agencies. Common competency models are essential for the interoperability of the information systems of these organizations. This in turn is required for them to provide better training and job matching services, which by their very nature require close interorganizational collaboration.

A first full version of DOGMA-MESS is operational now. We are currently experimenting with experts representing multiple educational organizations in the Dutch bakery sector. Pilot tests have been done, and a set of templates is currently being specialized by a number of domain experts representing various Dutch Bakery schools, under supervision of a core domain expert from the Dutch National Bakery Centre. Initial user experiences with tool and system are generally positive. The most serious limitation is complexity rather than lack of functionalities. One of our main development activities therefore is experimentally simplifying workflows and interfaces. For example, initially users had to define their concept types before they were going to create a definition. However, often they find out which concept type they need only *during* definition construction. We now allow for a type creation process to be spawned from a definition creation process, and afterwards to return to the right step in that originating process.

After a face-to-face demo and one hour practice session, most are able to create specializations online asynchronously. Task and competency templates have been defined for the hundreds of tasks that bakery students need to master by the core domain expert. A task template, for example, has the task as the *focus concept*, around which relations need to be specialized such as who is the person doing the task, what resources, materials, equipment, and tools are needed as inputs, what is the resulting output, and what quality aspects are required for task components.

The templates are currently being specialized by these experts into organizational specializations (see Fig. 3 for an example of such a specialization). With most start-up problems having been addressed, the initial knowledge base of dozens of definitions is now growing into hundreds of definitions. Ultimately, thousands, tens of thousands, or even more definitions will be obtained in a typical domain. Given that time of these experts is very expensive, they should only focus on the most relevant definitions at any moment in time. How to reach this definition convergence after the divergence of the organizational specialization process is the aim of the third stage of each ontology version. Ontology alignment is a key issue here.

4   Organizational Ontology Alignment in DOGMA-MESS

Ontology alignment means making -literally- common sense out of a set of individual ontologies. This is the most difficult ontology engineering process, with the most of degrees of freedom for operationalization. Formalization of the ontology engineering process is required to ensure the quality of the design of this process.

4.1   Characterizing Ontology Engineering Processes

An important class of ontology engineering processes concerns *ontology integration*. This process has been studied extensively in the literature (for a state-of-the-art survey, cf. [10,14]). Although different groups vary in their exact definition, ontology integration is generally considered to consist of four key subprocesses: the *articulation*, *mapping*, *alignment*, and *merging* of ontologies. Ontology articulation deals with the problem of how to define conceptualizations. Ontology mapping concerns how to link elements from different ontologies, alignment how meanings common to different ontologies can be identified, and merging focuses on how to generate completely new ontologies out of existing ones. Interorganizational ontology engineering to a large extent focuses on the first three subprocesses, merging being of less relevance.

All ontology integration methodologies use some combination of these *macro-ontology engineering processes*. However, in their operational implementation of these processes, which we call *micro-ontology engineering processes*, methodologies differ widely. We use the following (non-exhaustive) set of micro-process primitives: the creation, modification, termination, linking, and selection of ontological definitions. The *creation* of a definition generates a new definition, often from a template. A *modification* changes concepts and relations in an existing definition. A *termination* removes the definition from the ontology. These operations were described in detail in terms of pre and post conditions in [4]. Still lacking were formal definitions of the micro-processes of explicitly *linking* ontologies and elements from ontologies and *selecting* relevant definitions from a (growing) ontology. Many operationalizations of these processes have already been developed in the field. The question is how to apply and (re)combine them to increase the quality of real-world ontology engineering processes.

Contexts are important building blocks in our decomposition and linking of ontology engineering processes [2]. We define a context of an entity as the set of circumstances surrounding it. *Context dependencies* constrain the possible relations between the entity and its context. Many different types of context dependencies exist, within and between ontological elements of various levels of granularity, ranging from individual concepts of definitions to full ontologies. One of the best studied dependencies, which we focus on in this paper, are specialization dependencies. For instance, an organizational definition of a particular task (the entity) can have a specialization dependency with a task template (its context). The constraint in this case is that each organizational definition must be a specialization of the template. In conceptual graphs terms, this would mean that the template must have a projection into the organizational definition. We give an exhaustive analysis of such dependencies in interorganizational ontology engineering in [3]. In Sect. 4.3, we will only illustrate specialization dependencies, by formally describing and decomposing one type of ontology integration (macro)-process: the alignment of organizational ontologies.

4.2 A Formalization of Specialization Dependencies

We formalize the DOGMA-MESS methodology in terms of a set of ontologies and their (specialization) context dependencies. First, we define an ontology as a logical theory:

Ontology. An ontology is defined as a structure $O = \langle S, A \rangle$, where S is the signature and A is a set of ontological axioms. The signature typically consists of a set of concept symbols and relation symbols, the latter denotes relations whose arguments are defined over the concepts.The axiomatization specifies the intended interpretation of the signature. It essentially defines which relation symbol r in S is to be interpreted as subsumption relation. Formally, this requires that r defines a strict partial order (poset). Furthermore, A optionally defines a strict partial order on the relation symbols in S

(thus defining a specialization hierarchy on the relations), and a particular sort or class of axioms (or semantic constraints), depending on the kind of ontology.

Specialization Dependencies. These dependencies, illustrated by arrows in Fig. 2, are an important context dependency in DOGMA-MESS, and are used to connect the various interorganizational ontology entities. Conceptual graph theory is one of the most universal and powerful formalisms for dealing with specializations, with its concept and relation type hierarchies and generalization hierarchies of graphs. We use it here to further define and operationalize our methodology. A conceptual graph can be represented by a logical theory $\langle S, A \rangle$, where the signature consists of concept types, canonical relation types, and a set of ontological definitions (CGs). The axiomatization consists of a concept type hierarchy and a relation type hierarchy defined by partial orders in terms of concept types and relation types in S respectively. Most ontological entities, such as the definitions stored in an ontology (i.e. template, organizational specializations), can be represented as conceptual graphs, and the usual conceptual graph operations can be applied to reason about them.

- An interorganizational ontology IOO contains the following sub-ontologies: a meta-ontology MO, an upper common ontology UCO, a lower common ontology LCO, and a set of organizational ontologies OO_i, one for each member organization.
- Each ontology O contained in IOO consists of a concept type hierarchy CTH and a set of ontological definitions D. Each definition $d \in D$ is a well-formed conceptual graph.
- The meta-ontology MO consists of a meta-concept type hierarchy CTH_M, and an optional set of core canonical relations CR_M. Each relation $cr_m = \langle c_1, r, c_2 \rangle \in CR_M$, with $c_1, c_2 \in CTH_M$, and $r \in CR$, which is a standard set of relation types similar to the canonical relation types described in [16].
- The upper common ontology UCO consists of a upper common concept type hierarchy CTH_{UC}, an optional set of domain canonical relations CR_{UC}, and a non-empty set of templates T. $CTH_{UC} \leq CTH_M$ and $CR_{UC} \leq CR_M$, in the standard CG theory sense.
- The lower common ontology LCO consists of a lower common concept type hierarchy CTH_{LC}, and a set of common specializations D_{LC}. At the start of a version period $D_{LC} = \emptyset$. $CTH_{LC} \leq CTH_{UC}$ and $\forall d_{lc} \in D_{LC} : \exists t \in T \wedge d_{lc} \leq t$.
- Each organizational ontology OO_i consists of an organizational concept type hierarchy CTH_{Oi} and a set D_{Oi} of organizational specializations of templates from the UCO, with $CTH_{Oi} \leq CTH_{UC}$ and $\forall d_{Oi} \in D_{Oi} : \exists t \in T \wedge d_{Oi} \leq t$.
- Each of the constructs defined above is indexed by a version number v. For clarity, this index number is not shown in the definitions.

4.3   Selecting Relevant Organizational Specializations

In this section, we formalize the process of aligning organizational ontologies in DOGMA-MESS by selecting the most relevant organizational specializations as the common specializations (see Fig. 2). Such a process increases the *relevance* of definitions, since the community will focus on those definitions most in line with its (evolving) goals. The rationale is that templates at first will be coarse, as the community is still learning about its goals and interests of its members. Over time (and versions), however, templates can become more focused. This requires that only the most relevant definitions are passed on as templates (and possibly other domain definitions) to the UCO of the next version, since time and energy of domain experts are limited. The notion of relevance of ontological definitions in an evolving collaborative community is still almost unexplored in the ontology engineering literature.

As collaborative communities evolve and learn, their number of ontological definitions, often specializations of earlier definitions, grows. Furthermore, the collaborative goals and requirements become clearer over time, often leading to new types of definitions that need to be created. At the same time, certain older ones become less relevant because of the shifting collaborative focus. The process of creating and modifying ontological definitions is very expensive, since many experts are to be involved, who often have to consult their respective organizations before being able to make a commitment.

To increase relevance and efficiency of the interorganizational ontology engineering process, some way of selecting the most relevant definitions in a particular stage of the evolution of the interorganizational ontology is needed. DOGMA-MESS has currently implemented this selection process in the following way:

- The organizational ontology alignment process starts the moment the set of organizational ontologies OO_i has been updated for the current version v of the IOO.
- The community defines a set of *relevance definitions* D_R.
  Example: a group of bakers started of with making definitions of general baking tasks, but has now discovered that the real education gap is in baking sweet products:

  $$D_R = \{ \quad \boxed{\text{Baking}} \rightarrow \boxed{\text{produces}} \rightarrow \boxed{\text{Sweet\_Stuff}} \quad \}$$

- Each $d_r \in D_R$ is now *lexonized*, which means that it is automatically flattened into a set of lexons L_R. Lexons are similar to binary conceptual relations, with a role/co-role pair instead of a single connecting relation type. If no role/co-role mapping exists in the DOGMA knowledge base, the co-role is left empty. The co-role helps to find additional linguistic forms of the same conceptual relation. Using these lexons, the set of *relevance relations* R_R now is formed by all "surface forms" of relevant conceptual relations (i.e. creating a conceptual relation from a role, and another relation from the co-role, arrows inverted).

  Example: $R_R = \{ \quad \boxed{\text{Baking}} \rightarrow \boxed{\text{produces}} \rightarrow \boxed{\text{Sweet\_Stuff}} \quad , \quad \boxed{\text{Sweet\_Stuff}} \rightarrow \boxed{\text{is\_produced\_in}} \rightarrow \boxed{\text{Baking}} \quad \}$

- For each organizational specialization d_{Oi} in each organizational ontology OO_i, a relevance score s_r is now computed, by checking if the relevance relations project into the definition. $\forall d_{Oi} \in D_{Oi}$, with CTH_{Oi} :
  - $s_r(d_{Oi}) = 0$.
  - $\forall r_r \in R_R$: if $\exists \pi \ r_r$ in d_{Oi}, then $s_r(d_{Oi}) = s_r(d_{Oi}) + 1$.

  Example: assume Baker A is an expert in cakes, and always stresses that cakes should be just sweet enough. His organizational ontology OO_A therefore contains this organizational task specialization d_{OA}:

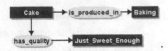

Also, in his concept type hierarchy $CT H_{OA}$, Cake $<$ Sweet_Stuff. Since the second relevance relation projects into this graph, its relevance score is increased by one.

– Now, all definitions have an associated relevance score. The scores can now be used to rank the relevance of the organizational definitions to the common interest. Once ranked, either the x highest ranked definitions can be selected, or those definitions that meet a certain threshhold. If the community is very busy, or has only very limited time, the threshold level can be set higher, so that less definitions need to be considered for lifting into the UCO of IOO version $v + 1$.

Of course, this is only a crude measure of relevance. Important is that a start has been made with operationalizing this fuzzy, but necessary idea and grounding it in a useful interorganizational ontology engineering methodology, from where it can evolve into more sophisticated approaches. The relevance assessment procedure is currently being experimented with in the CODRIVE project as a way of increasing focus and thus motivation of users.

5   Discussion

This paper has made several contributions. A conceptual model of interorganizational ontology engineering was presented, and the beginning of its formalization, which can be used to analyze and compare ontology engineering methodologies. DOGMA-MESS was introduced, as a prime example of an interorganizational ontology engineering methodology. One core IOE process, organizational ontology alignment, was described. This process explicitly addresses the fundamental problem of keeping ontological definition processes relevant, a necessary condition for these complex socio-technical processes to scale.

The aim of this paper was not to examine any particular ontology engineering process in depth, much work has been done in this respect in, for example, Semantic Web research. Neither was it to provide a full theory of ontological context dependency management. Substantial work already exists on this in the Conceptual Structures community, e.g. [16,15,11]. Also outside this community much relevant work already exists. For example, Guha et al. [13] adopt a notion of context primarily for scaling the management of the very large knowledge base Cyc. Our aim, however, was to introduce a framework and concrete methodology to apply this valuable theoretical work to a very pressing organizational and societal problem: making interorganizational ontology engineering work. In other words, our interest is how to apply such semantic techniques to community meaning negotiation goals [6].

In future work, we will refine our conceptual model of interorganizational ontology engineering and position related methodologies in our descriptive framework. Extensive intra and cross-case analyses will be done, providing both valuable data and extensions and refinements of the DOGMA-MESS methodology and system.

Having specialization and other context dependencies clearly defined and inform the design of knowledge definition processes is a necessary but not a sufficient condition for guaranteeing efficient and relevant interorganizational ontology engineering processes. A thorough grounding in socio-technical principles of community informatics, such as legitimate meaning negotiation, is required for interorganizational ontology engineering to succeed [5]. Systematically supporting this meaning negotiation process through a community portal and guided discussion facilities, tightly linked to the DOGMA-MESS knowledge base, is is one of our key research foci at the moment.

6 Conclusions

In this paper, literally speaking, we have put ontology engineering in context. We have characterized interorganizational ontology engineering as a process involving different domain experts, with well-defined roles, producing increasingly specialized versions of composite interorganizational ontologies.

Our goal is to efficiently produce relevant ontological definitions. To support this complex process, we introduced our DOGMA-MESS methodology. It combines a comprehensive socio-technical methodology with a formal framework of specialization context dependencies. With it, macro-processes like ontology alignment can be decomposed into managable combinations of micro-processes, such as template creation and specialization. Moreover, various ways of relevance scoring can support the subtle socio-technical dynamics in expert communities of practice. There is thus not one right way of designing these macro/micro processes. We aim to expand our work in intensive empirical evaluations of DOGMA-MESS in and across a range of high-impact cases.

Our aim was not to produce a comprehensive formal definition of macro and micro ontology engineering processes nor of context dependencies. This would have been far too ambitious, nor do justice to the large amount of related work. Rather, our methodology is in fact a meta-methodology that can accommodate many different ontological methodologies and technologies, synthesizing and tailoring them to the needs of of real communities of use.

By positioning related work using our conceptualization of interorganizational ontology engineering and its practical implementation in a working methodology and system, the factors influencing the relevance and efficiency of this extremely complex socio-technical process can be better understood. This understanding should considerably progress the identification of research gaps, alignment of research efforts, and applicability of results from ontology integration research. In the end, ontologies are not a goal in themselves, but instruments to facilitate collaborative *community dynamics*. We are confident that DOGMA-MESS will help improve this dynamics by discovering new ways for communities to find conceptual common ground.

References

1. F.-R. Aschoff, Schmalhofer, F., and L. van Elst. Knowledge mediation: A procedure for the cooperative construction of domain ontologies. In *Proc. of the ECAI 2004 Workshop on Agent-Mediated Knowledge Management*, pages 29–38, 2004.

2. P. De Leenheer and A. de Moor. Context-driven disambiguation in ontology elicitation. In P. Shvaiko and J. Euzenat, editors, *Context and Ontologies: Theory, Practice, and Applications. Proc. of the 1st Context and Ontologies Workshop, AAAI/IAAI 2005, Pittsburgh, USA, July 9, 2005*, pages 17–24, 2005.

3. P. De Leenheer, A. de Moor, and R. Meersman. Context dependency management in ontology engineering. Technical Report STAR-2006-03-01, VUB STARLab, Brussel, March 2006.

4. A. de Moor. *Empowering Communities: A Method for the Legitimate User-Driven Specification of Network Information Systems.* PhD thesis, Tilburg University, The Netherlands, 1999. ISBN 90-5668-055-2.

5. A. de Moor. Ontology-guided meaning negotiation in communities of practice. In P. Mambrey and W. Gräther, editors, *Proc. of the Workshop on the Design for Large-Scale Digital Communities at the 2nd International Conference on Communities and Technologies (C&T 2005), Milano, Italy, June 2005*, 2005.

6. A. de Moor. Patterns for the pragmatic web. In *Proc. of the 13th International Conference on Conceptual Structures, ICCS 2005, Kassel, Germany, July 17-22, 2005*, pages 1–18, 2005.

7. S. Decker, D. Fensel, F. van Harmelen, I. Horrocks, S. Melnik, M. Klein, and J. Broekstra. Knowledge representation on the Web. In *Proc. of the 2000 International Workshop on Description Logics (DL2000), Aachen, Germany*, 2000.

8. T. Edgington, B. Choi, K. Henson, T.S. Raghu, and A. Vinze. Adopting ontology to facilitate knowledge sharing. *Communications of the ACM*, 47(11):217–222, 2004.

9. J. Euzenat. Building consensual knowledge bases: Context and architecture. In N.J.I. Mars, editor, *Towards Very Large Knowledge Bases - Proceedings of the KB&KS '95 Conference*, pages 143–155. IOS Press, 1995.

10. J. Euzenat, T. Le Bach, J. Barrasa, et al. State of the art on ontology alignment. Knowledge Web Deliverable KWEB/2004/d2.2.3/v1.2, 2004.

11. B. Ganter and G. Stumme. Creation and merging of ontology top-levels. In *Proc. of the 11th International Conference on Conceptual Structures, ICCS 2003 Dresden, Germany, July 21-25, 2003*, pages 131–145, 2003.

12. T.R. Gruber. A translation approach to portable ontology specifications. *Knowledge Acquisition*, 5(2):199–220, 1993.

13. R. Guha and D. Lenat. Cyc: a midterm report. *AI Magazine*, 11(3):32–59, 1990.

14. Y. Kalfoglou and M. Schorlemmer. Ontology mapping: The state of the art. In *Proc. of the Dagstuhl Seminar on Semantic Interoperability and Integration (Dagstuhl, Germany)*, 2005.

15. G. Mineau and O. Gerbé. Contexts: A formal definition of worlds of assertions. In *Proc. of the 5th International Conference on Conceptual Structures, ICCS '97, Seattle, Washington, USA, August 3-8, 1997*, pages 80–94, 1997.

16. J.F. Sowa. *Conceptual Structures: Information Processing in Mind and Machine.* Addison-Wesley, 1984.

17. P. Spyns, R. Meersman, and M. Jarrar. Data modelling versus ontology engineering. *SIGMOD Record*, 31(4):12–17, 1998.

FCA-Based Browsing and Searching
of a Collection of Images

Jon Ducrou[1], Björn Vormbrock[2], and Peter Eklund[3]

[1] School of Information Technology and Computer Science, The University
of Wollongong, Northfields Avenue, Wollongong, NSW 2522, Australia
jrd990@uow.edu.au
[2] AG Algebra und Logik, FB Mathematik, Technische Universität Darmstadt,
Schloßgartenstr. 7, D–64289 Darmstadt, Germany
vormbrock@mathematik.tu-darmstadt.de
[3] School of Economics and Information Systems, The University of Wollongong,
Northfields Avenue, Wollongong, NSW 2522, Australia
peklund@uow.edu.au

Abstract. This paper introduces ImageSleuth, a tool for browsing and
searching annotated collections of images. It combines the methods of
Formal Concept Analysis (FCA) for information retrieval with the graphical information conveyed in thumbnails. In order to use thumbnails of
images to represent concept extents, line diagrams can not be efficiently
utilised and thus other navigation methods are necessary. In addition
to established methods like search and upper/lower neighbours, a query
by example function and the possibility to restrict the attribute set are
included. Moreover, metrics on conceptual distance and similarity are
discussed and applied to automated discovery of relevant concepts. This
paper describes the FCA base of ImageSleuth which formed the basis for
its design and the implementation which followed.

1 Motivation

Formal Concept Analysis (FCA) has been successfully applied in Information
Retrieval for browsing and searching text documents ([CS01], [KC00]). The richer
structure of the concept lattice has advantages over simple keyword search or
tree structures. For keyword search, the user has to remember or guess the
correct keywords. For searching in trees, the names of nodes serve as keywords,
but there is a unique path leading to the desired information. Moreover, once
a categorisation scheme for the documents is chosen, this hierarchy is enforced
for every search. In concept lattices multiple paths can lead to a result, so the
user may guide the search via the addition of required properties step by step
without the restriction imposed by a single inheritance hierarchy. The order of
these properties is irrelevant.

This paper illustrates how ImageSleuth uses FCA methods for information
retrieval within a collection of images. Any such approach has to take into consideration the graphical nature of this information. The established method for

H. Schärfe, P. Hitzler, and P. Øhrstrøm (Eds.): ICCS 2006, LNAI 4068, pp. 203–214, 2006.

browsing collections of images is to display all images as *thumbnails*. A thumbnail is a smaller version of the original image, small enough to view many images simultaneously but large enough to distinguish features of the full size image. Within a collection of thumbnails, each thumbnail is usually the same size and displayed in a two dimensional layout, sorted by a simple feature of the image (e.g. name, date, filesize, etc). The desired outcome is to combine thumbnails as the technique that best conveys the content of an image with the advantages of FCA information retrieval for the annotated information associated with the image. This requires that the concept lattice representation has a different presentation and navigation paradigm compared to that of text documents.

This paper contains four more sections. In Section 2, a description of the FCA-background of ImageSleuth is presented. Section 3 explains the implementation, while Section 4 describes an example. Finally, Section 5 contains concluding remarks.

2   Using FCA to Browse Images

In this section, the mathematical structure underlying ImageSleuth and the resulting search and browse options are described. We assume that the reader is familiar with the basic notions of Formal Concept Analysis such as context, formal concept and conceptual scaling. For an introduction to FCA we refer to [GW99].

Following the approach used for browsing and searching of text documents, ImageSleuth computes concept lattices of contexts having the collection of images as objects and their annotated features as attributes. These features may be information about the depicted object annotated by hand as well as automatically extracted graphical information. In contrast to most approaches for FCA document retrieval, no line diagram of the lattice is displayed. Instead, following [KC00], the user is always located at one concept of the concept lattice. This allows thumbnails of the images to be shown as the extent of the present concept and thus to convey most of the graphical information characterising this concept. The intent is represented as a list of attributes. As no line diagram of the lattice is shown, lists of upper and lower neighbours are the only representation of the lattice structure around the present concept. Searching and browsing in the image collection then corresponds to moving from concept to concept in the lattice. By including new attributes in the intent, the user moves to a smaller concept where all images in the extent have these features. ImageSleuth offers the following possibilities to navigate in the concept lattice:

- Restriction of the set of attributes in consideration
- Move to upper/lower neighbour
- Search by attributes
- Search for similar objects (Query by example)
- Search for similar concepts

The possibility to restrict the set of attributes in consideration allows focus on the features that are relevant for the current navigation needs of the user.

Otherwise large sets of irrelevant attributes would increase the number of concepts and make search unnecessarily complex. ImageSleuth offers predefined sets of attributes (called *perspectives*) covering different aspects of the images. The user may combine these perspectives and include or remove perspectives during the search. Scale attributes are natural candidates for such attribute sets but other sets are allowed (for example, overlapping perspectives and perspectives which are subsets of other perspectives).

The option to search for similar concepts requires a similarity measure. In order to use this similarity together with the normal search or query-by-example, (where the user may describe the searched concept with attribute or object sets which are not intent or extent of a concept) we want the similarity measure to be defined for semiconcepts as introduced in [LW91] as a generalisation of concepts:

Definition 1. *A semiconcept of a context* $\mathbb{K} := (G, M, I)$ *is a pair* (A, B) *consisting of a set of objects* $A \subseteq G$ *and a set of attributes* $B \subseteq M$ *such that* $A = B'$ *or* $B = A'$. *The set of all semiconcepts of* $\mathbb{K}$ *is denoted by* $\mathfrak{H}(\mathbb{K})$.

Note that every concept is a semiconcept. The underlying structure of ImageSleuth is thus:

1. A context $\mathbb{K} := (G, M, I)$ with a collection of images as object set G, possible features as attribute set M and an incidence relation I assigning features to objects.
2. A collection $\mathcal{P}$ of subsets of M called perspectives. Every subset $\mathcal{A} \subseteq \mathcal{P}$ defines a subcontext $\mathbb{K}_{\mathcal{A}} := (G, \bigcup \mathcal{A}, I_{\mathcal{A}})$ with $I_{\mathcal{A}} := I \cap (G \times \bigcup \mathcal{A})$ of $\mathbb{K}$.
3. A similarity measure

$$s : \bigcup_{\mathcal{A} \subseteq \mathcal{P}} \mathfrak{H}(\mathbb{K}_{\mathcal{A}})^2 \to [0, 1]$$

assigning to every pair of semiconcepts of a subcontext $\mathbb{K}_{\mathcal{A}}$ a value between 0 and 1 which indicates the degree of similarity.

Since for every $\mathcal{A} \subseteq \mathcal{P}$ the contexts $\mathbb{K}_{\mathcal{A}}$ and $\mathbb{K}$ have the same object set and every attribute of $\mathbb{K}_{\mathcal{A}}$ is an attribute of $\mathbb{K}$ it follows for every $m \in \bigcup \mathcal{A}$ that $m^I = m^{I_{\mathcal{A}}}$. Since for $(A, B) \in \mathfrak{B}(\mathbb{K}_{\mathcal{A}})$ we have

$$A = B^{I_{\mathcal{A}}} = \bigcap \{m^{I_{\mathcal{A}}} \mid m \in B\} = \bigcap \{m^I \mid m \in B\}$$

it follows that A is the extent of a concept of $\mathfrak{B}(\mathbb{K})$. Therefore, $\phi(A, B) := (A, A^I)$ defines a map $\phi : \mathfrak{B}(\mathbb{K}_{\mathcal{A}}) \to \mathfrak{B}(\mathbb{K})$ and the image of ϕ is a $\wedge$-subsemilattice of $\mathfrak{B}(\mathbb{K})$. In the following, the different navigation means based on this structure are described.

2.1   Restriction of the Attribute Set

By including different perspectives the user defines a subcontext of $\mathbb{K}$ in which all operations are performed. She may change this subcontext while browsing, thus obtaining at the present concept further information and search options. If at the concept (A, A^{I_A}) the perspective $S \in \mathcal{P}$ is included (i.e. the set of attributes in consideration is increased), then ImageSleuth moves to the concept $(A^{I_{A \cup \{S\}} I_{A \cup \{S\}}}, A^{I_{A \cup \{S\}}})$ of $\mathfrak{B}(\mathbb{K}_{A \cup \{S\}})$. Since for $\mathcal{A} \subseteq \mathcal{P}$ and $S \in \mathcal{P}$ the extent of every concept of $\mathbb{K}_{\mathcal{A}}$ is an extent of $\mathbb{K}_{A \cup \{S\}}$ we have $A = A^{I_{A \cup \{S\}} I_{A \cup \{S\}}}$ and the set of images shown does not need to be updated when a further perspective is included. This allows the addition of perspectives during the search without losing information. A similar strategy is known from Toscana (cp. [TJ02]) where the user moves through different scales. At every point the user may also remove a perspective S which takes her to the concept $(A^{I_{A \setminus \{S\}}}, A^{I_{A \setminus \{S\}} I_{A \setminus \{S\}}})$. If in this way an attribute of A^{I_A} is removed from the current subcontext then the extent may be increased since $A^{I_A} \subseteq A^{I_{A \setminus \{S\}}}$.

2.2 Moving to Upper and Lower Neighbours

ImageSleuth uses most of its interface to show thumbnails of images in the extent of the chosen concept. As a result the user never sees the line diagram of a lattice. Instead, the lattice structure around the current concept is represented through the list of upper and lower neighbours which allow the user to move to super- or subconcepts. For every upper neighbour (C, D) of the current concept (A, B) the user is offered to remove the set $B \setminus D$ of attributes from the current intent. Dually, for every lower neighbour (E, F) the user may include the set $F \setminus B$ of attributes which takes her to this lower neighbour. By offering the sets $B \setminus D$ and $F \setminus B$ dependencies between these attributes are shown. Moving to the next concept not having a chosen attribute in its intent may imply the removal of a whole set of attributes. In order to ensure that the extent of the given concept is never empty it is not possible to move to the minimal concept.

2.3 Search and Query-by-Example

Browsing of the image collection is achieved by moving to neighbouring concepts. In many cases the user will want to go directly to images having a certain set of attributes $B \subseteq \bigcup \mathcal{A}$. This is offered by the search function which computes, for the selected attributes, the concept $(B^{I_A}, B^{I_A I_A})$. Its extent is the set of all images having these attributes, its intent contains all attributes implied by B.

Another type of search is performed by the query-by-example function. Instead of defining a set of attributes, a set of objects A is defined as the sample set. The query-by-example function then computes the common attributes of these images (in the selected subcontext) and returns all other images having these attributes by moving to $(A^{I_A I_A}, A^{I_A})$. In this way, query-by-example is the dual of the search function. While the search for images having certain attributes is not affected by the removal or addition of perspectives to the subcontext, query-by-example depends strongly on the selected subcontext. The more attributes taken into consideration, the smaller the set of images that have exactly the same attributes as the examples.

2.4 Similarity

The aim of query-by-example is to find objects which are similar to the objects in a given sample set. This is a narrow understanding of similarity implying equivalence in the considered subcontext; for the query-by-example function two objects g, h are "similar" in a subcontext $\mathbb{K}_A$ if $g^{I_A} = h^{I_A}$. If the objects are uniquely described by the attributes in the chosen subcontext then query-by-example seldom yields new information. A more general approach is to define a similarity measure. In [Le99] several similarity measures on attribute sets are investigated. Similarity of two objects g and h is then described as the similarity of the attribute sets g' and h'. In order to use the grouping of objects provided by the formal concepts, ImageSleuth works with a similarity measure on semi-concepts which allows the return of a ranked list of similar concepts. We use semiconcepts since the set of sample images chosen by the user is not necessarily the extent of a concept. The similarity measure is derived from the following metric:

Definition 2. *On the set $\mathfrak{H}(\mathbb{K})$ of semiconcepts of a context $\mathbb{K} := (G, M, I)$ the metric $d : \mathfrak{H}(\mathbb{K}) \times \mathfrak{H}(\mathbb{K}) \to [0, 1]$ is defined as*

$$d((A, B), (C, D)) := \frac{1}{2} \left(\frac{|A \setminus C| + |C \setminus A|}{|G|} + \frac{|B \setminus D| + |D \setminus B|}{|M|} \right).$$

This definition formalizes the idea that two semiconcepts are close if there are few objects and attributes belonging to only one of them. In order to compare the number of objects and the number of attributes where they differ, these numbers are set in relation to the total number of objects or attributes. Semiconcepts with small distance are considered similar. ImageSleuth uses $1 - d((A, B), (C, D))$ as the similarity of (A, B) and (C, D).

For a similar purpose Saquer and Deogun introduced in [SD01] a related similarity measure as

$$s((A, B), (C, D)) := \frac{1}{2} \left(\frac{|A \cap C|}{|A \cup C|} + \frac{|B \cap D|}{|B \cup D|} \right).$$

This definition of similarity extends to semiconcepts (A, B), (C, D) if $A \cup C \neq \emptyset$ and $B \cup D \neq \emptyset$. In particular, the similarity $s((A, A'), (C, D)))$ is defined for every nonempty set A of objects and every concept $(C, D) \neq (G, \emptyset)$. For a sample set A of images, ImageSleuth uses a combination of both measures to return a ranked list of concepts similar to the semiconcept (A, A^{I_A}).

The given metric on semiconcepts has two advantages. First, it allows the return of a list of similar concepts rather than just a list of images. This provides a reasonable grouping of the similar images and, since the attributes of the concepts are displayed, it shows in which way the images relate to the sample set.

Second, in contrast to other approaches such as graph distance, the number of different objects of two concepts is taken into account. Instead of counting only the attributes in which two concept intents differ, we assume that the significance of this difference is reflected in the difference of their corresponding attribute sets. If (A, B) is a concept and (C, D), (E, F) are upper neighbours of (A, B) with $|C| \leq |E|$ then the attributes in $B \setminus F$ are considered as more characteristic for the concept (A, B) than the attributes in $B \setminus D$. Thus, if $|D| = |F|$ then (C, D) is closer to (A, B) than (E, F) even though they differ from (A, B) in the same number of attributes. In this way, even an incomparable concept may be the closest. This contradicts the intuition that, for a concept, its sub- and superconcepts should be closest. Yet upper and lower neighbours are directly accessible by other navigation means. The advantage of the search for similar concepts for a given concept is that it offers a selection of (in the lattice order) incomparable but close concepts which are otherwise invisible.

As the original query-by-example function described above is the dual of a search this approach can be used for the search function, too. If a search is carried out for a set of attributes B, and if B' is empty, then the concept (B', B'') contains only the information that these attributes do not occur together. No images are returned as a result of this search, since there are no images having the required attributes. In this case, the user may be shown a list of concepts similar to or with small distance to the semiconcept (B', B).

3   Implementation

This section introduces the application ImageSleuth. Focus is placed on the dataset used for testing, its history, navigation overview and a method for resolving the empty extent search result.

3.1   Image Collection

The dataset used is taken from the popular computer game *"The Sims 2"*. It features 412 objects of household funiture and fittings, described by 120 attributes which include in-game properties, suggestions for use and automatically extracted colour information. There are 7,516 concepts in the complete context. Each attribute of the context is assigned to one or more perspectives. In this dataset, 10 perspectives have been constructed.

3.2   History

The version of ImageSleuth presented here is the second version. The original prototype used concept neighbourhoods and include/remove attributes, but was limited to traversal between three mutually exclusive subcontexts via single objects. It underwent user-evaluation to test functionality and opinion of ImageSleuth's navigation paradigm. 29 honours level university students (from various disciplines) were asked to perform tasks and provide feedback on ImageSleuth v1. Results are overviewed in [DE05]. Results indicated that concept neighbourhoods offered a useful navigation method, users liked the *"grouping*

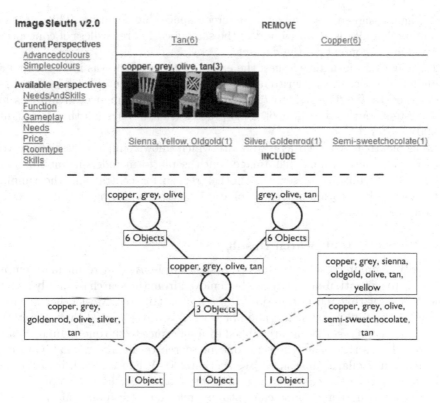

Fig. 1. An example screenshot of ImageSleuth and the lattice representation of the corresponding neighbourhood. The screenshot shows the four primary navigation functions of ImageSleuth. On the left is the listings of current and available perspectives (currently, advanced and simple colour perspectives are selected). Top and bottom show the remove and include functions respectively. The central pane shows the current concept; with intent listed as textual attributes and extent as thumbnailed images. The lattice neighbourhood shows the current concept at its centre.

of similar objects"[1] (concept extents) and the efficient searching by selection of defined attributes. Negative feedback included complaints about the interface and the systems performance. Analysis of the task results revealed the biggest problem: if a search included mutually exclusive attributes, it returned an empty extent, which left users confused. According to [Co99], making a user feel stupid is the worst possible software interaction fault.

The second version of ImageSleuth addressed the primary problems experienced by participants in the user testing sessions. These included interface layout, slow performance, inability to combine contexts and the empty extent search result problem. In the first version, *include* and *search* functionality was listed after the thumbnails, and users needed to scroll to the bottom of the page to continue navigation. This was repaired by partitioning the page into frames with

---

[1] A term used by more than one of the participants.

each frame assigned a set amount of screen space and function. This means a given functionality is always found in the same location regardless of conceptual position in, or content of, the dataset.

To address performance issues, the entire system (which was implemented as a single Perl script) was rewritten in C++ as a set of executables. The database was ported to PostGreSQL to take advantage of performance advantages for FCA systems outlined in [Ma06]. This process lead to a system that is roughly 10,000% faster.

ImageSleuth is accessed as a web site which allows simple access via a web browser. This also means that ImageSleuth is platform independent for users as all code is run on the server. Another reason for centralising the running of ImageSleuth is to allow logging of users' activities during usability testing sessions for analysis.

3.3 Empty Extent Search Result

The most common solution to concept searches in FCA, that result in an empty extent, is to offer attributes that can be removed from the search to supply a more general answer that meets a majority of search attributes. Most other forms of search (for example, text search) do not work this way - instead they supply the user with a list of results that are ranked by a relevance to the query. ImageSleuth tries to address this using the semiconcept search result and a combination of distance and similarity measures (see section 2.4). When a search is performed that would return the concept with an empty extent, the user can opt to allow the system to find and rank conceptually relevant concepts. This process is achieved by finding possible neighbours of the semiconcept and performing a bounded traversal which ranks the traversed concepts. These possible neighbours (Fig. 3, Line 3.) become the first concepts traversed. Each concept visited has its relevance calculated and stored. A test is applied to each concept visited to calculate whether it is to be used for further traversal. The test condition is based on the distance metric compared to a weighted average of the query concepts intent and extent size (Fig. 3, Line 8.). The condition is represented as:

$$Dist((A, B), (C, D)) \times SearchWidth < \tfrac{1}{2}(|A|/|G| + |B|/|M|)$$

where (A, B) is the query concept and (C, D) is the current concept of the traversal. $SearchWidth$ is a modifier to allow the search to be made wider or narrower. If the traversal is to continue, the concept's neighbourhood is added to the traversal list, the concept is marked as visited and the process continues (Fig. 3, Lines 9-11.).

Relevance is calculated as the average of the similarity scores which is presented to the user as a percentage.

4   Empty Extent Search Result Example

The following is a simple example of how ImageSleuth's semi-concept searching works. This example uses two perspectives, *Function* and *RoomType* which have

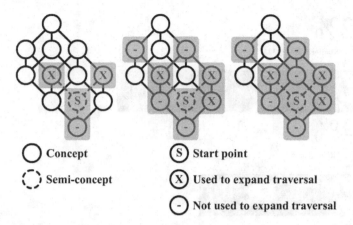

Fig. 2. An example of lattice traversal starting from a semi-concept. The traversal in this example is complete in 3 steps. The shaded area shows the computed concepts at each step.

```
1.     find_similar ( Concept: input, Number: width )
2.         input_size = size ( input.intent ) + size ( input.extent )
3.         candidate = upperNeigh ( input ) ∪ lowerNeigh ( input )
4.         exclude = ( input )
5.         while ( size ( candidate ) > 0 )
6.             concept = pop ( candidate )
7.             exclude = exclude ∪ concept
..
..             compute and store rank information for concept.
..
8.             if ( distance ( input , concept ) × width
                        < weightedAverage( input ) )
9.                 candidate = candidate ∪ upperNeigh ( concept )
10.                candidate = candidate ∪ lowerNeigh ( concept )
11.                candidate = candidate / exclude
12.            end if
13.        end while
14.    end
```

Fig. 3. Pseudocode representation of search traversal. Parameters are the starting concept or semiconcept (*input*) and a numeric value used to modify the width of the search (*width*).

20 attributes in total. The *Function* perspective is a simple nominal scale with each object having one *function* attribute. The *RoomType* perspective, on the other hand, is more complex with each object having zero or more *room type* attributes. With this context the complete lattice has 194 concepts.

64.92%

Distance: 0.965189 Similarity: 0.333333

Electronics, Study(7)

55.74%

Distance: 0.914985 Similarity: 0.2

Bedroom, Electronics, LivingRoom, Study(5)

54.42%

Distance: 0.921883 Similarity: 0.166667

Appliances(21)

Distance: 0.921883 Similarity: 0.166667

Electronics(21)

Fig. 4. Results of a concept traversal from the query *"Applications, Electronics, Study"* using the perspectives *"Function, Room Type"*

The query for this example will be *"Applications, Electronics, Study"*, the first two attributes from the *Function* perspective and the remaining one from *Room Type*. *Function* being nominally scaled, the inclusion of two attributes from this perspective means that if the concept was completed it would result in the empty extent concept or $(\emptyset, M)$. Although this result is technically correct, it does not suit the query's intention.

To identify a concept that is more representative, a concept traversal is started using the semiconcept, $(\emptyset, (Applications, Electronics, Study))$. In this example, the traversal visits 12 concepts, four of which are conceptually close enough to extend the traversal. Consequently, only 6.19% of the total lattice is computed. The first three of five rankings are shown in Fig. 4. Relevance is shown as a large percentage, while individual distance and similarity scores are displayed below. Each result is displayed as a list of attributes representing the intent and a collection of thumbnails representing the extent. The highest ranking concept, with relevance 64.92%, has the intent (*Electronics, Study*), which is two of the three original query attributes. Following that, at 55.74%, is the concept with the intent (*Bedroom, Electronics, LivingRoom, Study*). The third ranking, at 54.42% relevance, has two concepts, with the intents (*Applications*) and (*Electronics*), which represent the mutually exclusive elements of the original query.

5   Conclusion

Presented is an image based navigation paradigm combining the methods of Formal Concept Analysis for information retrieval with the graphical information conveyed as thumbnails. This paradigm is formalised and realised via the ImageSleuth application which uses a collection of images taken from the game, *The Sims 2*.

It was required that the concept lattice representation used in ImageSleuth had a different presentation and navigation paradigm compared to that of text documents; in contrast to most approaches for FCA document retrieval, no line diagram of the lattice is displayed. In our approach, the user chooses perspectives of interest and is always located at one concept of the concept lattice, with the extent of the current concept displayed as thumbnails. Query-by-example and a method for ranking attribute search results when an exact match is not to be found are also described and exemplified in ImageSleuth. Also shown is how ImageSleuth has been improved from the previous version after testing and user evaluation.

References

[Co99]   A. Cooper: The Lunatics are Running the Asylum, SAMS, 1999.

[CS01]   R. Cole, G. Stumme: CEM – A conceptual email manager. In: B. Ganter, G. W. Mineau (eds.): Conceptual structures: Logical, linguistic, and computational issues. Proc. ICCS 2000. LNAI **1867** Springer, Heidelberg 2000, 438–452.

[DE05]   J. Ducrou, P. Eklund: Browsing and Searching MPEG-7 Images using Formal Concept Analysis. To Be Published, Feb 06 in: ACTA: IASTED AIA.

[GW99]   B. Ganter, R. Wille: Formal concept analysis: mathematical foundations. Springer, Heidelberg 1999.

[KC00]   M. Kim, P. Compton: Developing a Domain-Specific Document Retrieval Mechanism. In: Proc. of the 6th pacific knowledge acquisition workshop (PKAW 2000). Sydney, Australia.

[Le99]   K. Lengnink: Ähnlichkeit als Distanz in Begriffsverbänden. In: G. Stumme, R. Wille (eds.): Begriffliche Wissensverarbeitung: Methoden und Anwendungen. Springer, Heidelberg 2000, 57–71.

[LW91]   P. Luksch, R. Wille: A mathematical model for conceptual knowledge systems. In: H. H. Bock, P. Ihm (eds.): *Classification, data analysis, and knowledge organisation*. Springer, Heidelberg 1991, 156 – 162.

[Ma06]   B. Martin, P. Eklund: Spatial Indexing for Scalability in FCA. In: Formal
          Concept Analysis: 4th International Conference (ICFCA 2006), Lecture
          Notes in Computer Science, Volume 3874, 2006, 205–220.
[SD01]   J. Saquer, J. S. Deogun: Concept aproximations based on rough sets and sim-
          ilarity measures. In: Int. J. Appl. Math. Comput. Sci., Vol.11, No.3, 2001,
          655 – 674.
[TJ02]   P. Becker, J. Hereth, G. Stumme: ToscanaJ - An Open Source Tool for Quali-
          tative Data Analysis,In: Advances in Formal Concept Analysis for Knowledge
          Discovery in Databases. Proc. Workshop FCAKDD of the 15th European
          Conference on Artificial Intelligence (ECAI 2002), 2002.
[TJ]   The ToscanaJ Homepage. <http://toscanaj.sourceforge.net>
[VW95]   F. Vogt, R. Wille: TOSCANA - a graphical tool for analyzing and exploring
          data. In: Proceedings of the DIMACS International Workshop on Graph
          Drawing (GD'94), 1995, 226 – 233.

Semantology: Basic Methods for Knowledge Representations

Petra Gehring and Rudolf Wille

Technische Universität Darmstadt,
Institut für Philosophie und Fachbereich Mathematik, D–64289 Darmstadt
gehring@phil.tu-darmstadt.de, wille@mathematik.tu-darmstadt.de

Abstract. In this paper, we introduce the term *"Semantology"* for naming the theory of semantic structures and their connections. Semantic structures are fundamental for representing knowledge which we demonstrate by discussing *basic methods of knowledge representation*. In this context we discuss why, in the field of knowledge representation, the term "Semantology" should be given preference to the term "Ontology" .

Contents

1 Introduction

In today's scientifically oriented world, *knowledge representations* are considered to be of great importance. Hence multifarious methods are offered for representing knowledge in an immeasurable variety of domains. Such methods are grounded consciously or unconsciously on *semantic structures* which carry the meaning of the represented knowledge. In this paper we want to discuss *basic methods of knowledge representation* constituted by structures of scientific semantics; the corresponding methodology is treated in the frame of *"Semantology"* which we understand as the general theory of semantic structures and their connection. The term "Semantology" may help to avoid naturalistic or essentialistic fallacies, which the term *"Ontology"* may suggest as, for instance, the naturalistic idea that scientific models can match existing realities of nature.

2 Semantic Structures

Semantic structures considered in this treatment obtain their meaning from some scientific semantics. For discussing semantic structures and their meanings in

H. Schärfe, P. Hitzler, and P. Øhrstrøm (Eds.): ICCS 2006, LNAI 4068, pp. 215–228, 2006.

general, it is useful to refer to *Peirce's classification of sciences*. This classification scales the sciences "in the order of abstractness of their objects, so that each science may largely rest for its principles upon those above it in the scale while drawing its data in part from those below it" ([Pe92]; p.114). We mainly activate the first level of Peirce's classification:

I. Mathematics   II. Philosophy   III. Special Sciences

where *Mathematics* is viewed as the most abstract science studying hypotheses exclusively and dealing only with potential realities, *Philosophy* is considered as the most abstract science dealing with actual phenomena and realities, while all *other sciences* are more concrete in dealing with special types of actual realities.

Since modern mathematics is essentially based on *set-theoretical semantics*, semantic structures having mathematical meaning can be represented by set structures. Mathematicians are developing those structures in great variety, many of which even in advance. Peirce already wrote that mathematicians are "gradually uncovering a great Cosmos of Forms, a world of potential being" ([Pe92], p.120).

	roof	ceiling	wall	fire-wall	staircases	stairwell	substructure	baseplate	chimney
BauONW §15	X	X	X	X	X	X	X	X	X
BauONW §16	X							X	X
BauONW §17		X	X	X	X	X	X		X
BauONW §18 Abs.1	X	X	X	X		X		X	X
BauONW §18 Abs.2	X	X	X	X	X	X	X		
BauONW §25			X	X					
BauONW §26			X	X					
BauONW §27			X	X					
BauONW §28			X	X					
BauONW §29				X					
BauONW §30			X						
BauONW §31	X								
BauONW §32					X	X			
BauONW §33						X			
BauONW §39								X	X
BauPG		X			X	X	X		
EnEG	X	X	X	X		X	X		
WHG									X
LWG									X
WärmeschutzV	X	X	X	X		X	X		
VGS									X
DIN 1054							X	X	X
DIN 1055			X	X	X	X	X	X	X
DIN 4102	X	X	X	X	X	X			
DIN 4108 Teil 1 u. 2	X	X	X	X		X	X		
DIN 4108 Teil 3	X			X	X		X		
DIN 4109	X	X	X	X	X	X			
DIN 18150									X
DIN 18160									X
DIN 18195	X		X	X		X	X	X	
DIN 18531	X								
DIN 68800	X								
ATV-Merkblätter									X

Fig. 1. Formal context concerning the shell of a one-family house

Semantic structures having philosophical meaning are based on *philosophic-logical semantics* which are grounded on networks of philosophical concepts. In traditional philosophical logic, concepts are viewed as the basic units of thought; they and their combinations to judgments and conclusions form "the three essential main functions of thinking" ([Ka88], p.6), which constitutes the logical semantics of philosophy. Semantic structures having their meaning with respect to special sciences are based on semantics which are grounded on networks of special concepts of those sciences.

Let us illustrate the described understanding of semantic structures by an *example*. For this we choose a project which was initiated by the Department of Building and Housing of the State of "Nordrhein-Westfalen". At the beginning of the 1990th, representatives of that department asked the Darmstadt Research Group on Concept Analysis whether members of the group could develop, in cooperation with them, a prototype of an *information system about laws and regulations concerning building construction*. The main purpose of that system was defined to be a support for the planning department and building control office as well as for people that are entitled to present building projects to the office in order to enable these groups to consider the laws and technical regulations in planning, controlling, and implementing building projects (cf. [KSVW94], [EKSW00], [Wi05c]).

The first question in starting the project was how to find an adequate knowledge representation for the desired information system. The natural idea to establish a slender-boned thesaurus about all relevant aspects of laws and building techniques turned out to be too extensive and therefore not manageable. Thus, more elementary semantic structures became desirable which could be represented by formal contexts as defined in Formal Concept Analysis [GW99a][1].

Now the main question was: What are the "information objects" users of the information system have to look for? It was a breakthrough when we finally identified the relevant text units of the laws and regulations as those information objects. With this understanding we needed only five hours to establish a comprehensive formal context having as objects the relevant text units and as attributes building components and requirements concerning a one-family house. A smaller subcontext of that context is shown in Fig. 1; its concept lattice depicted in Fig. 2 functioned later as a query structure of the information system.

---

[1] Let us recall that a *formal context* is mathematically defined as a set structure (G, M, I) in which G and M are sets and I is a binary relation between G and M; the elements of G and M are called *formal objects* and *formal attributes*, respectively. One says: a formal object g *has* a formal attribute m if g is in relation I to m. A *formal concept* of (G, M, I) is a pair (A, B) where A and B are subsets of G and M, respectively, and A is just the set of all objects having all attributes of B, and B is just the set of all attributes applying to all objects of A. A formal concept (A_1, B_1) is said to be a *subconcept* of a formal concept (A_2, B_2) if A_1 is contained in A_2 or, equivalently, if B_1 contains B_2. The set of all formal concepts of (G, M, I) together with the subconcept-relation always forms the mathematical structure of a complete lattice which is named the *concept lattice* $\underline{\mathfrak{B}}(G, M, I)$.

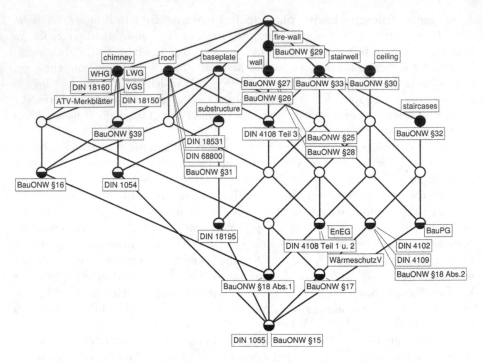

Fig. 2. Concept lattice as query structure "Shell of a one-family house"

Such a labelled concept lattice represents a semantic structure in a threefold manner: it can be understood as a set structure having purely *mathematical meaning*, as a general conceptual structure having *philosophic-logical meaning*, and as special conceptual structure having *purpose-oriented meaning* concerning planning, controlling, and implementing building projects. The *mathematical understanding* is important for developing a mathematical structure theory of concept lattices. This theory yielded, for instance, the means to establish and justify the TOSCANA software (cf. [KSVW94]), [BH05]) which makes possible to use suitable query structures to navigate purpose-oriented through the represented knowledge about laws and technical regulations. The *philosophic-logical understanding* allows in general to unify special conceptual structures on a more abstract level, still refering to actual realities; one might consider, for instance, the aggregation of the query structures "Shell of a one-family house" and "Operation and fire security" in [KSVW94], p.279. Since the philosophic-logical level is the most abstract level refering to actual realities, it may function also well as a transdisciplinary bridge between mathematical structures and (special) conceptual structures (cf. [Wi05a]).

Experiences have shown that labelled concept lattices, which purposefully represent semantic structures, usually stimulate the creation of knowledge caused by those semantic structures. This shall be indicated by just one event in the

Department of Building and Housing. After a longer period of collaboration in developing the information system, directors of the department wanted to see the progess of our collaboration. They understood fairly well how to read the labelled diagrams of the concept lattices and made many remarks and even helpful suggestions. When we showed the diagram of the query structure "Shell of a one-family house", suddenly, the head of directors shouted: "This is unbelievable! For building a chimney of a one-family house, one has to observe twelve laws and regulations! We really need law compression!" Indeed, since laws and regulations are usually created and revised over a long time, they are in danger to become too large and too many. Therefore it is helpful to represent laws and regulations in networks of concepts which allow to notice more connections and even improvements of them. In general, experts of the represented contents are very fast in grasping essential relationships within a labelled diagram of a concept lattice; in particular, they even recognize mistakes in the underlying data contexts.

3   Basic Methods for Knowledge Representation

Representations of knowledge about scientifically accessible domains should enable the reconstruction of the represented knowledge by users with a relevant scientific background, i.e., users who have internalized enough semantic structures of the corresponding special sciences. What are methods for establishing those knowledge representations? In this paper, answers to that question are concentrating on basic methods for knowledge representations (cf. [Wi06]) which allow the three-fold semantic understanding discussed in Section 2.

3.1. A *(formal) context*, as the one in Fig. 1, is a semantic structure which yields the most elementary representation of knowledge. Mathematically, such a context is a set structure which is usually called an *incidence structure* and investigated predominantly combinatorially (e.g. [De70]). Philosophically, such a context may be understood as a *logical structure* consisting of a collection of (general) objects and a collection of (general) attributes joined by relationships indicating which objects have which attributes. In special sciences, such a context is mainly viewed as an elementary *data table* representing relationships between (special) objects and (special) attributes.

In the example in Fig. 1, objects are text units of laws resp. regulations and attributes are building components; a cross in the data table indicates that the text unit whose name is heading the row of the cross is relevant for the component whose name is heading the column of the cross. For example, the crosses in the row headed by "DIN 1054" indicate that the standard "DIN 1054" is relevant for the components "substructure", "baseplate", "chimney", and not relevant for the other components. For reconstructing the represented knowledge, it might sometimes help to increase the readability of the data table by suitably permuting rows and columns.

3.2. A *concept lattice* presented by a *labelled line diagram*, as the one depicted in Fig. 2, yields a representation of conceptual knowledge which humans comprehend very well. Many experiences have shown that experts of the contents represented by a concept lattice reconstruct astonishingly fast relevant meanings of the presented concepts and relationships. Therefore, one can assume that depicted concept lattices are able to activate corresponding semantic structures in human thought.

Mathematically, concept lattices belong to the class of *lattice-ordered structures* which have been intensively studied since the late nineteenth century (cf. [DP02]). Today there exists a huge network of lattice-ordered structures which serve as mathematical semantic structures in theoretical and practice-oriented research. In particular, *Formal Concept Analysis* as mathematical theory of concept lattices benefits from this research.

Philosophically, labelled line diagrams of concept lattices are understood as representations of contextually founded *concept hierarchies* in which the concepts are constituted by their extensions and intensions (cf. [DIN79], [DIN80]). Those concept hierarchies function as logical semantic structures for the representation of knowledge. Thus, the *philosophical logic of concepts* has to play a basic role for the actual research on knowledge representation.

In special sciences, knowledge representations by concept lattices and their line diagrams are grounded on the *special semantics* of those sciences. Which semantic structures are activated, respectively, is usually dependent on aim and purposes of the knowledge representation. For example, the knowledge representation of the concept lattice depicted in Fig. 2 has the purpose to support architects in planning the shell of a one-family house subject to the relevant laws and regulations. For designing the connection of the chimney and the roof, the architect has therefore to observe all text units which are relevant for both. These text units form the extension of the largest subconcept of the two concepts generated by the attributes "chimney" and "roof". In the line diagram, those three concepts are represented by the non-labelled circle on the very left and the two circles with the labels "chimney" and "roof". The wanted text units are indicated by the labels to circles which can be reached by descending pathes starting from the non-labelled circle on the very left. Thus, the architect has to consider §15, §16, §17, and §18 Abs.1 of the Building Law of Nordrhein-Westfalen (BauONW) and the German standard DIN 1055.

3.3. A *system of concept lattices* based on subcontexts of a given formal context is desirable if the concept lattice of the whole context is too large. The general idea is that the concept lattice of the whole context are reconstructable from the concept lattices of the distinguished subcontexts. One might think of an atlas of maps where the maps are the concept lattices of the system (cf. [Wi85]). Knowledge representations by those systems could be metaphorically viewed as *conceptual landscapes of knowledge* where the navigation through such landscapes is supported by activated semantic structures (cf. [Wi97])

Mathematically, different types of concept lattice systems have been extensively investigated which is documented in [GW99a], mainly in the chapters on *decompositions* and *constructions* of concept lattices. A frequently used method of decomposition and construction leads to the so-called *nested line diagrams* which are often better readable than the usual line diagrams of concept lattices (cf. [Wi84]). In the basic case, the attribute set of a formal context is divided into two subsets which together with the object set form two subcontexts. Then, a line diagram of the concept lattice of one of the subcontexts is drawn with large rectangles for representing the formal concepts. Following that, the line diagram of the concept lattice of the other subcontext is copied in each of the large rectangles of the first diagram. Finally, in each copy of the second diagram those little circles are marked which represent the formal concepts of the original context (cf. [GW99a], p.90). To justify this construction, some basic mathematical argumentations are necessary (see [GW99a], p.75ff.).

Philosophically, a system of concept lattices visualized by labelled line diagrams is understood as a contextually founded *logical system of concept hierarchies* which can be elaborated to obtain useful knowledge representations. The most successful knowlege systems elaborated in that way are the so-called *TOSCANA-systems*, the design of which was inspired by the metaphor of conceptual landscapes of knowledge (cf. [KSVW94], [VW95], [Wi97]). Basically, a TOSCANA-system is founded on a formal context and a system of attribute collections covering all attributes of the given context. Each of those attribute collections together with the object collection of the context yields a subcontext, named a *conceptual scale*. The concept lattice of such a scale, visualized by a well-drawn line diagram, is named a *query structure* because it can be used to interrogate knowledge represented by the TOSCANA-system (see e.g. Fig. 2). Query structures can even be combined to show nested line diagrams which particularly support the navigation through the represented knowledge. The description of actual software for maintaining and activating TOSCANA-systems can be found in [BH05].

Special sciences with their special semantics and purposes give rise to *special systems of concept lattices*. This shall be exemplified by the following research project in *developmental psychology*. The psychologist Th. B. Seiler together with coworkers has investigated how the concept of "work" develops in the mind of children of the age of 5 to 13 (see [SKN92]). 62 children were interviewed about their understanding of "work", and the content of each of these interviews were represented by a concept lattice. Then the researchers put the line diagrams of those concept lattices up in a row according to the age of the children. That lining up made already clear to a large extent how to reconstruct the development of the concept of "work". Therefore no further analysis of connections between the 62 concept lattices was necessary. Nevertheless, a TOSCANA-system could have helped to represent an even more complete representation of the knowledge gained by the investigation. How much a special TOSCANA-system is formed according to the special contents and purposes of the desired knowledge system, this has been described in [EKSW00] for the already discussed *informa-tion system about laws and regulations concerning building construction.*

3.4. A *concept lattice with a collection of its attribute inferences*[2] extends the knowledge representations discussed in subsection **3.2** by offering knowledge about attribute inferences valid in the underlying context. This enlargement adds elements of an *inferential semantics based on (formal) attributes* to the structural semantics of the given concept lattice. Dually, one can enlarge the structural semantics by elements of an *inferential semantics based on (formal) objects*. Both enlargements can be unified in an *inferential semantics based on (formal) concepts*. The contextual foundation of those semantics allows to make structurally explicit the relationships between all those semantics.

Mathematically, a *theory of attribute implications* of formal contexts has been developed which, in particular, states that a concept lattice can be re-constructed as lattice, up to isomorphism, by the set of all attribute implica-tions valid in the underlying context (cf. [GW99a], Section 2.3). Furthermore, the *theory of attribute clauses* yields that a concept lattice with its object con-cepts as constants can be reconstructed as lattice with constants, up to iso-morphism, by the set of all attribute clauses valid in the underlying context (cf. [GW99b], Section 2). Both results show how close the structural seman-tics of concept lattices and the inferential semantics of attribute inferences are mathematically.

Philosophically, a concept lattice with a collection of attribute inferences can be understood as a *contextual logic of attribute inferences* based on a concept hierarchy. Such a logic is mathematically supported by the Contextual Con-cept Logic (cf. [Wi00a]), particularly by the Contextual Attribute Logic (cf. [GW99b]), both founded on Formal Concept Analysis. In the extended case of a system of concept lattices together with a collection of attribute inferences, the corresponding philosophical logic can be understood as a *logic of distributed systems* as developed in [BS97]. This logic has already remarkable applications as, for instance, to the construction and analysis of switching circuits (see e.g. [Kr99]).

In special sciences, a *contextual logic of attribute inferences* based on a concept hierarchy have a great variety of applications. Here, only the so-called *attribute exploration* (see [GW99a], p.85ff.) shall be mentioned as an appplication method. This method is used to complete knowledge representations making knowledge explicit which is implicitly coded in a specified universe of discourse. The key idea of the exploration is to ask step by step whether an attribute implication valid in the actual context is also valid in the universe. If yes, then the impli-cation is listed as valid in the universe. If not, then an object of the universe, which has all attributes of the implication premise but not all of the implication

---

[2] The most important attribute inferences of a formal context (G, M, I) are the *at-tribute implications* $B_1 \rightarrow B_2$ where B_1 and B_2 are subsets of M satisfying that each formal object having all formal attributes of B_1 has also all formal attributes of B_2. Further attribute inferences are the *attribute clauses* $B_1 \multimap B_2$ satisfying that each formal object having all formal attributes of B_1 has at least one attribute of B_2.

conclusion, has to be made explicit and added to the actual context (cf. [Wi06], method **M9.1**). For example, in exploring how adjectives can characterize musical pieces a typical question was: If the adjectives "dramatic", "transparent", and "lively" apply to a musical piece, do the adjectives "sprightly", "rhythmizing", and "fast" also apply to that piece? The answer was "No" and justified by naming as counterexample the third movement of Beethoven's moonlight sonata which was judged to be "dramatic", "transparent", and "lively", but not "sprightly" (cf. [WW06]).

3.5. A *power context family*[3] is a semantic structure which yields elementary representations of knowledge about connections between objects, attributes, basic and relational concepts. Mathematically, investigations of those connections may benefit from the research in the *Algebra of Relations* (cf. [PK79], [Poe04]). Philosophically, the multifarious work on the logic of relations yields a supporting background; in particular, Peirce's extensive work on the *logic of relatives* is a valuable source (for an introduction see [Pe92], Lecture Three). R. W. Burch has amalgamated various systems of logic, developed by Peirce over his long career, under the title *PAL* (Peircean Algebraic Logic) which extends the logic of relatives (see [Bu91]). In special sciences, a power context family might be viewed as a sequence of data tables representing relationships between (special) objects resp. object sequences and (special) attributes.

For creating knowledge representations by power context families, it is important to understand the close relationship between power context families and *relational databases* [AHV95]. In [EGSW00] it is shown how a representation of all flights inside Austria by a relational model in the sense of Codd can be transferred into a representation by a power context family. This power context family could even serve as basis for a contextual-logic extension of a TOSCANA-system, which was established by using PAL-term formations and their derivatives. Representations of the knowledge coded in the extended TOSCANA-system could then be activated by query graphs for retrieving flight information (see [EGSW00], Section 4).

3.6. *Concept graphs* of a power context family are semantic structures which represent (*formal*) *judgments* based on the knowledge represented by the underlying power context family and its concept lattices. Those judgments are understood in the sense of the traditional philosophical logic with its doctrines of concepts, judgments, and conclusion. This means that "the *matter* of judgment consists in given cognitions that are joined into unit of consciousness; in the determination of the manner in which various presentations as such belong to one consciousness consists the *form* of judgment" ([Ka88], p.106f).

---

[3] A *power context family* is mathematically defined a sequence $\vec{\mathbb{K}} := (\mathbb{K}_0, \mathbb{K}_1, \mathbb{K}_2, \ldots)$ of formal contexts $\mathbb{K}_k := (G_k, M_k, I_k)$ with $G_k \subseteq (G_0)^k$ for $k = 1, 2, \ldots$. The formal concepts of $\mathbb{K}_0$ are called *basic concepts*, those of $\mathbb{K}_k$ with $k = 1, 2, \ldots$ are called *relation concepts* because they represent k-ary relations on the basic object set G_0 by their extents (resp. extensions).

Mathematically, the form of judgment is set-theoretically defined as *concept graph* of a power context family[4]. The mathematical theory of concept graphs has been mainly developed to establisch a *Contextual Judgment Logic* which is understood as an extension of the Contextual Concept Logic (cf. [Wi00b]). This development does not only extend the structural semantics of formal concepts, but also the inferential semantics of concept graphs (cf. [Wi04]).

Philosophically, matter and form of judgments can be successfully represented by *conceptual graphs* which have been invented by J. F. Sowa [So84] and further developed by him and many other collaborators. In [So92], Sowa wrote: "Conceptual graphs are a system of logic based on the existential graphs of Charles Sanders Peirce and the semantic networks of artificial intelligence. The purpose of the system is to express meaning in a form that is logically precise, humanly readable, and computationally tractable." Conceptual graphs gave rise to their *mathematical abstraction* by concept graphs; therefore there is a very close connection between the representation of judgments by conceptual graphs and by concept graphs. A comparison of the philosophical foundations of conceptual graphs and of formal concept analysis in general is presented in [MSW99].

In special sciences, knowledge representations by concept(ual) graphs are frequent which could already be seen by papers published in the series of Springer lecture notes on conceptual structures. As example, only the TOSCANA representations, mentioned already under **3.5**, shall be discussed further. Such a TOSCANA-system offers flight information in the form of conceptual graphs. Those graphs occur as answers to requests which are given by constraints concerning flight connections, time intervals for departures and arrivals, possible flight days, etc. The answers are graphically presented as information networks which can be interactively changed, in particular, to smaller networks by increasing the inserted constraints and, finally, to a fixed flight schedule (cf. [EGSW00]).

4   Semantology and Ontology

It may be helpful to have a clear idea of the epistomological status of the structural entities which we call *"semantic structures"* in the field of knowledge representation and processing. How do we address the meaningful totality these structures are part (and taken) of? In which way are semantic structures related to the world? Is there a general methodological perspective on what knowledge representations are dealing with?

---

[4] A *concept graph* of a power context family $\vec{\mathbb{K}} := (\mathbb{K}_0, \mathbb{K}_1, \mathbb{K}_2, \ldots)$ with $\mathbb{K}_k := (G_k, M_k, I_k)$ for $k = 0, 1, 2, \ldots$ is a structure $\mathfrak{G} := (V, E, \nu, \kappa, \rho)$ for which
- (V, E, ν) is a relational graph, i.e. a structure (V, E, ν) consisting of two disjoint sets V and E together with a map $\nu : E \to \bigcup_{k=1,2,\ldots} V^k$,
- $\kappa : V \cup E \to \bigcup_{k=0,1,2,\ldots} \mathfrak{B}(\mathbb{K}_k)$ is a mapping such that $\kappa(u) \in \mathfrak{B}(\mathbb{K}_k)$ for all u with $u \in V$ if $k = 0$ or $\nu(u) = (v_1, \ldots, v_k) \in V^k$ if $k = 1, 2, \ldots$,
- $\rho : V \to \mathfrak{P}(G_0) \setminus \{\emptyset\}$ is a mapping such that $\rho(v) \subseteq Ext(\kappa(v))$ for all $v \in V$ and, furthermore, $\rho(v_1) \times \cdots \times \rho(v_k) \subseteq Ext(\kappa(e))$ for all $e \in E$ with $\nu(e) = (v_1, \ldots, v_k)$; in general, $Ext(\mathfrak{c})$ denotes the extent of the formal concept $\mathfrak{c}$.

We propose the term *"Semantology"* as a first step to answer those questions and as a basic methodological concept for knowledge representation research. "Semantology" is an existing expresion in linguistics. There it stands for "science of the meaning of words" or "science of semantics", but as far as we see without any important terminological function within the linguistic field. In a recent philosophical publication the word appears in the pejorative sense, meaning something like: thinking too much in semantical terms (cf. [Wa04]). In our context, "Semantology" can directly refer to the idea of a semantics implying, as semantics, meta-structure or universe or "archive" ([Fou69] p.103ff.) that can be explored by the means of different sciences of concepts (in the field of mathematics, e.g. the science of set-theoretic concepts versus the science of geometric concepts (cf. [Th71]); in the field of philosophy, one should, in particular, name the philosophical history of concepts; in special sciences, e.g. the science of musical concepts (cf. [Eg78])). More precisely, we understand Semantology as the *theory of semantic structures and their connections* which, in particular, make possible the creation of suitable methods for knowledge representations. Thus, Semantology should also cover the general methodology of representing information and knowledge.

In the discourse of computer sciences the term *"Ontology"* is used to indicate a certain complex or totality of meanings - an "entity" on a meta-level that semantic structures imply or on that they refer. Ontology in this general sense of exploring the relation of data to "world in general" or "reality" is also a keyword of todays artificial intelligence research. It aims at the idea of modelling how language is used to *specify our practical world*. So why do not speak of an ontology (or of ontologies in plural) to point out what may be the implicit and ultimate meta-structure/totality of "semantic structures" in the field of knowledge representation and processing?

The *concept of ontology* has a long and tangled history, although it has not at all, as an interpreter promoted it, "as old as philosophy itself" ([Bi03], p.632). The epistomological program named "Ontology" takes shape with the post-scholastical re-reading of Aristotle during the 17th Century. In the philosophy of enlightenment the idea of a *"science of being"* was part of the project of metaphysics as a general science of all that is possible to be thought insofar it as itself "is": in words of Leibniz: A Scientia Generalis "de Cogitabili in universum quatenus tale est" ([Lei03], p.511). Kant and Hegel criticized this sentence of "ontology" and replaced the term by more complex, at most indirectly ontological ("transcendental", "speculative") concepts.

Todays usage of the term "Ontology" in computer sciences seems to refer more or less vaguely to E. Husserl's idea of a *"formal ontology"* ([Hu50], p.27). According to P. Øhrstrøm et al. the analytic philosopher W. V. O. Quine must be considered as an author of great influence ([OAS05], p.433) what may be the case in the sphere of analytic philosophy. The claim of Husserl's theory (as of Quine's) is not a metaphysical one. Nevertheless, the pragmatic usage of the word "Ontology" in todays computer science is imprecise - and exactly in this point: it is indecisive in regard to possible metaphysical implications. "Ontology" leaves the question of the status of the background-idea of a "world" as a meta-structure of semantic structures either open or it shifts more or less involuntarily into ancient philosophical metaphysics: it indicates the existence or even truth of something like a "plain" reality in a quite naïve way. In other words: a tacid suggestion is lying in the term "Ontology", the suggestion that there may exist a plain reality which a scientific model can match.

From our point of view the coining of the term *"Semantology"* may help to avoid naturalistic or essentialistic fallacies. It is precisely a complex semanto-logical, and not an ontological totality, that is built-up (and reduced) by the semantic structures that knowledge representation addresses (and reveals). One may speak of a semantological "world" or better - as above - of a "universe". But neither a certain semantic structure nor a complex totality of them should be linked to the idea of being.

5   Further Research

Our idea of *Semantology* has to be elaborated further which, in particular, in-cludes to make the corresponding notion of reality more explicit. On such basis the *methods of knowledge representation* have to be widely extended and consol-idated; that research may be supported by the detailed analysis of the concept of concepts in [Wi05b] and the presented methods of conceptual knowledge pro-cessing in [Wi06]. The development of further graphical representation methods is particularly interesting. Generally, it would be scientifically important to in-vestigate methods developed in computer science by the program of *Ontology* (cf. [SS04]) whether they could also be understood semantologically.

References

[AHV95]   S. Abiteboul, R. Hull, V. Vianu: *Foundations of databases.* Adison-Wesley, Reading/Mass. 1995.

[BS97]   J. Barwise, J. Seligman: *Information flow: the logic of distributed systems.* Cambridge University Press, Cambridge/UK 1997.

[BH05]   P. Becker, J. Hereth Correia: The ToscanaJ Suite for implementing con-ceptual information systems. In: [GSW05], 324–348.

[Bi03]   D. Bielefeld: Ontology. In: *The Encyclopedia of Science and Religion.* Vol. 2 (2003).

[Bu91]   R. W. Burch: *A Peircean reduction thesis.* Texas Tech University Press, Lubbock 1991.

[DP02]   B. A. Davey, H. A. Priestley: *Introduction to lattices and order.* 2nd edition. Cambridge University Press, Cambridge/UK 2002.

[De70]   P. Dembowski: Kombinatorik. Bibliographisches Institut, Mannheim 1970.

[DIN79]   Deutsches Institut für Normung: DIN 2330 - Begriffe und Benennungen: Allgemeine Grundsätze. Beuth, Berlin-Köln 1979.

[DIN80]   Deutsches Institut für Normung: DIN 2331 - Begriffssysteme und ihre Darstellung. Beuth, Berlin-Köln 1980.

[Eg78]   H. H. Eggebrecht (Hrsg.): *Handbuch der musikalischen Terminologie.* Stuttgart 1978.

[EGSW00] P. W. Eklund, B. Groh, G. Stumme, R. Wille: A contextual-logic extension of TOSCANA. In: B. Ganter, G. Mineau (eds.): *Conceptual structures: logical, linguistic and computational issues.* LNAI **1867**. Springer, Heidelberg 2000, 453-467.

[EKSW00] D. Eschenfelder, W. Kollewe, M. Skorsky, R. Wille: Ein Erkundungssystem zum Baurecht: Methoden der Entwicklung eines TOSCANA-Systems. In: [SW00], 254–272.

[Fou69] M. Foucault: *L'archéologie du savoir.* Gallimard, Paris 1969.

[GSW05] B. Ganter, G. Stumme, R. Wille (eds.): *Formal Concept Analysis: foundations and applications.* State-of-the-Art Survey. LNAI **3626**. Springer, Heidelberg 2005.

[GW99a] B. Ganter, R. Wille: *Formal Concept Analysis: mathematical foundations.* Springer, Heidelberg 1999.

[GW99b] B. Ganter, R. Wille: Contextual Attribute Logic. In: W. Tepfenhart, W. Cyre (eds.): *Conceptual structures: standards and practices.* LNAI **1640**. Springer, Heidelberg 1999, 377-388.

[Hu50] E. Husserl: *Ideen zu einer reinen Phänomenologie und phänomenologischen Philosophie* (= Husserliana 3). Kluwer, Den Haag 1950.

[Ka88] I. Kant: Logic. Dover, Mineola 1988.

[Kr99] M. Karl: *Eine Logik verteilter Systeme und deren Anwendung auf Schaltnetzwerke.* Diplomarbeit. FB Mathematik, TU Darmstadt 1999.

[KSVW94] W. Kollewe, M. Skorsky, F. Vogt, R. Wille: TOSCANA - ein Werkzeug zur begrifflichen Analyse und Erkundung von Daten. In: [WZ94], 267–288.

[Lei03] G. W. Leibniz: Opuscules et fragmentes inédites. Edited by L. Coutural. Paris 1903. New print 1961.

[MSW99] G. Mineau, G. Stumme, R. Wille: Conceptual structures represented by conceptual Graphs and formal concept analysis. In: W. Tepfenhart, W. Cyre (eds.): *Conceptual structures: standards and practices.* LNAI **1640**. Springer, Heidelberg 1999, 423-441.

[OAS05] P. Øhrstrøm, J. Andersen, H. Schärfe: What has happened to ontology. In: F. Dau, M.-L. Mugnier, G. Stumme (eds.): *Conceptual structures: common semantics for sharing knowledge.* LNAI **3596**. Springer, Heidelberg 2005, 425–438.

[Pe92] Ch. S. Peirce: *Reasoning and the logic of things.* Edited by K. L. Ketner; with an introduction by K. L. Ketner and H. Putnam. Havard University Press, Cambridge 1992.

[Poe04] R. Pöschel: Galois connections for operations and relations. In: K. Denecke, M. Erné, S. L. Wismath (eds): *Galois connections and applications.* Kluwer, Dordrecht 2004, 231–258.

[PK79] R. Pöschel, L. A. Kaluznin: *Funktionen und Relationenalgebren.* VEB Verlag der Wissenschaften, Berlin 1979.

[SKN92] Th. B. Seiler, B. Koböck, B. Niedermeier: Rekonstruktion der Entwicklung des Arbeitsbegriffs mit Mitteln der Formalen Begriffsanalyse. Manuskript. TU Darmstadt 1992.

[So84] J. F. Sowa: *Conceptual structures: information processing in mind and machine.* Adison-Wesley, Reading 1984.

[So92] J. F. Sowa: Conceptual graph summary. In: T. E. Nagle, J. A. Nagle, L. L. Gerholz, P. W. Eklund (eds.): *Conceptual structures: current research and practice.* Ellis Horwood, 1992, 3–51.

[SS04] S. Staab, R. Studer (eds.): *Handbook on Ontologies.* Springer, Heidelberg 2004.

[SW00]   G. Stumme, R. Wille (Hrsg.): *Begriffliche Wissensverarbeitung: Methoden und Anwendungen.* Springer, Heidelberg 2000.

[Th71]   R. Thom: 'Modern' mathematics: an educational and philosophical error? *American Scientist* **59** (1971), 695–699.

[VW95]   F. Vogt, R. Wille: TOSCANA – A graphical tool for analyzing and exploring data. In: R. Tamassia, I. G. Tollis (eds.): *Graph drawing '94.* LNCS **894**. Springer, Heidelberg 1995, 226–233.

[Wa04]   D. von Wächter: Ontologie und Semantologie. In: M. Siebel, M. Textor (eds.): *Semantik und Ontologie. Beiträge zur philosophischen Forschung.* Ontos, Frankfurt a. M., London 2004.

[Wi84]   R. Wille: Liniendiagramme hierarchischer Begriffssysteme. In: H. H. Bock (Hrsg.): *Anwendungen der Klassifikation: Datenanalyse und numerische Klassifikation.* Indeks-Verlag, Frankfurt 1984, 32–51; English translation: Line diagrams of hierachical concept systems. *International Classification* **11** (1984), 77–86.

[Wi85]   R. Wille: Complete tolerance relations of concept lattices. In: G. Eigenthaler, H. K. Kaiser, W. B. Müller, W. Nöbauer (eds.): *Contributions to General Algebra 3.* Hölder-Pichler-Temsky, Wien 1985, 397–415.

[Wi97]   R. Wille: Conceptual landscapes of knowledge: a pragmatic paradigm for knowledge processing. In: G. Mineau, A. Fall (eds.): *Proceedings of the International Symposium on Knowledge Representation, Use, and Storage Efficiency.* Simon Fraser University, Vancouver 1997, 2–13; reprinted in: W. Gaul, H. Locarek-Junge (Eds.): *Classification in the Information Age.* Springer, Heidelberg 1999, 344–356.

[Wi00a]   R. Wille: Boolean Concept Logic. In: B. Ganter, G. Mineau (eds.): *Conceptual structures: logical, linguistic and computational issues.* LNAI **1867**. Springer, Heidelberg 2000, 317-331.

[Wi00b]   R. Wille: Contextual Logic summary. In: G. Stumme (ed.): *Working with conceptual structures. Contributions to ICCS 2000.* Shaker-Verlag, Aachen 2000, 265–276.

[Wi04]   R. Wille: Implicational concept graphs. In: K. E. Wolff, H. Pfeiffer, H. Delugach (eds.): *Conceptual structures at work.* LNAI **3127**. Springer, Heidelberg 2004, 52–61.

[Wi05a]   R. Wille: Allgemeine Wissenschaft und transdisziplinäre Methodologie. *Technikfolgenabschätzung - Theorie und Praxis* Nr. 2, 14. Jahrgang, Forschungszentrum Karlsruhe 2005, 57–62.

[Wi05b]   R. Wille: Formal Concept Analysis as mathematical theory of concepts and concept hierarchies. In: [GSW05], 1–33.

[Wi05c]   R. Wille: Conceptual Knowledge Processing in the field of economics. In: [GSW05], 226–249.

[Wi06]   R. Wille: Methods of Conceptual Knowledge Processing. In: R. Missaoui, J. Schmid (eds.): *Formal Concept Analysis. ICFCA 2006.* LNAI **3874**. Springer, Heidelberg 2006, 1–29.

[WW06]   R. Wille, R. Wille-Henning: Beurteilung von Musikstücken durch Adjektive: Eine begriffsanalytische Exploration. In: K. Proost, E. Richter (Hrsg.): *Von Intentionalität zur Bedeutung konventionalisierter Zeichen. Festschrift für Gisela Harras zum 65. Geburtstag.* Narr, Tübingen 2006, 453–475.

[WZ94]   R. Wille, M. Zickwolff (Hrsg.): *Begriffliche Wissensverarbeitung - Grundfragen und Aufgaben.* B.I.-Wissenschaftsverlag, Mannheim 1994.

The Teridentity and Peircean Algebraic Logic

Joachim Hereth Correia and Reinhard Pöschel

Technische Universität Dresden
Fakultät Mathematik und Naturwissenschaften, Institut für Algebra
D-01062 Dresden, Germany
{Joachim.Hereth_Correia, Reinhard.Poeschel}@tu-dresden.de

Abstract. A main source of inspiration for the work on Conceptual Graphs by John Sowa and on Contextual Logic by Rudolf Wille has been the Philosophy of Charles S. Peirce and his logic system of Existential Graphs invented at the end of the 19th century. Although Peirce has described the system in much detail, there is no formal definition which suits the requirements of contemporary mathematics.

In his book *A Peircean Reduction Thesis: The Foundations of topological Logic*, Robert Burch has presented the Peircean Algebraic Logic (PAL) which aims to reconstruct in an algebraic precise manner Peirce's logic system.

Using a restriction on the allowed constructions, he is able to prove the Peircean Reduction Thesis, that in PAL all relations can be constructed from ternary relations, but not from unary and binary relations alone. This is a mathematical version of Peirce's central claim that the category of thirdness cannot be decomposed into the categories of firstness and secondness.

Removing Burch's restriction from PAL makes the system very similar to the system of Existential Graphs, but the proof of the Reduction Thesis becomes extremely complicated. In this paper, we prove that the teridentity relation is – as also elaborated by Burch – irreducible, but we prove this without the additional restriction on PAL. This leads to a proof of the Peircean Reduction Thesis.

Introduction

The influence of Peirce's philosophy on the development of the theory of conceptual structures is visible in many areas. Both conceptual graphs (see [Sow84], [Sow92], [Sow00]) and the developments in contextual logic (see [Arn01], [Wil00], [Wil00b], [DaK05]) are influenced by his ideas in general and his system of existential graphs in particular.

Philosphical ideas and Peirce's work on formalizing logic converge on the Reduction Thesis: "The triad is the lowest form of relative from which all others can be derived." (MS 482 from [PR67]). This expresses both his philosophical believe that the categories of firstness, secondness and thirdness suffice and no category of fourthness etc. is needed. Also it is to be understood that all relatives (these correspond to relations in nowadays mathematical terminology) can be

H. Schärfe, P. Hitzler, and P. Øhrstrøm (Eds.): ICCS 2006, LNAI 4068, pp. 229–246, 2006.

generated from triads (ternary relations) but not from unary and binary relations alone. Peirce was conviced that at least on the mathematical level this thesis can be proven. According to Herzberger in [Her81] Peirce mentioned he found a proof, but no corresponding publication has been found.

In his article [Her81], Herzberger summarizes Peirce's understanding on the thesis and provides a first approach for an algebraic proof. In [Bur91], Burch gives a more extended and elaborated framework. He shows that his framework, the Peircean Algebraic Logic is able to represent the same relations as the existential graphs. However, to prove the Reduction Thesis, he imposes a restriction on the constructions in PAL. The juxtaposition of graphs (this corresponds to the product Def. 1.1(PAL1)) is only allowed as last or before the last operation. Removing this restriction makes PAL simpler (our version of PAL needs only one join-operator as opposed to two in [Bur91]) and probably more alike to the system of existential graphs. The proof of the reduction thesis in contrast becomes exceedingly difficult.

Many attempts have failed for non-obvious reasons. In fact, often the parts that seemed to be obvious turned out to be wrong afterwards. For this reason we present the complete mathematical proof of the difficult part of the reduction thesis. Due to space restrictions, we will not show the part that any relation can be constructed (in PAL) from ternary relations. For this, we refer to [Her81], [Bur91] or [HCP04].

Organization of This Paper

In the following section we present the various tools needed to describe the relations that can be generated from unary and binary relations. Each subsection will be introduced by a comment on the purpose of the subsequent definitions. Then the representation theorem for the relations that can be generated without ternary relations will be presented. The paper concludes with a short final section consisting of only the reduction thesis.

Mathematical Notations

To avoid disambiguities, we define some abbreviations used in this paper. The set of all m-ary relations over some set A is denoted by $\mathrm{Rel}^{(m)}(A) := \{\varrho \mid \varrho \subseteq A^m\}$ (and relations will be denoted by greek letters). The set of all relations is denoted by $\mathrm{Rel}(A) := \bigcup\{\mathrm{Rel}^{(m)}(A) \mid m \in \mathbb{N}\}$. Please note, that also empty relations have arities. Empty relations with different arities are considered to be different, that is for $n \neq m$ the empty relations $\emptyset^n \subseteq A^n$ and $\emptyset^m \subseteq A^m$ are considered to be different. Often we will talk about the places of a relation. If m is the arity of a relation, we will write $\underline{m}$ instead of $\{1, \ldots, m\}$. The empty set $\emptyset$ is identified with $\emptyset^1$. The arity of a relation ϱ is denoted by $\mathrm{ar}(\varrho)$.

A tuple $(a_1, \ldots, a_n)$ will be shortened to the notation $\underline{a}$ if the arity of the relation the tuple belongs to can be derived from the context. If not otherwise noted, A denotes an arbitrary set.

1   Peircean Algebraic Logic (PAL)

The operations of the *Peircean Algebraic Logic (PAL)* are closely related to the existential graphs that Peirce developed in the late 1890s. They have been identified by Burch in [Bur91] as the fundamental operations in Peirce's understanding of the manipulation of relations. For a detailed discussion of these operations we refer to [Bur91], for this paper we adopt Burch's operations.

1.1 Definition. Let $\varrho \in \mathrm{Rel}^{(m)}(A)$ and $\sigma \in \mathrm{Rel}^{(n)}(A)$. We define the following operations:

(PAL1) The *product* of relations:

$$\varrho \times \sigma := \{(a_1, \ldots, a_m, b_1, \ldots, b_n) \in A^{m+n} \mid \underline{a} \in \varrho, \, \underline{b} \in \sigma\},$$

(PAL2) for $1 \le i < j \le m$ the *join* of i and j of a relation is defined by

$$\delta^{i,j}(\varrho) :=$$
$$\{(a_1, \ldots, a_{i-1}, a_{i+1}, \ldots, a_{j-1}, a_{j+1}, \ldots, a_m) \in A^{m-2} \mid \exists \underline{a} \in \varrho : a_i = a_j\}$$

(PAL3) $\neg \varrho := \{\underline{a} \in A^m \mid \underline{a} \notin \varrho\}$ (the *complement* of ϱ),

(PAL4) if α is a permutation on $\underline{m}$, then
$$\pi_\alpha(\varrho) := \{(a_1, \ldots, a_m) \mid (a_{\alpha(1)}, \ldots, a_{\alpha(m)}) \in \varrho\}.$$

Remark 1. Let ϱ be an m-ary relation, let $1 \le i < j \le m$ and let α be the folowing permutation on $\underline{m}$:

$$\begin{pmatrix} 1, \ldots, i-1, i+1, \ldots, j-1, j+1, \ldots, m-1, & m, & i, & j \\ 1, \ldots, i-1, & i, & \ldots, j-2, j-1, \ldots, m-3, m-2, m-1, m \end{pmatrix}$$

π_α moves the i-th and j-th place of a relation to the $m-1$-th and m-th place. Then we have $\delta^{i,j}(\varrho) = \delta^{m-1,m}(\pi_\alpha(\varrho))$. For this reason we will only have to investigate the specific case $\delta^{m-1,m}(\varrho)$ as the general case can be derived together with the permutation operation.

Syntactically the terms of PAL are symbols combined by (symbols of) the operations of PAL. In this paper, the symbols will always stand for relations which will be naturally interpreted as the relations themselves. Formally, this is expressed by the following definition.

1.2 Definition. Let Σ be a set with $\mathrm{id}_3 \notin \Sigma$ and let $\mathrm{ar} : \Sigma \to \mathbb{N}$ be a mapping. Let $\Sigma_0 := \Sigma \cup \{\mathrm{id}_3\}$. The elements of Σ are called *atomic (Σ, ar)-PAL-terms* (briefly *atomic Σ-term*), id_3 is called *(syntactical) teridentity*. We set $\mathrm{ar}_0(t) := \mathrm{ar}(t)$ for all $t \in \Sigma$ and $\mathrm{ar}_0(\mathrm{id}_3) := 3$. We define recursively for all $i \in \mathbb{N}$ the sets

$$\begin{aligned}
\Sigma_{i+1} := \quad &\Sigma_i \\
&\cup \{(t \times s) \mid t, s \in \Sigma_i\} \\
&\cup \{\delta^{i,j}(t) \mid t \in \Sigma_i \text{ and } 1 \le i < j \le \mathrm{ar}_i(t)\} \\
&\cup \{\neg t \mid t \in \Sigma_i\} \\
&\cup \{\pi_\alpha(t) \mid \in \Sigma_i \text{ and } \alpha \text{ is permutation of } \underline{\mathrm{ar}_i(t)}\}
\end{aligned}$$

(note that the operations are considered as syntactical symbols, not as operations on relations) and correspondingly $\mathrm{ar}_{i+1}(u) := \mathrm{ar}_i(u)$ if $t \in \Sigma_i \cap \Sigma_{i+1}$ and for $u \in \Sigma_{i+1} \setminus \Sigma_i$ we define

$$\mathrm{ar}_{i+1}(u) := \begin{cases} \mathrm{ar}_i(t) + \mathrm{ar}_i(s) & \text{if } u = (t \times s), \\ \mathrm{ar}_i(t) - 2 & \text{if } u = \delta^{i,j}(t), \\ \mathrm{ar}_i(t) & \text{if } u = \neg t, \\ \mathrm{ar}_i(t) & \text{if } u = \pi_\alpha(t). \end{cases}$$

Obviously, we have $\mathrm{ar}_{i+1} \upharpoonright_{\Sigma_i} = \mathrm{ar}_i$. Instead of the (syntactially correct) $\delta^{i,j}((t \times s))$ we will write $\delta^{i,j}(t \times s)$. The set of (Σ, ar)-*PAL-terms* (or Σ-terms for short) is $\mathsf{T}_{\mathsf{PAL}}(\Sigma, \mathrm{ar}) := \bigcup_{i \in \mathbb{N}} \Sigma_i$. The mapping $\mathrm{ar}' := \bigcup_{i \in \mathbb{N}} \mathrm{ar}_i$ assigns the *arity* to each PAL-term.

An *interpretation of* $\mathsf{T}_{\mathsf{PAL}}(\Sigma, \mathrm{ar})$ *over* A is a mapping from $\mathsf{T}_{\mathsf{PAL}}(\Sigma, \mathrm{ar})$ into $\mathrm{Rel}(A)$, based on a mapping $[\![\ ]\!] : \Sigma \to \mathrm{Rel}(A)$ satisfying $\mathrm{ar}([\![t]\!]) = \mathrm{ar}(t)$, that is an n-ary atomic term has to be mapped to an n-ary relation. This mapping is then extended canonically to the set $\mathsf{T}_{\mathsf{PAL}}(\Sigma, \mathrm{ar})$ by translating the syntactial operation symbols into the corresponding operations on the relations, that is $[\![\mathrm{id}_3]\!] := \{(a, a, a) \mid a \in A\}$ and

$$[\![u]\!] := \begin{cases} [\![t]\!] \times [\![s]\!] & \text{if } u = (t \times s), \\ \delta^{i,j}([\![t]\!]) & \text{if } u = \delta^{i,j}(t), \\ \neg([\![t]\!]) & \text{if } u = \neg t, \\ \pi_\alpha([\![t]\!]) & \text{if } u = \pi_\alpha(t). \end{cases}$$

In the case $\Sigma \subseteq \mathrm{Rel}(A)$ the *natural interpretation* is given by $[\![\varrho]\!] := \varrho$ for $\varrho \in \Sigma$. In this case we deliberately blur the distinction between syntax and semantics which is clear from the context. Because the arity $\mathrm{ar}(\varrho)$ is canonically given for $\varrho \in \Sigma \subseteq \mathrm{Rel}(A)$ we write $\mathsf{T}_{\mathsf{PAL}}(\Sigma)$ instead of $\mathsf{T}_{\mathsf{PAL}}(\Sigma, \mathrm{ar})$. The set of relations which can be generated with PAL from the relations in Σ is denoted by $\langle \Sigma \rangle_{\mathsf{PAL}}^A := \{[\![t]\!] \mid t \in \mathsf{T}_{\mathsf{PAL}}(\Sigma)\}$.

Analogously we define the set of (Σ, ar)-PAL$\setminus\{\mathrm{id}_3\}$-terms which is denoted by $\mathsf{T}_{\mathsf{PAL}\setminus\{\mathrm{id}_3\}}(\Sigma, \mathrm{ar})$, and for $\Sigma \subseteq \mathrm{Rel}(A)$ the sets $\mathsf{T}_{\mathsf{PAL}\setminus\{\mathrm{id}_3\}}(\Sigma)$ and $\langle \Sigma \rangle_{\mathsf{PAL}\setminus\{\mathrm{id}_3\}}^A$ for PAL without teridentity by replacing the definition of Σ_0 by $\Sigma_0 := \Sigma$.

Remark 2. Different terms may be interpreted as the same relation. For instance, for relations $\varrho, \sigma, \tau \in \Sigma \subseteq \mathrm{Rel}(A)$ the following identity of interpretations $[\![((\varrho \times \sigma) \times \tau)]\!] = [\![(\varrho \times (\sigma \times \tau))]\!]$ is easy to see, but formally the terms $((\varrho \times \sigma) \times \tau)$ and $(\varrho \times (\sigma \times \tau))$ are different.

Connected Places

Associated with PAL is a graphical notation, as presented in [Bur91] and [HCP04]. In the graphical representation it is obvious how places (called hooks in [Bur91]) are connected with each other. As we need the notion of connectedness but will not introduce the graphical representation, we define connectedness formally following the constructions by PAL-terms.

1.3 Definition. Let $\Sigma \subseteq \mathrm{Rel}(A)$ and $t \in \mathsf{T_{PAL}}(\Sigma)$ be a term with $m := \mathrm{ar}(t)$. The places $k, l \in \underline{m}$ are said to be *t-connected* if one of the following conditions is satisfied:

   (i) $t \in \Sigma \cup \{\mathsf{id}_3\}$ or
   (ii) $t = (u \times v)$ and k, l are u-connected or $(k - m), (l - m)$ are v-connected or
   (iii) $t = \delta^{m-1,m}(u)$ and k, l are u-connected or
  (iii') $t = \delta^{m-1,m}(u)$ and $k, m - 1$ and l, m are u-connected or
  (iii") $t = \delta^{m-1,m}(u)$ and $l, m - 1$ and k, m are u-connected or
  (iv) $t = \neg u$ and k, l are u-connected or
   (v) $t = \pi_\alpha(u)$ and $\alpha^{-1}(k), \alpha^{-1}(l)$ are u-connected.

A set $P \subseteq \underline{m}$ is said to be *t-connected* if the elements of P are pairwise *t*-connected.

For the reduction thesis the relations generated by PAL without teridentity are very important. The following lemma is a first indication on a special property of these relations.

1.4 Lemma. *Let Σ be a set and* $\mathrm{ar} : \Sigma \to \mathbb{N}$ *with* $\max\{\mathrm{ar}(\sigma) \mid \sigma \in \Sigma\} \leq 2$. *Let* $t \in \mathsf{T}_{\mathsf{PAL} \setminus \{\mathsf{id}_3\}}(\Sigma, \mathrm{ar})$ *and let $X \subseteq \underline{\mathrm{ar}(t)}$ be t-connected. Then $|X| \leq 2$.*

Proof. We proceed by induction on the structure of terms. For atomic terms the assertion trivially holds. The case $t = \mathsf{id}_3$ is not possible because PAL is considered without teridentity. If $t = (u \times v)$, it is easy to verify that two places can be t-connected only if they are both less or equal to $\mathrm{ar}(u)$ or both strictly greater. This means that either $\max(X) \leq \mathrm{ar}(u)$ or $\min(X) > \mathrm{ar}(u)$, and consequently X is u-connected or $\{x - \mathrm{ar}(u) \mid x \in X\}$ is v-connected. By the induction hypothesis one concludes $|X| = |\{x - \mathrm{ar}(u) \mid x \in X\}| \leq 2$. Now let us consider the case $t = \delta^{m-1,m}(u)$ where $m := \mathrm{ar}(u)$. If there are $x, y \in X$ with $x \neq y$ such that x and y are u-connected, one conludes from the induction hypothesis that x and y cannot be u-connected to $m - 1$ or m, therefore the cases Def. 1.3(iii') and (iii") cannot apply for x and y and there can be no third element t-connected to x or y. If all $x, y \in X$ with $x \neq y$ are not u-connected then in order to be t-connected they must be u-connected to $m - 1$ or m. Therefore in this case $X \subseteq (\{k \in \underline{m} \mid k, m-1\ u\text{-connected}\} \cup \{k \in \underline{m} \mid k, m\ u\text{-connected}\}) \setminus \{m - 1, m\}$ and therefore $|X| \leq 2 + 2 - 2 = 2$. For $t = \neg u$ the set X is t-connected if and only if X is u-connected, therefore the assertion holds. For $t = \pi_\alpha(u)$ the assertion can easily be seen because α is a bijection and one can therefore apply the inverse mapping: X is t-connected $\Longleftrightarrow \{\alpha^{-1}(x) \mid x \in X\}$ is u-connected.

Essential Places

Later we shall introduce representations of relations as unions of intersections of special relations. Formally, these special relations have to have the same arity as the relation represented. However, they are essentially unary or binary relations. To make formally clear what "essentially" means, we introduce the notion of "essential places".

1.5 Definition. Let $\varrho \in \mathrm{Rel}^{(m)}(A)$ and $i \in \underline{m}$ a place of the relation. A place i is called a *fictitious place of* ϱ if

$$\forall \underline{a} \in \varrho \; \forall b \in A \implies (a_1, \ldots, a_{i-1}, b, a_{i+1}, \ldots, a_m) \in \varrho.$$

A non-fictitious place is called *essential place of* ϱ. The set of essential places of ϱ is denoted by $E(\varrho)$.

Essential places are the places of the relation, where one cannot arbitrarily exchange elements in the tuple.

1.6 Lemma. *For any relation* $\varrho \in \mathrm{Rel}^{(m)}(A)$ *holds*

$$\varrho \in \{\emptyset^m, A^m\} \iff E(\varrho) = \emptyset.$$

Proof. "$\implies$" is easy to see. For "$\impliedby$" let $\varrho \in \mathrm{Rel}^{(m)}(A) \setminus \{\emptyset^m\}$. Let $\underline{a} \in \varrho$ and $\underline{b} \in A^m$. Every $i \in \underline{m}$ is a fictitious place of ϱ, therefore a_i can be replaced by b_i and one gets $(a_1, \ldots, a_{i-1}, b_i, a_{i+1}, \ldots, a_m) \in \varrho$. Consecutively applying this exchange for all places shows $\underline{b} \in \varrho$ and consequently $\varrho = A^m$. □

The following lemmata are useful to show in the representation Theorem 2.1 that the special relations of the representations are essentially at most binary.

1.7 Lemma. *Let A be a set with at least two elements, let ϱ be an m-ary relation and let σ be another relation over A. Then*

   (i) $E(\varrho) \subseteq \underline{m}$,
   (ii) $E(\mathrm{id}_3) = \underline{3}$,
   (iii) $E(\varrho \times \sigma) = E(\varrho) \cup \{m + i \mid i \in E(\sigma)\}$,
   (iv) $E(\delta^{m-1,m}(\varrho)) \subseteq E(\varrho) \setminus \{m - 1, m\}$
   (v) $E(\neg\varrho) = E(\varrho)$
   (vi) $E(\pi_\alpha(\varrho)) = \{\alpha^{-1}(i) \mid i \in E(\varrho)\}$.

Proof. (i) and (ii) are trivial. For (iii) let $\varrho \in \mathrm{Rel}^{(m)}(A)$ and $\sigma \in \mathrm{Rel}^{(n)}(A)$. For $1 \leq i \leq m$ and $i \notin E(\varrho)$ one has the following equivalencies:

$$(a_1, \ldots, a_m, b_1, \ldots, b_n) \in \varrho \times \sigma, c \in A$$
$$\iff \underline{a} \in \varrho, \underline{b} \in \sigma, c \in A$$
$$\iff a_i \in A, (a_1, \ldots, a_{i-1}, c, a_{i+1}, \ldots, a_m) \in \varrho, \underline{b} \in \sigma$$
$$\iff a_i \in A, (a_1, \ldots, a_{i-1}, c, a_{i+1}, \ldots, a_m, b_1, \ldots, b_n) \in \varrho \times \sigma$$

and similarly for $1 \leq i \leq n$

$$(a_1, \ldots, a_m, b_1, \ldots, b_n) \in \varrho \times \sigma, c \in A$$
$$\iff \underline{a} \in \varrho, \underline{b} \in \sigma, c \in A$$
$$\iff b_i \in A, \underline{a} \in \varrho, (b_1, \ldots, b_{i-1}, c, b_{i+1}, \ldots, b_n) \in \sigma$$
$$\iff b_i \in A, (a_1, \ldots, a_m, b_1, \ldots, b_{i-1}, c, b_{i+1}, \ldots, b_n) \in \varrho \times \sigma.$$

Therefore $\neg E(\varrho \times \sigma) = (\underline{m} \setminus E(\varrho)) \cup \{m + i \mid i \in \underline{n} \setminus E(\sigma)\}$, consequently $E(\varrho \times \sigma) = E(\varrho) \cup \{m + i \mid i \in E(\sigma)\}$. (iv) Let $i \in \underline{m-2} \setminus E(\varrho)$, $c \in A$ and $(a_1, \ldots, a_{m-2}) \in \delta^{m-1,m}(\varrho)$, then there exists $b \in A$ with $(a_1, \ldots, a_{m-2}, b, b) \in \varrho$. Because i is fictitious we have $(a_1, \ldots, a_{i-1}, c, a_{i+1}, \ldots, a_{m-2}, b, b) \in \varrho$ and therefore $(a_1, \ldots, a_{i-1}, c, a_{i+1}, \ldots, a_{m-2}) \in \delta^{m-1,m}(\varrho)$. We deduce $\underline{m-2} \setminus E(\varrho) \subseteq \neg E(\delta^{m-1,m}(\varrho))$, that is $E(\delta^{m-1,m}(\varrho)) \subseteq E(\varrho) \setminus \{m-1, m\}$. (v) If $\varrho = \emptyset^m$ then this follows from Lem. 1.6. Otherwise let $\underline{a} \in A^m \setminus \varrho$ and $i \in \underline{m} \setminus E(\varrho)$. Let us assume that $i \in E(\neg \varrho)$. Then there must be some $c \in A$ such that $(a_1, \ldots, a_{i-1}, c, a_{i+1}, \ldots, a_m) \in \varrho$. But because $i \notin E(\varrho)$ and $a_i \in A$ this implies $\underline{a} \in \varrho$, contradiction. Therefore $\underline{m} \setminus E(\varrho) = \underline{m} \setminus E(\neg \varrho)$, that is $E(\neg \varrho) = E(\varrho)$. (vi) is easy to verify.

1.8 Lemma. *Let $S \subseteq \mathrm{Rel}^{(m)}(A)$. Then*

$$E(\bigcap S) \subseteq \bigcup_{\sigma \in S} E(\sigma).$$

Proof. Let $i \in \underline{m} \setminus (\bigcup_{\sigma \in S} E(\sigma))$, $\underline{a} \in \bigcap S$ and $c \in A$. Then for all $\sigma \in S$ holds $i \notin E(\sigma)$ and therefore $(a_1, \ldots, a_{i-1}, c, a_{i+1}, \ldots, a_m) \in \sigma$ and consequently $(a_1, \ldots, a_{i-1}, c, a_{i+1}, \ldots, a_m) \in \bigcap S$, therefore $i \in \underline{m} \setminus E(\bigcap S)$. Thus we get $E(\bigcap S) \subseteq \bigcup_{\sigma \in S} E(\sigma)$. $\qquad\square$

1.9 Lemma. *Let $S \subseteq \mathrm{Rel}^{(m)}(A)$ for some $m \in \mathbb{N}$. Then*

$$\delta^{m-1,m}(\bigcap S) = \left(\bigcap_{\substack{\sigma \in S \\ \{m-1,m\} \cap E(\sigma) = \emptyset}} \delta^{m-1,m}(\sigma) \right) \cap \delta^{m-1,m} \left(\bigcap_{\substack{\sigma \in S \\ \{m-1,m\} \cap E(\sigma) \neq \emptyset}} \sigma \right).$$

Proof. "$\subseteq$": Let $\underline{a} \in \delta^{m-1,m}(\bigcap S)$. Then there exists some $c \in A$ such that $\underline{b} := (a_1 \ldots, a_{m-2}, c, c) \in \bigcap S$, therefore for all $\sigma \in S$ also $\underline{b} \in \sigma$, consequently $\underline{a} \in \delta^{m-1,m}(\sigma)$. Because of $S' := \{\sigma \in S \mid \{m-1, m\} \cap E(\sigma) \neq \emptyset\} \subseteq S$ we have $\underline{b} \in \bigcap S'$ and therefore $\underline{a} \in \delta^{m-1,m}(\bigcap S')$.

"$\supseteq$": Let $\underline{a} \in (\bigcap \{\delta^{m-1,m}(\sigma) \mid \sigma \in S, \{m-1, m\} \cap E(\sigma) = \emptyset\}) \cap \delta^{m-1,m}(\bigcap \{\sigma \in S \mid \{m-1, m\} \cap E(\sigma) \neq \emptyset\})$. Then there exists some $c \in A$ such that $(a_1, \ldots, a_{m-2}, c, c) \in \bigcap \{\sigma \in S \mid \{m-1, m\} \cap E(\sigma) \neq \emptyset\}$, that is $(a_1, \ldots, a_{m-2}, c, c) \in \sigma$ for all σ with $\{m-1, m\} \cap E(\sigma) \neq \emptyset$. For every $\sigma \in S$ with $\{m-1, m\} \cap E(\sigma) = \emptyset$ there is some $d_\sigma \in A$ such that $(a_1, \ldots, a_{m-2}, d_\sigma, d_\sigma) \in \sigma$. Because of $m-1, m \notin E(\sigma)$ one can replace the d_σ by c and gets $(a_1, \ldots, a_{m-2}, c, c) \in \sigma$. As this tuple is in each $\sigma \in S$ one concludes $\underline{a} \in \delta^{m-1,m}(\bigcap S)$. $\qquad\square$

Core and Comparability

The proof of Thm. 2.1 became more complex because the join between relations does not preserve inclusions. They do in many cases but not in all. For the special case of domains with two elements the exceptions were investigated by classifying relations by separability (see [DHC06]). With the notions of the core $K(\varrho)$ of a relation ϱ and of comparability between relations this is incorporated into the relation $\preceq$, which basically checks if the projection of a binary relation (ignoring those elements which can not be separated) is included in a unary relaiton (or conversely).

1.10 Definition. Let $\varrho \in \mathrm{Rel}^{(2)}(A)$ be a binary relation. Then

$$K(\varrho) := \{c \in A \mid \forall a,b \in A : (a,b) \in \varrho \implies (a,c) \in \varrho\}.$$

is called the *core* of ϱ.

1.11 Corollary

$$\neg K(\neg \varrho) = \{c \in A \mid \exists a,b \in A : (a,b) \notin \varrho \text{ and } (a,c) \in \varrho\}.$$

1.12 Definition. Let $\sigma, \tau \in \mathrm{Rel}^{(1)}(A) \cup \mathrm{Rel}^{(2)}(A)$. We define

$$\sigma \preceq \tau : \Longleftrightarrow \begin{cases} \sigma \subseteq \tau & \text{if } \mathrm{ar}(\sigma) = \mathrm{ar}(\tau) = 1 \\ \sigma = \tau & \text{if } \mathrm{ar}(\sigma) = \mathrm{ar}(\tau) = 2 \\ \sigma \subseteq K(\tau) & \text{if } \mathrm{ar}(\sigma) = 1, \mathrm{ar}(\tau) = 2 \\ \neg K(\neg \sigma) \subseteq \tau & \text{if } \mathrm{ar}(\sigma) = 2, \mathrm{ar}(\tau) = 1 \end{cases}$$

To simplify notation we set $\varrho^{-1} := \varrho$ for any unary relation $\varrho \in \mathrm{Rel}^{(1)}(A)$. We say the relations σ and τ are *comparable* if

$$\sigma \preceq \tau \text{ or } \tau \preceq \sigma$$

and we say σ and τ are *inverted comparable* if σ^{-1} and τ^{-1} are comparable.

The following lemma shows that the comparability is stable under some PAL-operations. These are the operations we will need in Thm. 2.1.

1.13 Lemma. *Let $\varrho_1 \in \mathrm{Rel}^{(1)}(A)$ and let $\varrho_2 \in \mathrm{Rel}^{(2)}(A)$ such that ϱ_1 and ϱ_2 are comparable. Then:*

(i) *$\neg \varrho_1$ and $\neg \varrho_2$ are comparable.*
(ii) *ϱ_1 and $\delta^{1,2}(\tau_1 \times \varrho_2)$ are comparable for any $\tau_1 \in \mathrm{Rel}^{(1)}(A)$.*
(ii') *ϱ_1 and $\delta^{2,3}(\tau_2 \times \varrho_2)$ are comparable for any $\tau_2 \in \mathrm{Rel}^{(2)}(A)$.*

Proof. (i) follows trivially from Def. 1.12. (ii) We define $\sigma := \delta^{1,2}(\tau_1 \times \varrho_2) = \{c \in A \mid \exists a \in \tau_1 : (a,c) \in \varrho_2\}$. If $\sigma \in \{\emptyset^1, A^1\}$ the assertion holds. Otherwise, we have two possibilities for ϱ_1 and ϱ_2 to be comparable. (ii.a) If $\varrho_1 \preceq \varrho_2$, then we have for any $t \in \varrho_1 \subseteq K(\varrho_2)$ that from $s \in \sigma \neq \emptyset^1$ follows that there is some $a \in A$ such that $a \in \tau_1, (a,s) \in \varrho_2$ and by Def. 1.10 $a \in \tau_1, (a,t) \in \varrho_2$, consequently $t \in \sigma$. We deduce $\varrho_1 \subseteq \sigma$ which implies $\varrho_1 \preceq \delta^{1,2}(\tau_1 \times \varrho_2)$. (ii.b) The second possiblity for ϱ_1 and ϱ_2 to be comparable is $\varrho_2 \preceq \varrho_1$. Then exists for any $s \in \sigma$ some $a \in \tau_1$ such that $(a,s) \in \varrho_2$. From $\sigma \neq A^1$ we know that there is some $b \in A$ with $(a, b) \notin \varrho_2$. Therefore by Cor. 1.11 we deduce $s \in \neg K(\neg \varrho) \subseteq \varrho_1$ and therefore $\sigma \subseteq \varrho_1$. We conclude $\delta^{1,2}(\tau_1 \times \varrho_2) \preceq \varrho_1$.

The proof for (ii') is similar. We define analogously $\sigma := \delta^{2,3}(\tau_2 \times \varrho_2)$. If for σ holds $\forall a \in A : ((\{a\} \times A \subseteq \sigma)$ or $(\{a\} \times A \subseteq \neg\sigma))$, then $K(\sigma) = A^1$ and therefore $\varrho_1 \preceq \delta^{2,3}(\tau_2 \times \varrho_2)$. Otherwise we consider the following two cases: if $\varrho_1 \preceq \varrho_2$, then there exists for any $t \in \varrho_1$ and $(a, b) \in \sigma$ some $c \in A$ such that $(a, c) \in \tau_2, (c, b) \in \varrho_2$, and with $t \in \varrho_1 \subseteq K(\varrho_2)$ we deduce $(a, c) \in \tau_2$ and $(c, t) \in \varrho_2$ which implies $(a, t) \in \sigma$, therefore $t \in K(\sigma)$ and consequently $\varrho_1 \subseteq K(\sigma)$ and therefore $\varrho_1 \preceq \delta^{2,3}(\tau_2 \times \varrho_2)$. Otherwise, we have $\varrho_2 \preceq \varrho_1$. Then there are by Cor. 1.11 for any $c \in \neg K(\neg\sigma)$ elements $a, b \in A$ with $(a, b) \in \sigma$ and $(a, c) \notin \sigma$. From $(a, b) \in \sigma$ one deduces the existence of $d \in A$ with $(a, d) \in \tau_2$ and $(d, a) \in \varrho_2$. Let us assume $(d, c) \in \varrho_2$ then one has together with $(a, d) \in \tau_2$ that $(a, c) \in \sigma$, contradiction. Therefore $(d, c) \notin \varrho_2$ and together with $(d, a) \in \varrho_2$ one gets by Cor. 1.11 that $c \in \neg K(\neg\varrho_2)$, that is $\neg K(\neg\sigma) \subseteq \neg K(\neg\varrho_2) \subseteq \varrho_1$ and therefore $\delta^{2,3}(\tau_2 \times \varrho_2) \preceq \varrho_1$. $\square$

The following lemma was an important clue to find the proof of Thm. 2.1. It allows us to represent a connected graph (in the graphical representation, elements are denoted by lines and c connects the four relations ϱ_1, ϱ_2, σ_1 and σ_2)[1] by the intersection of four other graphs (each element $c_1, \ldots, c_4$ is connecting only two relations). Of course, this is not possible in general, but only for comparable relations.

1.14 Crux-Lemma. *Let* $\varrho_1, \sigma_1 \in \mathrm{Rel}^{(1)}(A)$ *and* $\varrho_2, \sigma_2 \in \mathrm{Rel}^{(2)}(A)$ *such that* ϱ_1 *and* ϱ_2 *are comparable and* σ_1 *and* σ_2 *are comparable. Then for any* $a, b \in A$

$$\exists c \in A : c \in \varrho_1 \cap \sigma_1, \qquad \text{(A)}$$
$$(a, c) \in \varrho_2, \qquad \text{(B)}$$
$$(b, c) \in \sigma_2 \qquad \text{(C)}$$

$$\Longleftrightarrow$$

$$\exists c_1, c_2, c_3, c_4 \in A : c_1 \in \varrho_1 \cap \sigma_1, \qquad \text{(a)}$$
$$(a, c_2) \in \varrho_2, \ c_2 \in \sigma_1, \qquad \text{(b)}$$
$$c_3 \in \varrho_1, \ (b, c_3) \in \sigma_2, \qquad \text{(c)}$$
$$(a, c_4) \in \varrho_2, \ (b, c_4) \in \sigma_2 \qquad \text{(d)}$$

Proof. "$\Longrightarrow$" is obvious. For "$\Longleftarrow$" we have to consider several cases.

(I) (*) $\varrho_1 \preceq \varrho_2$ and (**) $\sigma_1 \preceq \sigma_2$: Then we can set $c := c_1$. Condition (A) is then the same as (a), in particular it follows that $c_1 \in \varrho_1 \underset{(*)}{\subseteq} K(\varrho_2)$. Together with $(a, c_2) \in \varrho_2$ from (b) we get by the definition of the core that $(a, c_1) \in \varrho_2$, that is (B). Analogously we conclude from (a) that $c_1 \in \sigma_1$ and by the equations (**) and (b) that $(b, c_1) \in \sigma_2$, that is (C).

---

[1] In this paper, we do not introduce the graphical notation due to space restrictions. See [HCP04] for details.

(II) (*) $\varrho_1 \preceq \varrho_2$ and (**) $\sigma_2 \preceq \sigma_1$. There are two subcases: (II.i) $\forall c' \in A :$ $(b, c') \in \sigma_2$ and (II.ii) $\exists c' \in A : (b, c') \notin \sigma_2$. For (II.i), we chose $c = c_1$, as in (I) we conclude, that c_1 fulfills Ⓐ and Ⓑ. From the condition of (II.i) we obtain Ⓒ. For (II.ii), we can set $c = c_3$. We have by ⓒ that $c_3 \in \varrho_1$. From the condition of (II.ii) and $(b, c_3) \in \sigma_2$ (follows from ⓒ, this is Ⓒ) we deduce by Cor. 1.11 that $c_3 \in \neg K(\neg \sigma_2) \underset{(**)}{\subseteq} \sigma_1$, and therefore $c_3 \in \varrho_1 \cap \sigma_1$, that is Ⓐ. Due to $c_3 \in \varrho_1 \underset{*}{\subseteq} K(\varrho_2)$ and $(a, c_2) \in \varrho_2$ (by ⓑ) we conclude $(a, c_3) \in \varrho_2$, hence Ⓑ.

(III) The case $\varrho_2 \preceq \varrho_1$ and $\sigma_1 \preceq \sigma_2$ is handled analogously to (II). If $\forall c' \in A :$ $(a, c') \in \varrho_2$, we can set $c = c_1$, if $\exists c' \in A : (a, c') \notin \varrho_2$, we chose $c = c_2$.

(IV) Finally, we consider (*) $\varrho_2 \preceq \varrho_1$ and (**) $\sigma_2 \preceq \sigma_1$. Now we have four subcases:

(IV.i) (△) $\forall c' \in A : (a, c') \in \varrho_2$ and (▽) $\forall d' \in A : (b, d') \in \sigma_2$. For $c = c_1$ we get Ⓐ from ⓐ, Ⓑ from (△), and Ⓒ from (▽).

(IV.ii) (△) $\exists c' \in A : (a, c') \notin \varrho_2$ and (▽) $\forall d' \in A : (b, d') \in \sigma_2$. We show that $c = c_2$ is a possible choice. With (△) and $(a, c_2) \in \varrho_2$ we obtain by Cor. 1.11 that $c_2 \in \neg K(\neg \varrho_2) \underset{(*)}{\subseteq} \varrho_1$. Also from ⓑ we know $c_2 \in \sigma_1$ and therefore Ⓐ. Condition Ⓑ follows directly from ⓑ, while Ⓒ follows from (▽).

(IV.iii) $\forall c' \in A : (a, c') \in \varrho_2$ and $\exists d' \in A : (b, d') \notin \sigma_2$. This case is analogous to (IV.ii), we can set $c = c_3$.

(IV.iv) (△) $\exists c' \in A : (a, c') \notin \varrho_2$ and (▽) $\exists d' \in A : (b, d') \notin \sigma_2$. From (△) and $(a, c_4) \in \varrho_2$ (from ⓓ) we deduce (by Cor. 1.11) that $c_4 \in \neg K(\neg \varrho_2) \underset{(*)}{\subseteq} \varrho_1$, analogously from (▽) and $(b, c_4) \in \sigma_2$ (ⓓ again), that $c_4 \in \neg K(\neg \sigma_2) \underset{(**)}{\subseteq} \sigma_1$, therefore $c_4 \in \varrho_1 \cap \sigma_1$, that is Ⓐ. Conditions Ⓑ and Ⓒ follow from ⓓ. □

⊔⊓-representations of Relations

Now, we will introduce the notion of ⊔⊓-representation. It corresponds to the disjunctive-conjunctive (normal) form of first-order predicate logic formulas.

1.15 Definition. Let $\varrho \in \mathrm{Rel}^{(n)}(A)$. Then we say the set $\mathcal{S} \subseteq \mathfrak{P}(\mathrm{Rel}^{(n)}(A))$ is a ⊔⊓-*representation* of ϱ if

(i) $\varrho = \bigcup \{\bigcap S \mid S \in \mathcal{S}\}$ and
(ii) $\bigcup \mathcal{S}$ is finite.

For $\Sigma \subseteq \mathrm{Rel}(A)$ and a Σ-term t a ⊔⊓-*representation* $\mathcal{S} \subseteq \mathfrak{P}(\mathrm{Rel}^{(n)}(A))$ is said to be *consistent with* t if $\mathcal{S}$ is a ⊔⊓-representation of $\varrho := \llbracket t \rrbracket$ and for every $\sigma \in \bigcup \mathcal{S}$ we have:

(iii) $E(\sigma)$ is t-connected and
(iv) $\sigma \upharpoonright_{E(\sigma)} \in \langle \Sigma \rangle^A_{\mathrm{PAL}}$.

The following lemmata show how ⊔⊓-representations have to be transformed to provide a ⊔⊓-representation of the result of the PAL-operation under consideration.

1.16 Lemma. *Let* $\varrho_1, \varrho_2 \in \mathrm{Rel}(A)$ *and let* $\mathcal{S}_1$ *and* $\mathcal{S}_2$ *be* $\bigcup\bigcap$*-representations of* ϱ_1 *and* ϱ_2 *respectively. Then*

$$\mathcal{S} := \{\{\sigma_1 \times A^{\mathrm{ar}(\varrho_2)} \mid \sigma_1 \in S_1\} \cup \{A^{\mathrm{ar}(\varrho_1)} \times \sigma_2 \mid \sigma_2 \in S_2\} \mid S_1 \in \mathcal{S}_1, S_2 \in \mathcal{S}_2\}$$

is a $\bigcup\bigcap$*-representation of* $\varrho_1 \times \varrho_2$.

Proof. Let $m := \mathrm{ar}(\varrho_1)$ and $n := \mathrm{ar}(\varrho_2)$, let $\underline{a} \in A^m$ and $\underline{b} \in A^n$. It is easy to see that for any relation $\tau_1 \in \mathrm{Rel}^{(m)}(A)$ we have $(*)$ $\underline{a} \in \tau_1$ $\Longleftrightarrow$ $(a_1, \ldots, a_m, b_1, \ldots, b_n) \in \tau_1 \times A^n$ and analogously $(*')$ $\underline{b} \in \tau_2$ $\Longleftrightarrow$ $(a_1, \ldots, a_m, b_1, \ldots, b_n) \in A^m \times \tau_2$ for every $\tau_2 \in \mathrm{Rel}^{(n)}(A)$. Consequently

$$(a_1, \ldots, a_m, b_1, \ldots, b_n) \in \varrho_1 \times \varrho_2$$

$\xrightarrow{\text{Def. 1.1(1)}}$ $\underline{a} \in \varrho_1, \underline{b} \in \varrho_2$

$\xrightarrow{\text{Def. 1.15}}$ $\exists S_1 \in \mathcal{S}_1, S_2 \in \mathcal{S}_2 : \underline{a} \in \bigcap S_1$ and $\underline{b} \in \bigcap S_2$

$\xrightarrow{(*),(*')}$ $\exists S_1 \in \mathcal{S}_1, S_2 \in \mathcal{S}_2 :$

$$(a_1, \ldots, a_m, b_1, \ldots, b_n) \in ((\bigcap S_1) \times A^n) \cap (A^m \times (\bigcap S_2))$$

$\xrightarrow{\text{Def of } \mathcal{S}}$ $\exists S \in \mathcal{S} : (a_1, \ldots, a_m, b_1, \ldots, b_n) \in \bigcap S$

$\xrightarrow{\text{Def. 1.15}}$ $(a_1, \ldots, a_m, b_1, \ldots, b_n) \in \bigcup\{\bigcap S \mid S \in \mathcal{S}\}$. $\qquad\square$

The finiteness condition (ii) follows from the finiteness of $\bigcup \mathcal{S}_1$ and $\bigcup \mathcal{S}_2$.

1.17 Lemma. *Let* $\mathcal{S}_1$ *be a* $\bigcup\bigcap$*-representation of* $\varrho \in \mathrm{Rel}(A)$. *Then*

$$\mathcal{S} := \{\{\neg \tau(S_1) \mid S_1 \in \mathcal{S}_1\} \mid \tau : \mathcal{S}_1 \to \bigcup \mathcal{S}_1, \tau(S) \in S \text{ for all } S \in \mathcal{S}_1\}$$

is a $\bigcup\bigcap$*-representation of* $\neg\varrho$.

Proof. Basically we use de Morgan's law and the distributivity of $\cap$ and $\cup$, although in a generalized version. We show first, that every tuple not in ϱ is an element of the relation described by $\mathcal{S}$. Let $m := \mathrm{ar}(\varrho)$ and $\underline{a} \in \neg\varrho$. Because $\mathcal{S}_1$ is a $\bigcup\bigcap$-representation of ϱ and by Def. 1.15 one can conclude that for every $S \in \mathcal{S}_1$ there is some relation $\sigma_S \in S$ such that $\underline{a} \notin \sigma_S$ (otherwise $\underline{a} \in \bigcap S \subseteq \varrho$, contradiction). The mapping $\tau_{\underline{a}} : \mathcal{S}_1 \to \bigcup \mathcal{S}_1$ with $\tau_{\underline{a}}(S) := \sigma_S$ is obviously a choice function (as used in the definition of the $\bigcup\bigcap$-representation $\mathcal{S}$) and $\underline{a} \notin \tau_{\underline{a}}(S_1)$ (i.e. $\underline{a} \in \neg\tau_{\underline{a}}(S_1)$) for all $S_1 \in \mathcal{S}_1$ and consequently $\underline{a} \in \bigcap\{\neg\tau_{\underline{a}}(S)_1) \mid S_1 \in \mathcal{S}_1\} \subseteq \bigcup\{\bigcap S \mid S \in \mathcal{S}\}$.

After having shown that every tuple in $\neg\varrho$ is described by $\mathcal{S}$, one can similarly show that every element not in $\neg\varrho$, that is every $\underline{a} \in \varrho$ is not described by $\mathcal{S}$. By Def. 1.15 we see that there is some $S_{\underline{a}} \in \mathcal{S}_1$ such that $\underline{a} \in \bigcap S_{\underline{a}}$. Then for any choice function $\tau : \mathcal{S}_1 \to \bigcup \mathcal{S}_1$ one has $\underline{a} \in \tau(S_{\underline{a}})$, that is $\underline{a} \notin \neg\tau(S_{\underline{a}}) \supseteq \bigcap\{\neg\tau(S_1) \mid S_1 \in \mathcal{S}\}$, consequently $\underline{a} \notin \bigcup\{\bigcap S \mid S \in \mathcal{S}\}$.

We have $|S| \leq |S_1|$ for all $S \in \mathcal{S}$ and $|\mathcal{S}| \leq |\bigcup \mathcal{S}_1|^{|S_1|}$. Due to $|\bigcup \mathcal{S}| \leq$ $\max\{|S| \mid S \in \mathcal{S}\} \cdot |\mathcal{S}| \leq |S_1| \cdot |\bigcup \mathcal{S}_1|^{|S_1|}$ the finiteness condition (ii) for $\bigcup \mathcal{S}$ follows from the finiteness of $\bigcup \mathcal{S}_1$. □

1.18 Lemma. *Let $\mathcal{S}$ be a $\bigcup\bigcap$-representation consistent with some Σ-term t. Then there exists a $\bigcup\bigcap$-representation $\mathcal{S}'$ consistent with t, satisfying the following conditions:*

(i) $\forall \sigma_1, \sigma_2 \in S \in \mathcal{S}' : \sigma_1 \subseteq \sigma_2 \implies \sigma_1 = \sigma_2$,
(ii) $\forall \sigma \in \bigcup \mathcal{S}' : \sigma \neq \emptyset^{\mathrm{ar}(t)}$,
(iii) $\forall \sigma \in \bigcup \mathcal{S}' : \sigma \neq A^{\mathrm{ar}(t)}$,
(iv) $S \neq \emptyset$ for every $S \in \mathcal{S}'$ if $|\mathcal{S}'| > 1$ and
(v) $\bigcup \mathcal{S}' \subseteq \bigcup \mathcal{S}$.

Proof. (i): From Def. 1.15(ii) we deduce that every set $S \in \mathcal{S}$ is finite. For that reason there are minimal relations (w. r. t. inclusion) in S. Let $\widetilde{S}$ be the set of these minimal relations, then $\bigcap \widetilde{S} = \bigcap S$ and $\mathcal{S}_{(i)} := \{\widetilde{S} \mid S \in \mathcal{S}\}$ fulfills condition (i) and $\bigcup \mathcal{S}_{(i)} \subseteq \bigcup \mathcal{S}$.

(ii): The empty relation $\emptyset^{\mathrm{ar}(\varrho)}$ absorbes all other relations, in the sense that $\bigcap S = \emptyset^{\mathrm{ar}(\varrho)}$ if $\emptyset^{\mathrm{ar}(\varrho)} \in S$ and for this reason the $\bigcup\bigcap$-representation given by $\mathcal{S}_{(ii)} := \{S \mid S \in \mathcal{S}_{(i)}, \emptyset^{\mathrm{ar}(\varrho)} \notin S\}$ fulfills conditions (i) and (ii) and also $\bigcup \mathcal{S}_{(ii)} \subseteq \bigcup \mathcal{S}_{(i)} \subseteq \bigcup \mathcal{S}$.

(iii): The full relation $A^{\mathrm{ar}(\varrho)}$ has no influence on the intersection of relations, that is for all $S \in \mathcal{S}$ holds $\bigcap S = \bigcap (S \setminus \{A^{\mathrm{ar}(\varrho)}\}$. Therefore the $\bigcup\bigcap$-representation $\mathcal{S}_{(iii)} := \{S \setminus \{A^{\mathrm{ar}(\varrho)} \mid S \in \mathcal{S}_{(ii)}\}$ fulfills conditions (i)–(iii) and also $\bigcup \mathcal{S}_{(iii)} \subseteq \bigcup \mathcal{S}_{(ii)} \subseteq \bigcup \mathcal{S}$. Finally, because of $\bigcap \emptyset = A^{\mathrm{ar}(\varrho)}$ it is either $\varrho = A^{\mathrm{ar}(\varrho)}$ and $\mathcal{S}' := \{\emptyset\}$ fulfills conditions (i)–(v) or $\emptyset \notin \mathcal{S}_{(iii)}$ and $\mathcal{S}' := \mathcal{S}_{(iii)}$ fulfills conditions (i)–(v).

From the inclusion property (v) also follow the finiteness and consistency properties Def. 1.15(ii)–(iv). □

1.19 Definition. A $\bigcup\bigcap$-representation fulfilling the conditions (i), (ii), (iii) and (iv) of Lem. 1.18 is said to be *normalized*.

1.20 Lemma. *Let $\varrho_1 \in \mathrm{Rel}^{(1)}(A)$ and $\varrho_2 \cup \mathrm{Rel}^{(2)}(A)$ be arbitrary relations. Then ϱ_1 and ϱ_2 are comparable with $\emptyset^1$ and A^1, and ϱ_1 is comparable with $\emptyset^2$ and A^2.*

Proof. For $\emptyset^1$ and A^1 this is trivial. From $K(\emptyset^2) = K(A^2) = A^1$. this follows also for $\emptyset^2$ and A^2. □

2   The Representation Theorem

After the mathematical tools have been prepared, we can now prove the first central result, the representation theorem for the relations generated from unary and binary relations in PAL without the teridentity. Many parts of the proof are rather technical because many subcases have to be distinguished. The most difficult case is the join operation, where the Crux-Lemma is needed to show that property (iii) of Thm. 2.1 is preserved, which states that we do not need the teridentity to construct the essential relations of the representation.

2.1 Theorem. *Let* $\Sigma := \mathrm{Rel}^{(1)}(A) \cup \mathrm{Rel}^{(2)}(A)$. *Then for every* $\Sigma\text{-}PAL\backslash\{\mathrm{id}_3\}$*-term* t *there is a* $\bigcup\bigcap$*-representation* $\mathcal{S}$ *consistent with* t *such that*

(i) $|E(\sigma)| \leq 2$ *for all* $\sigma \in \bigcup\mathcal{S}$,

(ii) $\forall \sigma_1, \sigma_2 \in \bigcup\mathcal{S} : E(\sigma_1) \cap E(\sigma_2) \neq \emptyset \implies \sigma_1\lceil_{E(\sigma_1)}$ *and* $\sigma_2\lceil_{E(\sigma_2)}$ *are comparable or inverted comparable and*

(iii) $\{\sigma\lceil_{E(\sigma)} \mid \sigma \in \bigcup\mathcal{S}\} \subseteq \langle\Sigma\rangle^A_{\mathrm{PAL}\backslash\{\mathrm{id}_3\}}$.

Proof. The proof works by induction over the possible constructions of a Σ-term (see Def. 1.2). When checking the consistency of a $\bigcup\bigcap$-representation with a term, we do not have to consider the condition Def. 1.15(iv) because condition (iii) of the theorem is stronger. Condition (i) of the theorem holds by Lem. 1.4 for all $t \in \mathsf{T}_{\mathrm{PAL}\backslash\{\mathrm{id}_3\}}(\Sigma)$. We have stated this condition explicitly to make clear that the essential relations are at most binary and the notion of comparability (see Def. 1.12) can therefore be applied as in condition (ii).

(I) If t is atomic then $t = \sigma \in \Sigma$. We can therefore simply set $\mathcal{S} := \{\{\sigma\}\}$. Conditions (ii) and (iii) hold trivially. Obviously, $\mathcal{S}$ is a $\bigcup\bigcap$-representation of $\sigma = [\![t]\!]$. Due to $E(\sigma) \subseteq \mathrm{ar}(\sigma)$ the representation is consistent with t by Def. 1.3(i).

(II) The case $t = \overline{\mathrm{id}_3}$ is not possible because $\mathrm{id}_3 \notin \Sigma = \Sigma_0$.

(III) If $t = (t_1 \times t_2)$ then by the induction hypothesis there exist $\bigcup\bigcap$-representations $\mathcal{S}_1$ and $\mathcal{S}_2$ consistent with t_1 and t_2 respectively. Let $n_1 := \mathrm{ar}(t_1)$ and $n_2 := \mathrm{ar}(t_2)$. We set $\mathcal{S}$ as in Lem. 1.16. By this lemma we see that $\mathcal{S}$ is a $\bigcup\bigcap$-representation of $[\![t_1]\!] \times [\![t_2]\!] = [\![t]\!]$.

It is easy to see that $E(\sigma_1 \times A^{n_2}) = E(\sigma_1)$ and $E(A^{n_1} \times \sigma_2) = \{n_1 + i \mid i \in E(\sigma_2)\}$ for all $\sigma_1, \sigma_2 \in \mathrm{Rel}(A)$. Using the induction hypothesis we can deduce $(\triangle)$ $\sigma_1 \times A^{n_2}\lceil_{E(\sigma_1 \times A^{n_2})} = \sigma_1\lceil_{E(\sigma_1)} \in \langle\Sigma\rangle^A_{\mathrm{PAL}\backslash\{\mathrm{id}_3\}}$ and $A^{n_1} \times \sigma_2\lceil_{E(A^{n_1} \times \sigma_2)} = \sigma_2\lceil_{E(\sigma_2)}$ and therefore condition (iii) of the theorem holds.

The consistency of $\mathcal{S}$ with t follows by Def. 1.3(ii). Now let $\tau_1, \tau_2 \in \bigcup\mathcal{S}$. If τ_1 is of the form $\sigma_1 \times A^{n_2}$ and τ_2 of the form $A^{n_1} \times \sigma_2$ (or vice-versa) for some $\sigma_1 \in \bigcup\mathcal{S}_1$ and $\sigma_2 \in \bigcup\mathcal{S}_2$ then $E(\tau_1) \subseteq \{1, \ldots, n_1\}$ and $E(\tau_2) \subseteq \{n_1 + 1, \ldots, n_1 + n_2\}$, that is $E(\tau_1) \cap E(\tau_2) = \emptyset$ and they do not fulfill the premise of condition (ii). If $\tau_1 = \sigma_1 \times A^{n_2}$ and $\tau_2 = \sigma_2 \times A^{n_2}$ for $\sigma_1, \sigma_2 \in \bigcup\mathcal{S}_1$ then $\tau_1\lceil_{E(\tau_1)}$ and $\tau_2\lceil_{E(\tau_2)}$ are comparable by $(\triangle)$ and the induction hypothesis. Analogously we can show that they are comparable if they are both of the form $A^{n_1} \times \sigma$ for some $\sigma \in \bigcup\mathcal{S}_2$. This proves condition (ii) of the theorem.

(IV) The most complicated case is $t = \delta^{i,j}(t_1)$. Let $\mathcal{S}_1$ be a $\bigcup\bigcap$-representation consistent with t_1 and $m := \mathrm{ar}(t_1)$. There are three subcases: (IV.i) if every place $p \in \underline{m}$ is t_1-connected to i or j then by Def. 1.3(iii-iii") we have that $\underline{m}$ is t-connected. Then $\{\{[\![t]\!]\}\}$ is trivially a $\bigcup\bigcap$-representation consistent with t. As for the atomic case we see that conditions (ii) and (iii) hold.

In the cases following now (the subcases (IV.ii) and (IV.iii)), there is some place in $\underline{\mathrm{ar}(t_1)}$ which is not t_1-connected to i and j. Without loss of generality we assume $i = m$ and $j = m - 1$ and that 1 is not t_1-connected to these two places. Let $\mathcal{S}_1$ be a normalized $\bigcup\bigcap$-representation consistent with t_1 with the properties given in the theorem (which exists by the induction hypothesis and Lem. 1.18) and let $S_1 \in \mathcal{S}_1$.

(IV.ii) If $m - 1$ and m are t_1-connected, then we have by Lem. 1.8 and condition (i) that $E(\tau) \subseteq \{m-1, m\}$ for $\tau := \bigcap\{\sigma \mid \sigma \in S_1, E(\sigma) \cap \{m-1, m\} \neq \emptyset\}$. By Lem. 1.7(iv) and Lem. 1.6 we see that $\delta^{m-1,m}(\tau) \in \{\emptyset^{m-2}, A^{m-2}\}$.

By Lem. 1.9, $\delta^{m-1,m}(\bigcap\{\sigma \in S_1\}) = \bigcap\{\delta^{i,j}(\sigma) \mid \sigma \in S_1 \text{ and } m-1, m \notin E(\sigma)\} \cap \delta^{m-1,m}(\tau)$. If $\delta^{m-1,m}(\tau) = \emptyset^{m-2}$ we set simply $S_1' := \{\emptyset\}$, otherwise $S_1' := \{\delta^{m-1,m}(\sigma) \mid \sigma \in S_1 \text{ and } m-1, m \notin E(\sigma)\}$ and have in both cases $\delta^{m-1,m}(\bigcap\{\sigma \in S_1\}) = \bigcap S_1'$. Due to $S_1' \subseteq S_1$ we also have $\{\sigma' \restriction_{E(\sigma')} \mid \sigma' \in S_1'\} \subseteq \{\sigma \restriction_{E(\sigma)} \mid \sigma \in S_1\} \cup \{\emptyset^{m-2}\}$ and therefore conditions (i) and (iii) hold.

Let $\mathcal{S} := \{S' \mid S \in \mathcal{S}_1\}$, applying the same construction to all elements of $\mathcal{S}_1$. It is easy to see that this is a $\bigcup\bigcap$-representation of $[\![t]\!]$. We have $\bigcup \mathcal{S} \subseteq \bigcup \mathcal{S}_1 \cup \{\emptyset^{m-2}\}$ and by the induction hypothesis (condition ii) and Lem. 1.20 we see that condition (ii) holds and that $\mathcal{S}$ is consistent with t.

(IV.iii) Finally we consider the case, that 1 is not t_1-connected to $m - 1$ or m, but some place other than these three is connected to $m - 1$ or m.

If there is some relation $\widetilde{\varrho}_1 \in S$ with $E(\widetilde{\varrho}_1) = \{m\}$ we set $\varrho_1 := \widetilde{\varrho}_1 \restriction_{\{m\}}$ (because S_1 is normalized and by condition (ii) there can be at most one such relation), otherwise we set $\widetilde{\varrho}_1 := A^m$ and $\varrho_1 := A^1$ (and therefore $\varrho_1 = \widetilde{\varrho}_1 \restriction_{\{m\}}$)). In the second case we have $E(\widetilde{\varrho}_1) = \emptyset \subseteq \{m\}$. Analogously, if there is some relation $\widetilde{\sigma}_1 \in S$ with $E(\widetilde{\sigma}_1) = \{m - 1\}$ we set $\sigma_1 := \widetilde{\sigma}_1 \restriction_{\{m-1\}}$, and otherwise $\widetilde{\sigma}_1 := A^m$ and $\sigma_1 := A^1$.

If there is some place $k \in \underline{m} \setminus \{m - 1, m\}$ which is t_1-connected to m and some relation $\widetilde{\varrho}_2 \in S$ with $E(\widetilde{\varrho}_2) = \{k, m\}$ ($(\lhd)$ will denote this condition) we set $\varrho_2 := \widetilde{\varrho}_2 \restriction_{\{k,m\}}$. By condition (i) there can be at most one such place, and by condition (ii) and Lem. 1.18(i) there can be at most one such relation. If there is no such relation or no such place we set $k := 1$, $\widetilde{\varrho}_2 := A^m$ and $\varrho_2 := A^2$ (denoted by $(\blacktriangleleft)$). In all these cases we have $\widetilde{\varrho}_2 \restriction_{\{k,m\}} = \varrho_2$ and that ϱ_1 and ϱ_2 are comparable (the latter by condition (ii) or Lem. 1.20).

Similarly, if there is some place $l \in \underline{m} \setminus \{m - 1, m\}$ which is t_1-connected with $m - 1$ and some relation $\widetilde{\sigma}_2 \in S$ with $E(\widetilde{\sigma}_2) = \{l, m - 1\}$ we set $\sigma_2 := \widetilde{\sigma}_2 \restriction_{\{l,m-1\}}$ (denoted by $(\rhd)$). If no such place or no such relation exists we set $l := 1$, $\widetilde{\sigma}_2 := A^m$ and $\sigma_2 := A^2$ (condition $(\blacktriangleright)$). We have $E(\widetilde{\sigma}_2) \subseteq \{l, m - 1\}$ and $\widetilde{\sigma}_2 \restriction_{\{l,m-1\}} = \sigma_2$ in all cases and that σ_1 and σ_2 are comparable.

Because S_1 is normalized we know that $\emptyset^m \notin \{\widetilde{\varrho}_1, \widetilde{\varrho}_2, \widetilde{\sigma}_1, \widetilde{\sigma}_2\}$ and that the places $m - 1, m \notin E(\widetilde{\varphi})$ for all $\widetilde{\varphi} \in S \setminus \{\widetilde{\varrho}_1, \widetilde{\varrho}_2, \widetilde{\sigma}_1, \widetilde{\sigma}_2\}$.

Let $\tau_{11} := \varrho_1 \times \sigma_1$ and $\widetilde{\tau}_{11} := A^{m-2} \times \tau_{11}$. Then we have $\widetilde{\tau}_{11} \restriction_{E(\widetilde{\tau}_{11})} = \tau_{11} \in \langle \Sigma \rangle^A_{\mathsf{PAL} \setminus \{\mathsf{id}_3\}}$ (if $\widetilde{\varrho}_1, \widetilde{\sigma}_1 \in S$) or $\widetilde{\tau}_{11} \restriction_{E(\widetilde{\tau}_{11})} = \varrho_1 \in \langle \Sigma \rangle^A_{\mathsf{PAL} \setminus \{\mathsf{id}_3\}}$ (if $\widetilde{\varrho}_1 \in S, \widetilde{\sigma}_1 = A^m$) or $\widetilde{\tau}_{11} \restriction_{E(\widetilde{\tau}_{11})} = \sigma_1 \in \langle \Sigma \rangle^A_{\mathsf{PAL} \setminus \{\mathsf{id}_3\}}$ (if $\widetilde{\varrho}_1 = A^m, \widetilde{\sigma}_1 \in S$) or $\widetilde{\tau}_{11} \restriction_{E(\widetilde{\tau}_{11})} = A^0 \in \langle \Sigma \rangle^A_{\mathsf{PAL} \setminus \{\mathsf{id}_3\}}$ (if $\widetilde{\varrho}_1 = \widetilde{\sigma}_1 = A^m$), that is in all cases $(\Diamond_{11})$ $\widetilde{\tau}_{11} \restriction_{E(\widetilde{\tau}_{11})} \in \langle \Sigma \rangle^A_{\mathsf{PAL} \setminus \{\mathsf{id}_3\}}$.

We set $\widetilde{\tau}_{21} := \{\underline{a} \in A^m \mid (a_k, a_m) \in \varrho_2, a_{m-1} \in \sigma_1\}$ and $\tau_{21} := \varrho_2 \times \sigma_1$. We get $\widetilde{\tau}_{21} \restriction_{\{k,m-1,m\}} = \pi_{(23)}(\tau_{21})$ in any case and $\widetilde{\tau}_{21} \restriction_{E(\widetilde{\tau}_{21})} = \pi_{(23)}(\tau_{21}) \in \langle \Sigma \rangle^A_{\mathsf{PAL} \setminus \{\mathsf{id}_3\}}$ (if $(\lhd)$ and $\widetilde{\sigma}_1 \in S$), $\widetilde{\tau}_{21} \restriction_{E(\widetilde{\tau}_{21})} = \varrho_2 \in \langle \Sigma \rangle^A_{\mathsf{PAL} \setminus \{\mathsf{id}_3\}}$ (if $(\lhd)$ and $\widetilde{\sigma}_1 = A^m$), $\widetilde{\tau}_{21} \restriction_{E(\widetilde{\tau}_{21})} = \sigma_1 \in \langle \Sigma \rangle^A_{\mathsf{PAL} \setminus \{\mathsf{id}_3\}}$ (if $(\blacktriangleleft)$ and $\widetilde{\sigma}_1 \in S$) and $\widetilde{\tau}_{21} \restriction_{E(\widetilde{\tau}_{21})} =$

$A^0 \in \langle \Sigma \rangle^A_{\text{PAL} \setminus \{\text{id}_3\}}$ (if ($\blacktriangleleft$) and $\widetilde{\sigma}_1 = A^m$), consequently in each of these cases ($\Diamond_{21}$) $\widetilde{\tau}_{21} \restriction_{E(\widetilde{\tau}_{21})} \in \langle \Sigma \rangle^A_{\text{PAL} \setminus \{\text{id}_3\}}$.

Analogously we set $\widetilde{\tau}_{12} := \{\underline{a} \in A^m \mid a_m \in \varrho_1, (a_l, a_{m-1}) \in \sigma_2\}$ and get $\widetilde{\tau}_{12} \restriction_{\{l, m-1, m\}} = \pi_{(132)}(\varrho_1 \times \sigma_2)$ and ($\Diamond_{12}$) $\widetilde{\tau}_{12} \restriction_{E(\widetilde{\tau}_{12})} \in \langle \Sigma \rangle^A_{\text{PAL} \setminus \{\text{id}_3\}}$.

In this subcase (IV.iii) we know that some place in $\underline{m} \setminus \{m-1, m\}$ is t_1-connected to $m-1$ or m, therefore we can deduce $k \neq l$. We set $\tau_{22} := \varrho_2 \times \sigma_2$ and $\widetilde{\tau}_{22} := \{\underline{a} \in A^m \mid (a_k, a_m) \in \varrho_2, (a_l, a_{m-1}) \in \sigma_2\}$. We get $\widetilde{\tau}_{22} \restriction_{\{k, l, m-1, m\}} = \pi_{(243)}(\tau_{22})$ if $k < l$, $\widetilde{\tau}_{22} \restriction_{\{k, l, m-1, m\}} = \pi_{(1243)}(\tau_{22})$ if $l < k$, for the restriction on the essential places we have:

	($\triangleleft$)		($\blacktriangleright$)	($\triangleright$)	($\blacktriangleright$)
	($\triangleright$)				
	$k < l$	$l < k$			
$\widetilde{\tau}_{22} \restriction_{E(\widetilde{\tau}_{22})} =$	$\pi_{(243)}(\tau_{22})$	$\pi_{(1243)}(\tau_{22})$	ϱ_2	σ_2	$\emptyset^2$

and in all cases ($\Diamond_{22}$) $\tau_{22} \restriction_{E(\tau_{22})} \in \langle \Sigma \rangle^A_{\text{PAL} \setminus \{\text{id}_3\}}$.

As noted before, ϱ_1 and ϱ_2 are comparable as are σ_1 and σ_2. Therefore we can apply the Crux-Lemma. We will use the fact that the essential places of $\widetilde{\varrho}_1$, $\widetilde{\varrho}_2$, $\widetilde{\sigma}_1$ and $\widetilde{\sigma}$ alre all contained in $\{k, l, m-1, m\}$ (this will be marked by $(*)$).

$$\exists a_1, \ldots, a_{m-2} : (a_1, \ldots, a_{m-2}) \in \delta^{m-1,m}\left(\bigcap \{\widetilde{\varrho}_1, \widetilde{\varrho}_2, \widetilde{\sigma}_1, \widetilde{\sigma}_2\}\right)$$

$$\Longleftrightarrow \exists a_1, \ldots, a_{m-2}, c \in A : (a_1, \ldots, a_{m-2}, c, c) \in \bigcap \{\widetilde{\varrho}_1, \widetilde{\varrho}_2, \widetilde{\sigma}_1, \widetilde{\sigma}_2\}$$

$$\overset{(*)}{\Longleftrightarrow} \exists a_k, a_l, c \in A : c \in \varrho_1 \cap \sigma_1, (a_k, c) \in \varrho_2, (a_l, c) \in \sigma_2$$

$$\overset{\text{Crux-Lemma}}{\Longleftrightarrow} \exists a_k, a_l, c_1, c_2, c_3, c_4 \in A : c_1 \in \varrho_1 \cap \sigma_1,$$
$$(a_k, c_2) \in \varrho_2, c_2 \in \sigma_1, c_3 \in \varrho_1, (a_l, c_3) \in \sigma_2,$$
$$(a_k, c_4) \in \varrho_2, (a_l, c_4) \in \sigma_2$$

$$\Longleftrightarrow \exists a_k, a_l, c_1, c_2, c_3, c_4 \in A : (c_1, c_1) \in \tau_{11},$$
$$(a_k, c_2, c_2) \in \tau_{21}, (c_3, a_l, c_3) \in \tau_{12}, (a_k, c_4, a_l, c_4) \in \tau_{22}$$

$$\overset{(*)}{\Longleftrightarrow} \exists a_1, \ldots, a_{m-2}, c_1, \ldots, c_4 \in A : (a_1, \ldots, a_{m-2}, c_1, c_1) \in \widetilde{\tau}_{11},$$
$$(a_1, \ldots, a_{m-2}, c_2, c_2) \in \widetilde{\tau}_{21}, (a_1, \ldots, a_{m-2}, c_3, c_3) \in \widetilde{\tau}_{12},$$
$$(a_1, \ldots, a_{m-2}, c_4, c_4) \in \widetilde{\tau}_{22}$$

$$\Longleftrightarrow \exists a_1, \ldots, a_{m-2} \in A :$$
$$(a_1, \ldots, a_{m-2}) \in \bigcap \{\delta^{m-1,m}(\widetilde{\tau}_{xy}) \mid x, y \in \{1, 2\}\}.$$

Let $S' := \{\delta^{m-1,m}(\sigma) \mid \sigma \in S, m-1, m \notin E(\sigma)\} \cup \{\delta^{m-1,m}(\widetilde{\tau}_{x,y}) \mid x, y \in \{1, 2\}\}$. By Lem. 1.9 we get $\delta^{m-1,m}(\bigcap S) = \bigcap S'$. For $\sigma \in S$, $m-1, m \notin E(\sigma)$ we have $\delta^{m-1,m}(\sigma) \restriction_{E(\delta^{m-1,m}(\sigma))} = \sigma \restriction_{E(\sigma)}$ and together with ($\Diamond_{11}$), ($\Diamond_{21}$), ($\Diamond_{12}$) and ($\Diamond_{22}$) we see that conditions (i) and (iii) of the theorem hold for all relations in S'.

Now we set $\mathcal{S} := \{S' \mid S \in \mathcal{S}_1\}$, applying the same construction to all elements in $\mathcal{S}_1$. As before it is easy to see that conditions (i) and (iii) hold for all relations in $\bigcup \mathcal{S}$ and that $\mathcal{S}$ is a $\bigcup\bigcap$-representation of $[\![t]\!] = \delta^{m-1,m}(t_1) = \delta^{m-1,m}([\![t_1]\!]) = \delta^{m-1,m}(\bigcup\{\bigcap S \mid S \in \mathcal{S}_1\}) = \bigcup\{\delta^{m-1,m}(\bigcap S \mid S \in \mathcal{S}_1) = \bigcup\{\bigcap S' \mid S \in \mathcal{S}_1\}$, which is due to Def. 1.3(iii) consistent with t.

To show that condition (ii) holds requires again to consider several cases. Let $\widetilde{\varphi}_1, \widetilde{\varphi}_2 \in \bigcup\mathcal{S}$. (IV.iii.a) If there is no place in $E(\widetilde{\varphi}_1) \cap E(\widetilde{\varphi}_2)$ which is t_1-connected to $m - 1$ or m, then we see by Def. 1.3(iii-iii") and the foregoing construction that there are relations $\widetilde{\psi}_1, \widetilde{\psi}_2 \in \bigcup\mathcal{S}$ such that $\widetilde{\varphi}_1 = \delta^{m-1,m}(\widetilde{\psi}_1)$, $\widetilde{\varphi}_2 = \delta^{m-1,m}(\widetilde{\psi}_2)$ and $m - 1, m \notin E(\widetilde{\psi}_1) \cup E(\widetilde{\psi}_2)$. Consequently, $\widetilde{\varphi}_1{\restriction}_{E(\widetilde{\varphi}_1)} = \widetilde{\psi}_1{\restriction}_{E(\widetilde{\psi}_1)}$ and $\widetilde{\varphi}_2{\restriction}_{E(\widetilde{\varphi}_2)} = \widetilde{\psi}_2{\restriction}_{E(\widetilde{\psi}_2)}$ are comparable according to the induction hyposthesis.

(IV.iii.b) If there is some place in $E(\widetilde{\varphi}_1) \cap E(\widetilde{\varphi}_2)$ which is t_1-connected to $m - 1$ or m, then due to property (i) we have $E(\widetilde{\varphi}_1), E(\widetilde{\varphi}_2) \subseteq \{k, l\}$. There are several possibilities.

(IV.iii.b.1) $\widetilde{\varphi}_1 = \delta^{m-1,m}(\widetilde{\psi}_1)$ and $\widetilde{\varphi}_2 = \delta^{m-1,m}(\widetilde{\psi}_2)$ with $E(\widetilde{\varphi}_1) = E(\widetilde{\varphi}_2) \in \{\{k\}, \{l\}\}$. Then $\widetilde{\psi}_i{\restriction}_{E(\widetilde{\psi}_i)} = \widetilde{\varphi}_i{\restriction}_{E(\widetilde{\varphi}_i)}$ for $i \in \{1, 2\}$ which are comparable due to the induction hypothesis.

(IV.iii.b.2) $\widetilde{\varphi}_1 = \delta^{m-1,m}(\widetilde{\psi}_1)$ with $E(\widetilde{\psi}_1) = E(\widetilde{\varphi}_2) = \{k\}$, but $\widetilde{\varphi}_2$ is of the form $\widetilde{\varphi}_2 = \delta^{m-1,m}(\widetilde{\tau}_{21})$ or $\widetilde{\varphi}_2 = \delta^{m-1,m}(\widetilde{\tau}_{22})$. From $k \in E(\widetilde{\varphi}_2) \underset{\text{L. 1.7(iv)}}{\subseteq}$
$E(\widetilde{\psi}_2)$ we deduce $\widetilde{\tau}_{21}{\restriction}_{E(\widetilde{\tau}_{21})} \in \{\pi_{(23)}(\tau_{21}), \varrho_2, \pi_\alpha(\tau_{22})\}$ with $\alpha = (243)$ if $k < l$ and $\alpha = (1243)$ otherwise. Therefore $\widetilde{\psi}_2{\restriction}_{E(\widetilde{\psi}_2)} \in \{\delta^{2,3}(\varrho_2 \times \sigma_1), \delta^{2,3}(\varrho_2 \times A^1), \delta^{2,3}(\varrho_2 \times \sigma_2), \delta^{2,3}(\sigma_2^{-1} \times \varrho_2^{-1})\}$ (if $k < l$ we have $\delta^{2,3}(\tau_{22})$ and if $l < k$ it is the join of the inverted relations). By the induction hypothesis are $\varphi_2{\restriction}_{E(\varphi_2)} = \widetilde{\psi}_2{\restriction}_{E(\widetilde{\psi}_2)}$ and $\widetilde{\varrho}_2{\restriction}_{E(\widetilde{\varrho}_2)} = \varrho_2$ inverted comparable, by applying Lemma 1.13(ii) we see that $\widetilde{\varphi}_1{\restriction}_{E(\widetilde{\varphi}_1)}$ and $\widetilde{\varphi}_2{\restriction}_{E(\widetilde{\varphi}_2)}$ are inverted comparable (or comparable if $l < k$).

(IV.iii.b.3) The case $\widetilde{\varphi}_1 = \delta^{m-1,m}(\widetilde{\psi}_1)$ with $E(\widetilde{\psi}_1) = \{l\}$ and $\widetilde{\varphi}_2 = \delta^{m-1,m}(\widetilde{\tau}_{12})$ or $\widetilde{\varphi}_2 = \delta^{m-1,m}(\widetilde{\tau}_{22})$ is handled analogously to case (IV.iii.b.2).

(IV.iii.b.4) Finally, if $E(\widetilde{\varphi}_1) = E(\widetilde{\varphi}_2) = \{k, l\}$ then both must be of the form $\widetilde{\varphi}_1 = \delta^{m-1,m}(\widetilde{\tau}_{22})$ and $\widetilde{\varphi}_2 = \delta^{m-1,m}(\widetilde{\tau}'_{22})$ with $\widetilde{\tau}_{22}{\restriction}_{E(\widetilde{\tau}_{22})} = \varrho_2 \times \sigma_2$ and $\widetilde{\tau}'_{22}{\restriction}_{E(\widetilde{\tau}'_{22})} = \varrho'_2 \times \sigma'_2$ (they may come from different sets $S_1, S_2 \in \mathcal{S}_1$). However, by induction hypothesis ϱ_2 and ϱ'_2 are comparable and therefore $\varrho_2 = \varrho'_2$ and for the same reason $\sigma_2 = \sigma'_2$ and consequently $\widetilde{\tau}_{22} = \widetilde{\tau}'_{22}$ and therefore $\widetilde{\varphi}_1$ and $\widetilde{\varphi}_2$ are equal and comparable (and inverted comparable).

Therefore, property (ii) of the theorem also holds for $\mathcal{S}$ in the case (IV.iii).

(V) If $t = \neg t_1$ then there exists by the induction hypothesis a $\bigcup\bigcap$-representation $\mathcal{S}_1$ consistent with t_1. By Lem. 1.17 the set $\{\{\neg\tau(S_1) \mid S_1 \in \mathcal{S}_1\} \mid \tau : \mathcal{S}_1 \to \bigcup\mathcal{S}_1, \tau(S) \in S$ for all $S \in \mathcal{S}_1\}$ is a $\bigcup\bigcap$-representation of $\neg[\![t_1]\!] = [\![t]\!]$ which is consistent with t by Def. 1.3(iv) and Lem. 1.7(v). Condition (i) follows from Lem. 1.7(v), condition (ii) follows from Lem. 1.13(i).

(VI) If $t = \pi_\alpha(t_1)$ for some permutation α on $\underline{ar}\, t$, then there exists by the induction hypothesis $\bigcup\bigcap$-representations $\mathcal{S}_1$ consistent with t_1. Obviously $\mathcal{S} := \{\{\pi_\alpha(\sigma) \mid \sigma \in S\} \mid S \in \mathcal{S}_1\}$ is a $\bigcup\bigcap$-representation consistent with $[\![t]\!]$ (by Lem. 1.7(vi)). It is easy to see that conditions (i) and (iii) hold. For relations $\pi_\alpha(\sigma), \pi_\alpha(\tau) \in S$ with $\alpha(i) \in E(\pi_\alpha(\sigma)) \cap E(\pi_\alpha(\tau))$. If $|E(\sigma) \cup E(\tau)| = 1$ this trivially holds, if $|E(\sigma)| = |E(\tau)| = 2$ we can deduce $E(\sigma) = E(\tau)$ using Lem. 1.4 and even $\sigma = \tau$ because σ and τ are comparable by the induction hypothesis and therefore $\pi_\alpha(\sigma)$ and $\pi_\alpha(\tau)$ are equal, in particular comparable and inverted comparable. Otherwise, we can assume without loss of generality that $|E(\sigma)| = 2$ and $|E(\tau)| = 1$. Let $E(\sigma) = \{i, j\}$ and $i \in E(\tau)$. By the induction hypothesis we know that $\sigma{\upharpoonright}_{E(\sigma)}$ and $\tau{\upharpoonright}_{E(\tau)}$ are comparable if $i < j$ and inverted comparable if $j < i$. If α inverts the order of i and j we get that $\pi_\alpha(\sigma){\upharpoonright}_{E(\pi_\alpha(\sigma))} = (\sigma{\upharpoonright}_{E(\sigma)})^{-1}$. We see that then either $\pi_\alpha(\tau){\upharpoonright}_{E(\pi_\alpha(\tau))}$ and $\pi_\alpha(\sigma){\upharpoonright}_{E(\pi_\alpha(\sigma))}$ are comparable (if $\alpha(i) < \alpha(j)$) or inverted comparable (otherwise). Therefore condition (ii) holds in all cases. $\qquad\square$

3   The Peircean Reduction Thesis

The representation theorem presented in the last section allows to prove the difficult part of the Peircean Reduction Thesis. It is easy to see that all relations can be generated from the unary and binary relations – if the teridentity can be used. Thus the Peircean Reduction Thesis is true if and only if it is impossible to generate the teridentity itself from unary and binary relations. In fact, using the properties of the $\bigcup\bigcap$-representation provided by Theorem 2.1 we can prove that the teridentity cannot be generated. Therefore at least one ternary relation is needed besides unary and binary relations to generate all relations.

3.1 Theorem. Let $|A| \geq 2$ and $\Sigma := \mathrm{Rel}^{(1)}(A) \cup \mathrm{Rel}^{(2)}(A)$. Then

$$\langle \Sigma \rangle^A_{\mathsf{PAL}\setminus\{\mathrm{id}_3\}} \subsetneqq \langle \Sigma \rangle^A_{\mathsf{PAL}}.$$

Proof. By definition we have $\mathrm{id}_3 \in \langle \Sigma \rangle^A_{\mathsf{PAL}}$. Let $\varrho \in \langle \Sigma \rangle^A_{\mathsf{PAL}\setminus\{\mathrm{id}_3\}}$ be a ternary relation. Let us assume $\mathrm{id}_3 = \varrho$. Let $t \in \mathsf{T}_{\mathsf{PAL}\setminus\{\mathrm{id}_3\}}(\Sigma)$ be a term describing ϱ. Then by Thm. 2.1 and Lem. 1.18 there exists a normalized $\bigcup\bigcap$-representation $\mathcal{S}$ consistent with t. From Lem. 1.4 we can deduce that there is some place in $\underline{3}$ which is not t-connected to the other two places. Without loss of generality we assume that 1 is not t-connected to 2 or 3. Let $S \in \mathcal{S}$. Because $\mathcal{S}$ is normalized, there is some tuple $(x, x, x) \in S$. Let $y \in A$ with $x \neq y$. For any relation $\sigma \in S$ with $1 \notin E(\sigma)$ we have $\sigma = A^1 \times \sigma{\upharpoonright}_{\{2,3\}}$ and therefore $(y, x, x) \in \sigma$. If we have $1 \in E(\sigma)$ for no $\sigma \in S$, then we have $(y, x, x) \in \bigcap S \subseteq \bigcup\{\bigcap S \mid S \in \mathcal{S}\} = \varrho$ but $(y, x, x) \notin \mathrm{id}_3$, a contradiction.

Therefore, for every $S \in \mathcal{S}$ there has to be a relation ϱ_S with $1 \in E(\varrho_S)$, and because 1 is not t-connected to any other place we have $E(\varrho_S) = \{1\}$. Let $a, b \in A$ be two distinct elements. Let S_a be an element of $\mathcal{S}$ with $(a, a, a) \in \bigcap S_a$ (and therefore $a \in \varrho_{S_a}{\upharpoonright}_{E(\varrho_{S_a})}$), and analogously S_b such that $(b, b, b) \in \bigcap S_b$ and therefore $b \in \varrho_{S_b}{\upharpoonright}_{E(\varrho_{S_b})}$. By condition (ii) of Thm. 2.1 we know that ϱ_{S_a} and ϱ_{S_b} are comparable and therefore $a \in \varrho_{S_a} \subseteq \varrho_{S_b}$ or $b \in \varrho_{S_b} \subseteq \varrho_{S_a}$. The set $\mathcal{S}$ is normalized and by condition (ii) of Thm. 2.1 we know that ϱ_{S_a} is the only relation $\sigma \in S_a$ with $1 \in E(\sigma)$ (and likewise for ϱ_{S_b}). Therefore we can deduce either $(a, b, b) \in \bigcap S_b \subseteq \varrho$ or $(b, a, a) \in \bigcap S_a \subseteq \varrho$, in both cases a contradiction to $\varrho = \mathrm{id}_3$. Thus the teridentity is not representable in PAL without teridentity. $\square$

References

[Arn01]   M. ARNOLD, *Einführung in die kontextuelle Relationenlogik.* Diplomarbeit,
          Technische Universität Darmstadt, 2001.

[Bur91]   R. W. BURCH, *A Peircean Reduction Thesis: The Foundations of Topolog-
          ical Logic.* Texas Tech University Press, 1991.

[DHC06]   F. DAU, J. HERETH CORREIA: Two Instances of Peirce's Reduction Thesis.
          In: R. Missaoui and J. Schmid (eds.): *Formal Concept Analysis,* Proc. of the
          4th Int. Conf. on Formal Concept Analysis, vol. 3874 of *LNAI,* Springer,
          Berlin–Heidelberg 2006, pp. 105-118.

[DaK05]   F. DAU, J. KLINGER: From Formal Concept Analysis to Contextual Logic.
          In B. Ganter, G. Stumme, R. Wille (Eds.): *Formal Concept Analysis: Foun-
          dations and Applications,* vol. 3626 of *LNAI,* Springer, Berlin–Heidelberg
          2004, pp. 81–100.

[HCP04]   J. HERETH CORREIA, R. PÖSCHEL: The Power of Peircean Algebraic Logic
          (PAL). In: P. Eklund (ed.): *Concept Lattices,* Proc. 2nd Int. Conf. on Formal
          Concept Analysis, vol. 2961 of *LNAI,* Springer, Berlin–Heidelberg 2004, pp.
          337–351.

[Her81]   HANS G. HERZBERGER, *Peirce's Remarkable Theorem.* In: L. W. SUMNER,
          J. G. SLATER, AND F. WILSON (eds.), *Pragmatism and Purpose: Essays
          Presented to Thomas A. Goudge,* University of Toronto Press, Toronto, 1981.

[PR67]   C. S. PEIRCE: *The Peirce Papers in the Houghton Library, Harvard Univer-
          sity.* Catalogue by R. Robin: Annotated Catalogue of the Papers of Charles
          S. Peirce. University of Massachusetts Press, Massachusetts 1967.

[Sow84]   J. F. SOWA: *Conceptual structures: Information processing in mind and
          machine.* Addison-Wesley, Reading 1984.

[Sow92]   J. F. SOWA, *Conceptual graphs summary.* In: T. NAGLE, J. NAGLE, L. GER-
          HOLZ, AND P. EKLUND (eds.), *Conceptual structures: Current Research and
          Practice,* Ellis Horwood, 1992, pp. 3–52.

[Sow00]   J. F. SOWA: *Knowledge Representation: Logical, Philosophical, and Com-
          putational Foundations.* Brooks Cole Publishing Co., Pacific Grove 1999.

[Wil00b] R. WILLE: Contextual Logic summary. In: G. Stumme (ed.): *Working with
          conceptual structures. Contributions to ICCS 2000.* Shaker-Verlag, Aachen
          2000, pp. 265-276.

[Wil00]   R. WILLE, *Lecture notes on Contextual Logic of Relations.* Preprint, Darm-
          stadt University of Technology 2000.

Transaction Agent Modelling: From Experts to Concepts to Multi-Agent Systems

Richard Hill, Simon Polovina, and Dharmendra Shadija

Web and Multi-Agent Research Group,
Sheffield Hallam University, Sheffield, UK
{r.hill, s.polovina, d.shadija}@shu.ac.uk

Abstract. Whilst the Multi-Agent System (MAS) paradigm has the potential to enable complex heterogeneous information systems to be integrated, there is a need to represent and specify the nature of qualitative conceptual transactions in order that they are adequately comprehended by a goal-directed MAS. Using the Transaction Agent Model (TrAM) approach we examine the use of Conceptual Graphs to model an extension to an existing MAS in the community healthcare domain, whereby the existing agent capabilities are augmented with a robust set of behaviours that provide emergency healthcare management. We illustrate how TrAM serves to enrichen the requirements gathering process, whilst also supporting the definition and realisation of quantitative measures for the management of qualitative transactions.

1   Introduction

Complex information systems are often rich with convoluted interactions, involving repetitive and bespoke transactions that frequently consist of many other, smaller, heterogeneous sub-systems. The Multi-Agent System (MAS) paradigm is attractive to the systems designer in that these complex organisational operations can be mapped to 'agents' in a way that corresponds better to the real-world view. Accordingly a MAS designer typically assembles, iterates and refines a series of design using the Unified Modelling Language (UML) or, more recently, Agent-oriented UML (AUML) [1]. However whilst the emerging methods that assist the conversion of initial requirements into program code are proliferating there still remains a significant intellectual distance between MAS model and implementation.

As we have identified elsewhere [11] [12], [13], [19] Agent Oriented Software Engineering (AOSE) methodologies such as Gaia [23], Prometheus [17], Zeus [16] and MaSE [6] have attempted to provide a unifying development framework. Except for Tropos [4] however, little work has been published that encompasses the whole cycle from initial requirements capture through to implementation of MAS. To prevent significant disparities between program code and the more abstract models, Tropos facilitates the modelling of systems at the knowledge level

H. Schärfe, P. Hitzler, and P. Øhrstrøm (Eds.): ICCS 2006, LNAI 4068, pp. 247–259, 2006.
© Springer-Verlag Berlin Heidelberg 2006

and highlight the difficulties encountered by MAS developers, especially since notations such as UML force the conversion of knowledge concepts into program code representations [4]. As a methodology Tropos seeks to capture and specify 'soft' and 'hard' goals during an 'Early Requirements' capture stage, in order that the Belief-Desire-Intention (BDI) architectural model [10] of agent implementation can be subsequently supported. Whilst model-checking is provided through the vehicle of Formal Tropos [8], this is an optional component and is not implicit within the MAS realisation process.

Therefore we describe an extension to an existing demonstrator in the community healthcare domain, whereby qualitative transactions in emergency healthcare management are elucidated and expressed in order that an information system can be designed and developed as a MAS. As such we explain how an improved MAS design framework that places a greater emphasis upon the initial requirements capture stage, by supplementing the modelling process with Conceptual Graph notation [21] can assist the realisation of complex information systems.

2   The Problem Domain Addressed

High quality medical care depends upon prompt, accurate recording, communication, and retrieval of patient data and medical logistics information. In emergency medicine, such information can make the difference between life and death [15] or severe permanent injury. To elaborate; this vital information enables better planning and scheduling of medical resources [15]. Furthermore, a hospital can assemble the appropriate team of specialists and prepare the necessary equipment at the earliest opportunity, before they encounter the patient.

Prior experience with the Intelligent Community Alarm (InCA) [2] has illustrated the potential of information systems designed using the MAS approach. Home-based community care requires a variety of healthcare services to be delivered within the recipient's own home, allowing them to continue to live independently, maintaining the best possible quality of life. Healthcare services include help with domestic cleaning, 'meals-on-wheels', medical treatment and assistance with basic bodily functions. Whilst the aims of community care address humanitarian issues, other services such as emergency healthcare are an example of the challenging and expensive tasks of managing the organisation, logistics and quality assurance [3].

As each element of care is often delivered by independent agencies, the number of autonomous command and control systems quickly increases, leaving the overall managers of the care (the Local Authority) to protect the individual agencies from disclosing sensitive and irrelevant information. Although information technology is established in emergency healthcare management, it is clear that in many instances the vast quantities of disparate heterogeneous information repositories can lead to the undermining of effective system operations [15].

Collaborative intelligent agents, which overcome the difficulties of integrating disparate hardware and software platforms, enables queries to be mediated to
the most appropriate information source, with the potential to build effective, co-ordinated emergency healthcare management systems. As such the benefits of multi-agent systems, and examples of the types of improvements offered for healthcare management, are documented by Zambonelli et al. [24] and Beer et al. [2].

Human agents regularly engage in economic transactions and Beer et al. [3] demonstrated the transactions involved when processing emergency alarm conditions in a community care environment. These apparently innocuous transactions proved much more difficult to deploy than first envisaged, though there were distinct advantages in favour of the agent approach. Upon raising the alarm, the message from the Elderly Person required brokering to determine the correct type of service, as well as locating the nearest, most available source of help. Subsequent work has enabled the further development of the Intelligent Community Alarm (InCA) to include the management of information in response to emergency healthcare requests.

3   Transaction Agent Model (TrAM)

Prior work has established that there is a stage whereby the elicitation of qualitative issues within a problem domain can assist the MAS development process [11], [12], [13], [20]. The combination of requirements capture and analysis with Conceptual Graphs (CG), and its transformation into the Transaction Model (TM), based upon the work of Geerts and McCarthy [9], provides a much more rigorous input artefact in readiness for design specification with UML [22]. We have accordingly defined this combination as the Transaction Agent Model or TrAM. The use of TrAM enriches the requirements capture stage and serves as a precursor to existing AOSE methodologies that require a design representation such as AUML as an input.

3.1   Modelling the System

Initial requirements models such as use case diagrams can be mapped to a CG [19] as in Figure 1.

The TM denotes that the emergency care is a transaction that arises due to the occurrence of two complementary economic events, namely '999 call' and 'Life-threatening Incident' (Figure 3). These are considered economic events because they denote the demand upon a limited resource. The '999 call' requires limited resources of the UK health authority, who have to make provision for this support at the potential expense of other resources (e.g. community care) upon which their finances could be spent.

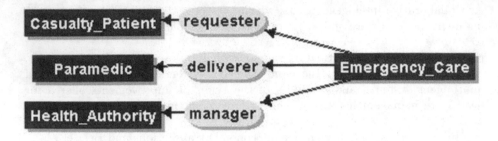

Fig. 1. Conceptual representation of use case model

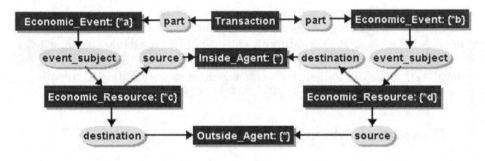

Fig. 2. Generic Transaction Model (TM)

For TrAM we then review this CG to determine how it fits in to the generic TM (Figure 2).

Similarly the life-threatening incident calls upon the paramedics' priorities, in terms of potentially being required elsewhere or more simply the time it takes the paramedic to arrive at the scene. Hence the health authority needs to provide (i.e. be the source of) some optimally cost-effective estimated time of arrival (ETA).

Clearly if money was no object then an infinite number of paramedics could be employed and the ETA would be zero thus the most lives would be saved. However as the health authority's finances are limited by what it receives from the government, plus competing prioritising demands for that finance to be spent elsewhere, money is an economic resource that is scarce. This in turn makes an unrestricted number of paramedics impossible, thus also making ETA an economic resource. This is demonstrated by the very fact that ETA denotes a delay caused by competing demands (e.g. another accident being attended to by the paramedic) or the physically remote location of the paramedic when the 999 call is made (as there is only a limited amount of money to pay for paramedics). The corresponding benefit for these 'costs' is the economic resource of the saved

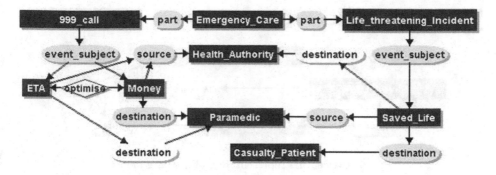

Fig. 3. Emergency Care Transaction Model

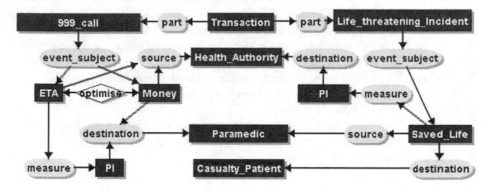

Fig. 4. Emergency Care transaction model with added 'measure' concept

life of the casualty patient. This concept of life being an economic resource may sound callous, but as we see from the cost-benefit trade-off, there are sadly some lives that the health authority cannot afford to be saved. Once again an optimal ETA vs. monetary cost must be found, and efficiencies continually found to improve this optimisation further for the benefit of saving more lives.

Accordingly, following the TrAM process the following CG for the TM can be derived as in Figure 3. In accordance with TrAM the CG of Figure 3 should also show a transaction between the casualty patient and the health authority or the paramedic. But this cannot as yet be defined because, unlike the money destination (e.g. salary paid) to the paramedic, the above TM has the rather odd situation of the ETA destination being the paramedic.

This clearly does not model the relationship between the concepts in a realistic manner, therefore it is necessary to represent the economic resource with a meaningful measure such as a Performance Indicator (PI). This PI would then be used to measure the effectiveness of the paramedic in attending life-threatening situations.

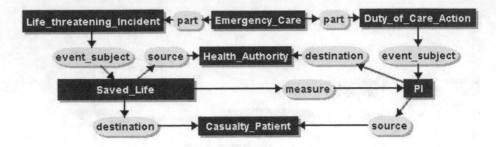

Fig. 5. Conceptual representation of use case model

Likewise the health authority is shown as the destination of the saved life when, once again, it would better be informed by some effective measure. Hence we find the need for an expressive PI. These measures thereby offer the relevant, quantifiable information upon which the health authority and paramedic can make the most informed decisions, as indeed would their software agents. The following CG in Figure 4 thus captures these dimensions. We thus require that some economic resources need to have a characteristic of being measurable.

Figure 5 illustrates how we can proceed to capture the transaction of the casualty patient, whilst the type hierarchy deduced after transformation with the TM is illustrated in Figure 6.

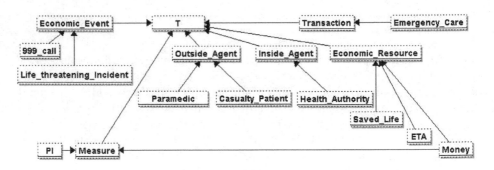

Fig. 6. Emergency Care Type Hierarchy

Once again, a PI enables the casualty patient to be most informed about the duty of care that he or she was given (and consequently the aggregate quality of 'duty of care' that a person can expect, perhaps against some benchmark, should that person be in the unfortunate situation of being a casualty patient).

3.2   Designing the Agent Classes

Our exemplar emergency healthcare transaction illustrates the how TrAM can be used to develop a complex multi-agent system for the management of an emergency healthcare information system.

Emergency healthcare management exposes a vast number of qualitative issues in the widest sense, and it is apparent that AUML has limitations when attempting to elicit these conceptual issues. Our experiences with CG illustrate that this notation appears to offer the multi-agent system designer a considerable advantage when it comes to assembling a specification of requirements that captures the essence of real-world representations, particularly when used in conjunction with AUML. It is also apparent that CG models lack the detail necessary to specify agent program code, unlike the comprehensive representations that can be expressed with AUML.

CG	Class Diagram	Description
Transaction	TransactionMaker Class	Initiates the Transaction
EconomicEvent	EEAction Class	Represents the EconomicEvent
EconomicResource	EResource Class	Represents the EconomicResource
OutsideAgent	OutsideAgent Class	Represents the OutsideAgent
InsideAgent	InsideAgent Class	Represents the InsideAgent
	ERList Class	Represents a group of Economic Resources
Relationship		**Type of Relationship**
EventSubject	EventSubject relation between EEAction and ERList	Association
Part	'Part of' relationship between EEAction and TransactionMaker	Aggregation
Source	Owns relationship between Agent class and EResource	Association
Destination	Owns relationship between Agent class and EResource	Association
	Specialisation relationship between Agent and InsideAgent	Inheritance
	Specialisation relationship between Agent and OutsideAgent	Inheritance

Fig. 7. Deriving the agent classes and relationships from the Transaction Model

Once the initial requirements have been captured and transformed with the TM, it is necessary to convert the resulting models into a design specification in readiness for program code generation. Figure 7 describes the mappings between the TM concepts, relationships, and the resulting classes, leading to the model in Figure 8.

3.3   Building the Demonstrator

The TrAM approach to MAS construction has enabled disparate emergency services to be integrated in order to share information, whilst retaining loose coupling. The basic HES architecture is illustrated in Figure 9, and contains details specific to the deployed demonstrator:

- *External Agent*: This agent provides an interface to the user to access the emergency services. It executes outside the platform main container and makes remote request to the Broker agent, enhancing system security.
- *Broker Agent*: this agent control the access to emergency services agents; it is also be used to verify the identity of the user in terms of role-based access control to sensitive patient data.
- *Casualty Department and Ambulance Call Operator Agent*: both agents are an example of health emergency service agents. These agents can allow the user to contribute or access information about a specific incident. Such agents cooporate and collaborate to enrich the information captured, facilitating proactive behaviours.
- *Buffer Agent*: This agent retains information if the platform is not reachable due to infrastructure problems.
- *Casualty Department and Ambulance Control Operator database wrappers*: agents that control access to heterogeneous data sources.

The process of eliciting agents using the TrAM model has exposed both singular agents and agent roles that are performed by a MAS. The details of this process are omitted for brevity.

4   Conclusions

Emergency healthcare management systems are an example of the complexity often encountered in many problem domains. They incorporate many islands of both information and functionality, therefore the MAS paradigm seems relevant, providing the necessary design abstraction.

There are higher order issues and these affect the eventual system. If we consider the individual goals of an agent, they aspire to satisfy some rules predetermined by their owner (the human agent). Typically these goals arecaptured as qualitative concepts such as: 'To better serve the public', 'to improve the standard of living', 'to make money now and in the future', 'to improve quality of care'.

Thus if we elect to model concepts then it immediately opens up the possibility of modelling at much greater levels of abstraction. The need to capture and express qualitative concepts demands a means of representation that is rich and flexible. If these qualitative concepts are to be represented later as quantitative measures then it is important that the thought processes and design iterations are documented and rigorously evaluated. Without such a

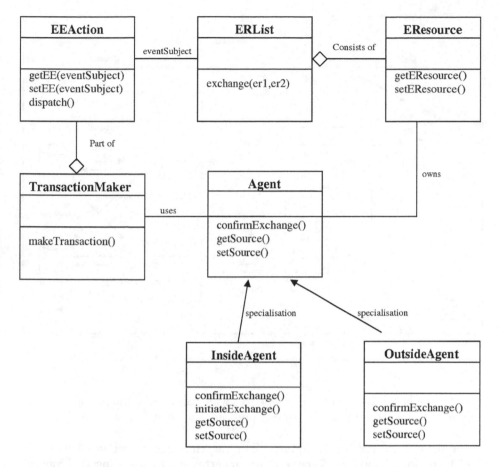

Fig. 8. Class diagram derived from the Transaction Model

representation medium, the quantification would not be documented and key ideas and opportunities may have been missed.

There is a need to represent the complexity of multi-agent based systems and it is becoming increasingly common to embody software with human character-istics such as beliefs, desires and intentions (BDI) [10]. Since the CG notation is flexible, it can be used to quickly capture the high level requirements of a sys-tem. Additionally, the in-built type hierarchy (where every concept inherits from the universal supertype 'T') enables the concept types elicited to be scrutinised both in terms of the name by which the concept will be referred to as well as the relationships between two or more concepts with respect to 'T'.

Since agents in an organisation will be required to conduct transactions, it seems appropriate to consider the notion of transactions and how they might offer a possible representation for agent system design. Geerts and McCarthy [9] offer 'event accounting', upon which the work of Polovina [18], as well as Sowa [21][pp109-111], is based. The attractions of the TM are:

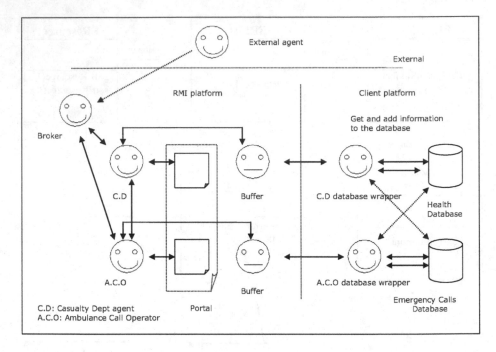

Fig. 9. Deployment architecture of the Emergency Healthcare MAS

- the model is generic and applicable to a wide variety of domain transactions;
- models are straightforward to comprehend and relate to standard business vocabulary;
- a hierarchy of types is introduced into the requirements gathering process;
- the notion of 'balance' forces discussion pertaining to the respective concept names (or 'slot' names as they will inevitibly become later).

Balanced transactions are particularly powerful when qualitative concepts are recorded. Rather than force the analyst to quantify the qualitative concepts, these can be left in the model to achieve a balance. Subsequent analysis focuses attention upon the qualitative concepts that need to be resolved, whilst reinforcing the importance of the inclusion of the concept, albeit qualitative and therefore difficult to deal with. Aware of this, the system design participants are motivated to select the most appropriate domain term for the concept. Copnsequently, we believe that the TrAM route from abstract requirements gathering through to design specification contributes towards the development of complex information systems by:

- ensuring that a conceptual representation of the information system has captured qualitative transactions that can be decomposed into quantitative measures prior to verification;
- permitting the deduction of performance indicators from the 'measure' concept. Relationships between the concepts in the TM force questions to be

asked, enabling terms and rule for the ontology to be derived, before a balance is achieved;
- using Conceptual Graphs and the Transaction Model to elicit an ontology that agents can utilise as part of a knowledge base;
- providing a mapping from conceptual model to AUML design specification.

The mappings identified so far have proved repeatable across different scenarios and assist agent code generation.

5 Further Work

Since the key mappings have been identified, we are now automating the translation of the iterated Transaction Model into a design specification such as UML, in order to facilitate the use of TrAM in many other disparate domains.

Acknowledgements

Programming of the Health Emergency System MAS demonstrator was carried out by Ahmed Mohammed and Owen Lobb as part of their Masters Degree studies. This project is in receipt of an AgentCities Deployment Grant from the European Union AgentCities.rtd Project (IST-2000-28385).

References

1. Bauer, B., Muller, J. P., and Odell, J., (2000), 'Agent UML: A Formalism for Specifying Multi-agent Software Systems', in Agent-Oriented Software Engineering, vol. 1957, Ciancarini, P. and Wooldridge, M. J., Eds. Springer-Verlag, 2000, pp. 91-104.
2. Beer, M. D., Bench-Capon, T. J. M. and Sixsmith, A., (1999) 'Some Issues in Managing Dialogues between Information Agents', In Proceedings of Database and Expert Systems Applications '99, Lecture Notes in Computer Science 1677, Springer, Berlin, 1999, pp. 521-530.
3. Beer, M. D, Huang, W., and Sixsmith, A. (2002), 'Using Agents to Build a Practical Implementation of the InCA (Intelligent Community Alarm) System', in L. C. Jain, Z. Chen, & N. Ichalkaranje, 'Intelligent Agents & their Applications', Springer, pp320-345.
4. Bresciani, P., Giorgini, P., Giunchiglia, F., Mylopoulos, J., and Perini, A., (2004) 'TROPOS: An Agent-Oriented Software Development Methodology', Journal of Autonomous Agents and Multi-Agent Systems, vol. 8, pp. 203-236, 2004.
5. Dau F., (2003) Lecture Notes in Computer Science 2892: 'The Logic System of Concept Graphs with Negation: And Its Relationship to Predicate Logic', Heidelberg: Springer-Verlag, 2003
6. DeLoach, S., (1999) 'Multi-Agent Systems Engineering: A Methodology and Language for Designing Agent Systems', citeseer.ist.psu.edu/deloach99multiagent.html
7. FOUNDATION FOR INTELLIGENT PHYSICAL AGENTS, 'FIPA Iterated Contract Net Interaction Protocol Specification'. Accessed: 2005, 11/21. 2000. http://www.fipa.org/specs/fipa00030/PC00030D.html

8. Fuxman, A., Kazhamiakin, R., Pistore, M., and Roveri, M., (2003) 'Formal Tropos: language and semantics (Version 1.0)'. Accessed: 2005, 4th November. 2003. http://www.dit.unitn.it/ ft/papers/ftsem03.pdf

9. Geerts, G., and McCarthy, W., 'Database Accounting Systems', in Information Technology Perspectives in Accounting: and Integrated Approach, Eds. B. Williams and B. J. Sproul, Chapman and Hall Publishers, 1991, pp. 159-183.

10. Georgeff, M. P., Pell, B., Pollack, M. E., Tambe, M., and Wooldridge, M., (1999) 'The Belief-Desire-Intention Model of Agency', in ATAL '98: Proceedings of the 5th International Workshop on Intelligent Agents V, Agent Theories, Architectures, and Languages, 1999, pp. 1-10.

11. Hill, R., Polovina, S., Beer, M. D., (2005) 'From Concepts to Agents: Towards a Framework for Multi-Agent System Modelling', Fourth International Joint Conference on Autonomous Agents and Multi-Agent Systems, AAMAS '05, 25-29 July, Utrecht University, Netherlands, ACM Press, in press.

12. Hill, R., Polovina, S., Beer, M. D., (2005) 'Improving AOSE with an Enriched Modelling Framework', 6th International Workshop on Agent Oriented Software Engineering (AOSE-2005), Fourth International Joint Conference on Autonomous Agents and Multi-Agent Systems, AAMAS '05, 25 July, Utrecht University, Netherlands, ACM Press, in press.

13. Hill, R., Polovina, S., Beer, M. D., (2005) 'Managing Community Healthcare Information in a Multi-Agent System Environment', First International Workshop on Multi-Agent Systems for Medicine, Computational Biology, and Bioinformatics (BIOMED), Fourth International Joint Conference on Autonomous Agents and Multi-Agent Systems, AAMAS '05, 25 July, Utrecht University, Netherlands, ACM Press, in press.

14. Hill, R., Polovina, S., Beer, M. D., (2004) 'Towards a Deployment Framework for Agent-Managed Community Healthcare Transactions', The Second Workshop on Agents Applied in Health Care, 23-24 Aug 2004, Proceedings of the 16th European Conference on Artificial Intelligence (ECAI 2004), Valencia, Spain, IOS Press, 13-21.

15. Holzman, T. G. (1999). Computer-human interface solutions for emergency medical care. Interactions, 6 (3), 13-24.

16. Nwana, H., Ndumu, D., Lee, L., and Collis, J., (1999) 'ZEUS: A Tool-Kit for Building Distributed Multi-Agent Systems', Applied Artifical Intelligence Journal, vol. 13, no. 1 pp129-186.

17. Padgham, L., Winikoff, M., (2002) 'Prometheus: A Methodology for Developing Intelligent Agents', In: Proceedings of the Third International Workshop on Agent-Oriented Software Engineering, at AAMAS 2002.

18. Polovina, S., 'The Suitability of Conceptual Graphs in Strategic Management Accountancy (PhD Thesis)', 1993, Available at http://www.polovina.me.uk/phd

19. Polovina, S., Hill, R., Crowther, P., Beer, M. D., (2004) 'Multi-Agent Community Design in the Real, Transactional World: A Community Care Exemplar', Conceptual Structures at Work: Contributions to ICCS 2004 (12th International Conference on Conceptual Structures), Pfeiffer, H., Wolff, K. E., Delugach, H. S., (Eds.), Shaker Verlag (ISBN 3-8322-2950-7, ISSN 0945-0807), 69-82.

20. Polovina, S., Hill, R., (2005) 'Enhancing the Initial Requirements Capture of Multi-Agent Systems through Conceptual Graphs', Proceedings of 13th International Conference on Conceptual Structures (ICCS '05): Conceptual Structures: Common Semantics for Sharing Knowledge, July 18-22, 2005, Kassel, Germany, Springer, 439-452.

21. Sowa, J. F., 'Conceptual Structures: Information Processing in Mind and Machine', Addison-Wesley, 1984.
22. OMG, 'Unified Modeling Language Resource Page', vol. 2004, http://www.uml.org/
23. Wooldridge, M., Jennings, N., Kinny, D.(2000), 'The Gaia Methodology for Agent-Oriented Analysis and Design', In: Autonomous Agents and Multi-Agent Systems 3, pp. 285-312.
24. Zambonelli, F., Jennings, N. R., and Wooldridge, M., (2003), 'Developing multiagent systems: The Gaia methodology', ACM Trans.Softw.Eng.Methodol., vol. 12, pp. 317-370.

Querying Formal Contexts with Answer Set Programs*

Pascal Hitzler and Markus Krötzsch

AIFB, University of Karlsruhe, Germany

Abstract. Recent studies showed how a seamless integration of formal concept analysis (FCA), logic of domains, and answer set programming (ASP) can be achieved. Based on these results for combining hierarchical knowledge with classical rule-based formalisms, we introduce an expressive common-sense query language for formal contexts. Although this approach is conceptually based on order-theoretic paradigms, we show how it can be implemented on top of standard ASP systems. Advanced features, such as default negation and disjunctive rules, thus become practically available for processing contextual data.

1 Introduction

At the heart of formal concept analysis (FCA) lies the formation of formal concepts from formal contexts. As such, formal concept analysis is a powerful tool for extracting conceptual hierarchies from raw data. The resulting lattices structure the knowledge hidden in the raw data, i.e. formal contexts, in a way which appeals to human experts, and allows them to navigate the data in a new way in order to understand relationships or create new hypotheses.

Formal concept analysis thus has a commonsense knowledge representation aspect. This also becomes apparent by its multiple uses in the creation of ontologies for the semantic web. It serves as a basic tool for the conceptualization of data, and the conceptual hierarchies obtained from it can often form a modelling base for ontologies in more expressive logical languages.

In this paper, we address the question of querying the conceptual knowledge hidden within a formal context. We present a query language based on commonsense reasoning research in artificial intelligence. It allows to query formal contexts by means of logic programs written over attributes and objects, and features default negation in the sense of the knowledge representation and reasoning systems known as answer set programming (ASP).

Our results will also show how the resulting query system can be implemented on top of standard ASP systems, which allows to utilize the highly optimized systems currently available. Our work also sheds some foundational light on ASP

* The authors acknowledge support by the German Federal Ministry of Education and Research (BMBF) under the SmartWeb project, and by the European Commission under contract IST-2003-506826 SEKT and under the KnowledgeWeb Network of Excellence.

H. Schärfe, P. Hitzler, and P. hrstrm (Eds.): ICCS 2006, LNAI 4068, pp. 260–273, 2006.

itself, whose theoretical underpinnings, in particular in relation to order-theoretic perspectives, are not yet understood in a satisfactory way.

The paper will be structured as follows. We first review the foundational results from [1] which serve as a base for our contribution. Section 2 introduces the logical formalism of reasoning on hierarchical knowledge that we will build upon, and recalls some basics of formal concept analysis. Section 3 discusses default negation for conceptual hierarchies. Section 4 describes the query language which we introduce. In Section 5 we present the theoretical results which enable the implementation of the querying system on top of the dlv ASP system. We close with the discussion of related and future work in Section 6.

2 Logic of Domains and FCA

We need to establish a certain amount of formal terminology in order to be able to motivate our contribution. This will be done in this and the next section. Following [2], we first introduce the logic of domains, and then recall the basics of formal concept analysis.

We assume the reader to be familiar with the basic notions of order theory, and recall only the relevant notions of domain theory. Thus let $(D, \sqsubseteq)$ be a partially ordered set. A subset $X \subseteq D$ is *directed* if, for all x, $y \in X$, there is $z \in X$ with $x \sqsubseteq z$ and $y \sqsubseteq z$. We say that D is a *complete partial order* (*cpo*) if every directed set $X \subseteq D$ has a least upper bound $\bigsqcup X \in D$. Note that we consider the empty set to be directed, and that any cpo thus must have a least element $\bigsqcup \emptyset$ that we denote by $\bot$.

An element $c \in D$ is *compact* if, whenever $c \sqsubseteq \bigsqcup X$ for some directed set X, there exists $x \in X$ with $c \sqsubseteq x$. The set of all compact elements of D is written as $\mathsf{K}(D)$. An *algebraic cpo* is a cpo in which every element $d \in D$ is the least upper bound of the – necessarily directed – set $\{c \sqsubseteq d \mid c \in \mathsf{K}(D)\}$ of compact elements below it.

A set $O \subseteq D$ is *Scott open* if it is upward closed, and *inaccessible by directed suprema*, i.e., for any directed set $X \subseteq D$, we have $\bigsqcup X \in O$ if and only if $O \cap X \neq \emptyset$. The *Scott topology* on D is the collection of all Scott open sets of D, and a Scott open set is *compact*, if it is compact as an element of the Scott topology ordered under subset inclusion. A *coherent algebraic cpo* is an algebraic cpo such that the intersection of any two compact open sets is compact open. Coherency of an algebraic cpo implies that the set of all minimal upper bounds of a finite number of compact elements is finite, i.e. if $c_1, \ldots, c_n$ are compact elements, then the set $\mathsf{mub}\{c_1, \ldots, c_n\}$ of minimal upper bounds of these elements is finite. As usual, we set $\mathsf{mub}\, \emptyset = \{\bot\}$, where $\bot$ is the least element of D.

In the following, $(D, \sqsubseteq)$ will always be assumed to be a coherent algebraic cpo. We will also call these spaces *domains*. All of the above notions are standard and can be found e.g. in [3].

We can now define the basic notions of domain logic. The following is taken from [2], where further details can be found.

Definition 1. *Let D be a coherent algebraic cpo with set $K(D)$ of compact elements. A* clause *of D is a finite subset of $K(D)$. Given a clause X over D, and an element $m \in D$, we write $m \models X$ if there exists $x \in X$ with $x \sqsubseteq m$, i.e. X contains an element below m. In this case, we say that m is a* model *of X.*

The clausal logic introduced in Definition 1 will henceforth be called the *Logic RZ* for convenience.

Example 1. In [2], the following running example was given. Consider a countably infinite set of propositional variables $\mathcal{V}$, and the set $\mathbb{T} = \{\mathbf{f}, \mathbf{u}, \mathbf{t}\}$ of truth values ordered by $\mathbf{u} < \mathbf{f}$ and $\mathbf{u} < \mathbf{t}$. This induces a pointwise ordering on the space $\mathbb{T}^{\mathcal{V}}$ of all interpretations (or *partial truth assignments*). The partially ordered set $\mathbb{T}^{\mathcal{V}}$ is a coherent algebraic cpo[1] and has been studied, e.g., in [4] in a domain-theoretic context, and in [5] in a logic programming context. Compact elements in $\mathbb{T}^{\mathcal{V}}$ are those interpretations which map all but a finite number of propositional variables to $\mathbf{u}$. We denote compact elements by strings such as $pq\bar{r}$, which indicates that p and q are mapped to $\mathbf{t}$, and r is mapped to $\mathbf{f}$. Clauses in $\mathbb{T}^{\mathcal{V}}$ can be identified with formulae in disjunctive normal form, e.g. $\{pq\bar{r}, \bar{p}q, r\}$ translates to $(p \wedge q \wedge \neg r) \vee (\neg p \wedge q) \vee r$.

The Logic RZ provides a framework for reasoning with disjunctive information. However, it is also possible to encode conjunctive information: given a finite set X of compact elements of a domain D, the "conjunction" of the elements of D can be expressed by the clause $\mathsf{mub}(D)$. Indeed, whenever an element models all members of X, it is greater or equal than one of the minimal upper bounds of X.

Example 2. Consider the domain $\mathbb{T}^{\mathcal{V}}$ of Example 1. The set of minimal upper bounds of every finite set of compact elements in $\mathbb{T}^{\mathcal{V}}$ is either singleton or empty. For instance, the only minimal (and therefore least) upper bound of $p\bar{r}$ and pq is $pq\bar{r}$.

The Logic RZ enables logical reasoning with respect to a background theory of hierarchical knowledge that is encoded in the structure of the domain. Formal concept analysis (FCA), in contrast, provides techniques for representing data in form of conceptual hierarchies, that allow for simple relational descriptions. We quickly review the basic notions of FCA, and refer to [6] for an in-depth treatment.

A *(formal)* context $\mathbb{K}$ is a triple (G, M, I) consisting of a set G of *objects*, a set M of *attributes*, and an *incidence relation* $I \subseteq G \times M$. Without loss of generality, we assume that $G \cap M = \emptyset$. For $g \in G$ and $m \in M$ we write $g \ I \ m$ for $(g, m) \in I$, and say that g has the attribute m.

For a set $O \subseteq G$ of objects, we set $O' = \{m \in M \mid g \ I \ m \text{ for all } g \in O\}$, and for a set $A \subseteq M$ of attributes we set $A' = \{g \in G \mid g \ I \ m \text{ for all } m \in A\}$. A *(formal)* concept of $\mathbb{K}$ is a pair (O, A) with $O \subseteq G$ and $A \subseteq M$, such that $O' = A$ and $A' = O$. We call O the *extent* and A the *intent* of the concept (O, A).

---

[1] In fact it is also bounded complete.

The set $\mathcal{B}(\mathbb{K})$ of all concepts of $\mathbb{K}$ is a complete lattice with respect to the order defined by $(O_1, A_1) \leq (O_2, A_2)$ if and only if $O_1 \subseteq O_2$, which is equivalent to the condition $A_2 \subseteq A_1$. $\mathcal{B}(\mathbb{K})$ is called the *concept lattice* of $\mathbb{K}$.

The mappings $(\cdot)'$ are closure operators, which, under appropriate conditions, can be regarded as logical closures of theories in the Logic RZ. Details on this relationship between Logic RZ and formal concept analysis can be found in [7].

3   Logic Programming in the Logic RZ

In this section, we discuss how the Logic RZ can be extended to a logic programming paradigm, thus adding disjunctive rules and default negation to the expressive features of the formalism. In addition, we review the classical approach of answer set programming that will be related to logic programming in the Logic RZ in Section 5.

Following [2], we first explain how the Logic RZ can be extended naturally to a disjunctive logic programming paradigm.

Definition 2. *A (disjunctive logic) program over a domain D is a set P of rules of the form $Y \leftarrow X$, where X, Y are clauses over D. An element $w \in D$ is a model of P if, for every rule $Y \leftarrow X$ in P, if $w \models X$, then $w \models Y$. We write $w \models P$ in this case. A clause Y is a logical consequence of P if every model of P satisfies Y. We write $\mathsf{cons}(P)$ for the set of all clauses which are logical consequences of P.*

Note that the condition $w \models P$ is strictly stronger than $w \models \mathsf{cons}(P)$. For an example, consider the domain $\{p, q, \bot\}$ defined by $\bot < p$ and $\bot < q$, and the program P consisting of the single rule $\{p\} \leftarrow \{q\}$. Then $\mathsf{cons}(P)$ contains only tautologies, and thus is modelled by the element q. Yet q is not a model for P.

In [1], a notion of *default negation* was added to the logic programming framework presented above. The extension is close in spirit to mainstream developments concerning knowledge representation and reasoning with nonmonotonic logics. It will serve as the base for our query language.

Since the following definition introduces a nonmonotonic negation operator $\sim$ into the logic, we wish to emphasize that the negation $\dot{\,\,\,}$ from Example 1 is not a special symbol of our logic but merely a syntactical feature to denote the elements of one particular example domain. Similar situations are known in FCA: in some formal contexts, every attribute has a "negated" attribute that relates to exactly the opposite objects, but FCA in general does not define negation. Analogously, by choosing appropriate domains, Logic RZ can be used to model common monotonic negations. In contrast, the semantic extension introduced next cannot be accounted for in this way.

Definition 3. *Consider a coherent algebraic domain D. An extended rule is a rule of the form $H \leftarrow X, \sim Y$, where H, X, and Y are clauses over D. An extended rule is trivially extended if $Y = \{\}$, and we may omit Y in this case. We call the tuple (X, Y) the body of the rule and H the head of the rule. An (extended disjunctive) program is a set of extended rules.*

Informally, we read an extended rule $H \leftarrow X, \sim Y$ as follows: if X holds, and Y does not, then H shall hold. As usual in logic programming, the formal semantics of $\sim$ is defined by specifying the semantics of logic programs in which $\sim$ is contained. But in contrast to classical negation, semantics is not defined by specifying the effect of $\sim$ on logical interpretations, e.g. by using *truth tables*. The reason is that we want to enrich our reasoning paradigm with nonmonotonic features, which are characterized by the fact that previously drawn conclusions might become invalid when adding additional knowledge (i.e. program rules). But such semantics clearly cannot be defined *locally* by induction on the structure of logical formulae – the whole program must be taken into account for determining logical meaning. We consider the following formal definition, akin to the answer set semantics that will be introduced later on in this section.

Definition 4. *Consider a coherent algebraic domain D, an element $w \in D$, and an extended disjunctive program P. We define P/w to be the (non-extended) program obtained by applying the following two transformations:*

1. *Replace each body (X, Y) of a rule by X whenever $w \not\models Y$.*
2. *Delete all rules with a body (X, Y) for which $w \models Y$.*

An element $w \in D$ is an answer model *of P if it satisfies $w \models \mathsf{cons}(P/w)$. It is a* min-answer model *of P if it is minimal among all v satisfying $v \models \mathsf{cons}(P/w)$.*

Note that every min-answer model is an answer model. We do not require answer models w to satisfy the rules of the program P/w. However, one can show the following lemma.

Lemma 1. *Consider a coherent algebraic domain D and a disjunctive program (i.e. without default negation) P over D. If $w \in D$ is minimal among all elements v with property $v \models \mathsf{cons}(P)$, then $w \models P$.*

Proof. This was shown in [2, Lemma 5.3].   □

The proof of the following statement refers to [2, Lemma 5.1] which uses Zorn's Lemma (or, equivalently, the Axiom of Choice).

Lemma 2. *Consider a coherent algebraic domain D, and a disjunctive program P over D. If $w \in D$ is such that $w \models \mathsf{cons}(P)$, then there is an element $w' \sqsubseteq w$ such that $w' \models P$ and which is minimal among all v that satisfy $v \models \mathsf{cons}(P)$.*

Proof. By [2], $\mathsf{cons}(P)$ is a logically closed theory. By [2, Proof of Theorem 3.2], the set M of all models of $\mathsf{cons}(P)$ is compact and upwards closed. By [2, Lemma 5.1] we have that M is the upper closure of its finite set $C(M)$ of minimal compact elements. Consequently, for any $w \models \mathsf{cons}(P)$ we have $w \in M$ and there is a $w' \in C(M)$ with the desired properties.   □

Intuitively, the above lemma enables us to conclude that there is a minimal model below any model of the program P. The rationale behind the definition of min-answer model is that it captures the notion of answer set as used in answer set programming which we will introduce next.

Answer set programming (ASP) is a reasoning paradigm in artificial intelligence which was devised in order to capture some aspects of commonsense reasoning. We now briefly review the basic concepts of ASP so that we can make the relationship to nonmonotonic logic programming in the Logic RZ explicit in Section 5.

ASP is based on the observation that humans tend to *jump to conclusions* in real-life situations, and on the idea that this imprecise reasoning mechanism (amongst other things) allows us to deal with the world effectively. Formally, *jumping to conclusions* can be studied by investigating supraclassical logics, see [8], where *supraclassicality* means, roughly speaking, that under such a logic more conclusions can be drawn from a set of axioms (or knowledge base) than could be drawn using classical (e.g. propositional or first-order) logic. Answer set programming, as well as the related default logic [9], is also *nonmonotonic*, in the sense that a larger knowledge base might yield a smaller set of conclusions.

We next describe the notion of answer set for extended disjunctive logic programs, as proposed in [10]. It forms the heart of answer set programming systems like dlv[2] or smodels[3] [11,12], which have become a standard paradigm in artificial intelligence.

Let $\mathcal{V}$ denote a countably infinite set of propositional variables. An *ASP-rule* is an expression of the form

$$L_1, \ldots, L_n \leftarrow L_{n+1}, \ldots, L_m, \sim L_{m+1}, \ldots, \sim L_k,$$

where each L_i is a literal, i.e. either of the form p or $\neg p$ for some propositional variable $p \in \mathcal{V}$. Given such an ASP-rule r, we set $\text{Head}(r) = \{L_1, \ldots, L_n\}$, $\text{Pos}(r) = \{L_{n+1}, \ldots, L_m\}$, and $\text{Neg}(r) = \{L_{m+1}, \ldots, L_k\}$.

In order to describe the answer set semantics, or stable model semantics, for extended disjunctive programs, we first consider programs without $\sim$.

Thus, let P denote an extended disjunctive logic program in which $\text{Neg}(r)$ is empty for each ASP-rule $r \in P$. A set $W \subseteq \mathcal{V}^{\pm} = \mathcal{V} \cup \neg\mathcal{V}$ is said to be *closed by rules* in P if, for every $r \in P$ such that $\text{Pos}(r) \subseteq W$, we have that $\text{Head}(r) \cap W \neq \emptyset$. W is called an *answer set* for P if it is a minimal subset of $\mathcal{V}^{\pm}$ such that the following two conditions are satisfied.

1. If W contains complementary literals, then $W = \mathcal{V}^{\pm}$.
2. W is closed by rules in P.

We denote the set of answer sets of P by $\alpha(P)$. Now suppose that P is an extended disjunctive logic program that may contain $\sim$. For a set $W \subseteq \mathcal{V}^{\pm}$, consider the program P/W defined as follows.

1. If $r \in P$ is such that $\text{Neg}(r) \cap W$ is not empty, then we remove r i.e. $r \notin P/W$.

---

[2] http://www.dbai.tuwien.ac.at/proj/dlv/

[3] http://www.tcs.hut.fi/Software/smodels/

```
object(1).          object(2).          ...
attribute(a).       attribute(b).       ...
incidence(1,b).     ...

in_extent(G)  :- object(G), not outof_ext(G).
outof_ext(G)  :- object(G), attribute(M), in_intent(M), not incidence(G,M).

in_intent(M)  :- attribute(M), not outof_int(M).
outof_int(M)  :- object(G), attribute(M), in_extent(G), not incidence(G,M).
```

Fig. 1. Computing formal concepts using answer set programming

2. If $r \in P$ is such that $\mathtt{Neg}(r) \cap W$ is empty, then the ASP-rule r' belongs to P/W, where r' is defined by $\mathtt{Head}(r') = \mathtt{Head}(r)$, $\mathtt{Pos}(r') = \mathtt{Pos}(r)$ and $\mathtt{Neg}(r') = \emptyset$.

The program transformation $(P, W) \mapsto P/W$ is called the *Gelfond-Lifschitz transformation* of P with respect to W.

It is clear that the program P/W does not contain $\sim$ and therefore $\alpha(P/W)$ is defined. We say that W is an *answer set* or *stable model* of P if $W \in \alpha(P/W)$. So, answer sets of P are fixed points of the operator GL_P introduced by Gelfond and Lifschitz in [10], where $\mathsf{GL}_P(W) = \alpha(P/W)$.[4] We note that the operator GL_P is in general not monotonic, and call it the *Gelfond-Lifschitz operator* of P.

Example 3. We illustrate answer set programming by means of a program due to Carlos Damasio [13] given in Fig. 1. It computes all formal concepts for a given formal context. The program consists of declarations of the objects, attributes, and incidence relation in the form of facts, hinted at in the first three lines of Fig. 1. The remaining four lines suffice to describe the problem – run in an answer set programming system, the program will deliver several answer sets, which coincide with the formal concepts of the given context if restricted to the predicates `in_extent` and `in_intent`. Note that "`not`" stands for default negation $\sim$, and "`:-`" stands for $\leftarrow$. Variables are written uppercase.

We follow common practice in allowing variables to occur in programs, but we need to explain how the syntax of Fig. 1 relates to the answer set semantics given earlier. This is done by *grounding* the program by forming all ground instances of the rules by making all possible substitutions of variables by the constants occurring in the program. Variable bindings within a rule have to be respected. For example, the first rule

```
in_extent(G)  :- object(G), not outof_ext(G).
```

has the ASP-rules

```
in_extent(1)  :- object(1), not outof_ext(1).    and
in_extent(b)  :- object(b), not outof_ext(b).
```

---

[4] GL_P being a multi-valued map, we speak of W as a fixed point of GL_P if $W \in \mathsf{GL}_P(W)$.

as two examples of ground instances. The resulting ground atoms, i.e. atomic formulae such as `object(b)` and `outof_ext(1)`, can then be understood as propositional variables, and the semantics given earlier can be derived.

It was shown in [1] that extended disjunctive programs over the domain $\mathbb{T}^{\mathcal{V}}$ from Example 1 can be closely related to classical answer set programming over a set of ground atoms $\mathcal{V}$. We will generalize this result in Theorem 1 below, where we incorporate information from formal contexts as well.

4   Querying Formal Contexts

In this section, we integrate hierarchical background knowledge specified by a formal context with the generalized programming paradigm for the Logic RZ. The result can be considered as a query language for formal contexts.

Definition 5. *Let $\mathbb{K}$ be a finite formal context with concept lattice $\mathcal{B}(\mathbb{K})$, and let $\mathbb{T}^{\mathcal{V}}$ be the domain from Example 1. A query over $\mathbb{K}$ is any extended disjunctive program over the domain $\mathcal{B}(\mathbb{K}) \times \mathbb{T}^{\mathcal{V}}$.*

The fact that $\mathcal{B}(\mathbb{K}) \times \mathbb{T}^{\mathcal{V}}$ is a domain follows since both factors of this product are domains as well. That $\mathcal{B}(\mathbb{K})$ is a domain is ensured by restricting the above definition to finite contexts. Concepts of $\mathbb{K}$ can be viewed as conjunctions of attributes of $\mathbb{K}$ (or, similarly, as conjunctions of objects), thus allowing for a straightforward intuitive reading of the rules of a query. When formulating a rule, however, the restriction to concepts can be unwanted, and one might prefer to state arbitrary conjunctions over attributes and objects. To this end, we now develop a more convenient concrete syntax for queries.

Definition 6. *Given a context $\mathbb{K} = (G, M, I)$, a literal over $\mathbb{K}$ is either an element of $G \cup M$, or a formula of the form $p(t_1, \ldots, t_n)$ or $\neg p(t_1, \ldots, t_n)$, where p is an n-ary predicate symbol and $t_1, \ldots, t_n$ are terms over some first-order language. A rule over $\mathbb{K}$ then is of the form*

$$L_1, \ldots, L_n \leftarrow L_{n+1}, \ldots, L_m, \sim L_{m+1}, \ldots, \sim L_k,$$

where each of the L_i is a literal over $\mathbb{K}$. A simplified query for $\mathbb{K}$ is a set of rules over $\mathbb{K}$.

The intention is that rules over a context allow for an intuitive reading which is similar to that for classical ASP-rules, and that finite sets of rules unambiguously represent queries over the given context. We consider rules with variables as a short-hand notation for the (possibly infinite) set of all ground instances (with respect to the considered first-order signature), and thus can restrict our attention to rules that do not contain first-order variables.

When relating Definition 6 to Definition 5, we have to be aware that both definitions are somewhat implicit about the considered logical languages. Namely, the notion of a simplified query depends on the chosen first order language, and, similarly, queries employ the domain $\mathbb{T}^{\mathcal{V}}$ that depends on choosing some concrete set of propositional variables $\mathcal{V}$. Given standard cardinality constraints, the choice of these invariants is not relevant for our treatment, but notation is greatly simplified by assuming that $\mathcal{V}$ is always equal to the set of ground atoms (i.e. atomic logic formulae without variables) over the chosen language. Thus, we can also view ground atoms as elements of $\mathbb{T}^{\mathcal{V}}$ mapping exactly the specified element of $\mathcal{V}$ to true, and leaving everything else undetermined.

Moreover, note that both $\mathcal{B}(G, M, I)$ and $\mathbb{T}^{\mathcal{V}}$ have least elements (M', M'') and $\perp$, respectively. We exploit this to denote elements of $\mathcal{B}(G, M, I) \times \mathbb{T}^{\mathcal{V}}$ by elements of $G \cup M \cup \mathcal{V} \cup \overline{\mathcal{V}}$. Namely, each element $o \in G$ denotes the element $(((\{o\}'', \{o\}'), \perp)$, $a \in M$ denotes $(((\{a\}', \{a\}''), \perp)$, and $p \in \mathcal{V} \cup \overline{\mathcal{V}}$ denotes $((M', M''), p)$. This abbreviation is helpful since the atomic elements are supremum-dense in $\mathcal{B}(G, M, I) \times \mathbb{T}^{\mathcal{V}}$ and thus can be used to specify all other elements.

Definition 7. *Consider a context* $\mathbb{K} = (G, M, I)$, *and a rule over* $\mathbb{K}$ *of the form*

$$L_1, \ldots, L_n \leftarrow L_{n+1}, \ldots, L_m, {\sim}L_{m+1}, \ldots, {\sim}L_k.$$

The associated *extended disjunctive rule over* $\mathcal{B}(\mathbb{K}) \times \mathbb{T}^{\mathcal{V}}$ *is defined as*

$$\{L_1, \ldots, L_n\} \leftarrow \bigsqcup \{L_{n+1}, \ldots, L_m\}, {\sim}\{L_{m+1}, \ldots, L_k\}.$$

Given a simplified query P, *its associated query* $\widehat{P}$ *is obtained as the set of rules associated to the rules of* P *in this sense.*

Conversely, it is also possible to find an appropriate simplified query for arbitrary queries. The problem for this transformation is that the simplified syntax does not permit disjunctions in the bodies of rules, and generally restricts to atomic expressions. It is well-known, however, that disjunctions in bodies do usually not increase expressiveness of a rule language. Indeed, consider the extended disjunctive rule

$$\{l_1, \ldots, l_n\} \leftarrow \{l_{n+1}, \ldots, l_m\}, {\sim}\{l_{m+1}, \ldots, l_k\},$$

where l_i are elements of $\mathcal{B}(\mathbb{K}) \times \mathbb{T}^{\mathcal{V}}$ as in Definition 5. Using a simple form of the so-called Lloyd-Topor transformation, we can rewrite this rule into the set of rules

$$\{\{l_1, \ldots, l_n\} \leftarrow \{l_j\}, {\sim}\{l_{m+1}, \ldots, l_k\} \mid j \in \{n+1, \ldots, m\}\}.$$

It is straightforward to show that this transformation preserves answer models and min-answer models, and we omit the details.

Similarly, it is possible to treat heads $\{l_1, \ldots, l_n\}$, where the l_i are not necessarily literals. Indeed, each l_i can be written as $l_i = \bigsqcup A_i$, where $A_i = \{a_{i1}, \ldots, a_{in_i}\}$ is a finite set of literals, and a rule of the form

$$\{l_1, \ldots, l_n\} \leftarrow \{l\}, {\sim}\{l_{m+1}, \ldots, l_k\}$$

can thus be transformed into the set of rules

$$\{\{e_1, \ldots, e_n\} \leftarrow \{l_j\}, \sim\{l_{m+1}, \ldots, l_k\} \mid e_i \in A_i\}.$$

Intuitively, this transformation is obtained by bringing the head into conjunctive normal form (using a distributivity law) and subsequent splitting of conjunctions into different clause heads. This constitutes another Lloyd-Topor transformation, and it is again straightforward – but also quite tedious – to show that the transformation preserves answer models and min-answer models.

Similar techniques could be applied to transform complex expressions in the default negated part of the body. Therefore, we can restrict our subsequent considerations to simplified queries without loss of generality.

5 Practical Evaluation of Queries

Based on the close relationship to answer set programming discussed in Section 3, we now present a way to evaluate queries within standard logic programming systems. This has the huge advantage that we are able to employ highly optimized state of the art systems for our reasoning tasks. Furthermore, the connection to standard answer set programming creates further possibilities to combine contextual knowledge with other data sources that have been integrated into answer set programming paradigms.

Our goal is to reduce queries to (classical) answer set programs. For this it is necessary to translate both the rules of the program and the data of the underlying formal context into the standard paradigm. On a syntactic level, we already established a close correspondence based on the notion of a simplified query. We now show how to take this syntactic similarity to a semantic level.

Definition 8. *Given a simplified query P for a context $\mathbb{K} = (G, M, I)$, consider the syntactic representation of P as an extended disjunctive logic program over the set of variables $G \cup M \cup \mathcal{V}$. Furthermore, define a program $\mathsf{ASP}(\mathbb{K})$ over this set of variables to consist of the union of*

1. *all rules of the form $o \leftarrow a_1, \ldots, a_n$, with $a_1, \ldots, a_n \in M$, $o \in \{a_1, \ldots, a_n\}'$,*
2. *all rules of the form $a \leftarrow o_1, \ldots, o_n$, with $o_1, \ldots, o_n \in G$, $a \in \{o_1, \ldots, o_n\}'$.*

By $\mathsf{ASP}(P)$, we denote the extended disjunctive logic program $P \cup \mathsf{ASP}(\mathbb{K})$ that is thus associated with the simplified query P.

Obviously, $\mathsf{ASP}(\mathbb{K})$ (and therefore also $\mathsf{ASP}(P)$) will in general contain redundancies, which could be eliminated if desired by using stem base techniques (see [6]). We will not discuss this optimization in detail as it is not necessary for our exhibition.

On the semantic level of min-answer models and answer sets, the relationship between queries and logic programs is described by the following function.

Definition 9. *Consider the domain $\mathbb{T}^{\mathcal{V}}$ and a context $\mathbb{K} = (G, M, I)$. A mapping ι from elements of $\mathcal{B}(\mathbb{K}) \times \mathbb{T}^{\mathcal{V}}$ to subsets of $G \cup M \cup \mathcal{V}^{\pm}$ is defined by setting $\iota(w) = \{p \mid w \models p\}$.*

Note that the above definition is only meaningful when employing our convention of using elements $p \in G \cup M \cup \mathcal{V}^{\pm}$ to denote (not necessarily atomic) elements of $\mathcal{B}(\mathbb{K}) \times \mathbb{T}^{\mathcal{V}}$, as discussed in the previous section. This relationship need not be injective, but elements from $\mathcal{V}^{\pm}$ and $G \cup M$ are never associated with the same element of $\mathcal{B}(\mathbb{K}) \times \mathbb{T}^{\mathcal{V}}$, and this suffices to eliminate any confusion in our following considerations.

Lemma 3. *Consider a simplified query P for a context $\mathbb{K}$ with associated query $\widehat{P}$. For any element $w \in \mathcal{B}(\mathbb{K}) \times \mathbb{T}^{\mathcal{V}}$, we find that $\mathsf{ASP}(P)/\iota(w) = P/\iota(w) \cup \mathsf{ASP}(\mathbb{K})$.*

Furthermore, an ASP-rule $L_1, \dots, L_n \leftarrow L_{n+1}, \dots, L_m$ is in $P/\iota(w)$ iff the corresponding rule $\{L_1, \dots, L_n\} \leftarrow \bigsqcup \{L_{n+1}, \dots, L_m\}$ is in $\widehat{P}/w$.

Proof. The first part of the claim is immediate, since there are no default negated literals in $\mathsf{ASP}(\mathbb{K})$. For the second part, note that any rule in either $P/\iota(w)$ or $\widehat{P}/w$ stems from a rule of the form $L_1, \dots, L_n \leftarrow L_{n+1}, \dots, L_m, {\sim}L_{m+1}, \dots, {\sim}L_k$ in P. To finish the proof, we just have to observe that $w \models \{L_{m+1}, \dots, L_k\}$ iff $\iota(w) \models L_i$ for some $i = m+1, \dots, k$. □

Restricting to non-extended disjunctive programs only, the next lemma establishes the basis for the main result of this section.

Lemma 4. *Consider a simplified query P over some context $\mathbb{K} = (G, M, I)$, such that no default negation appears in P. If $w \in \mathcal{B}(\mathbb{K}) \times \mathbb{T}^{\mathcal{V}}$ is such that $w \models \widehat{P}$, then $\iota(w)$ is closed under rules of $P \cup \mathsf{ASP}(\mathbb{K})$.*

Conversely, assume there is a consistent set $W \subset (G \cup M \cup \mathcal{V})^{\pm}$ closed under rules of $P \cup \mathsf{ASP}(\mathbb{K})$. Then $W \supseteq \iota(w)$ for some element $w \in \mathcal{B}(\mathbb{K}) \times \mathbb{T}^{\mathcal{V}}$ with property $w \models \widehat{P}$. In particular, if W is minimal among all sets closed under said rules, then W is equal to $\iota(w)$.

Proof. For the first part of the claim, let $w \models \widehat{P}$ be as above, and consider an ASP-rule $r \in P \cup \mathsf{ASP}(\mathbb{K})$ given by $L_1, \dots, L_n \leftarrow L_{n+1}, \dots, L_m$. First consider the case $r \in P$. By Definition 7, $\{L_1, \dots, L_n\} \leftarrow \bigsqcup \{L_{n+1}, \dots, L_m\}$ is in $\widehat{P}$. If $L_{n+1}, \dots, L_m \in \iota(w)$, then $w \models L_i$, $i = n+1, \dots, m$ by the definition of ι. Hence $w \models \bigsqcup \{L_{n+1}, \dots, L_m\}$ and thus $w \models L_j$ for some $j \in \{1, \dots, n\}$. But then $L_j \in \iota(w)$ and so $\iota(w)$ satisfies the rule r.

On the other hand, that $\iota(w)$ satisfies any ASP-rule $r \in \mathsf{ASP}(\mathbb{K})$ is obvious from the fact that $\iota(w) \cap G$ is an extent with corresponding intent $\iota(w) \cap M$. This finishes the proof that $\iota(w)$ is closed under rules of $\mathsf{ASP}(P)/\iota(w)$.

For the second part of the claim, consider a set W as in the assumption. First note that elements of $\neg G \cup \neg M$ do not occur in the rules of $P \cup \mathsf{ASP}(\mathbb{K})$. Consequently, whenever W is closed under rules of $P \cup \mathsf{ASP}(\mathbb{K})$, we find that $V = W \cap (G \cup M \cup \mathcal{V}^{\pm})$ has this property as well.

Now consider sets $O = V \cap G$ and $A = V \cap M$. We claim that (O, A) is a concept of $\mathbb{K}$, i.e. $O' = A$. Since V is closed under $\mathsf{ASP}(\mathbb{K})$, whenever $a \in O'$ for some $a \in M$, we find $a \in W$, and hence $A \supseteq O'$. Analogously, we derive $O \supseteq A'$.

Using standard facts about concept closure $(\cdot)'$ [6], we obtain $O'' \subseteq A' \subset O$ and $A'' \subseteq O' \subset A$ which establishes the claim.

Given that $V \cap (G \cup M)^{\pm}$ is the (disjoint) union of the extent and intent of a concept, it is obvious that $V = \iota(w)$ for some $w \in \mathcal{B}(\mathbb{K}) \times \mathbb{T}^{\mathcal{V}}$. Here we use the assumed consistency of W and V to ensure that the $\mathbb{T}^{\mathcal{V}}$-part of V can indeed be expressed by an appropriate w.

We still have to show that $w \models \widehat{P}$. For any rule $\{L_1, \ldots, L_n\} \leftarrow \bigsqcup \{L_{n+1}, \ldots, L_m\}$ in $\widehat{P}$, there is an ASP-rule $L_1, \ldots, L_n \leftarrow L_{n+1}, \ldots, L_m$ in P. By the definition of ι, it is clear that $\iota(w) = V$ models this ASP-rule iff w models the corresponding rule, which establishes the claim, since V models all ASP-rules. $\square$

Theorem 1. *Consider a simplified query P with associated query $\widehat{P}$ and program $\mathsf{ASP}(P)$. If $\widehat{P}$ has any min-answer models, then the function ι from Definition 9 is a bijection between min-answer models of P and answer sets of $\mathsf{ASP}(P)$.*

Proof. Consider an answer set W of $\mathsf{ASP}(P)$ such that $W \neq (G \cup M \cup \mathcal{V})^{\pm}$. Considering P/W as a simplified query, we can apply Lemma 4, to find some element $w \models \widehat{P/W}$ such that $W = \iota(w)$. By Lemma 3, an ASP-rule is in P/W iff a corresponding rule is in $\widehat{P}/w$, and we conclude $w \models \widehat{P}/w$. We claim that w additionally is a min-answer model. Indeed, by Lemma 2, there is an element $w' \sqsubseteq w$ such that $w' \models \widehat{P}/w$ and which is minimal among all $v \models \mathsf{cons}(P)$. For a contradiction, suppose that $w \neq w'$, i.e. w is not a min-answer model. Using Lemmas 3 and 4, we find that $w' \models \widehat{P}/w$ implies that $\iota(w')$ is closed under rules of $P/\iota(w)$. Closure of $\iota(w')$ under rules of $\mathsf{ASP}(\mathbb{K})$ is immediate since the first component of w' is required to be a concept. Thus, we obtain a model $\iota(w')$ for $\mathsf{ASP}(P)$ which is strictly smaller than $\iota(w)$. This contradicts the assumed minimality of $\iota(w) = W$, so that w must be a min-answer model.

Conversely, we show that $\iota(w)$ is an answer set of $\mathsf{ASP}(P)$ whenever w is a min-answer model of $\widehat{P}$. Combining Lemmas 3 and 4, we see that $\iota(w)$ is closed under rules of $\mathsf{ASP}(P)/\iota(w)$. For a contradiction, suppose that $\iota(w)$ is not minimal, i.e. there is some model $V \subseteq W$ that is also closed under rules of $\mathsf{ASP}(P)$. Then, by Lemma 4, we have $V \supseteq \iota(v)$ for some v for which $v \models \widehat{P}/w$. Clearly, $\iota(v) \subseteq V \subseteq W$ implies $v \sqsubset w$, thus contradicting our minimality assumption on w. $\square$

The above result can be compared to the findings in [1], where programs over $\mathbb{T}^{\mathcal{V}}$ were related to answer set programs. The corresponding result can be obtained from Theorem 1 by restricting to the empty context $(\emptyset, \emptyset, \emptyset)$.

6   Conclusions and Further Work

We have shown how artificial intelligent commonsense reasoning in the form of answer set programming can be merged with conceptual knowledge in the sense of formal concept analysis. We have utilized this in order to develop a commonsense query answering system for formal contexts, which features the full strength of disjunctive answer set programming.

Based on our results, it is straightforward to implement this e.g. on top of the dlv system[5] [11], which has recently been supplemented to support extensions like ours in a hybrid fashion [14].

The dlv system also provides modules for interfacing with conceptual knowledge in other paradigms like OWL [15] or RDF [16], resulting in a hybrid reasoning system. These features become available for us by way of dlv and thus allow for an integrated querying and reasoning paradigm over heterogeneous knowledge. This could also be further enhanced with query-based multicontext browsing capabilities in the sense of [17].

Finally, let us remark that our work sheds some light on recently discussed issues concerning the interplay between conceptual knowledge representation and rule-based reasoning.[6] Indeed, our approach realizes a strong integration between paradigms, with the disadvantage that it is restricted to hierarchical conceptual knowledge in the sense of formal concept analysis. It may nevertheless be a foundation for further investigations into the topic from an order-theoretic perspective.

References

1. Hitzler, P.: Default reasoning over domains and concept hierarchies. In Biundo, S., Frühwirth, T., Palm, G., eds.: Proceedings of the 27th German conference on Artificial Intelligence, KI'2004, Ulm, Germany, September 2004. Volume 3238 of Lecture Notes in Artificial Intelligence., Springer, Berlin (2004) 351–365
2. Rounds, W.C., Zhang, G.Q.: Clausal logic and logic programming in algebraic domains. Information and Computation **171** (2001) 156–182
3. Abramsky, S., Jung, A.: Domain theory. In Abramsky, S., Gabbay, D., Maibaum, T.S., eds.: Handbook of Logic in Computer Science. Volume 3. Clarendon, Oxford (1994)
4. Plotkin, G.: T^ω as a universal domain. Journal of Computer and System Sciences **17** (1978) 209–236
5. Fitting, M.: A Kripke-Kleene-semantics for general logic programs. The Journal of Logic Programming **2** (1985) 295–312
6. Ganter, B., Wille, R.: Formal Concept Analysis – Mathematical Foundations. Springer, Berlin (1999)
7. Hitzler, P., Wendt, M.: Formal concept analysis and resolution in algebraic domains. In de Moor, A., Ganter, B., eds.: Using Conceptual Structures – Contributions to ICCS 2003, Shaker Verlag, Aachen (2003) 157–170
8. Makinson, D.: Bridges between classical and nonmonotonic logic. Logic Journal of the IGPL **11** (2003) 69–96
9. Reiter, R.: A logic for default reasoning. Artificial Intelligence **13** (1980) 81–132
10. Gelfond, M., Lifschitz, V.: Classical negation in logic programs and disjunctive databases. New Generation Computing **9** (1991) 365–385
11. Eiter, T., Leone, N., Mateis, C., Pfeifer, G., Scarcello, F.: A deductive system for nonmonotonic reasoning. In Dix, J., Furbach, U., Nerode, A., eds.: Proceedings

---

[5] http://www.dbai.tuwien.ac.at/proj/dlv/

[6] See e.g. the work of the W3C Rule Interchange Format working group at http://www.w3.org/2005/rules/wg.html.

of the 4th International Conference on Logic Programming and Nonmonotonic Reasoning (LPNMR'97). Volume 1265 of Lecture Notes in Artificial Intelligence., Springer, Berlin (1997)

12. Simons, P., Niemelä, I., Soininen, T.: Extending and implementing the stable model semantics. Artificial Intelligence **138** (2002) 181–234

13. Damasio, C.: Personal communication (2004)

14. Eiter, T., Ianni, G., Schindlauer, R., Tompits, H.: A uniform integration of higher-order reasoning and external evaluations in answer set programming. In Kaelbling, L.P., Saffiotti, A., eds.: Proceedings of the 19th International Joint Conference on Artificial Intelligence (IJCAI-05). (2005)

15. Smith, M.K., McGuinness, D.L., Welty, C.: OWL Web Ontology Language Guide. W3C Recommendation 10 February 2004 (2004) available at `http://www.w3.org/TR/owl-guide/`.

16. Manola, F., Miller, E.: Resource Description Framework (RDF) Primer. W3C Recommendation 10 February 2004 (2004) available at `http://www.w3.org/TR/rdf-primer/`.

17. Tane, J.: Using a query-based multicontext for knowledge base browsing. In: Formal Concept Analysis, Third International Conf., ICFCA 2005-Supplementary Volume, Lens, France, IUT de Lens, Universite d'Artois (2005) 62–78

Towards an Epistemic Logic of Concepts

Tanja Hötte* and Thomas Müller

Universität Bonn, Institut für Philosophie, Lennéstr. 39, 53113 Bonn, Germany
thoette@uni-bonn.de, thomas.mueller@uni-bonn.de

Abstract. What does it take to possess a concept? Behaviour of various degrees of complexity is based on different levels of cognitive abilities. Concept possession ranges between mere stimulus-response schemes and fully developed propositional representations. Both biological and artifical systems can be described in terms of these levels of cognitive abilities, and thus we can meaningfully ask whether a given system *has concepts*. We regard that question not in terms of behavioural criteria, but from a formal point of view. We focus on the interrelation between a given objective structure of concepts and a subject's representation of that structure. The main question is how much of the structure of the objective side needs to be mirrored subjectively in order to grant possession of concepts. Our approach shows a strong parallel to epistemic logic. There, the objective side can be represented by an algebra of true propositions, and an epistemic subject can represent some of these propositions as what she believes to be true. As in propositional epistemic logic, in an epistemic logic of concepts the main issue is finding adequate closure conditions on the subjective set of representations. We argue that the appropriate closure conditions can be stated formally as closure under witnesses for two types of relationships among concepts: in order for a subject to possess a concept c she has to represent both a *sibling* and a *cousin* of c. We thus arrive at a first formally perspicious candidate for a psychologically adequate epistemic logic of concepts.

1 Introduction

Biological systems show behaviour of various degrees of complexity, some of which is based on cognition. Some types of behaviour clearly do not presuppose cognition; among them are reflexes, but also mere stimulus-response behaviour such as that exhibited by Pavlov's dogs who slobber when they hear a bell ring. On the other hand, some types of human behaviour are clearly based on propositional representations and thus, on a high form of cognitive abilities; reading this article is among them. In between there are types of behaviour that presuppose conceptual representations without demanding full propositional cognition. Such behaviour is not confined to human beings: there are also some types of animal behaviour that fall within that range, e.g., the behaviour of parrots that Pepperberg describes in her book *The Alex Studies* [8]. Once these different levels of

---

* Research project "Wissen und Können", working group Stuhlmann-Laeisz, funded by the VolkswagenStiftung. Project website: www.wuk.uni-bonn.de

H. Schärfe, P. Hitzler, and P. Øhrstrøm (Eds.): ICCS 2006, LNAI 4068, pp. 274–285, 2006.

cognitive abilities have been recognised, it becomes possible to use them for describing not just biological, but also artificial systems. Thus we can meaningfully ask whether a given biological or artificial system *has concepts*.

When does a system have concepts? We will address this question from a formal point of view: we will presuppose that we know what a system *represents*, and we will try to describe minimal structural conditions that have to be met by the set of those representations in order to call them *conceptual*. Thus we leave the empirical question of how to derive representations from behaviour aside. The main link between behaviour and representation ascriptions will clearly have to be via various types of classificatory tasks. — We will also not consider the problem of concept acquisition, but aim at describing fully competent epistemic subjects. The concepts we focus on are perception-based ones like colour concepts or concepts of natural kinds, not theoretical concepts.

Our approach shows a strong parallel to *epistemic logic*. In that well-established branch of modal logic, one assumes that there is an objective side (that which can be represented) in the form of an algebra of *true propositions*, and an epistemic subject can represent some of these propositions as what she *believes to be true*. Whether the subject is granted knowledge normally depends on a number of side conditions, most of which are *closure conditions* for the set of propositions believed to be true. E.g., many systems of epistemic logic presuppose closure under logical consequence, so that a subject will only be granted knowledge of a proposition p if she also represents all logical consequences of p as believed-true. There is a long-standing debate about the adequacy of such closure conditions under the heading of "the problem of logical omniscience": empirical subjects just cannot represent all logical consequences of any proposition as believed-true, since empirical subjects are finite, but there are infinitely many logical consequences of any given proposition (cf. [2], [3]).

In developing an epistemic logic of concepts, we have to face a similar problem. In parallel to propositional epistemic logic, we distinguish between an objective side of *concepts* and a subjective side of *representations*. Our main question can be phrased in terms of closure conditions: which conditions does a subjective set of representations have to meet in order to be truly conceptual? Just like for propositions, a trivial condition which would allow for representing only a single concept won't do. On the other hand, imposing strict closure conditions runs the risk of ruling out too many empirical subjects for concept possession at all. Our task will be to find a good balance between these two extremes.

In Section 2, we will first describe the objective side of how concepts are structured. In Section 3, we then address the question of how much of that structure must be mirrored on the subjective side in order to grant subjective possession of concepts.

2   The Structure of Sets of Concepts

In this section we will regard relations between the elements of a set of concepts. But let us first say a word about the kind of such elements: we focus on perception based concepts and we do not assume that the epistemic subject possesses a language. This restricts the concepts that we will be dealing with to rather simple ones. In particular, we do not consider second-order concepts, i.e., concepts that are themselves sets of concepts. Also, as we are interested in minimal conditions of concept possession, we presuppose only a thin notion of concept, not one that has rich internal structure. In this respect, our approach differs from, e.g., the detailed account of human concepts proposed by Kangassalo [6].

In line with this approach, we do not wish to presuppose too fine-grained a structure among the concepts that we are dealing with. We will focus exlusively on the *subconcept* relation and on relations definable in terms of it. The subconcept relation is the most basic relation on any set of concepts, and it will have to be considered in any case. There are further relations on a set of concepts that might be interesting. E.g., each concept could have a *complement*, or we could consider an *incompatibility* relation among concepts. However, in this paper we stick to positively defined, perception based concepts. With respect to these, negation appears to presuppose an additional layer of theory. Finally, having opted for first-order concepts only, the *element* relation among such a set of concepts is empty.

The fact that we do not require language capability as an essential ingredient of concept possession means that we must not assume that the epistemic subject can individuate concepts by name. Thus, the fact that languages can have synonymous terms referring to one and the same concept, does not pose a problem for our approach.

We will only consider finite structures in this paper. This will allow us to sidestep a number of technical issues, but there is also a philosophical basis for this restriction: we are interested in ascribing concept possession to actual (biological or artificial) epistemic subjects, i.e., to finite beings.

We will denote a set of concepts by $\mathcal{C}$. Single concepts will be denoted by lower case sans serif expressions, e.g., a, b, cornflower etc.

2.1   The Subconcept Relation

$\mathcal{C}$ is structured by the (reflexive) subconcept relation, usually dubbed ISA — e.g., as in bird ISA animal.

Now let us regard how the subconcept relation orders a set C of concepts. If C is a singleton there is not much to order, except that ISA is reflexive on the element. If C contains two concepts, we have the following possibilities: one concept can be a subconcept of the other or they can have nothing to do with each other. In case C contains three or more elements, two concepts can have a common sub- or superconcept, and they can be otherwise connected by the subconcept relation via a chain of intermediate concepts. It never happens that a concept is both a proper sub- and a proper superconcept of another concept. This suggests that, formally, the set of concepts $\mathcal{C}$ is at least partially ordered: ISA is reflexive, antisymmetric and transitive on $\mathcal{C}$.

Does the notion of a partial order really characterise ISA? The alternatives would be either to drop or to add formal requirements. Dropping requirements appears implausible: ISA is certainly reflexive. It is also transitive: a subconcept of a subconcept of c is itself a subconcept of c. How about antisymmetry? Could there be concepts a and b such that a is a subconcept of b and b a subconcept of a while a ≠ b? We frankly do not understand what that would mean. Thus, ISA is at least a partial order.

Maybe ISA satisfies additional formal requirements? Two intuitions pull in different directions at this point. On the one hand, in the tradition of formal concept analysis [5], it is customary to suppose that a set of concepts forms a lattice. This means that any two concepts have a common subconcept and a common superconcept, which is a lot of structure. On the other hand, in a philosophical tradition stemming from Aristotle, concepts can be thought of as ordered in a taxonomic tree. Formally, ISA on a taxonomic tree is a partial ordering that fulfills the additional formal requirement of backward linearity. In a taxonomic tree, concepts are distinguished by their genus proximum and their differentia specifica. Ordering concepts in such trees is intuitively appealing — think of zoology. However, such an order is unable to account for many of our intuitive uses of concepts, as taxonomy is always just with respect to *one* hierarchy of classification. We do however mix different classifying hierarchies. For example, peas are both vegetables and green. In a taxonomic tree, it follows that vegetable must be a subconcept of green or vice versa. Apparently, this is not the case. One can force concepts into a tree by adding multiple copies of nodes, as in Fig. 1. This is, however, both inelegant and yields an exponential overhead of new nodes.

Faced with the task of deciding between general partial orders, lattices or trees, we do not see overwhelming, univocal arguments in favour of one of the more specific

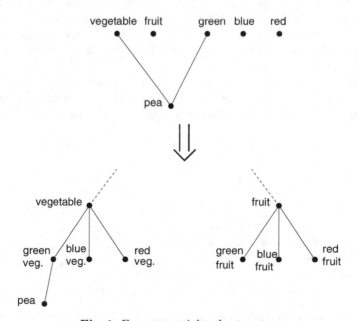

Fig. 1. From a partial order to a tree

structures. Thus we will only presuppose that $\langle \mathcal{C}, \mathsf{ISA} \rangle$ is a partial order. This does not exclude the more specific cases, while keeping our theory general.

Note that we do not require ISA to connect all of $\mathcal{C}$. Let $con(\mathsf{a})$ be the set of concepts connected to a concept a via the ISA relation (i.e., the reflexive transitive closure of ISA and its converse). There could be some $\mathsf{b} \in \mathcal{C}$ such that $\mathsf{b} \notin con(\mathsf{a})$, in which case $con(\mathsf{a}) \cap con(\mathsf{b}) = \emptyset$. We do not exclude the possibility of $\mathcal{C}$ consisting of several of such ISA-clusters, isolated from each other. In fact, we do not think it likely that this is the case for an objective ordering of concepts, but in this paper we will maintain full generality. In Sect. 3.3 we will argue that it is useful to allow a *subjective* ordering of concepts to contain disconnected elements.

In the finite case (as well as in 'friendly' infinite cases), the ISA relation gives rise to a *covering relation* that relates just closest subconcepts in the partial order. Let $\mathsf{a}, \mathsf{b} \in \mathcal{C}$. Then a is covered by b, written $\mathsf{a} \prec \mathsf{b}$, iff (i) a ISA b, (ii) $\mathsf{a} \neq \mathsf{b}$, and (iii) on the assumption that a ISA c and c ISA b we have $\mathsf{c} = \mathsf{a}$ or $\mathsf{c} = \mathsf{b}$.

2.2 Siblings and Cousins

Let us introduce two useful notions with the help of $\prec$: the sibling relation and the cousin relation.

The intuition for the *sibling* relation comes from trees: in a tree, a sibling of a node is a node with the same mother. Here, we generalise this notion to partial orders, excluding the reflexive case.

$$sib(\mathsf{a}, \mathsf{b}) \Leftrightarrow_{df} \mathsf{a} \neq \mathsf{b} \land \exists \mathsf{c}(\mathsf{a} \prec \mathsf{c} \land \mathsf{b} \prec \mathsf{c}).$$

This relation is irreflexive and symmetric, but not transitive (in view of $\mathsf{a} \neq \mathsf{b}$). Sharing a covering node ($\exists \mathsf{c}(\mathsf{a} \prec \mathsf{c} \land \mathsf{b} \prec \mathsf{c})$) is not necessarily transitive (though on trees it is). Staying in the family metaphor, one could say that partial orders allow for half-siblings. And the half-sister of Eve's half-brother need not be her sister at all (see Fig. 2 (a)).

The second relation that we will introduce is the *cousin* relation. Two concepts are cousins if they have immediate superconcepts that are siblings. Cousins are more independend than siblings, but still connected via the ISA relation.

$$cousin(\mathsf{a}, \mathsf{b}) \Leftrightarrow_{df} \mathsf{a} \neq \mathsf{b} \land \exists \mathsf{c}, \mathsf{d}(\mathsf{a} \prec \mathsf{c} \land \mathsf{b} \prec \mathsf{d} \land sib(\mathsf{c}, \mathsf{d}))$$

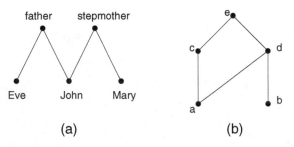

(a)   (b)

Fig. 2. (a) patchwork families, (b) a is a candidate for being one's own cousin

Again, reflexivity it excluded. We have to put in this condition explicitly because otherwise in the case of the partial order of Fig. 2 (b), a would be its own cousin. Note that *cousin* is a symmetric relation.

3   Possession of a Concept

After we have made some observations on the form of sets of concepts in general and introduced relations on them, we now turn to the subjective side of concept possession. We thus consider the epistemic view that an animal or artificial system has of the objective set of concepts. The question is how much of that set has to be grasped in order to ascribe concept possession sensibly. In the following we will first discuss the extreme cases of requiring no structure at all and of demanding so much that even humans cannot be ascribed concept possession any more. Then we will present an intermediate position, based on results of cognitive science as well as on solutions to the problem of omniscience (see Sect. 1).

To state the question formally, let i be an epistemic subject. Now let us introduce a set $\mathcal{P}_i \subseteq \mathcal{C}$ for each i. $\mathcal{P}_i$ shall denote the set of concepts i possesses. Now we can state the question as follows: given $\mathsf{a} \in \mathcal{P}_i$, what other concepts must $\mathcal{P}_i$ contain? And what structure does $\mathcal{P}_i$ need to have?

Relations on $\mathcal{P}_i$

Each $\mathcal{P}_i$ might just be a small subset of $\mathcal{C}$. Still, $\mathcal{P}_i$ can have an internal structure. In the following, we presuppose that the epistemic subject does not err in the concepts nor in their relations (cf. Sect. 1). A similar presupposition is operative in propositional epistemic logic. In general, we would not ascribe possession of some cognitive content to a subject who mixes up the inferential relations in which that content stands. Thus, for $\mathsf{a}, \mathsf{b} \in \mathcal{P}_i$ we have

$$\mathsf{a} \ \mathsf{ISA}_i \ \mathsf{b} \ \text{iff} \ \mathsf{a} \ \mathsf{ISA} \ \mathsf{b},$$

where ISA_i denotes the ISA relation defined on $\mathcal{P}_i$ instead of on $\mathcal{C}$. From the epistemic point of view this is reasonable, since if an epistemic subject possesses both a concept and one of its superconcepts, he has to treat them as concept and superconcept. If he treated them differently, e.g., as siblings, our intuition would not grant him possession of both concepts. Formally, $\mathcal{P}_i$ is thus a subordering of $\mathcal{C}$.

This perfect matching need not hold for the covering relation. $\mathcal{P}_i$ is likely to be less fine-grained than $\mathcal{C}$. In the examples of Fig. 3, all nodes belong to $\mathcal{C}$, but just the circled ones belong to $\mathcal{P}_i$. Let $\prec_i$ denote the covering relation of ISA_i. We have plum $\prec_i$ fruit, but not plum $\prec$ fruit. Similarly with all the other nodes at the bottom level. This does not conflict with our use of concepts. What we use extensively is the ISA relation, but we would always be reluctant to label a concept as a direct superconcept absolutely — there is always the possibility that we do not know enough of the area, or that we have missed some in-between concept.

Regarding *sib* and *cousin*, which depend on the covering relation, even more can change if we determine them via $\mathcal{P}_i$. Let sib_i and $cousin_i$ denote the relations defined on $\mathcal{P}_i$ instead of on $\mathcal{C}$, i.e., in terms of $\prec_i$ instead of $\prec$, via the respective definitions from Sect. 2.2. Now regard the example of Fig. 3(a) once more. We have $sib(\mathsf{plum}, \mathsf{apricot})$ and $cousin(\mathsf{plum}, \mathsf{grape})$, but no $cousin_i$ relation holds in $\mathcal{P}_i$, and we have derived relations like $sib_i(\mathsf{plum}, \mathsf{grape})$ etc. For this special example, sib_i is even larger than *sib*, while $cousin_i$ gets trivialised on $\mathcal{P}_i$. Using a different example, for $\mathcal{P}_j = \mathcal{P}_i \setminus \{\mathsf{fruit}\}$, $sib_j = \emptyset$, we can enlarge $cousin_i$ as well (see Fig. 3(b)). These examples illustrate the fact that sib_i and $cousin_i$ generally are only weakly related to *sib* and *cousin*. If we use sibling or cousin relationships in specifying closure conditions, it therefore seems advisable to employ the objective relation, i.e., the restrictions of *sib* and *cousin* to $\mathcal{P}_i$. This means that these relations cannot in general be defined internally (in terms of $\mathcal{P}_i$ and $\prec_i$): $sib(\mathsf{a}, \mathsf{b})$ can hold in $\mathcal{P}_i$ even if the superconcept c witnessing the sibling relation is not in $\mathcal{P}_i$. The same holds for *cousin*. We therefore adopt *sib* and *cousin* as primitive relations on $\mathcal{P}_i$: $sib(\mathsf{a}, \mathsf{b})$ holds in $\mathcal{P}_i$ if and only if $\mathsf{a}, \mathsf{b} \in \mathcal{P}_i$ and $sib(\mathsf{a}, \mathsf{b})$ holds in $\mathcal{C}$.

This way, concepts not connected via ISA_i can still stand in the sibling or cousin relation in $\mathcal{P}_i$.

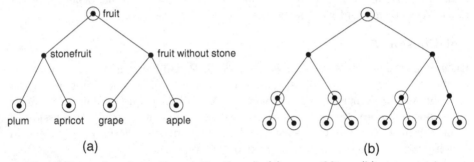

(a)   (b)

Fig. 3. Examples of two $\mathcal{P}_i$s smaller than $\mathcal{C}$: (a) more siblings (b) more cousins

3.1   Atomism

The absolutely minimal requirement for possession of a concept is just possession of that very concept. The approach to concept possession that specifies no additional requirement is called atomism. That view appears to be endorsed by Fodor [4, p. 121]. Atomism yields maximal independence of concepts. In particular, there are no conditions on the complexity of the structure of the set of concepts an epistemic subject can possess. Since every concept can be possessed on its own, the test for concept possession cannot go far beyond testing the capability to classify objects with respect to the concepts. This opens up the possibility of ascribing concepts to small children and animals. However, it turns out that atomism is untenable.

Seen from the point of cognitive science, the problem with atomism lies exactly in its generous openness. Since there is no complexity of the structure of concepts required, there is no distinction between concept possession and stimulus-response behaviour. As discussed in Sect. 1, this is not adequate.

3.2 Holism and Other Closure Conditions

Now we address the issues on the other side of the scale: demanding as much structure as there is.

Holism. According to a stricly holistic approach, an epistemic subject needs to possess the whole set of concepts in order to be granted possession of even a single concept. This is too harsh a requirement even for humans. Even if the objective set should be finite, it is probably very large; too large for the capacity of a single mortal being. Recall from Sect. 2.1 that $con(a)$ is the set of concepts connected to a concept a via the ISA relation. Then the holism condition reads as follows:

$$\forall a, b[(a \in \mathcal{P}_i \wedge b \in con(a)) \rightarrow b \in \mathcal{P}_i].$$

Note that this condition does not necessarily require possession of all concepts from $\mathcal{C}$. It just requires possession of all concepts that are connected to the concept in question via ISA. Should there be two or more isolated structures of concepts (cf. Sect. 2.1) all but one could be left out completely.

Closure under Superconcepts. Closure under superconcepts is a bit less demanding.

$$\forall a, b[(a \in \mathcal{P}_i \wedge a \text{ ISA } b) \rightarrow b \in \mathcal{P}_i].$$

In this case, one does not need to possess the whole set of concepts, but just everything along the order relation starting with the concept in question. This recursive abstraction and generalisation is very natural for humans (a dog is a mammal, a mammal is an animal, an animal is a living being, ...), but it is not clear why it should be essential for concept possession. In particular, it will be next to impossible to find an animal capable of this level of abstraction. On the other hand, there are clever animals to whom we want to ascribe concept possession, e.g., the parrot Alex (see Sect. 1). So this condition demands too much.

Closure under Subconcepts. Closure under subconcepts is formally similar, but has a different motivation.

$$\forall a, b[(a \in \mathcal{P}_i \wedge b \text{ ISA } a) \rightarrow b \in \mathcal{P}_i].$$

Again, the order relation is followed, but this time downwards in the set of concepts. It is not a process of abstraction but of specialisation that is carried out. Once more, this is natural for humans (some dogs are large dogs, some large dogs are St. Bernhards, some St. Bernhards are trained for rescue in the mountains, ...), but might rule out concept possession for animals.

Closure under Sub- and Superconcepts. For the sake of completeness, let us regard the conjunction of the closures of the above, closure under both super-concepts and subconcepts.

$$\forall a, b\, [(a \in \mathcal{P}_i \wedge (a\ \mathsf{ISA}\ b \vee b\ \mathsf{ISA}\ a)) \rightarrow b \in \mathcal{P}_i].$$

This closure is subject to both objections raised above. In addition, as this condition has to hold for every concept in $\mathcal{P}_i$, this type of closure yields the whole set of concepts connceted to a: we are back at holism.

Closure under Siblings and Cousins. A closure condition that requires less abstraction is closure under sibling and cousin concepts.

$$\forall a, b[(a \in \mathcal{P}_i \wedge (sib(a, b) \vee cousin(a, b))) \rightarrow b \in \mathcal{P}_i].$$

Note that we talk here about the *sib* and *cousin* relations on the objective set of concepts (cf. the discussion before Sect. 3.1). But again, do we have to possess all sibling concepts in order to possess one single concept? In the Indo-European language family, we have eleven basic colour concepts. There are other languages that have considerably less basic colour concepts (down to just two concepts, for black and white). If closure under siblings were a necessary condition, people with just the concepts of black and white could not even have the same colour concepts for these two colours as we have. But they do, as Berlin and Kay have shown in [1]. So this condition is not adequate either.

Closure under Witnesses. Another approach towards more realistic closure conditions consists in weakening the universal quantification to an existence quantification. Instead of requiring all concepts of a certain kind (e.g., all superconcepts) one can just require (at least) one concept of that kind. This one concept witnesses that the condition does not run empty (which might be another danger for all-quantified conditions). We will call this kind of closure *closure under witnesses*. For example, one can demand closure under the following condition:

$$\forall a\, [(a \in \mathcal{P}_i \wedge \exists b(a\ \mathsf{ISA}\ b)) \rightarrow \exists b(a\ \mathsf{ISA}\ b \wedge b \in \mathcal{P}_i)],$$

which means that if i possesses a concept a and there is a superconcept of a at all, i possesses at least one superconcept of a as well. This condition is met if $\mathcal{P}_i$ contains just one chain of concepts connected through ISA. Analogously, one could formulate a definition using subconcepts, going down in the order instead of up.

The problem of this approach lies again in our goal to leave a possibility for animals to possess concepts. As we discussed already in the case of closure under all superconcepts, humans do not appear to have a problem with iterated generalisation. Neither do they have a problem with iterated specialisation. But there is no reason why these faculties should be essential for concept possession. There is even a reason against that. From psychology we know that there is a level of perceptually based prototype concepts that are learned easiest by children. Learning how to generalise or to specialise such concepts comes much later in the development. Still our intuition tells us that children possess these concepts as soon as they show the appropriate classification behaviour.

So far we have seen that requiring no structure of $\mathcal{P}_i$ does not give enough complexity for concept possession. On the other hand, stipulating closure conditions that are universally quantified amounts to requiring more cognitive abilities than are necessary for mere concept possession. So the adequate condition must lie somewhere in between. We have discussed weakening the closure under super-concepts and subconcepts to existential quantification. This still runs into the problem of requiring the capability of repeated generalisation or specification for concept possession. In the next section we will regard a weakened condition on closure under siblings and cousins, which appears to be psychologically adequate.

3.3 Closure Under Witnesses for Siblings and Cousins

In their paper [7], Newen and Bartels propose an account of what it takes to possess a concept based on psychology and animal studies. They argue that there are certain levels of complexity of behaviour that can be related to levels of cognitive ability. Concepts are located on an intermediate level. Concept possession requires flexible behaviour more complex than fixed stimulus-response schemata, but less complex than, e.g., planning based on propositional knowledge.

Based on their assessment of the complexity of behaviour typically thought to require conceptual representations, Newen and Bartels propose two structural requirements for concept possession. In order to possess a concept, an epistemic subject should first possess another concept of the same dimension of classification, e.g., not just red, but blue, too. We formalise this in terms of the (objective) sibling relation

$$\forall a[a \in \mathcal{P}_i \rightarrow \exists b(sib(a, b) \wedge b \in \mathcal{P}_i)].$$

The other requirement is to possess another concept on the same level of abstraction that is not, however, a sibling. The intuition behind this goes back to classifying objects: an epistemic subject shall be capable of sorting objects under different dimensions of classifications, with respect to different aspects. This, of course, need not work for an arbitrary object — just for those that fall under more than one concept that the epistemic subject possesses. Again, this requirement is symmetric. So far, the cousin relation is our best candidate for a formalisation.

$$\forall a[a \in \mathcal{P}_i \rightarrow \exists b(cousin(a, b) \wedge b \in \mathcal{P}_i)].$$

This condition does not exactly match the above intuition. What we have is that in a tree, if the condition is fulfilled, we can ascribe concept possession to the epistemic subject. So the condition is sufficient, but it is not necessary: there can be nodes on the same level that are related more distantly. On general partial orders, the condition is not even sufficient. We encounter two issues: a concept can be both a cousin and a sibling of another concept, with respect to different superconcepts. One would have to exclude that this one concept is used to fulfill both conditions. The second issue is with the intuition of levels of abstraction: it can happen that two concepts on different levels are cousins. In this case it is not possible to ascribe levels unambiguously to $\mathcal{C}$.

Some partial orders admit the definition of levels, while others do not. Formally, we can capture this distinction as follows: If a ISA b, then $c_1, \ldots, c_n \in \mathcal{C}$ form a *direct conection* from a to b if $a = c_1 \prec \cdots \prec c_n = b$. We call n the *length* of the connection. Let $dc(a, b)$ be the set of all direct connections from a to b. Then we can say that $\mathcal{C}$ *respects levels* if for any elements a and b, all direct connections are of the same length, i.e.,if the following holds:

$$\forall a, b \in \mathcal{C} \, \forall \langle c_1, \ldots, c_n \rangle, \langle d_1, \ldots, d_m \rangle \in dc(a, b) \, [a \text{ ISA } b \rightarrow n = m] .$$

That is, in a partial order that respects levels, any two nodes have a fixed distance. In trees, this is obvious: if a ISA b, then $dc(a, b)$ has exactly one element.

If $\mathcal{C}$ respects levels, we can define a level function: Let $\mathcal{L} : \mathcal{C} \to \mathbb{Z}$ be a function, assigning every concept in $\mathcal{C}$ an integer indicating its level of abstraction. If a is a perceptually based prototype concept, $\mathcal{L}(a) = 0$ shall hold. In addition, we require $a \prec b \Leftrightarrow \mathcal{L}(a) + 1 = \mathcal{L}(b)$.

With this terminology, we can state the following condition for possessing concepts on the same level of abstraction.

$$\forall a \, [a \in \mathcal{P}_i \rightarrow \exists b (\mathcal{L}(a) = \mathcal{L}(b) \land a \neq b \land \neg sib(a, b) \land b \in \mathcal{P}_i)] .$$

That is, to every concept possessed by the epistemic subject i there is a different concept possessed by i that is on the same level of abstraction, but not in the same dimension of classification (i.e., not a sibling).

Further work will be required to weigh carefully the pros and cons of demanding the additional structure embodied in $\mathcal{L}$. So far, we have the condition on the existence of siblings and we search for another symmetric condition requiring the existence of one more element of $\mathcal{P}_i$. By closure under these conditions, we get a set of at least four concepts.

4   Conclusion

When does a biological or artifical system have concepts? In this paper, we addressed this question from a formal point of view. We focused on the interrelation between a given objective structure of concepts, which we argued is a partial ordering, and a subject's representation of that structure, which we argued should be a subordering. The main question was how much of the structure of the objective side needs to be mirrored subjectively in order to grant possession of concepts. We thus discussed a number of closure conditions for subjective sets of representations. Based on results from cognitive science, we argued that the appropriate closure condition that strikes a balance between atomism and holism is closure under witnesses for siblings and cousins.

As we argued in Sect. 3.3, we believe that closure under witnesses for siblings is both formally and psychologically adequate. Closure under witnesses for cousins is a formally precise condition that psychologically at least points in the right direction. In order to further develop the epistemic logic of concepts we hope to benefit from discussions with both cognitive scientists and scientists working in the field of AI.

Acknowledgements

We would like to thank Albert Newen and our anonymous referees for helpful comments. T.H. acknowledges support by the VolkswagenStiftung.

References

1. Berlin, B. and Kay, P.: *Basic Colour Terms: Their Universality and Evolution*, University of California Press, Berkeley 1969
2. Cursiefen, S.: *Formen des logischen Allwissens. Eine problemorientierte Darstellung modallogischer Systeme*, PhD thesis, Bonn 2006
3. Fagin, R., Halpern, J., Moses, Y., and Vardi, M.: *Reasoning about Knowledge*, MIT Press, 1995
4. Fodor, J. A.: *Concepts: Where Cognitive Science Went Wrong*, Oxford University Press, Oxford 1998
5. Ganter, B. and Wille, R.: *Formal Concept Analysis: Mathematical Foundations*, Springer, Heidelberg 1999
6. Kangassalo, H., The concept of concept. In Ohsuga, S., Kangassalo, H., Jaakkola, H., Hori, K., and Yonezaki, N. (eds.), *Information Modelling and Knowledge Bases III: Foundations, Theory and Applications*, IOS Press, Amsterdam 1992, pp. 17–58
7. Newen, A. and Bartels, A.: Conditions of Possessing Concepts, submitted
8. Pepperberg, I.: *The Alex Studies: Cognition and Communicative Abilities of Grey Parrots*, Harvard, Cambridge (Mass.) [4]2002

Development of Intelligent Systems and Multi-Agents Systems with Amine Platform

Adil Kabbaj

INSEA
Rabat, Morocco, B. P 6217
akabbaj@insea.ac.ma
http://sourceforge.net/projects/amine-platform

Abstract. Amine is a Java open source multi-layer platform dedicated to the development of intelligent systems and multi-agents systems. This paper and companion papers [2, 3] provide an overview of Amine platform and illustrate its use in the development of dynamic programming applications, natural language processing applications, multi-agents systems and ontology-based applications.

1   Introduction

Amine is a Java open source multi-layer platform and a modular Integrated Development Environment, dedicated to the development of intelligent systems and multi-agents systems. Amine is a synthesis of 20 years of works, by the author, on the development of tools for various aspects of Conceptual Graph theory.

This paper shows how dynamic programming, natural language processing and multi-agents systems can be developed using Amine. The companion paper [2] illustrates the use of Amine in the development of ontology based applications. A forthcoming paper will describe the use of Amine in problem-solving applications; and especially in the development of a strategic card game called Tijari (which has some similarity with Bridge card game).

We urge the reader to consult Amine web site[1] for more detail. Source code, samples and documentation can be downloaded from sourceforge site[2].

The paper is organized as follows: section 2 introduces briefly Amine platform. A more detailed description of Amine architecture is provided in the companion paper [3]. Section 3 introduces the re-engineering, extension and integration of Prolog+CG language [4, 5] in Amine. Section 4 introduces the re-engineering, extension and integration of Synergy language [6, 7] in Amine. It shows also how Synergy has been extended to enable dynamic programming. Section 5 discusses the use of the new version of Prolog+CG in the development of natural language processing applications. Section 6 illustrates briefly the use of Amine, in conjunction with Jade[3] (Java

---

[1] amine-platform.sourceforge.net
[2] sourceforge.net/projects/amine-platform
[3] jade.tilab.com/

H. Schärfe, P. Hitzler, and P. Øhrstrøm (Eds.): ICCS 2006, LNAI 4068, pp. 286–299, 2006.
© Springer-Verlag Berlin Heidelberg 2006

Agent Development Environment), in the development of a multi-agents system called Renaldo. A forthcoming paper will describe the development of Renaldo in more detail.

Section 7 provides a comparison of Amine with other CG tools. Section 8 outlines some current and future work. Section 9 concludes the paper.

2 An Overview of Amine Platform

Amine is a modular integrated environment composed of four hierarchical layers: a) *ontology layer* provides "structures, processes and graphical interfaces" to specify the "conceptual vocabulary" and the semantic of a domain, b) *algebraic layer* is build on top of the ontology layer: it provides "structures, operations and graphical interfaces" to define and use "conceptual" structures and operations, c) *programming layer* is build on top of the algebraic layer: it provides "programming paradigms/languages" to define and execute "conceptual" processes and, d) *multi-agent layer* provides plugs-in to agent development tools, allowing for the development of multi-agent systems.

More specifically:

1. *Ontology layer:* It concerns the creation, edition and manipulation of *multi-lingua ontology.* The companion paper [2] presents this layer in detail (including ontology meta-model and ontology related processes).

2. *Algebraic layer:* this layer provides several types of structures and operations: elementary data types (AmineInteger, AmineDouble, String, Boolean, etc.) and structured types (AmineSet, AmineList, Term, Concept, Relation and Conceptual Graph). In addition to operations that are specific to each kind of structure, Amine provides a set of basic common operations (clear, clone, toString, etc.) and various common matching-based operations (match, equal, unify, subsume, maximalJoin and generalize). Structures can be generic; they can contain variables and the associated operations take into account *variable binding* (the association of a value to a variable) and *binding context* (the programming context that determines how variable binding should be interpreted and resolved, i.e. how to associate a value to a variable and how to get the value of a variable).

    Amine structures and operations (including CG structure and CG operations) are APIs that can be used by any Java application. They are also "inherited" by the higher layers of Amine.

    The companion paper [3] provides more detail on the algebraic layer of Amine. It highlights also the use of Java interfaces to enhance the genericity of Amine. See also Amine web site for mode detail on this basic feature of Amine.

3. *Programming layer:* Three complementary programming paradigms are provided by Amine: a) *pattern-matching and rule-based programming paradigm,* embedded in PROLOG+CG language which is an object based and CG-based extension of PROLOG language, b) *activation and propagation-based programming paradigm,* embedded in SYNERGY language, and c) *ontology or memory-based programming paradigm* which is concerned by incremental and automatic integration of knowledge in an ontology (considered as an agent

memory) and by information retrieval, classification and other related ontology/memory-based processes.

4. *Agents and Multi-Agents Systems layer:* Amine can be used in conjunction with a Java Agent Development Environment to develop multi-agents systems. Amine does not provide the basic level for the development of multi-agents systems (i.e. implementation of agents and their communication capabilities using network programming) since this level is already offered by other open source projects (like Jade). Amine provides rather plugs-in that enable its use with these projects in order to develop multi-agents systems.

Amine provides also several graphical user interfaces (GUIs): Ontology GUI, CG Notations editors GUI, CG Operations GUI, Dynamic Ontology GUI, Ontology processes GUI, Prolog+CG GUI and Synergy GUI. *Amine Suite Panel* provides an access to all GUIs of Amine, as well as an access to some ontology examples and to several tests that illustrate the use of Amine structures and their APIs. Amine has also a web site, with samples and a growing documentation.

Amine four layers form a *hierarchy*: each layer is built on top of and use the lower layers (i.e. the programming layer inherits the ontology and the algebraic layers). However, a lower layer can be used by itself without the higher layers: the ontology layer (i.e. with the associated APIs) can be used directly in any Java application without the other layers. Algebraic layer (i.e. with the associated APIs) can be used directly too, etc. Among the goals (and constraints) that have influenced the design and implementation of Amine was the goal to achieve a higher level of modularity and independence between Amine components/layers.

Amine platform can be used as a modular integrated environment for the development of intelligent systems. It can be used also as the kernel; the basic architecture of an *intelligent agent*: a) the ontology layer can be used to implement the dynamic memory of the agent (agent's ontology is just a perspective on the agent's memory), b) the algebraic layer, with its various structures and operations, can be used as the "knowledge representation capability" of the agent, c) the programming layer (i.e. dynamic ontology engine, Prolog+CG and Synergy) can be used for the formulation and development of many inference strategies (induction, deduction, abduction, analogy) and cognitive processes (reasoning, problem solving, planning, natural language processing, dynamic memory, learning, etc.), d) Synergy language can be used to implement the reactive and event-driven behaviour of the agent.

One long-term goal of the author is to use Amine, in conjunction with Java Agent Development Environments (like Jade), to build various kinds of intelligent agents, with multi-strategy learning, inferences and other cognitive capabilities.

The group of Peter Ohrström and Henrik Scharfe has developed on-line course that covers some parts of Amine Platform[4]. Amine is used by the author to teach Artificial Intelligence (AI) courses. Amine is suited for the development of projects in various domains of AI (i.e. natural language processing, problem solving, planning, reasoning, case-based systems, learning, multi-agents systems, etc.).

---

[4] www.huminf.aau.dk/cg/

3 Re-engineering, Extension and Integration of Prolog+CG in Amine

Prolog+CG has been developed by the author as a "stand-alone" programming language [4, 5]. The group of Peter Ohrström developed a very good on-line course on some aspects of Prolog+CG. Let us recall three key features of previous versions of Prolog+CG:

- CG (simple and compound CGs) is a basic and primitive structure in Prolog+CG, like list and term. And like a term, a CG can be used as a structure and/or as a representation of a goal. Unification operation of Prolog has been extended to include CG unification. CG matching-based operations are provided as primitive operations.
- By a supplementary indexation mechanism of rules, Prolog+CG offers an object based extension of Prolog.
- Prolog+CG provides an interface with Java: Java objects can be created and methods can be called from a Prolog+CG program. Also, Prolog+CG can be activated from Java classes.

The interpreter of Prolog+CG, that takes into account these features (and others) has been developed and implemented in Java by the author.

The above three key features are still present in the new version of Prolog+CG but the re-engineering of Prolog+CG, which was necessary for its integration in Amine platform, involved many changes in the language (and its interpreter). Five main changes are of interest (see Amine Web Site for more details):

- Type hierarchy and Conceptual Structures (CSs) are no more described in a Prolog+CG program. Prolog+CG programs are now interpreted according to a specified ontology that includes type hierarchy and CSs. Also, a Prolog+CG program has the current ontology as support: Prolog+CG interpreter attempts first to interpret each identifier in a program according to the current lexicon of the current ontology. If no such identifier is found, then the identifier is considered as a simple identifier (without any underlying semantic).
- The notion of project is introduced: user can consult several programs (not only one) that share the same ontology.
- Prolog+CG inherit the first two layers of Amine: all Amine structures and operations are also Prolog+CG structures and operations. And of course, Prolog+CG user can manipulate the current ontology and the associated lexicons according to their APIs.
- The interface between the new version of Prolog+CG and Java is simpler and "natural" in comparison with previous interfaces (see Amine Web site for more detail).
- Interoperability between Amine components: Prolog+CG can be used in conjunction with the other components of Amine (i.e. dynamic ontology engine and Synergy can be called/used from a Prolog+CG program).

4   Re-engineering, Extension and Integration of Synergy in Amine

In [6, 7] we proposed *CG activation-based mechanism* as a computation model for *executable conceptual graphs*. Activation-based computation is an approach used in visual programming, simulation and system analysis where graphs are used to describe and simulate sequential and/or parallel tasks of different kinds: functional, procedural, process, event-driven, logical and object oriented tasks. Activation-based interpretation of CG is based on *concept lifecycle, relation propagation rules* and *referent/designator instantiation*. A concept has a state (which replaces and extends the notion of control mark used by Sowa) and the concept lifecycle is defined on the possible states of a concept. Concept lifecycle is similar to process lifecycle (in process programming) and to active-object lifecycle (in concurrent object oriented programming), while relation propagation rules are similar to propagation or firing rules of procedural graphs, dataflow graphs and Petri Nets.

SYNERGY is a visual multi-paradigm programming language based on CG activation mechanism. It integrates functional, procedural, process, reactive, object-oriented and concurrent object-oriented paradigms. The integration of these paradigms is done using CG as the basis knowledge structure, without actors or other external notation. Previous versions of Synergy have been presented [6, 7]. The integration of Synergy in Amine required re-engineering work and some changes and extensions to the language and to its interpreter. New features of Synergy include:

- Long-term memory introduced in previous definitions of Synergy corresponds now to ontology that plays the role of a support to a Synergy "expression/ program",
- Previous versions of Synergy did not have an interface with Java. The new version of Synergy includes such an interface; Java objects can be created and methods activated from Synergy. This is an important feature since user is not restricted to (re)write and to define anything in CGs. Also, primitive operations are no more restricted to a fixed set of operations.
- The new version of Synergy has an access to the two first layers of Amine. Also, since Prolog+CG, Synergy and dynamic ontology formation process are integrated in the same platform and share the same underlying implementation; it is now possible to develop applications that require all these components. We provide an example of this synergy in the next section.
- Another new feature is the possibility to perform dynamic programming, i.e. dynamic formation-and-execution of the program. We focus on this feature in the rest of this section.

4.1   Dynamic Programming with Synergy

To illustrate what we mean by "dynamic programming", let us start with the idea of database inference proposed by Sowa [9, p. 312] that combines the user's query with background information about the database to compute the answer. Background information is represented as type definitions and schemata. Sowa stressed the need for an inference engine to determine what virtual relations to access. By joining schemata and doing type expansions, the inference engine expands the query graph to a working graph (WG) that incorporates additional background information. Actors bound to the schemata determine which database relations to access and which functions and procedures to execute. According to Sowa, his inference engine can support a dynamic way of deriving dataflow graphs [9, p. 312]. In other words, his inference engine can be considered as a basis for *a dynamic programming approach* (recall that dataflow graphs, Petri Nets, executable CG and other similar notations have been used to develop visual programming languages). Indeed, his inference engine is not restricted to database, it can be extended to other domains and be considered as an approach to dynamic programming.

Our task was to adapt, generalize and integrate the inference engine of Sowa to Synergy. The new version of Synergy includes the result of this integration. Figure 1 illustrates the implementation of Sowa's example in Synergy. Background information (procedural knowledge in terms of strategies, methods, procedures, functions, tasks, etc.) is stored in ontology as situations associated to concept types (Figure 1.a). During the interpretation/execution of the working graph (WG) (Figure 1), if a concept needs a value that can not be computed from the actual content of the WG (Figure 1.b), then Synergy looks, in the ontology, for the best situation that can compute the value (i.e. the descriptor) of the concept. The situation is then joined to the WG (Figure 1.c) and Synergy resumes its execution. In this way, the program (i.e. the WG) is dynamically composed during its execution (Figure 1).

This simple example illustrates the advantage of Amine as an Integrated Development Environment (IDE); it illustrates how various components of Amine (ontology, CG operations, Prolog+CG, Synergy) can be easily used in one application: semantic analysis of the request can be done by a Prolog+CG program. The result (an executable CG), will be provided to Synergy which illustrates the visual execution of the "dynamic program". After the termination of the execution, the final CG will be an input for a text generation program (that can be implemented in Prolog+CG) to provide a text that paraphrases the composed "program" responsible for the result. See Amine Web Site for more detail on dynamic programming with Synergy.

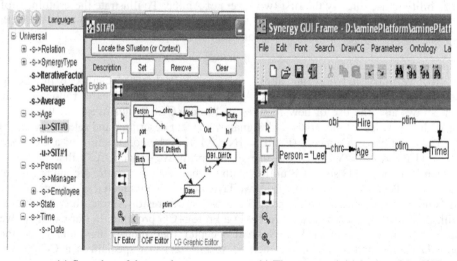

(a) Snapshot of the ontology   b) The request: initial state of the WG

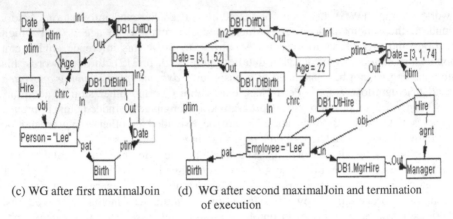

(c) WG after first maximalJoin   (d) WG after second maximalJoin and termination
of execution

Fig. 1. Example of Dynamic programming with Synergy (adapted from Sowa [9])

5 Natural Language Processing with Prolog+CG

As stressed in a previous paper [5], several features of Prolog+CG makes it a suitable
language for the development of natural language processing (NLP) applications: a)
Prolog+CG is an extension of Prolog which is suited for NLP, b) CG, both simple and
compound, is provided as a basic data structure, c) Prolog+CG allows CG with
variables (variable as concept type, concept designator, concept descriptor, or as
relation type), d) several CG matching-based operations are provided (maximalJoin,
generalize, subsume, contract, expand, analogy, etc.), e) CG basic operations are
available (find a concept or a relation in a CG that verify some constraints, etc.), f) the
possibility to construct and update a CG (by adding more concepts and relations).

All these features (and others) are made simpler with the integration of Prolog+CG
in Amine platform. Note that with the new version of Prolog+CG, there is also the
possibility to use directly the first two layers of Amine. To illustrate the usefulness of
all these features for NLP, let us consider briefly their use in three sub-tasks of NLP:
semantic analysis, question/answering and phrase/text generation.

Semantic analysis with Prolog+CG
In [5], we illustrated how the above features of Prolog+CG can be exploited to
develop a semantic analysis process. As a recall, let us consider the following rule
that shows also the use of new features of Prolog+CG. It illustrates : a) the use of
variables as concept type, concept designator, concept descriptor and relation type, b)
the construction of a concept (E_NP = [N : A1]), c) the construction of a CG (G =
[N: A1]-R1->[T1 = V1]), d) the use of the primitive branchOfCG that locates a
branch B in the CG G so that B unifies with the pattern given as the second argument
of branchOfCG, e) the use of the first two layers of Amine: branch (i.e. a relation with
its source and target concepts) and CG are two structures of Amine, these structures
with their methods can be used directly in a Prolog+CG program. In our example, we
have a call to the method getSourceConcept() that returns the source of the
branch/relation and a call to the method specialize() that specializes a CG by the
maximal join of another CG.

```
stativePart([A|P1], P1, G_NP, E_NP, G) :-
    Adj(A, R1, T1, V1), !,
    E_NP = [N : A1],
    G = [N : A1]-R1->[T1 = V1],
    branchOfCG(B, [N : A1]-R1->[T1 = V1], G),
    E_N is B:getSourceConcept(),
    G:specialize(E_N, G_NP, E_NP).
```

Let us consider now the change in the formulation of the lexicon: in previous versions of Prolog+CG, the semantic of the words should be specified in the Prolog+CG program itself. For instance, consider the word "open" with some of its different meanings:

```
lexicon("open",verb,
            [Human]<-agnt-[Open]-obj>[OpenableObject]).
lexicon("open", verb, [Key]<-agnt-[Open]-obj->[Door]).
lexicon("open", verb, [Open_Box]<-agnt-[Open]-obj->[Box]).
lexicon("open", verb, [Shop]<-pat-[Open]-
                                        -obj->[Door],
                                        -ptime->[Time]).
```

With the new version of Prolog+CG, another formulation is now possible: the above different meanings can be considered as background information, stored in the used ontology as situations associated to the type Open. User can access the ontology to get background information (definition, canon, situation, etc.) for a specific type or individual. These changes in the formulation of a lexicon in Prolog+CG lead to the following reformulation:

```
lexicon("open", verb, Open). // one entry for the word "open"

lexicon(_verb, verb, _type, _sem) :-
    lexicon(_verb, verb, _type),
    getSemantic(_type, _sem).
```

Definition of the goal getSemantic/2 is provided below. It searches, from the ontology, the background information for a specific type or individual. Note the call to the method getCanon() that returns the canon of the type (or returns null if the type has no canon) and the call to the method getSituationsDescription() that returns, in a list, all situations descriptions that are associated to the specified type.

```
getSemantic(_Type, _Sem) :-
    _Sem is _Type:getCanon(),
    dif(_Sem, null).

getSemantic(_Type, _Sem) :-
    _EnumSitDescr is _Type:getSituationsDescription(),
    dif(_EnumSitDescr, null),
    _ListSitDescr is
            "aminePlatform.util.AmineObjects":
                        enumeration2AmineList(_EnumSitDescr),
    member(_Sem, _ListSitDescr).
```

Word disambiguation is performed in the current version of our semantic analysis process by using the backtracking of Prolog+CG: if the maximal join of the word's semantic with the working graph fails, Prolog+CG backtracks and resatisfies the goal getSemantic/2 which returns another meaning (i.e. another conceptual structure) for the current word.

Question/Answering

Semantic analysis of a (short) story would produce a compound CG (see the fragment below). Let us call it CGStory. In our example, CGStory is a fusion of three networks: a) temporal network composed by "after" relations that specify the temporal succession of actions, events, and states, b) causal network composed by "cause" relations, and c) intentional network composed by "motivationOf" and "reason" relations:

```
story(
[Action #act1 =
      [Time : Early]<-time-[WakeUp]-pat->[Man: John]]-after->
[State #stt1 = [Hungry]-pat->[Man: John]]-after->
...
[State #stt1]<-reason-[Goal=
            [Action=[Food]<-obj-[Eat]-agnt->[Man:John]]]-
                              <-reason-[Action #act2],
                              <-reason-[Action #act5],
                              <-reason-[Action #act7],
                              <-reason-[Action #act8]
[Action #act3]<-motivationOf-[Goal =
      [Action = [Man:John]<-dest-[Greet]-agnt->[Woman: Mary]]
                          ]<-reason-[Action #act4]
[Event #evt1]-
            -cause->[State #stt2 = [ParkingLot]<-pat-[Slick]],
            <-cause-[Event #evt2]
                        ).
```

Semantic analysis process is applied also to questions and for each type of question; there is a specific strategy responsible for the search and the composition of the answer [1]. Here is the formulation in Prolog+CG of the strategy for answering "why" question. It concerns the intentional network: the strategy locates in CGStory the branch/relation with relation type "reason" or "motivationOf" and the branch's source concept should unify with the content of the request. The recursive definition of the goal reason/2 provides the possibility to follow an "intentional path" to get the reason of the reason, etc.

```
answerWhy(A, Y)  :-
    story(_story),
    member(R, [reason, motivationOf]),
    branchOfCG(B, [T = G]<-R-[T2 = A], _story),
    reason([T = G], Y).

reason(X, X).
reason([T = G], Y)  :-
    story(_story),
    member(R, [reason, motivationOf]),
    branchOfCG(B, [T1 = G1]<-R-[T = G], _story),
    reason([T1 = G1], Y).
```

For instance, to the question "why did john drive to the store ?", the question/answering program returns:

```
?- questionAnswering("why did john drive to the store ?",
                                                 _answer).
{_answer = [Goal = [Action = [Eat #0] -
                                      -agnt->[Man :John],
                                      -obj->[Food]
                    ]
            ]};
{_answer = [State = [Hungry]-pat->[Man :John]]};
 no
?-
```

Of course, the above definition of "why-strategy" is simplistic, but the aim of the example is to show how Prolog+CG, in the context of Amine, constitutes a suitable programming environment for CG manipulation and for the development of NLP applications.

Phrase generation
Nogier [8] proposed a phrase generation process that is based on: a) word selection, b) transformation of the input CG to a "syntactic CG" using semantic/syntactic corresponding rules, c) and then linearization of the "syntactic CG" using syntactic and morphological rules. All these rules can be implemented in Prolog+CG. To produce a concise and precise sentence, the generation process has to select the most specific words for the concepts in the input CG [8]. The approach proposed by Nogier can be implemented in Amine as follows: use the dynamic ontology engine of Amine to classify the input CG according to its concepts. The result of the classification, for each concept, is a list of "Conceptual Structures (CS) nodes" in the ontology that are the most close to the input CG. Select from these CS nodes, those that correspond to type definitions. Compute the correlation coefficient proposed by Nogier on the selected definitions to get the "best" words for the current concepts in the input CG. We are developing a phrase generation process that is based on the work of Nogier [8] and that uses the above implementation for the "word selection" procedure.

6   Multi-Agents Systems (MAS) with Amine and Jade: The Case of Renaldo

Instead of developing a specific multi-agents layer for Amine, we decided to use available open-source "Java Agent Development Environments" in conjunction with Amine. In her DESA, Kaoutar ElHari explored the use of Amine and Jade[5]. Jade allows the creation of Agents (i.e. it offers a Java class *Agent* and manages the underlying network processing), the use of different kinds of behaviours and the communication with ACL according to FIPA specification. Currently, we use Jade to handle the lower level of the MAS (i.e. creation and communication between agents) and Amine for the higher level (i.e. cognitive and reactive capabilities of the agents are implemented using Amine).

---

[5] jade.tilab.com

Currently, the MAS layer of Amine contains one plug-in (Amine/Jade) implemented as a package: *amineJade*. Other plugs-in (i.e. other packages) could be added as other combinations of Amine and "Java Agent Development Environments" are considered (for instance Amine and Beegent[6]). The package "amineJade" offers basically two classes:

- The class *PPCGAgent* that extends the class Agent (provided by Jade) with Prolog+CG interpreter (as its main attribute) and with other attributes and methods (like *send*, *sendAndWait* and *satisfyGoal*).
- The class *JadeMAS* that offers the possibility, via the method *createMAS*, to create and initiate a multi-agents system.

Let us consider briefly the case of *Renaldo*; a MAS that concerns the simulation of a child story. The setting of the story is a forest; it corresponds to the environment of Renaldo. The characters of the story (the bear John, the bird Arthur, the bee Betty, etc.) are the agents of Renaldo. Each type of agents (bear, bird, bee, etc.) has a set of attributes, knowledge, goals, plans and actions that are specified as a Prolog+CG program. A specific agent can have, in addition, specific attributes, knowledge, goals, plans and actions, specified also as a Prolog+CG program.

The MAS Renaldo is implemented as a Prolog+CG program/file: "Renaldo.pcg". The execution of the MAS Renaldo is initiated by calling the goal "renaldo" which is defined as follows:

```
renaldo :-
    "aminePlatform.mas.amineJade.JadeMAS":createAgents(
                        [John, Arthur, Betty, Environment]),
    John:satisfyGoal(
        goal([Bear: John]<-pat-[SatisfyNeed]-obj->
                 [Hungry]-Intensity-> [Intensity = High]))).
```

The argument of the method createAgents() is a list of agents identifiers. From the identifier of each agent (i.e. John), the method gets the associated Prolog+CG program/file that specifies the agent (i.e. "John.pcg"). It then locates the header fact of the agent to get the ontology and names of other Prolog+CG programs associated to the agent. For instance, the header of the agent John (from the program "John.pcg") is:

```
header("RenaldoOntology.xml", ["Bear.pcg"]).
```

The method createAgents() will then create an instance of PPCGAgent class for each agent and initiates the associated Prolog+CG interpreter with the specified Prolog+CG files and ontology file. For instance, createAgents() will create an instance of PPCGAgent class for John and will initiate its Prolog+CG interpreter with the files "John.pcg" and "Bear.pcg", and with the ontology "RenaldoOntology.xml".

Note: the environment is implemented as an agent that manages the access, by the agents, to shared objects (resources like foods, water, river, etc.) and it is responsible also for the treatment of events.

---

[6] www2.toshiba.co.jp/beegent/index.htm

After the creation and initiation of the agents (due to the execution of the method createAgents()), "renaldo" assigns to John the goal "satisfy hungry with intensity high". The method satisfyGoal() of PPCGAgent calls the Prolog+CG interpreter of the agent (recall that each agent has its own Prolog+CG interpreter) to resolve the specified goal.

Renaldo in particular and Amine's MAS layer in general will be described in more detail in a forthcoming paper.

7 Related Works

Philip Martin[7] provides a detailed comparison of several available CG tools[8] (Amine, CharGer, CGWorld, CoGITaNT, Corese, CPE, Notio, WebKB). CGWorld is a Web based workbench for joint distributed development of a large KB of CG, which resides on a central server. CGWorld is no more developed. Corese is a semantic web search engine based on CG. WebKB is a KB annotation tool and a large-scale KB server. CoGITaNT is an IDE for CG applications. CharGer is a CG editor with the possibility to execute primitive actors and to perform matching operation. Notio is not a tool but a Java API specification for CG and CG operations. It is no more developed. It is re-used however by CharGer and Corese. CPE has been developed as a single standalone application. Currently, CPE is being upgraded to a set of component modules (to render CPE an IDE for CG applications). Its author announces that CGIF and basic CG operations (projection and maximal join) are coming soon. The new upgraded version of CPE is underway and it is not yet available. CGWorld, Notio and CPE will not be considered in our comparison of available (and active) CG tools.

In his comparison, Philip focuses mainly on the "ontology-server dimension" which is specific to his tool (WebKB); he did not consider other dimensions, i.e. other classes of CG tools. Indeed, CG tools can be classified under at least 8 categories of tools: CG editors, executable CG tools, algebraic tools (tools that provides CG operations), KB/ontology tools, ontology server tools, CG-based programming languages, IDE tools for CG applications and, agents/MAS tools.

The category "IDE for CG applications" means a set of APIs and hopefully of GUIs that allow user to construct and manipulate CGs and to develop various CG applications. Only Amine and CoGITaNT belong to this category. The category "CG-based programming language" concerns any CG tools that provide a programming language with CG and related operations as basic construct. Only Amine belongs to this category, with its two programming languages: Prolog+CG and Synergy. The category "Agents/MAS Architecture" concerns CG tools that allow the construction and execution of intelligent agents (with cognitive and reactive capabilities) and multi-agents systems (MAS). As illustrated in this paper, Amine, in conjunction with a Java Agent Development Environment, can be classified under this category.

---

[7] en.wikipedia.org/wiki/CG_tools
[8] These tools are listed also in www.conceptualgraphs.org/

Symbols used in the table:

"++": the tool offers different features concerning the associated category. For instance, Amine provides multi-lingua and multi-notations CG editors. The same for executable CG: it offers not only the equivalent of actors, as CharGer does, but a programming language based on executable CG. This paper illustrates in addition a new feature of Synergy: dynamic programming. The same for "KB/Ontology" category: Amine provides a rich ontology API, ontology editors and various basic ontology processes. And the same for "Programming" category: Amine provides two CG based programming languages (i.e. Prolog+CG and Synergy).

"+": the tool can be classified under the associated category.

"-": the tool can not be classified under the associated category.

"/": the tool is not intended to belong to the specified category but it uses some aspects of the category. For instance, "web ontology tools" like Corese and WebKB are not intended to be used as "algebraic tools" even if they use some CG operations (like projection and generalization).

	Amine	CharGer	CoGITaNT	Corese	WebKB
CG Editor(s)	++	++	++	-	-
Exec. CG	++	+	-	-	-
Algebraic	++	+	+	/	/
KB/Ontology	++	-	?	+	+
Ont. Server	-	-	-	-	++
IDE	++	/	+	-	-
Programming	++	-	-	-	-
Multi-Agent	+	-	-	-	-

Fig. 2. Comparison of available CG tools

8 Current and Future Work

Current and future works concern all layers of Amine as well as the development of applications in various domains:

a)   Development of the ontology layer: development of interfaces with ontologies that use RDF/OWL, development of Web services so that Amine ontologies can be used from the Web, development of an ontology server, enhance the current ontology drawing module, enhance the basic ontology processes, etc.

b)   Development of the algebraic layer: enhance the implementation of CG operations, consider other implementations, enhance the CG drawing module, etc.

c)   Development of the programming layer: enhance the debugger of Prolog+CG as well as its interpreter and its GUI, complete the implementation of all the features of Synergy, etc.

d)   Development of inference and learning strategies, that will be considered as memory-based strategies to provide an *operational memory-based and multi-strategy learning programming paradigm,*

e)   Development of several applications in various areas: reasoning, expert systems, ontology-based applications, natural language processing, problem solving and planning, case-based systems, multi-strategy learning systems, multi-agents systems, intelligent tutoring systems, etc.

9   Conclusion

Amine platform can be used to develop different types of intelligent systems and multi-agents systems, thanks to its architecture; a hierarchy of four layers (ontology, algebraic, programming and agents layers) and to the "openness" of Amine to Java. This paper illustrates the use of Amine in the development of dynamic programming applications, natural language processing, and in multi-agents systems applications. The companion paper illustrates the use of Amine in ontology-based applications.

We hope that Amine will federate works and efforts in CG community (and elsewhere) to develop a robust and mature platform for the development of intelligent systems (including semantic web) and multi-agents systems.

References

[1]   Graesser A. C., S. E. Gordon, L. E. Brainerd, QUEST: A Model of Question Answering, in Computers Math. Applic. Vol 23, N° 6-9, pp. 733-745, 1992
[2]   Kabbaj A., K. Bouzouba, K. ElHachimi and N. Ourdani, Ontologies in Amine platform: Structures and Processes in Amine, submitted to the 14th ICCS 2006.
[3]   Kabbaj A., Amine Architecture, submitted to the Conceptual Structures Tool Interoperability Workshop, hold in conjunction with the 14th ICCS 2006.
[4]   Kabbaj A. and M. Janta, From PROLOG++ to PROLOG+CG: an object-oriented logic programming, in Proc. Of the 8th ICCS'00, Darmstadt, Allemagne, Août, 2000.
[5]   Kabbaj A. and al., Uses, Improvements and Extensions of Prolog+CG: Case studies, in Proc. Of the 9th ICCS'01, San Francisco, August, 2001.
[6]   Kabbaj A., Synergy: a conceptual graph activation-based language, in Proc. Of the 7th ICCS'99, 1999.
[7]   Kabbaj A., Synergy as an Hybrid Object-Oriented Conceptual Graph Language, in Proc. Of the 7th ICCS'99, Springer-Verlag, 1999.
[8]   Nogier J-F., Génération automatique de langage et graphes conceptuels, Hermès, Paris, 1991.
[9]   Sowa J. F., Conceptual Structures: Information Processing in Man and Machine, 1984, Addison-Wesley.

Ontologies in Amine Platform: Structures and Processes

Adil Kabbaj[1], Karim Bouzouba[2], Khalid El Hachimi[1], and Nabil Ourdani[1]

[1] INSEA, B.P. 6217, Rabat, Morocco
akabbaj@insea.ac.ma
[2] EMI, Mohamed V University, Avenue Ibnsina B.P. 765 Rabat Morocco
karim.bouzoubaa@emi.ac.ma
http://sourceforge.net/projects/amine-platform

Abstract. The theoretical part of this paper presents and discusses the conflict between Ontology and Commonsense Knowledge (CK) in the context of Computer-Human Interaction/Communication. A resolution of this conflict is proposed, where Ontology is considered as a special case of CK (i.e. a formal CK). By this "rehabilitation" of Ontology to "CK Base", we extend "by the way" the meaning and scope of Ontology to include "CK-based Ontology". This new kind of ontology considers both definitional and schematic/situational knowledge. Moreover, we propose for this new kind of ontology an organizational scheme based on the use of Conceptual Structures (like Definition, Canon and Schematic Clusters).

The technical part presents the "implementation" of this theoretical framework in Amine platform [14] and especially in its *Ontology layer*. Amine's ontology meta-model is presented as well as various Ontology/CK-Base related processes.

1 Introduction

In Metaphysic and Philosophy, Ontology is not concerned by our "Commonsense Knowledge about the World", but by "the quest of what is necessary true about the Physical World" [20-22, 4]. Knowledge should be discarded in favour of a truthful and formal description of the "essence" of the World.

In the last decade, Artificial Intelligence (AI) and Information Sciences scientists started with "an analogical use" of the term Ontology; viewed as a Terminological/Definitional Knowledge Base that guarantees an efficient sharing of (Large) Knowledge Bases (KB). Ontology is therefore viewed as a "terminological" support to (large) KB [10-12, 21, 7, 3, and 24].

This "analogical use" of the term/concept Ontology has been criticized by philosophers [20-22, 10-12, 3, 7, 24] who propose instead the use of "Formal Ontology" and a methodology suited for ontology design [21, 10-12].

"Applied Ontology", based on Formal Ontology, is now emerging as a new multi-disciplinary field, with Semantic Web as the application domain "par excellence" [9, 3].

In general, Formal Ontology is proposed as a new approach to Computer-Human Interaction/Communication (CHI/C). However, "Full/Natural" (i.e. uncontrolled,

H. Schärfe, P. Hitzler, and P. Øhrstrøm (Eds.): ICCS 2006, LNAI 4068, pp. 300–313, 2006.

unrestricted, unlimited) CHI/C is also a central topic in Cognitive Science in general and in AI in particular. It is well established, from these sciences, that Commonsense Knowledge (CK) is a (if not the) vital aspect, in this "big/hard" problem.

There is therefore a conflict between two opposite approaches to the same domain (CHI/C): Formal Ontology, which is (in principle) knowledge-free, and Commonsense Knowledge based approach.

In the theoretical part of this paper (sections 2-3), the first author presents this problematical conflict and proposes the following solution: a/an (formal) ontology is considered as a special case of Commonsense Knowledge –CK-; formal ontology is a "formal CK": a rigorous, systematic, definitional and truthful CK.

By this "rehabilitation" of Ontology to "CK based perspective", we extend "by the way" the meaning and scope of Ontology to include "CK-based Ontology". This new kind of ontology considers both definitional and schematic/situational knowledge. Moreover, we propose for this new kind of ontology an organizational scheme based on the use of Conceptual Structures –CS- (like Definition, Canon and Schematic Clusters). More specifically, CSs are used to organize the meaning of (i.e. the CK related to) concept types and relations inside an Ontology.

The technical part of this paper (sections 4-5) presents the "implementation" of this theoretical framework in Amine platform [14] and especially in its Ontology layer. Amine's ontology meta-model is presented as well as various Ontology/CK-Base related processes.

2   Ontology vs Commonsense Knowledge (CK)

2.1   Ontology, Formal Ontology and "Applied Ontology"

As an inquiry in Metaphysic and Philosophy, Ontology is not concerned by our "CK about the World", but by "the quest of what is necessary true about the Physical World". It is concerned by "What are the types of objects that exist in the World?, What are the types of relations between these objects?" etc. [20-22, 10-12]. A type corresponds to the form/essence of a set/category of objects. These objects, with their types and relations, "exist" even if there is no human to observe them (Trees, Animals, Bus ... exist even if no human observes them) [20-22]. In this sense, these objects (and their forms/essences) are independent of human mind and they are different from concepts as considered in Cognitive Psychology, Artificial Intelligence (AI) and Cognitive Science [20-22]. Also, CK should not be considered in THIS context since the purpose of developing World Ontology is to specify the essence of the objects of the World, not to specify our CK about the World [20-22]. For instance, in THIS context, Ontology should contain the definition of the Type Bus but not the following schema: "typically, a Bus contains a set of about 50 passengers, it is the instrument of travel by those passengers at a speed less than or equal to 55 miles per hour, and it is the object of driving by some driver" [25 p. 129]. This schema represents a "generic/typical situation"; a CK about the type Bus. Several other schemata can be related to the type Bus (the same holds for any other type). A schema is not (necessarily) a domain-specific knowledge and not (necessarily) a linguistic construct (i.e. a "context of use" of a word); it may represent a generic situation or CK, as illustrated by the above schema.

However, for the Metaphysician/Philosopher, this CK about the type Bus, even if it is used by human in his everyday life/activity and in his interaction/communication, is not relevant for his Ontology and should not be included in it, because it is not about the essence (the essential form) of "what is the type" Bus.

Conclusion: if the purpose is the construction/development of a World Ontology that should contain only the "essence" of the physical objects, then yes, CK should be discarded and ignored.

To complete this brief introduction to "Metaphysical/Philosophical perspective of Ontology", let us consider the following points:

1. In Metaphysic/Philosophy, World Ontology is considered along two perspectives [4, 20, 21]: (i) "Realist" perspective (*revisionary metaphysics*) which considers only the Physical World; it is concerned only by types of the physical objects, not by abstract types. Informally, we can say that "realist" perspective is interested by the "World without Human" (or Human as a simple physical object, among the other physical objects)! (ii) "Phenomenological" perspective (*descriptive metaphysics*) which considers the "Commonsense World" including physical objects created (and actions performed) by human and abstract types ("Commonsense World" should not be confused with the "CK about the Commonsense World").

2. Metaphysicians/Philosophers have developed various "Universal Ontologies" for the World in totality. They have developed also "Domain-specific Ontologies" like "Ontology for Physics", "Ontology for Biology", "Ontology for Medicine", "Ontology for Math", etc. as well as "Ultra-domain-specific Ontologies" like "Ontology for Emotion", "Ontology for Human Action", etc.

3. Metaphysicians/Philosophers/Logicians have used/developed formal tools (formal logic, formal semantic, set theory, mereology, topology, etc.) for the formulation of their ontologies. They have developed "Formal Ontology" [20, 21, 10, and 11]. The philosopher Cocchiarella defines "Formal Ontology" as "the systematic, formal, axiomatic development of the logic of all *forms* and modes of being" [11]. Guarino insists that "formal" in this definition means "rigorous" and "related to the *forms* of being". Guarino concludes that Formal Ontology is concerned by the rigorous description of the forms (i.e. essence) of the physical Objects. He adds that in practice, Formal Ontology can be intended as the theory of *a priori distinctions*: (i) among the entities of the world (physical objects, events, regions, quantities of matter...); (ii) among the meta-level categories used to model the world (concepts, properties, qualities, states, roles, parts...).

    Guarino proposed "ontological principles" that "operationalize" the above "a priori distinctions" [10, 11]. He developed also a methodology for "Ontology design" based on these principles. He proposed these principles as a foundation for the "Ontology Level" of any Knowledge Representation system [12].

1. Metaphysicians/Philosophers/Logicians have used computer science (and some AI tools and techniques) for the implementation of their ontologies. The Basic Formal Ontology (BFO) framework, in which the philosopher B. Smith (and other philosophers) is involved, is an example of this "Applied Ontology".

2. With this new interest to Ontology, it became clear that any domain could/should have its ontology formally defined and implemented (using hopefully the methodology proposed by Guarino). Computer scientists, involved in Information Sciences (Data Bases, Information Systems, Semantic Web [3, 9], etc.), as well as AI scientists, involved in the development of Large/Sharable Knowledge Bases, have developed several ontologies, producing a new field that can be called "Applied Ontology" (this is the name of a new Journal edited by Guarino).

In conclusion, Formal Ontology should be used to develop ontologies for Computer-Computer Interaction/Communication (CCI/C) or when human is involved in a very limited, controlled and restricted way.

2.2 Commonsense Knowledge (CK) in Cognitive Psychology/Science

Aristotle defined Human as "Animal with Rationality". Maybe a better definition would be "Animal with Knowledge". Human intelligence and cognition is based on his Conceptual System, Knowledge, Memory and, of course, the related cognitive processes. This is the main conclusion from Cognitive Science and AI ([6, 8, 15-19, 23-25] ... it is nonsensical to cite in this paper the huge literature that confirms this conclusion). Natural language, Human behaviour, thinking and reasoning, Human interaction and communication ... can't be treated adequately without taking into account (and seriously) this centrality of Knowledge.

Following Kant, Peirce, Whitehead and the mainstream of AI and Cognitive Science, we may say that "Human IS Process/Knowledge", the World for Human IS his "World of Processes/Knowledge" which includes his CK about the "commonsense World". And his "direct perception" of this "commonsense World" may be a "faithful image" of the Physical World [20-22], but an image that is included (immersed) in his CK.

At a more concrete level, "Knowledge about the commonsense World" can be viewed as a subset (a fragment) of Human Knowledge (even if this view is very simplistic; Knowledge about Commonsense World can't be clearly separated from the other Knowledge "parts").

CK describes generic (and typical) knowledge about the "commonsense World". CK is composed of definitional Knowledge AND situational/typical knowledge. A CK Base may contain for instance, the definition of the concept Bus AND a "schematic cluster" for concept Bus. Schematic cluster is a collection of schemata like the schema for Bus presented before.

A schema represents (in general) a generic situation. It provides (in general) default knowledge and it is an important source of expectation and prediction. It may provide causal knowledge, explanation knowledge, axiomatic/constraint and rule based knowledge, procedural knowledge, strategic knowledge, emotional knowledge, etc.

Often, human is unable to provide a precise definition of a type. An ontologist will consider such a type as a primitive type and no related knowledge (especially CK) is specified in the Ontology. From the knowledge-perspective view, this position is too restrictive. CK, in terms of schematic clusters related to concept types, are a valuable source of knowledge, for primitive types as well as for defined types.

Schema Theory is a central component in Cognitive Science and AI [6, 8, 15-19, 23-25]. It is however outside the scope of this paper.

The important point here is that "What is the World?" for "ordinary" Human (not for the Metaphysician/Philosopher) IS the CK. In other words: what is refereed by Metaphysician/Philosopher as "Ontology" is refereed by "ordinary" Human as CK. Human, in his everyday life and activity, is not concerned by the "essence" of the objects or by "what is necessary true about the World or about a specific domain". He is concerned by his knowledge about the World (or knowledge about a specific domain). In his use/practice of natural language, in his activity, in his reasoning, in his problem solving activity, in his interaction and communication, etc., "ordinary" human uses his CK, not a Metaphysical/Philosophical ontology.

In Human-Human Interaction/Communication, the situation is very clear: the shared knowledge is not a "metaphysical/formal ontology" but CK.

The situation becomes problematic however, when the application domain concerns Computer-Human Interaction/Communication (CHI/C): Which approach to consider: "Formal Ontology" or CK?

2.3 CHI/C: Between Formal Ontology and Commonsense Knowledge

Human, in his communication and interaction with human or with computer, uses his CK. Therefore, the above question becomes: In the context of "real/natural" CHI/C, does the Computer (the intelligent system or agent) needs a "Formal Ontology Base" or a "CK Base"? Computer and Human should use/share similar Knowledge. It is therefore clear that the Computer should have also a CK Base (not a "Formal Ontology").

CK Base has a long history in AI. The importance of CK for (all) intelligent activities is strongly established. Different approaches have been (and are being) developed in AI to deal with (build and develop) CK. Among these approaches: (a) acquisition from natural language, (b) learning, (c) building a huge CK Base.

Cyc[1] is the best example of building a huge CK Base. It is build however with an "ad hoc" approach [11]. In Cyc, CK is represented by a partitioned hierarchy and a flat collection of axioms. Our proposition is to organize CK in terms of Conceptual Structures (especially Definition and Schematic Cluster) around concept types and relations. This is crucial for a more modular AND "cognitively-inspired" organization.

Another problematic treatment of CK is encountered in the very used KL-ONE-like family (and Description Logic) [2, 1]. For instance, our schema for Bus should not be inserted in TBox since this Knowledge Base concerns only intensional/definitional knowledge. And it should not be inserted in ABox since this Knowledge Base concerns only assertional/extensional knowledge; knowledge which is specific to a particular problem [1, p. 12] (or a particular "state of affairs"). As noted before, our schema for Bus (and many others) is not specific to a particular problem; it is a general knowledge. In fact, Brachman uses this same expression ("general knowledge") to refer to TBox [2]. However, the most part of "general knowledge" belongs to CK, which has no (clear) places in TBox/ABox scheme. Related critics to KL-ONE-like family (and to Description Logic) have been formulated by Doyle and Patil [5].

---

[1] http://www.cyc.com/cyc

Let us recapitulate: we started this section with the question "Which approach to consider for 'real/natural' CHI/C: 'Formal Ontology' or CK?" The answer is "the use of CK". But is it sufficient to share the same (or similar) content of CK (between Computer and Human) in order to guarantee a "natural-like" Interaction/ Communication? For instance, is it sufficient to use Cyc to guarantee a "natural-like" Interaction/Communication with Human? It is not enough because it is well known in computer science and in AI that the "Organization of Information/Knowledge" is as central and fundamental as the information/knowledge itself [15-19, 23]. If Human (cognition) organizes his CK in a way that is totally different from the organization used by Cyc (or by other similar axiomatic commonsense-based Ontology) then, the access, the retrieval, the use and the update of the knowledge base will not be the same, and the Interaction/Communication will not be "natural" (human will be frustrated by the behaviour of the computer). Communication/Interaction between humans is possible and efficient because humans share similar CK AND similar 'Knowledge Organization". "Knowledge Organization" is an inescapable and a fundamental phenomenon. From Cognitive Science and AI, we retain the basic conclusion that Conceptual Structures (like Definition, Canon and Schematic Clusters) are good candidates for CK Organization at the level of concept types.

2.4   A Solution to the Conflict Between Formal Ontology and CK in the Context of CHI/C

In definitive, there isn't a sharp distinction between "Formal Ontology" and "CK Base": unlike the belief of some philosophers (like B. Smith [20-22]), an Ontology is Knowledge "after all". It is a formalized CK, with a focus on Necessity-Truth and a discard of schematic/situational/typical knowledge. It is argued in previous sections that Knowledge is more important than Necessity-Truth in the context of "full/natural" CHI/C. Therefore, in this context, Formal Ontology can be considered as a special case of CK Base (CK where only definitional/axiomatic knowledge is considered). For CHI/C domains that require more CK, i.e. that require schematic/situational/typical knowledge, the domain ontology should correspond to a "full" CK base.

Amine's Ontology Layer allows the creation and use of ontologies that belong to this Ontology-CK base continuum.

3   Amine's Ontology Meta-model

Amine does not provide a specific ontology, but it provides "epistemological constructs" that determine the organization of "multi-lingua conceptual ontology". It is the responsibility of the user/designer to use "adequately" these constructs to model her/his ontology according to her/his domain semantics.

A "multi-lingua conceptual ontology", or "CK-based ontology" is a generic hierarchical knowledge base that "defines" the conceptual vocabulary (concept types, relations and individuals) of a domain. Several lexicons (for different languages) can be associated to a CK-based ontology.

The "definition" of a concept type is considered in Amine in a large sense: it corresponds not only to the (classical) type definition, but to the whole CK associated to the type. The CK for each type is organized in terms of different Conceptual Structures (CS). More specifically, Amine's ontology meta-model proposes three *kinds* of CS for three *kinds* of CK (CK that is related to each type):

a) *Definition:* this CS enables the expression of definitional knowledge.
b) *Canon:* this CS enables the expression of constraint and axiomatic knowledge. It is currently limited to the expression of canonical constraints [25].
c) *Schematic cluster (schemata):* schematic cluster is an open-end extendible and dynamic set of schemata. Concept, Relation and Schema are the three basic, fundamental and essential constituents of Knowledge. In most cases, we can't reduce the meaning of a concept to a closed and situation-free definition (assuming that such a definition is possible). A concept is like a "coin with two sides": a possible closed-fixed-static side (the definition) and an open-extendible-dynamic side (the schematic cluster). Schematic cluster constitutes a fundamental part in the specification of concept meaning. Schema is strongly related to other fundamental cognitive notions like "Patterns", "Conceptual Network", "Mental Models", "Situations", "Contexts" and "Experiences" (chunk of Knowledge) and to various implementations of schema in AI (frames, scripts, situation, schema, MOP, TOP, TAU, etc.). Schemata are high level discrete units of Knowledge (concepts represent the low level discrete units). Schemata are dynamic, interconnected, continually combined in higher schemata that are themselves combined, etc., forming a continuous, dynamic and extendible web of schemata.

At the structural level: an ontology in Amine is a generalization graph where nodes represent Conceptual Structures (CSs). Currently, there are four types of nodes (the first author is working on a fifth type of node: metaphor. See Amine Web Site for more detail). The fourth type of nodes; context node, is not presented in this paper:

a. *Type and RelationType nodes:* nodes that represent concept type and relation type respectively. These nodes contain the *definition* of the type (if provided) and/or the *canon* of the type (if provided).
b. *Individual node:* a node that represents an individual (an instance) of a concept type. This node contains the *description of the individual* (if provided).
c. *Schema/Situation node:* a node that contains the description of a schema/situation.

There are three types of links used in the composition of a "CK-based ontology":

a. *Specialization link (s):* A type (node) can be specialized by other types (related to them by the (s)pecialization/subtype link). Also, a schema/situation can be specialized by other schemas/situations.
b. *Instantiation link (i):* the (i)ndividual link relates an individual (an instance) to its type.
c. *Utilisation link (u):* In general, a schema/situation is not indexed under all types contained in it, but only to some of them (determined by the user or by a process that interact with the ontology). The schema is related to these types with (u)se links.

None of the above Conceptual Structures are mandatory: a "minimal" definition of a type is the specification of its super-types (the equivalent of primitive concepts in KL-ONE family). Then, if a definition is available, it can be specified. Also, schemata/situations can be added to the ontology as more CK is acquired and required. Also, the content of a canon can be augmented by more constraints and axiomatic rules as needed.

By constraining what kind of CK to consider, Amine's user can define and use different *kinds* of ontologies. For instance, the simple kind of ontologies is "taxonomic ontology"; ontology composed of primitive-types hierarchy. Another kind of Ontology is "terminological ontology" where the hierarchy is enriched by type definitions (and optionally axioms). Lastly, user can define and use "Commonsense Knowledge-based Ontology", where "full" CK is considered; with the use of schematic clusters.

Amine Ontology Layer can be used to develop all these kinds of ontologies.

Ontology modelling language: Amine ontology Layer is not committed to a specific modelling language; any language (CG, KIF, XML, RDF, Frame-Like, etc.) that is interfaced with Java may be used as a modelling language for the description of CS. Indeed, at the implementation level, a description of a CS is declared as "Object" (the root class in Java). However, since we adopt Conceptual Graph (CG) as the basic modelling language in the other layers of Amine (algebraic, programming and multi-agents layers), we use also CG as a modelling language in our ontology examples (see next sections and Amine Web Site for more information about this point).

To be able to create/edit/update ontology in Amine, the developer may either (i) use ontology layer's APIs, from a Java program, or (ii) use directly Amine's ontology GUI. Amine Web Site provides examples of ontology creation and update using related APIs from Java programs. The next section presents Amine's ontology GUI/Editor.

4 Amine's Ontology GUI

Amine's ontology GUI is a multi-view editor (a tree view, a drawing view and a browser view) that allows the creation, consultation, browsing, edition and update of ontology. This section briefly presents these three views. Figure 1 is a snapshot of an ontology edited using the tree view editor. The ontology is visualized in the main frame as a hierarchy. Tree nodes represent CSs nodes of the ontology (each type of node is represented with a different colour). As noted before, a type node can be specialized by other types (and related to them by the (s)pecialization link). For instance, PetrolEngine is a specialization of Engine and Gasoline. Also, a type can be related to several individuals (by (i)ndividual link) and it can be related to several schemata/situations (by (u)se link). For instance, situation SIT#7 is associated to the type Container. Also, a situation can be specialized by other situations. For instance, SIT#7 is specialized by SIT#9.

The user can select a specific language (from the languages list associated to the ontology) to get the ontology according to the selected language. If required by the

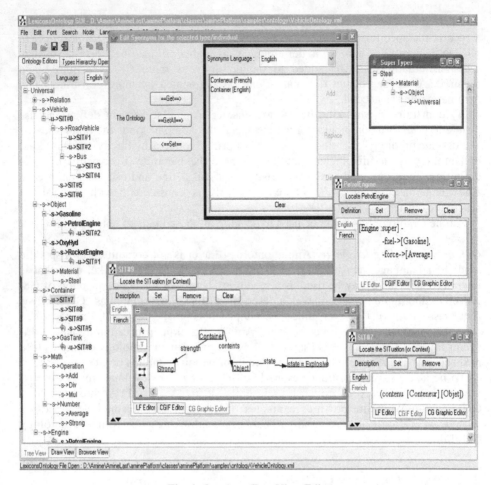

Fig. 1. Ontology Tree View Editor

user, the content of a node is visualized in an auxiliary frame. Some auxiliary frames
are shown in Figure 1: one frame shows the content of the type node "PetrolEngine"
and two other frames show contents of two situations. A Type node contains type's
definition (if provided) and canon (if provided). For instance, "PetrolEngine" has only
a definition. The description of a CS (see the three auxiliaries' frames in Figure 1) is a
CG that is displayed in multi-lingua multi-notations CG editor (Figure 1): CG can be
visualized according to the selected language (one tab for every language) and
according to the selected CG notation (one tab for every notation: Linear form, CGIF
and Graphical form).

Ontology's GUI provides also the possibility to display the super types of a given
type. For example, the frame "Super Types" shows the super types of the type
"Steel". It is also possible to edit synonyms in all available languages for a given type.
For instance, the Frame "Edit Synonyms for the type: Container" shows synonyms of
the type "Container" in both English and French.

The second possibility is to edit/consult ontology in "Draw View" Editor. Using graphical components (graphical nodes for CSs and arrows for links), this view allows to use the same functions as those available in tree view mode. Of course, the classical graphical operations are available (edition, selection, move, cut/copy/past, etc.) and any change (addition, suppression, modification of nodes and/or links) made on either modes affect directly the ontology and thus is visible in both modes. Also, user can zoom in/out the ontology, she/he can fix the vertical spacing between ontology nodes and she/he can locate the position of a specific type or an individual node in the ontology.

To produce an automatic drawing view of an ontology from its tree view, we use the basic Sugiyama algorithm[2] that enables automatic drawing of a hierarchical graph. An improvement of automatic drawing of ontology is underway.

For a large ontology, the user may have trouble reading the ontology from the drawing view. To accommodate this problem, a "Browser View Editor" is provided that allows the user to focus on a particular area of the ontology: a) the Browser Editor allows getting the neighbourhood of a particular node in the ontology, b) then, functions "expand" and "collapse" allow user to explore neighbourhood of any node in the Browser view. With expand/collapse functions, user can perform selective browsing of her/his ontology.

5 Ontology Related Processes

This section presents a short description of Amine's ontology related processes (see Amine Web Site for a detailed description). Unlike most of the "ontology dedicated tools/systems", Ontology processes in Amine are not limited to Classification (and related processes). Indeed, since Amine considers Ontology as a special case of CK Base, several CK Base processes are also offered (Classification, Information Retrieval, Dynamic Knowledge Integration, Elicitation and Elaboration). Here is a brief description of these processes:

- *Classification process* uses subsumption operation to classify a description in an ontology.
- *Information retrieval process (IR)* uses the classification process and searches to know if the specified description is contained or not in the ontology/memory. The aim of IR is not to answer by "yes" or "no", but rather to situate the specified description in the ontology; to determine its neighbourhood: which nodes are "fathers" (minimal generalizations of the specified description), which nodes are "children" (minimal specialization of the specified description) and which node is equal to the specified description. Here is an example that illustrates the use of IR from Prolog+CG program/console (see the companion paper for more detail on Prolog+CG [14]):

```
?- ask([Robot]<-agnt-[Wash]-thme->[Inanimate], [Wash]).

The description is : EQUAL to/than the known
[Wash #0] -
```

---

[2] plg.uwaterloo.ca/~itbowman/CS746G/Notes/Sugiyama1981_MVU/

```
        -Agnt->[Robot],
        -Thme->[Inanimate]

The description is : MORE_GENERAL to/than the known
[Wash #0] -
        -Thme->[Truck],
        -Agnt->[Robot]

The description is : MORE_GENERAL to/than the known
[Wash #0] -
        -Thme->[Car],
        -Agnt->[Robot]
 yes
?-
```

- *Dynamic integration process* concerns especially the integration of a description of type definition or of schema/situation. Dynamic integration process (or knowledge acquisition) performs automatic and incremental integration of new information in the ontology. The use of this integration process involves a similarity and generalization based construction and re-organization of the ontology. Contrary to classification, dynamic integration process is a "constructive learning task": it may create a new concept as a result of comparison between other concepts. Dynamic integration process is related to concept learning and machine learning in general, and to dynamic memory models in particular [15, 19].

- *Elicitation process* is an interactive process that helps a user to make his description D more precise and more explicit. For instance, assume that D is: [Vehicle]-engn->[Engine]. The user is asked if, by using Engine, he intends PetrolEngine or RocketEngine. The user may ask for the definition of the specified subtype (to decide if the proposed subtype is indeed the intended type). If the user selects one subtype (PetrolEngine for instance), the description is updated (Engine will be replaced by PetrolEngine): [Bus]-engn->[PetrolEngine]. The process will continue iteratively: it considers all the types used in the current description including the new types (PetrolEngine could be replaced by a more specific type). Beside this *type-directed elicitation process*, Amine provides a *situation-directed elicitation process:* while type-directed elicitation operates at the concept level (change of concept types), situation-directed elicitation operates at the structural level: the current description is integrated in the ontology, using the classification process, in order to situate the description in the ontology and to identify its neighbourhood; to determine situations that are more specific to the current description. The user is then asked if one of these situations fits his/her intended description. For instance, situation-directed elicitation may start from the result of the above type-directed elicitation ([Bus]-engn->[PetrolEngine]). The user gets then a request from Amine:

```
Does this situation tally with your description? :
       [Average]<-mass-[Bus]-engn->[PetrolEngine]
```

If the user's response is yes, the current description is replaced by the selected situation and the process continues with situations that are more specific than the selected situation. In this sense, the user is involved in an elicitation process.

- *Elaboration process* is introduced below.

See Amine Web Site for more detail about these processes. Classification, Information Retrieval and Dynamic Knowledge Integration will be described in more detail in a forthcoming paper.

Elaboration process uses the inheritance mechanism to provide more information about a description D. We differentiate between "deductive elaboration" which uses inheritance of type definitions and "inductive/plausible elaboration" which uses inheritance of schemata/situations ("plausible elaboration" because a situation provides only typical/plausible information about a type, contrary to a definition). A mixture of the two kinds of elaboration is possible.

Elaboration process is an interactive process: user provides a description D (of a proposition or of a situation). Then she/he asks the system to "elaborate/explicitate" D; "Can you provide more information about D?" The system "elaborates" by applying the inheritance mechanism to join "relevant information", from the current ontology, to D. For instance, if a concept type used in D has a definition, then this definition is joined to D. After that, user can ask the system to elaborate D further involving other joins (specialization), etc. Elaboration process is useful for users that are not familiar with (or expert in) a domain (and its conceptual vocabulary).

Let us consider the following simple example (Figure 2): the process starts with the user's description D (for instance, D = [Bus]-engn->[PetrolEngine]). The system locates, in the ontology, the definition of PetrolEngine and joins it to D (Figure 2.a). If the system is asked by the user to elaborate further, the process continues with the

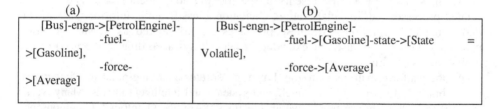

(a)	(b)
[Bus]-engn->[PetrolEngine]--fuel->[Gasoline], -force->[Average]	[Bus]-engn->[PetrolEngine]--fuel->[Gasoline]-state->[State =Volatile], -force->[Average]

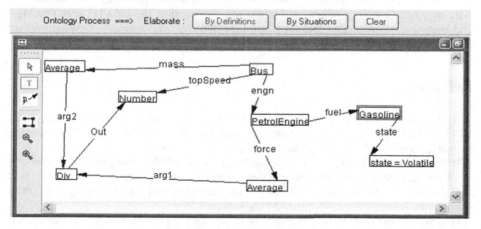

Fig. 2. Elaboration process

join of the Gasoline definition (Figure 2.b). The elaboration process will terminate after that because all superTypes of types present in the (new) formulation of D have no definition.

A similar treatment is performed by "plausible elaboration" which uses inheritance of schemata/situations. For instance, Figure 2.c shows the result of the specialization of the current description (Figure 2.b) by the join of an inherited schema/situation (schema that is associated to a superType of a type specified in the description).

Note that an inherited situation that can't be joined with the current description is ignored.

6 Conclusion, Current and Future Work

Formal Ontology is a formalized CK, with a focus on Necessity-Truth and a discard of schematic/situational/typical knowledge (which corresponds to a broader view and approach to CK). It is argued in this paper that Knowledge is more important than Necessity-Truth in the context of "full/natural" CHI/C. Therefore, in this context, Formal Ontology can be considered as a special case of CK Base (CK where only definitional/axiomatic knowledge is considered). For CHI/C domains that require more CK, i.e. that require schematic/situational/typical knowledge, the domain ontology should correspond to a "full" CK base.

Amine's Ontology Layer allows the creation and use of ontologies that belong to this Ontology-CK base continuum.

Amine's ontology layer presents a specific ontology meta-model and various related processes (edition, elaboration, elicitation, classification, information retrieval and dynamic integration). All these processes are considered as "basic" ontology processes; they can be used or extended in many ways according to the need of the application's domain.

Amine platform, with its Ontology Layer, constitutes a starting point for "integrative development" of ontology-based intelligent systems and intelligent agents. Many works are still to be done however, in several directions ("Applied Ontology", Commonsense KB, improving ontology editors and processes, investigating the links between ontology and metaphoric knowledge, between ontology and dynamic memory, development of intelligent agent with ontology/memory, etc.).

References

[1] Baader, F., et al. (eds.): *Description Logic Handbook*, Cambridge University Press, 2002
[2] Brachman R. J. and J. G. Schmolze, An Overview of the KL-ONE Knowledge Representation System, Cognitive Science 9, 171-216, 1985
[3] Brewter C. and K. O'Hara, Knowledge Representation with Ontologies: The Present and Future, p. 72-81, IEEE Intelligent Systems, January/February 2004
[4] Dölling J., Commonsense Ontology and Semantics of Natural Language, 1993 //www.uni-leipzig.de/~doelling/publikationen.html
[5] Doyle J. and R. S. Patil, Two theses of knowledge representation, Artificial Intelligence 48, pp. 261-297, 1991

[6] Eysenck M. W., *A Handbook of Cognitive Psychology*, Lawrence Erlbaum Associates, 1984

[7] Farrar S. and J. Bateman, General Ontology Baseline, 2004, //www.sfbtr8. uni-bremen.de/project.html

[8] Fauconnier G. and M. Turner, *The Way We Think*, Basic Books, 2002

[9] Fensel D. and al., OIL: An Ontology Infrastructure for the Semantic Web, IEEE Intelligent Systems, p. 38-45, March/April 2001

[10] Guarino N. Formal Ontology and Information Systems. 1998 www.ladseb.pd.cnr.it/ infor/Ontology/ontology.html

[11] Guarino N., Formal Ontology, Conceptual Analysis and Knowledge Representation. 1995 see URL [10]

[12] Guarino N. The Ontological Level, 1993. see URL [10]

[13] Kabbaj A., Development of Intelligent Systems and Multi-Agents Systems with Amine Platform, 2006 (in this volume).

[14] Kolodner J. L. and C. K. Riesbeck (eds.), *Experience, Memory, and Reasoning*, Lawrence Erlbaum Associates, 1986.

[15]   Lakoff G., *Women, Fire, and Dangerous Things: what categories reveal about the Mind*, University of Chicago Press, 1987.

[16] Luger G. F. and al., *Cognitive Science*, Academic Press, 1995

[17] Murphy, G. L., *The big book of concepts*, MIT Press, 2002

[18] Schank R. C., *Tell Me a Story: A new look at real and artificial memory*,MacMillan, 1991.

[19] Smith B., Formal Ontology, Common Sense and Cognitive Science, 1995 http://ontology.buffalo.edu/smith/articles/

[20] Smith B., Ontology, 2003. See URL in [22]

[21] Smith B., Beyond Concepts: Ontology as Reality Representation, 2004. See URL in [21]

[22] Sternberg R. J. (Ed.), *Thinking and Problem Solving*, Academic Press, 1994

[23] Sowa J. F., *Knowledge Representations*, Brooks Cole Publishing, 2000. See also www.jfsowa.com

[24] Sowa J. F., *Conceptual Structures: Information Processing in Man and Machine*, Addison-Wesley, 1984

Building a Pragmatic Methodology
for KR Tool Research and Development

Mary A. Keeler[1] and Heather D. Pfeiffer[2]

[1] Center for Advanced Research Technology in the Arts and Humanities
University of Washington, Seattle, Washington 98117, USA
mkeeler@u.washington.edu
[2] Department of Computer Science
New Mexico State University, Las Cruces, New Mexico 88003-8001, USA
hdp@cs.nmsu.edu

Abstract. Reviewing the evidence of CG-tool research and development in conference papers, we find little attention devoted to the issues and institution of scientific methodology, and only vague awareness of how that deficiency impairs progress. To focus attention and improve awareness, we briefly delineate the evolution of C.S. Peirce's theory of inquiry toward Eduard Hovy's general methodology for research, tracing from Peirce's early pragmatism to his "conditional idealism." We claim that methodological theory suggests a pragmatic method for KR research and tool advancement, in the form of an open-ended game somewhat like a child's game of building blocks, in which the forms of the "blocks" would be propositional rather than physical, with conditional propositions establishing the "dimensions," in place of the physical dimensions of blocks. The constraints would be logical and evidential (factual) rather than geometrical and gravitational (forceful). We challenge the entire Conceptual Structures community to help build a truly pragmatic methodology.

1 Introduction

Over the 14 years of ICCS meetings, some participants have wondered why work done in the Conceptual Graphs (CG) community exhibits so little scientific conduct, and what Peirce himself might recommend to improve its "pragmatic progress." Eduard Hovy informally expressed his disappointment after hearing CG papers presented, when he was invited to speak on methodology at the 2005 meeting [see 1]. He complained that the CG researchers presented insular work, making no comparisons with other techniques in similar applications outside the CG community or evaluation of tools within. His paper critiques the progress of several research communities, including the CG, and urges that their poor record of ontology-building success could most readily improve if at least two conditions are met: good methodologies for building and evaluating ontologies are developed, and those ontologies prove their utility in real applications [see 1: 91]. He identifies the problem for knowledge representation (KR), in general, after assessing its accomplishments: "KR work has been excellent in developing formalisms for representation, and for investigating the properties and requirements of various classes of deductive systems. But for practical applications such formalisms need *content;* the deductive systems need to work *on* something"

H. Schärfe, P. Hitzler, and P. Øhrstrøm (Eds.): ICCS 2006, LNAI 4068, pp. 314–330, 2006.

[1: 92]. His summary observation is that KR researchers are not yet able to build large, general-purpose semantic theories or semantic resources required for practical use on the scale of NLP: "such semantic data and theories as do exist are almost always limited to small-scale (or toy) applications ... [and] no accepted standard set of relations exists either" [1: 92].

Hovy stresses that the most troublesome aspect of work underway (such as in CGs) is "the near-complete absence of methodological discussion and emerging 'methodological theory' that would provide to the general enterprise of ontology building and relation creation the necessary rigor, systematicity, and eventually, methods for verification that would turn this work from an art to a science. ... Without at least some ideas about how to validate semantic resources, both the semantics builder and the eventual semantics user are in trouble" [1: 92-93].

In this paper, we review attempts to consider methodology within the CG community. We then investigate (with the aid of a host of Peirce scholars who have blazed productive paths through the maze of Peirce's writings) what methodological insights and guidance can be derived from Peirce's theory of inquiry. We conclude that both pragmatic methods and methodology might be instituted in the creation of an open-ended game.

2 Looking for Methodology in the CG Community

In the history of the CG community, the Peirce Workbench (also called "Peirce project") was an early attempt (1992) to establish some sort of context for at least comparing CG tools in some application domains. It was launched at the 7th workshop on Conceptual Graphs as an attempt to build cooperation among some 40 researchers, by coordinating their 27 tools into one test environment of 11 types [see 2]. Each participant group used different formatting and storing operations for graphs, and there was no well-defined Application Program Interface (API) for communication among all tools, which were built on different platforms in different programming languages. Nevertheless, the project marked the beginning of work on the Conceptual Graphs Interchange Format (CGIF), with the objective of making it possible to translate tools into exchangeable languages such as Java. Although the project itself did not continue, it demonstrated some need for developing collaborative methods among those in the community, but no apparent recognition of the need for methodological discussion of the sort Hovy describes.

The first obvious attempt came at the 1998 conference, when Tepfenhart outlined some of the fundamental technical ideas that form the basis of research efforts across the conceptual structures community. He concluded, "The variety of approaches, processing styles, and assumptions make it difficult for one author to apply the results of another. The same problem, framed in different language, is being solved many times by researchers who do not realize that it is the same problem" [see 3: 345].

2.1 Promising Efforts in 1999

Then came the Sisyphus I (SCG-1) initiative in 1999, devised by researchers at the Knowledge Science Institute, which challenged CG tool developers to solve a room-allocation task [see LNAI 1640, pp. 272-355]. Many hoped this testbed would "pour life into the CG community" by providing the opportunity: to compare various CG-tool approaches and determine the "CG state of the art," and to distinguish CG from other knowledge representation formalisms according to pertinent criteria [4: 213]. And yet, only the group testing the CoGITaNT tool even mentioned using an "experimental methodology": beginning with a theoretically formal model, then building software tools to implement the model, followed by a real-world application (presumably a prototype), and finally evaluating the system built—and reiterating this four-step process [see 5: 374]. Unfortunately, these researchers apparently attenuated that methodological plan: "This prototype is not yet a really usable tool. Further developments are needed in order to enable communication with the user in a friendly way, give the user the possibility to intervene at several stages of the solving process, and improve computational complexity of the solving process" [5: 375].

2.2 Assessment in 2000

No further obvious effort was attempted to achieve methodological procedures or standards until 2000, when Chein and Genest surmised that perhaps only a handful of operational CG applications remained. At conferences, they observed: "A program may be presented in a session [called] "applications" ... but it does not become, *ipso facto*, an AI or CGs application. ... External characteristics, those which consider the software from outside, like a black box, are essential for a software which claims to be an application" [6: 128]. In their paper ("CG Applications: Where Are We 7 Years After the First ICCS?"), they also discuss "the essential Internal characteristics" for an application: "[it] must have precise specifications, its conditions of use must be described, it must be reliable and robust, it must be validated, documentation must zbe available, its evolution must be anticipated, and so on" [6: 128]. And they identify four basic additional components specifically required for AI applications [see 6: 131-32].

In summary, Chein and Genest stressed: "In order to build applications, efficient tools are needed, and, what is rather distressing, we could deliver exactly the same discourse that Bob Levinson and Gerard Ellis had done when they launched the Peirce project in 1992!" They concluded that the CG community has difficulty "analyzing, synthesizing, and accumulating its knowledge," and further that "we have made numerous experiments, most of the time without drawing serious conclusions" [6: 138]. Nowhere in this paper is the question of methodology explicitly raised, but that is the question thoroughly begged throughout.

Guy Mineau's invited talk that year offered twenty-two recommendations for good practice in CG-based systems engineering, conceived in a three-layer architecture, as a "first set of guidelines" toward the common goal of a widely-supported, large-scale CG-based platform [see 7: 154]. But not until 2004, was a whole conference session devoted to frameworks for applications [see section "Conceptual Frameworks for Applications" in LNAI 3127, pp. 242-332], only two of which explicitly addressed the need for KR tool-development methodology [see 8 and 9], although it had been previously urged in 1999 and 2000 [see 10 and 11].

2.3 Recent Appraisement

Without solid methodological requirements, Hovy explains, "the reality is that most people build ontologies to support their knowledge representation needs in some practical application, and ... are more concerned about the computational effectiveness and correctness of their application than about the formal completeness, correctness, or consistency of the ontology *per se*." While theoretically, completeness, consistency, etc., would ensure that ontologies will avoid unwelcome surprises in system behavior, he observes that in development the strictures these requirements introduce are usually "so onerous that they make adopting the requirements tantamount to placing a severe limitation on the eventual scope of the ontology and hence of the whole practical enterprise" [1: 93]. Consequently, he explains, ontologies are built as relatively simple-term taxonomies with some inheritance inference, and do not enforce stricter logical requirements.

Hovy concludes that KR (among other techniques) lacks "a systematic and theoretically motivated methodology that guides the builder and facilitates consistency and accuracy, at all levels." In fact, he finds no evidence of an adequate theory on which to base such a methodology, and furthermore: "It *is not even clear how one would begin to approach the problem of designing theoretically motivated procedures from a suitably general point of view*" (our emphasis). He finds not one of the current builders of ontologies able "to provide a set of operationalizable tests that could be applied to every concept and relation to inform in which cases his or her choices were wrong" [1: 94].

Indicating how to approach such a methodology, Hovy considers what ontology builders actually *do*, in core operation(s) and in justifying their actions. He stresses that builders perform an *act of creation*, every time they generate a new (candidate) ontology item: "[they] decide *whether* to create a term, and if so, *how to place it* with regard to the other existing terms" [1: 94]. This portion of the act of defining the term begins the process of additional specification and definition. Hovy then traces this decision process as it plays out for the five "personality types" of builders he identifies, including "the philosophers," where he classifies those using the CG technique [1: 95]. He then describes a general methodology for building domain models [see 1: 97-98].

We eventually identify Hovy's approach to a general methodology as implicitly *pragmatic*, by first explicating (or clarifying) the motivation for such a methodology in terms of Peirce's theory of inquiry, then indicating how a particular sort of game can institute Peirce's pragmatic methodology systematically in KR tool development.

3 Pursuing a General Methodology in Peirce's Theory of Inquiry

When Peirce's *Collected Papers* was published in the early twentieth century, scholars were discouraged to find that Peirce never coherently stated his theory of inquiry. Thankfully, their persistent investigations provide substantial advantage in deciphering Peirce's writings, and the hope of addressing the concerns Hovy raises. The studies of early scholars such as Bronstein, Chisholm, Weiss, and Feibleman agree that Peirce considered his pragmatism to be a maxim (or general rule) for inquiry, and that his theory explained the course of inquiry in three elementary stages: abduction, deduction, and induction[1]. Abduction, the creative foundation of inquiry (which is presupposed by all induction), he came to identify as the essence of the pragmatic method [see 13: 166].

The puzzle that drove Peirce's pursuit of a theory of inquiry was the difficulty of explaining *how man could have managed to guess so many fruitful hypotheses in such a short historical time.* He argued that there must be an affinity between our minds and the course of the universe giving us insight into the general elements of nature. But he knew this alone could not enable the progress humans have made; it could not reduce the number of possible hypotheses to manageable numbers — and would not explain the countless useless and perverse ones tried in the history of science. His "better theory," says Weiss, is that our efforts to guess are *self-corrective*, "no matter where we start we arrive at results which we can modify as the circumstances demand, until they have the shape which satisfies the entire community" [13: 173]. Still, Peirce was not convinced that this was adequate to explain how science could progress so quickly.

We all know that in any development, some starting points are better and some results are more significant, some techniques are more effective than others and some means of production more efficient. How can we explain the selection of good enough guesses, throughout this process, that humans could progress so swiftly to civilization, and their science even more swiftly? According to Peirce, any such explanation is and always will be hypothetical (in fact, history is entirely hypothetical [see *CP* 2.511 fn.]). Any such hypothesis then must explain how, from the countless possibilities in any situation, do we formulate and select good hypotheses to test (or good hunches to try)? All of Peirce's philosophical work comes together in this challenge, which he eventually tried to explain as "the economics of research" [see *CP* 7.83-90].

Weiss gives us an account of the scope of deficiencies that Peirce suggested any *economic methodology* might address: "All men do not make signalized discoveries because they do not all adopt promising initial positions, are not familiar with new provocative items, do not have the requisite technical skill to express [themselves] to the satisfaction of technical arbiters whatever discoveries they might happen to make, and do not have the patience or time to add inference to inference so as to move forward steadily" [13: 178]. These are all possible errors or inefficiencies that any method at its core must help investigators routinely self-correct — by forming self-corrective *habits*.

Judged on this self-corrective requirement, all of Peirce's well-known, early formulations of a theory fail to be convincing, according to the scholarly literature. Clearly he does not maintain in his later writings the "guess" proposed in his early work, "The Fixation of Belief": that inquiry is a struggle to escape the irritation of doubt by attaining belief, true or false, and that the settlement of opinion is its sole aim [see, for best evidence, *CP* 6.485 (1908)]. The consequence, he realized, would be that *any* inquiry which settles belief is as good as any other, with no way of showing that some beliefs more likely conform to the facts than others.

---

[1] Note: Sowa and others have used these terms to discuss *types* of reasoning [12], but Peirce related these *ideal types* in his theory to account for the evolution of human knowledge.

3.1 Belief, Error, and Continuing Hypothetical Inference

Peirce's mature theory of inquiry abandons his early conclusions, on good grounds: If the settlement of opinion, by establishing belief that relieves the irritation of doubt, is the sole object of inquiry, and if belief is of the nature of habit, why should we not attain the desired end by taking anything we may fancy as answer to a question, and constantly reiterating it to ourselves, dwelling on all which conduce to that belief, and learning to turn with contempt and hatred from anything that might disturb it? [see *CP* 5.337]. Bronstein thinks Peirce revised his theory of inquiry because he realized that what we should demand of a method of attaining belief is that it should minimize *error*, not doubt. The urge to remove the irritation of doubt, is a biological fact which may cause us to adopt one or another method of fixing or re-establishing belief, but that fact tells us nothing about the validity of any method. It will not tell us *how any self-correction is determined* in the sustained inquiry that Peirce says is required to achieve true knowledge [see 13: 40-41].

In his later work Peirce proposes a more "fundamental hypothesis" to explain the successful method of science by inquiry that must lead to true knowledge, rather than merely conveniently settled opinion. He explicitly guesses: "there are Real things whose characters are entirely independent of our opinions about them; these Reals affect our senses according to regular laws" [*CP* 5.384]. We initiate inquiry when we confront a puzzling situation, then attempt to resolve the puzzle in the construction of a hypothesis that will enable us to anticipate the course of our experience (or not be surprised). A hypothesis is a *tentative* belief which may become firmer if it improves our anticipation. Theoretical as well as practical beliefs must involve expectation: when certain conditions are fulfilled, we obtain the consequence we expect. The general form of any hypothesis answers the pragmatic question: "What would be the consequences in my conduct (remembering that *thinking* is a form of conduct), if my conception of some observed phenomenon turned out to be true?" In other words, "what would I *do (or think)* differently if that conception explained what I observe of that phenomenon?" [see *CP* 5.534, 8.209] Any answer to this question is a hypothesis, whether implicit or explicitly expressed.

3.2 Beliefs Asserted as Conditional Propositions

Any belief can serve as an explicit hypothesis, if it is formulated in a conditional proposition whose antecedent specifies a course of action to be performed and whose consequent describes certain consequences to be expected. The hypothesis can be judged correct when we have perceived a correspondence between the *description* of these consequences and their *occurrence*. This, stresses Bronstein, is not a test of truth: "Rather, it is an attempt to explain what we mean when we say that a statement or belief is *true*" [13: 42]. In his 1903 "Lectures on Pragmatism," Peirce first clearly claims there is an "antecedent reality" and dismisses his earlier view that truth should be defined as the end of inquiry [see *CP* 5.211]. "What an inquiry presupposes," as Bronstein interprets Peirce, "is that there are phenomena subject to *law*, over which our thinking has no control," and which "we perceive directly in the course of inquiry," but which are "not introduced into the operation of knowing" [13: 43; our italics and punctuation].

Bronstein concludes that the principal weakness in Peirce's early theory was his failure to distinguish between something being true and our *knowing* that it is true. What results from inquiry is not that the belief becomes true, but that we gain some additional knowledge that we didn't have before the inquiry: that the belief is *justified* [13: 44]. Peirce clearly identified this confusion between truth and justification in his later work, but other pragmatists have perpetuated it. The confusion reaches to the depths of metaphysics, as Peirce discovered in reading Aristotle's works: "the first thing that strikes [the reader] is the author's [Aristotle's] unconsciousness of any distinction between grammar and metaphysics, between modes of signifying and modes of being" [*CP* 2.384; see 14]. The problem comes to focus in the question of identification and classification: "But identity ... is not a relation between two things, but between two representamens of the same thing" [*CP* 4.464].

Peirce then began to distinguish his theory of inquiry from the many contemporary theories known as "pragmatism" by calling it *conditional idealism*, expressed in a conditional proposition that aligned his theory with experimentalism: If a certain experiment were performed, then an experience of a given description would ensue. [*CP* 5.494]

3.3   What Will Be the Facts?

When we make a statement of fact, according to Peirce, we are asserting a *real possibility:* an imaginable event that would be realized under certain describable conditions, which must be specified in an explicit hypothesis. Bronstein reminds us that, on Peirce's view, "only individuals have actual existence (Firstness), ... [but] the world would be a mere chaos if these individuals were not related to each other and also subject to laws which govern their behavior" [13: 48]. The facts of science, then, are the *conditional* discoveries of those relations and behaviors (or habits). There is no such thing as an isolated fact, and the relations we discover and call facts are all conditionally dependent on how we perceive and conceive them. Any statement of fact (or assertion of something's reality) relies on the *assumed truth* of some *general conditional proposition* [see *CP* 5.457]. An assertion of scientific fact *means* that under certain conditions something *would be* true, whether the assertion explicitly (using the subjunctive conditional form) includes reference to those conditions or not. (This view underlies Peirce's theory of induction and distinguishes his philosophy, fundamentally, from some claimed followers, such as Popper, whose falsificationism is strictly deductive [see Haack (1995) *Evidence and Inquiry*, p. 131]. Putnam tells us that Popper even rejected the very idea of inductive logic [see Putman (2002) *The Collapse of the Fact/Value Dichotomy*, p. 141]).

Peirce's later theory distinguishes between a *proposition* and the *assertion* of that proposition, and insists that when you assert a proposition, you become responsible for it, as though you had placed a wager on it. "Now, if in the future it shall be ascertained that your assertion is correct, you win the wager. The rational meaning, then, or intellectual purport of your assertion lies in the future" [*CP* 5.543].

4 Genuine Inquiry and the Growth of Knowledge

Bronstein contends that Peirce's "signal contribution," in his revision of what we will continue to refer to as "pragmatism," was to realize "the importance of subjunctive conditionals for science and their irreducibility to material implications" [see 13: 49; *CP* 4.580]. Along with Bronstein, Chisholm concludes that Peirce's account of inquiry still relies on the account of belief as habit, but adds what might be called "outlook," expectations that anticipate imposed real constraints. A belief-habit is a law connecting behavior and possible experience, and a symbol (such as a statement or a sentence) is meaningful only if it can express a belief. But if it purports to concern matters that "transcend the limits of possible experience," then it cannot mean anything: any *meaningful* hypothesis must be *verifiable* (see expanded explanation, below, in section 4.3) [see 13: 93-4; *CP* 5.536; 5.597]. His theory of inquiry then becomes also a theory of meaning [see 13: 50].

Peirce's remark, "That belief gradually tends to fix itself under the influence of inquiry is, indeed, one of facts with which logic sets out," encouraged Chisholm to attempt a formulation of Peirce's basic theory of inquiry ("independently of the complications of his account of probability and induction") [*CP* 2.693; see also 3.161]. Chisholm compiled ten distinct tenets, which he entitled "The Conduct of Inquiry," from "the bewildering collection of Peirce's statements concerning doubt, surprise, habits of action, judgments of perception, common-sense, indubitability, fallibility, and truth" in the *Collected Papers* [see 13: 93]. We have modified Chisholm's list to serve in tracing the evolution of inquiry through Peirce's three stages (abduction, deduction, and induction) [see 13: 95-99]. We include abbreviated relevant discussion and comments from Peirce after each tenet; but we found it necessary to enrich the tenets we list under "Deduction," since Chisholm almost entirely neglects what Peirce describes must occur in that stage.

4.1 Abduction: Belief, Surprise, and Conjecture

"Abduction merely suggests that something may be" [CP 5.171].

1) The inquirer must have some beliefs to begin with; for inquiry does not begin until experience, by shock or surprise, breaks in upon some belief-habit. An empty mind cannot be surprised [see *CP* 5.512].

2) The inquirer should be guided by those personal beliefs which have survived the shock, many of which are indubitable. There is no genuine "complete doubt," as Cartesian theory presumes [see *CP* 5.416, 5.265].

3) As the inquirer ponders the surprising phenomena in relation to beliefs already held, conjectures or hypotheses instinctively suggest themselves. Each conjecture or hypothesis furnishes a possible explanation: "a syllogism exhibiting the surprising fact as necessarily consequent upon the circumstances of its occurrence together with the truth of the credible conjecture, as premises" [*CP* 6.469]; each comes in a flash of *insight* [see *CP* 5.173].

4) The relative plausibility of these hypotheses is considered in terms of what the inquirer happens to believe already. The hypothesis ought first, to be "entertained interrogatively," for there are many hypotheses "in regard to which knowledge already in our possession may, at once, quite justifiably either raise them to the rank of opinions, or even positive beliefs, or cause their immediate rejection" [*CP* 6.524]. "Accepting the conclusion that an explanation is needed when facts contrary to what we should expect emerge, it follows that the explanation must be such a proposition as would lead to the prediction of the observed facts, either as necessary consequences or at least as very probable under the circumstances" [*CP* 7.202].

4.2 Deduction: Conditional Prediction of Possible Consequences

"Deduction proves that something must be" (under what we can imagine would be the ideal conditions) [*CP* 5.171].

5) Most of the hypotheses which any inquirer is thus led to adopt will be false, but many of them will be true, at least under ideal conditions. Unfortunately, the faculty which originates conjectures supplies us with more false conjectures than with true ones. (The terms "true" and "false" are explained below, under "Induction."). Our insight is "strong enough not to be over-whelmingly more often wrong then right" [*CP* 5.173]. But in order to prevent future surprises and the ensuing misfortunes and irritable states, we need a way of sifting out the bad guesses. "Deduction ... relates exclusively to an ideal state of things. A hypothesis presents such an ideal state of things, and asserts that it is the icon, or analogue of an experience" [*CP* 7.205]. Experimentation requires preliminary "logical analysis," which renders the hypothesis as distinct as possible and *deduces* experimental predictions from the conjunction of the hypothesis and our other beliefs [see *CP* 6.472; 6.527]. The hypothesis has two relations to experience, one to the facts [induction] but the other to the hypothesis, and what effect that hypothesis, if embraced, must have in modifying our expectations in regard to future experience. Peirce clarifies in a footnote, "we infer by Deduction that if the hypothesis be true, any future phenomena of certain descriptions must present such and such characters" [*CP* 7.115 fn. and see 7.115].

6) Experience can eliminate a false hypothesis by virtually predicting the results of possible experiment. As soon as a hypothesis is adopted, deduction must trace out its necessary and probable experiential consequences [see *CP* 7.202]. Deduction draws virtual predictions. A virtual prediction is "an experiential consequence deduced from the hypothesis, and selected from among possible consequences independently of whether it is known, or believed, to be true, or not; so that at the time it is selected as a test of the hypothesis, we are either ignorant of whether it will support or refute the hypothesis, or, at least, do not select a test which we should not have selected if we had been so ignorant" [*CP* 2.96]. "The Deductions which we base upon the hypothesis produce conditional predictions concerning our future experience" [*CP* 7.115 fn.]. "'Conditional' is the right appellation, and not 'hypothetical,' if the rules of [my] Ethics of Philosophical Terminology are to be followed" [*CP* 2.316 fn.]. Experience, if given the opportunity, will perform a sifting or "pruning" function. In fact, this is the main service which experience renders — "to precipitate and filter off the false ideas, eliminating them" [*CP* 5.50].

4.3 Induction: Conditional Truth, Verifiability, Essential Fallibility

"Induction shows that something actually is operative" [*CP* 5.171].

7) If experience causes surprise, the new surprise will be accompanied by a series of events similar to those which accompanied the first surprise. "We now institute a course of quasi-experimentation in order to bring these predictions to the test, and thus to form our final estimate of the value of the hypothesis, and this whole proceeding I term Induction. ... By quasi-experimentation I mean the entire operation either of producing or of searching out a state of things to which the conditional predictions deduced from hypothesis shall be applicable and of noting how far the prediction is fulfilled" [*CP* 7.115 fn.]. If experience surprises, there will be new doubt leading to a new struggle to recover belief with a new hypothesis. If experience confirms, it will not inspire further inquiry, which is particularly unfortunate when the confirmed hypothesis happens to be false. Even when the confirmed hypothesis is true, complacency may result, which can impair the ultimate success of inquiry [see *CP* 5.168-69]. In so far as they greatly modify our former expectations of experience and in so far as we find them, nevertheless, to be fulfilled, we accord to the hypothesis a due weight in determining all our future conduct and thought. Even if the observed conformity of the facts to the requirements of the hypothesis has been fortuitous, we have only to persist in this same method of research and we shall gradually be brought around to the truth [see *CP* 7.115].

8) If given sufficient opportunity, experience would eliminate all false beliefs and leave us with none but true beliefs; this follows from Peirce's definitions of "true" and "false." In the long run "if inquiry were sufficiently persisted in" and experience given every opportunity to prune out the unstable beliefs, "the community of inquirers" would reach an agreement and all would share the same perfectly stable beliefs; some beliefs would continued to be re-affirmed and some would be denied [see *CP* 5.384, 5.311]. And the more we thus persist, particularly if we work together in *community*, the closer we come to this ideal. Peirce defines a *true* belief as a belief which would thus be an "ultimate conclusion" of inquiry. "The truth" is that to which belief would tend if it were to tend indefinitely toward absolute fixity [see *CP* 5.416]. These definitions become the foundation of his "conditional idealism," in which the concept of *truth* becomes the ideal (or hope) that motivates inquiry to persist indefinitely [see *CP* 5.494, 5.358 fn.]. It is much more important to frustrate the false hypothesis, than to confirm true ones, since we make more false ones than true ones. Hence, although it is not *solely* by surprise, it is *primarily* by surprise that experience teaches us [*CP* 5.51].

9) In order for experience to perform this function most efficiently, the inquirer should endeavor to submit all of his hypotheses and belief-habits to constant *experimental* test. We should make efforts to expose our hypotheses to the effects of experience, and reject those methods of fixing belief which attempt to shield our beliefs from those effects. The best effort is to conduct planned experimentation. An explanatory hypothesis must be *verifiable*; which is to say that it is "little more than a ligament of numberless possible predictions concerning future experience, so that if they fail, it fails" [*CP* 5.597]. We then submit these predictions to active test, by means of these actions: first, the act of choice by which the experimenter singles out certain identifiable objects to be operated upon; next, the external (or quasi-external) *act* by which the experimenter modifies those objects; and next, the subsequent *reaction* of the world upon the experimenter in a perception; and finally, the experimenter's recognition of the teaching of the experiment. "While the two chief parts of the event itself are the action and the reaction, yet the unity of essence of the experiment lies in its purpose and plan" [*CP* 5.424].

4.4 Self-correction as the Inquirer's Habit of Mind

10) This endeavor requires, in turn, that all inquirers have a "will to learn" and a constant dissatisfaction with their state of opinion at any time. Finally, "inquiry of every type, fully carried out, has the vital power of self-correction and of growth," but only if the inquirer has an intense desire to learn: a "dissatisfaction with one's present state of opinion," and a "sense that we do not know something" [*CP* 5.584; see 5.582; 5.583; also see 6.428]. Here can be the danger of the complacency when we get confirmation rather than surprise. An "indispensable ingredient" in any experiment is "sincere doubt in the experimenter's mind" concerning the truth of any hypothesis [see *CP* 5.424]. We should desire to know and be willing to work to find out [see *CP* 5.584]. While opinions that naturally suggest themselves are more often wrong than right, they can be corrected provided that we have a genuine dissatisfaction with them. Consequently we should not regard any as finally settled or certain, which is the basis of Fallibilism. "The first condition of learning is to know that we are ignorant. A man begins to inquire and to reason with himself as soon as he really questions anything and when he is convinced he reasons no more" [*CP* 7.322].

5 The Essential Open Mind

Peirce's persistent questioning took him well beyond his original pragmatism, as we have seen. And yet, his theory maintains the initial distinction that he says logic supposes in any investigation, between doubt and belief, which later became the distinction between a *question* and a *proposition* [see *CP* 7.313]. He explicitly relates these distinct elements in various expressions as his theory advances, beginning in his 1878 essays: "The object of reasoning is to find out, from the consideration of what we already know, something else which we do not know" [*CP* 5.365]. Meanwhile, his progress in developing relative logic gave him increasingly more insight into the strategic fine points of this relationship in the process of inquiry and, as Weiss concludes in one example, the critical balances to be maintained.

The most satisfactory inference [according to Peirce's theory] is one which enables us with the least risk to make the furthest advance beyond our present position. It must evidently move as far from the initial position as possible, but only to what is relevant. If we violate the first condition, we unnecessarily limit the range of application of the result; if we violate the second, we risk losing control over whatever material or truth we originally had. The combination of these two considerations enables us to preserve the past in the future, to extend the range, enrich the meaning while preserving the achievements of our inherited position. [13: 179]

By 1905, after he changed "pragmatism" to "pragmaticism," Peirce's maxim explicitly incorporated his *realist* insight, to balance the earlier *idealist* insight that true ideas are revealed by an "*a priori* evolutionism of the mind" (or because mind has evolved to be attuned to nature). According to pragmaticism, for the ultimate meaning of concepts to be true, they must represent their objects in the form of conditional resolutions consisting of conditional propositions and their antecedents, which he concludes amounts to saying "that possibility is sometimes of a real kind" [*CP* 5.453]. This realist *outlook* was the basis of his prolonged harangue against *nominalism*. From his early studies in the history of philosophy, he understood that nominalism had historical alliance with idealism and, by 1893, he had formulated his realist criticism of the nominalists' subtle problem: "they merely restate the fact to be explained under another aspect; or, if they add anything to it, add only something from which no definite consequences can be deduced. A scientific explanation ought to consist in the assertion of some positive matter of fact, other than the fact to be explained, but from which this fact necessarily follows; and if the explanation be hypothetical, the proof of it lies in the experiential verification of predictions deduced from it as necessary consequences" [see *CP* 8.30; 6.273]. Accepting a nominalist explanation as *sufficient* "would be to block the road of inquiry" [*CP* 6.273].

By 1906, he realized: "According to the nominalistic view, the only value which an idea has is to represent the facts, and therefore the only respect in which a system of ideas [or a generalization] has more value than the sum of the values of the ideas of which it is composed is that it is compendious" [*CP* 4.1]. But then, while insisting that hypothetical generalizations should be "submitted to the minutest criticism before being employed as premises," he declares, "It appears therefore that in scientific method the nominalists are entirely right. Everybody ought to be a nominalist at first, and to continue in that opinion until he is driven out of it by the *force majeure* of irreconcilable facts. Still he ought to be all the time on the lookout for these facts, considering how many other powerful minds have found themselves compelled to come over to realism" [*CP* 4.1]. From all this, we conclude that though a nominalist explanation is not *sufficient*, it could be at least *necessary*?

5.1 Nominalism as a Game of Inquiry

Peirce formulated a question to distinguish realists and nominalists by their opposite answers: "Do names of natural classes (such as man and horse) correspond with anything which all men, or all horses, really have in common, independent of our thought, or are these classes constituted simply by a likeness in the way in which our minds are affected by individual objects which have in themselves no resemblance or relationship whatsoever" [*CP* 8.13]. Pursuing inquiry with the latter, nominalist, form of investigation would result in reducing knowledge to a sort of "map" for the "unknowable territory" of existential being, and then accepting *that* representation as all we *can* know. Inquiry then becomes a sort of "game," in which we "take the map to *be* the territory." The advantage of "map-making" is that we can construct coherent accounts or stories to satisfy some conceived purposes, and submit those representations to "the minutest criticism" with respect to those purposes before admitting them as serious hypotheses to be tested [see 15].

We propose that KR researchers, instead of being unconsciously nominalistic in their conduct [see 11], should adopt nominalism consciously, as a formal method of inquiry by which to continue building and submitting to minutest criticism "their maps," (or what Chein and Genest have referred to as "essential internal characteristics" of software), while remaining "on the outlook for the facts," just as Peirce recommends (which are external to what Chein and Genest call that software "black box"). The strategic refinement made possible by Peirce's logic, especially his Existential Graphs (EG), suggests to us what sort of game could be constructed for instituting that method [see details in 15]. As Peirce says, "relative logic shows that from any proposition whatever, without a second, an endless series of necessary consequences can be deduced; and it very frequently happens that a number of distinct lines of inference may be taken, none leading into another" [*CP* 3.641]. The object of the game would be to "prune, filter, and select" the worthy hypotheses to test (or the essential internal characteristics to validate), among all the possible ones players might "wager" in the game.

A game method would be particularly appropriate for creating test-worthy hypotheses collaboratively. First, in explicitly (but sportively) formalizing the process of inquiry; second, by encouraging participants to make strategic contributions responsive to the progress of collaboratively formulating verifiable hypotheses that can be reconciled into testable predictions. The game must be "open-ended," as Peirce's theory mandates, and as simple to play as possible, leaving the challenge of building the "fail-proof ligaments in the construction" to KR processing. The analogy of a children's game of building blocks, comes to mind. In the game of inquiry, the forms of the "building blocks" would be propositional rather than physical, with conditional propositions establishing the "dimensions," in place of the physical dimensions of blocks. Rather than geometrical and gravitational (forceful) constraints, they would be logical and evidential (factual). These, conditionally-related, building blocks would "behave" as complex systems adapting to an "environment," in which fallibility would serve as "gravity" does in physical systems [see 15].

Peirce's fallibilism (identified with all three stages of inquiry, unlike Popper's falsificationism) reminds us that nothing is known for certain, that we should conduct inquiries so that sooner or later experience will catch up with any unstable (that is, invalid or unreliable) belief-habits and eliminate them. And since possible predictions are "numberless," fallibilism entails that no hypothesis can be completely verifiable" [see *CP* 1.149, 1.404, 2.663]. If inquiry's purpose is to find belief-habits that are reliable enough to serve as stable strategies in its evolution, then within the game context for investigation, complexity theory could apply. The building blocks for evolving the stable strategies of complex adaptive systems, composed of interacting agents, are described in terms of rules. Holland explains: "a major part of the environment of any given adaptive agent consists of other adaptive agents, so that a portion of any agent's efforts at adaptation is spent adapting to other adaptive agents" [16: 10]. Agents adapt their behavior by changing their rules as experience accumulates, just as improving hypotheses must do. We are exploring this framework for the game, in future work.

If, as Peirce predicts, "belief gradually tends to fix itself under the influence of inquiry," then a game strategy in which players are "always on the outlook for the facts" should prove successful [*CP* 2.693]. The nominalist game of inquiry must therefore be played in context where the facts can be found.

5.2   Realism in the Context for Inquiry

Returning to Hovy's paper on methodology, we find that his analysis and examples of current methodological deficiencies pinpoint the need for both minute criticism of ideas and for expanded scope in responding to external realities. He describes an *ideal* methodological objective for ontology creation that would reconcile the identified concept-creation procedures, and even assign relative priorities to their various methods and justification criteria *a priori*. But Hovy would agree with Peirce's caution that although there are universal presuppositions operative in the thinking of scientific investigators which may be essentially valid, "the history of science illustrates many of the stumbling blocks created by *a priori* dicta" [*CP* 1.114]. We have argued that something like the "nominalist game" should be instituted for such *idealistic* abduction and deduction, to accomplish collaborative reconciliation, but also *as a game* to remind us that the reconciled hypotheses and their predictions must not "block the road of inquiry."

Hovy then describes a more *realistic* approach.   His general methodology: *continual graduated refinement* [CGR] for building domain models proceeds in seven steps [1: 97-98]. CGR builders would begin by selecting "the principal criteria" for concept-creation and justification methods, and specify their order — which he insists *must be determined by the specific task or enterprise to be served*. He suggests basic questions to be answered in this preliminary step, with refinements to follow in a two-step process. These include: "Is this domain model to be used in a computational system? Or is it a conceptual product for domain analysis and description? Who are the intended users/readers of the ontology? What is their purpose for it? What justification criteria will they find most understandable? Do these criteria match the purpose or task in mind?" [1: 98]

These could be formulated as *pragmatic* questions, but typically in computer system development they are answered in the manner of traditional realism, rather than investigated in the manner that Peirce's critical realism demands. The traditional realist would be concerned with logical self-consistency and a "cognitive reality" of answers to these questions, rather than experientially by scientific induction. The result would be a *logical construction* of what the builder's finite intellect can apprehend as answers, rather than an outcome of the process of scientific inquiry.

Peirce identified this as the "nominalistic tendency" that distinguished traditional realism from his thorough-going realism [*CP* 1.29 (1869)].

We have previously described a *testbed* context [see 15, 17, 18] in which the answers to such questions could be experimentally obtained, where tool versions would be proposed as hypothetical solutions and these *predictions* tested on real tasks to solve. As Peirce says: "the entire meaning of a hypothesis lies in its conditional experiential predictions," to the extent that its predictions are true, the hypothesis is true [*CP* 1.29]. Why not build a context in which tools can be evaluated according to his scientific methodology? The very task of representing and managing the knowledge in such a context would be a healthy challenge for KR tool development. We think it is the appropriate challenge for any technology based in Peirce's logical theory, especially in his Existential Graphs (EG).

Peirce's EG have a central function in his philosophy, as instruments for observing experiments in reasoning. Greaves explains that they were designed to fill that role: "they would make both the logical structure and the entailments of propositions directly observable, in the same way that ... molecular diagrams make both the atomic structure and possible combinations of organic compounds observable" [19: 172]. Computer technology has given us Geographical Information Systems (GIS), we now need comparable technology for visualizing Peirce's pragmatic conduct of inquiry, and the CG community needs to pursue such an *outlook*.

6 Conclusions

At the beginning of Peirce's work on pragmatism, he briefly outlined what he conceived as the evolution of ways to establish beliefs, which he later advanced in his famous essays of 1877-8.

> Men's opinions will act upon one another and the method of obstinacy will infallibly be succeeded by the method of persecution and this will yield in time to the method of public opinion and this produces no stable result. Investigation differs entirely from these methods in that the nature of the final conclusion to which it leads is in every case destined from the beginning, without reference to the initial state of belief. ... But this will not be true for any process which anybody may choose to call investigation, but only for investigation which is made in accordance with appropriate rules. Here, therefore, we find there is a distinction between good and bad investigation. This distinction is the subject of study in logic. [*CP* 7.318-20 (1873)]

In the end, his theory of inquiry retains the objective of explaining a method for finding stable belief, but not at the cost of settling for what we call a "nominalist game." And yet, he came to realize that what is most wonderful about the mind is the ability to create ideas for which there is yet no existing prototype and "by means of this utter fiction it should manage to predict the results of future experiments and by means of that power should during the nineteenth century have transformed the face of the globe?" [*CP* 7.686] Clearly, the open-ended creation of ideas is also essential.

In Feibleman's evaluation of the lessons to be learned from Peirce's philosophy for those who might carry on his work (but not be blinded by it), the final lesson is something that Peirce was first to see: "the possibility of constructing an *open* system." All systems conceived by philosophers before Peirce had been *ipso facto* closed, he explains: "there is no real reason why there must be a limit to the size of our hypotheses. ... to maintain a single proposition tentatively should be no easier than to maintain a consistent set" [13: 325, 334].

We think that tools created with technology that is theoretically based in Peirce's work should be developed as open-ended experiments, relying on Peirce's minutely-critical logical instruments and his realistic outlook, for a suitably pragmatic (self-corrective) methodology that can cope with the challenges of their necessarily collaborative development.

References

General Notes: For all *CP* references, *Collected Papers of Charles Sanders Peirce,* 8 vols., ed. Arthur W. Burks, Charles Hartshorne, and Paul Weiss (Cambridge: Harvard University Press, 1931-58).

1. Hovy, E. [2005]. "Methodology for the Reliable Construction of Ontological Knowledge." In F. Dau, M-L. Mugnier, and G. Stumme (Eds.): *Lecture Notes in Artificial Intelligence,* Vol 3596. Berlin: Springer, pp. 91-106.
2. Ellis, G and Levinson, R. [1992]. "The Birth of PEIRCE: A Conceptual Graph Workbench." In H.D. Pfeiffer and T.E. Nagle (Eds), *Proceedings of the 7th Workshop on Conceptual Graphs,* LNAI 754. Berlin: Springer, pp. 219-228.
3. Tepfenhart, W. M. [1998]. "Ontologies and Conceptual Structures." In M-L. Mugnier, and M. Chein (Eds.): *Lecture Notes in Artificial Intelligence,* Vol 1453. Berlin: Springer, pp. 334-48.
4. Damianova, S. and Tpitampva, K. [1999]. "Using Conceptual Graphs to Solve a Resource Allocation Task." In W. Tepfenhar and W. Cyre, (Eds.): *Lecture Notes in Artificial Intelligence,* Vol 1640. Berlin: Springer, pp. 297-314.
5. Baget, J.F., Genest, D., and Mugnier, M.L. [1999]. "A Pure Graph-Based Solution to the SCG-1 Initiative." In W. Tepfenhar and W. Cyre, (Eds.): *Lecture Notes in Artificial Intelligence,* Vol 1640. Berlin: Springer, pp. 355-76.
6. Chein, M. and Genest, D. [2000]. "CGs Applications: Where Are We 7 Years After the First ICCS? " In B. Ganter, and G. Mineau (Eds.): *Lecture Notes in Artificial Intelligence,* Vol.1876. Berlin: Springer, pp. 127-39.
7. Mineau, G. [2000]. "The Engineering of a CG-Based System: Fundamental Issues." In B. Ganter, and G. Mineau (Eds.): *Lecture Notes in Artificial Intelligence,* Vol.1876. Berlin: Springer, pp. 140-56.
8. Keeler, M. [2004]. "Using Brandom's Framework to Do Peirce's Normative Science." In K.E. Wolff, H.D. Pfeiffer, and H.S. Delugach (Eds.): *Lecture Notes in Artificial Intelligence,* Vol. 3127, Springer-Verlag, pp. 37-53.
9. de Moor, A [2004]. "Improving the Testbed Development Process in Collaboratories." In K.E. Wolff, H.D. Pfeiffer, and H.S. Delugach (Eds.): *Lecture Notes in Artificial Intelligence,* Vol. 3127, Springer-Verlag, pp. 261-73.
10. Kayser, D. [1998]. "Ontologically Yours." In M-L. Mugnier, and M. Chein (Eds.): *Lecture Notes in Artificial Intelligence,* Vol 1453. Berlin: Springer, pp. 35-47.
11. Keeler, M. [2000]. "Pragmatically Yours." In B. Ganter, and G. Mineau (Eds.): *Lecture Notes in Artificial Intelligence,* Vol.1876. Berlin: Springer, pp. 82-99.
12. Sowa, J. [1990]. "Crystallizing Theories Out of Knowledge Soup." In Z. W. Ras and M. Zemankova (Eds.): *Intelligent Systems: State of the Art and Future Directions.* Ellis Horwood Ltd.: London, pp. 456-487. http://www.jfsowa.com/pubs/challenge

13. Wiener, P. P. and Young, F. H. [1952]. *Studies in the Philosophy of Charles S. Peirce.* Cambridge, MA: Harvard University Press.
14. Keeler, Mary [2003]. "Hegel in a Strange Costume: Reconsidering Normative Science in Conceptual Structures Research." In A. de Moor, W. Lex, and B. Ganter (Eds.): *Lecture Notes in Artificial Intelligence*, Vol. 2746. Berlin: Springer, pp. 37-53.
15. Keeler, M. and Pfeiffer, H.D. [2005]. "Games of Inquiry for Collaborative Concept Structuring." In F. Dau, M-L. Mugnier, and G. Stumme (Eds.): *Lecture Notes in Artificial Intelligence*, Vol 3596. Berlin: Springer, pp. 396-410.
16. Holland, J. H. [1995]. Hidden Order: How Adaptation Builds Complexity. Basic Books.
17. de Moor, A., Keeler, M. and Richmond, G. [2002]. "Towards a Pragmatic Web." In U. Priss, D. Corbett, and G. Angelova (Eds.): *Lecture Notes in Artificial Intelligence*, Vol. 2393, Springer-Verlag, pp. 235-249.
18. Pfeiffer, H.D. [2004]. "An Exportable CGIF Module from the CP Environment: A Pragmatic Approach." In K.E. Wolff, H.D. Pfeiffer, and H.S. Delugach (Eds.): *Lecture Notes in Artificial Intelligence*, Vol. 3127, Springer-Verlag, pp. 319-332.
19. Greaves, Mark [2002]. *The Philosophical Status of Diagrams.* Stanford University: CSLI Publications.

Simple Conceptual Graphs with Atomic Negation and Difference

Michel Leclère and Marie-Laure Mugnier

LIRMM, CNRS - Université Montpellier 2,
161, rue Ada, F-34392 Montpellier cedex, France
{leclere, mugnier}@lirmm.fr

Abstract. This paper studies the introduction of atomic negation into simple conceptual graphs. Several semantics of negation are explored w.r.t. the deduction problem and the query answering problem. Sound and complete algorithm schemes based on projection (or coref-projection) are provided in all cases. The processing of equality/inequality is added to the framework.

1 Introduction

Simple conceptual graphs (SGs) form the keystone of conceptual graphs (CGs). They are equivalent to the positive conjunctive existential fragment of first-order-logic [BM02]. The issue tackled in this paper is their extension with a restricted form of negation, namely atomic negation (in logical terms, negation of form $\neg p$, where p is an atom). Atomic negation allows to express knowledge as "this kind of relation does not hold between these entities", "this entity does not have this property" or "this entity is not of this type". This issue is studied both from semantic and computational viewpoints.

The framework. The reader is assumed to be familiar with basic notions about SGs. For further details about definitions and results used in this paper please see [CM04]. SGs are defined w.r.t. a vocabulary, called a *support* and denoted by $\mathcal{S}$. A support includes partially ordered sets of concept types and relations T_C and T_R. In the first sections we consider SGs without explicit coreference links ; coreference will be introduced as the same time as difference (section 4). Note however that a SG may include several concept nodes with the same individual marker, which is a case of implicit coreference. A SG is denoted by $G = (C, R, E, l)$, where C and R are respectively the concept and relation nodes, E is the family of edges and l is a mapping labeling nodes and edges (edges incident to a relation node are labeled from 1 to the arity of the relation). $r(c_1...c_k)$ is a short notation for a relation node with type r and argument list $(c_1...c_k)$, where $c_1...c_k$ are (not necessarily distinct) concept nodes. Φ is the classical translation from SGs (and the support) into FOL and $\vdash$ denotes classical FOL deduction. Given two SGs Q and G, it is known that when G is in normal form, there is a *projection* from Q to G if and only if $\Phi(\mathcal{S}), \Phi(G) \vdash \Phi(Q)$. With natural conditions on coreference, a SG always possesses a unique normal form (which

H. Schärfe, P. Hitzler, and P. Øhrstrøm (Eds.): ICCS 2006, LNAI 4068, pp. 331–345, 2006.

we note $nf(G)$ for a SG G). However, in case the normal form does not exist or cannot be computed, *coref-projection* can be used instead of projection [CM04]. In the sequel we use projection as the basic notion (knowing that it can be replaced by coref-projection if necessary). We note $G \preceq Q$ (or $Q \succeq G$) if there is a projection from Q to G. The deduction problem is then defined as follows.

Definition 1 (SG Deduction Problem). *The SG deduction problem takes two SGs G and Q as input and asks whether $Q \succeq G$.*

Another important problem is *query answering*. This problem takes as input a knowledge base (KB) composed of SGs representing facts and a SG Q representing a query, and asks for all answers to Q in the KB. The query Q is seen as a "pattern" allowing to extract knowledge from the KB. Generic nodes in the query represent variables to instantiate with individual or generic nodes in the base. With this interpretation, each projection from Q to G defines an answer to Q. An answer can be seen as the projection itself, or it can be seen as the subgraph of G induced by this projection. We call it the *image graph* of Q by π.

Definition 2 (Image graph). *Let π a projection from Q to G. The image graph of Q by π, denoted by $Image(Q, \pi)$, is the subgraph of G induced by the images of the nodes in Q by π.*

Distinct projections from Q to G may produce the same image graph, thus defining answers as image graphs instead of projections induces a potential loss of information. One advantage however of this answer definition is that the set of answers can be seen as a SG. We thus have the property that the results returned by a query are in the same form as the original data. This property is mandatory to process complex queries, i.e. queries composed of simpler queries.

Definition 3 (SG query answering problem). *Let Q be a query and G be a KB. The query answering problem asks for the set of image graphs of Q by all projections to G.*

If we consider the query answering problem in its decision form ("is there an answer to Q in the KB?") we obtain the deduction problem ("is Q deducible from the KB?").

Results. Several understandings of negation are explored in this paper, which are all of interest in real world applications. Briefly, when a query asks "find the x and y such that *not* $r(x, y)$", "not" can be understood in several ways. It might mean "the knowledge $r(x, y)$ cannot be proven" or "the knowledge *not* $r(x, y)$ can be proven". The first view is consistent with the closed-world assumption, the second one with the open-world assumption. In turn, the notion of proof can have several meanings. We point out that, as soon as negation is introduced, the deduction problem is no longer equivalent with the query answering problem in its decision form. Indeed, there are cases where classical deduction can be proven but no answer can be exhibited. These situations exactly correspond to cases where the law of excluded middle ("either A is true or *not* A is true") is used in the proof. This observation shifts the attention to logics in which the law of excluded middle does not hold. We have chosen to consider one of these logics, intuitionistic logic [Fit69]. It is shown that intuitionistic deduction exactly captures the notion of an answer. Furthermore, we establish that projection is sound and complete with respect to intuitionistic deduction in the logical fragment corresponding to SGs with atomic negation. It follows that atomic negation can be introduced in SGs with no overhead cost for the query answering problem. We also give a projection-based algorithm scheme for deduction with classical interpretation of negation. Finally, the processing of inequality is added to this framework.

Related works. One may wonder why bother about atomic negation, as general CGs (obtained from SGs by adding boxes representing negation, lines representing equality, and diagrammatic derivation rules, following Peirce's existential graphs) include general negation [Sow84] [WL94] [Dau03]. The main point is that checking deduction becomes an undecidable problem. Another important point is that, notwithstanding the qualities general CGs might have for diagrammatic logical reasoning, they are not at the application end-user level (see f.i. [BBV97]). Indeed, most applications are based on SGs and extensions that keep their intuitive appeal such as nested graphs, rules and constraints (f.i. the SG-family in [BM02]). Note that these latter extensions do not provide negation.

Few works have considered SGs with atomic negation. Simonet exhibited examples showing that projection is not complete anymore and proposed an algorithm based on an adaptation of the resolution method (unpublished note, 1998; see also [Mug00]). In [Ker01] simpler examples were exhibited and it was shown that projection remains complete in a very particular case (briefly when positive and negative relations are separated into distinct connected components). [Kli05] gave examples of problems related to the introduction of negation on relations (including equality) in the framework of protoconcept graphs (which can be translated into the conceptual graphs considered in the present paper). Moreover, as far as we know the problem of atomic negation in relationship with query problems had never been explored.

Paper organization. Section 2 introduces atomic negation into SGs, leading to *polarized* SGs. In section 3, several meanings of negation are discussed and related with the projection notion. Algorithm schemes for solving all the deduction and query problems are provided. In section 4, the results are extended to the processing of inequality. Due to space limitations, proofs are not included in this paper. The reader is referred to [ML05].

2   Polarized SGs

In this paper, we define *atomic negation* on relations only, but as explained below the results can easily be translated to concept types. Besides positive relation nodes, we now have negative relation nodes. A positive node is labeled by (r) or

$(+r)$, and a negative one by $(-r)$, where r is a relation type. As in [Ker01], we call *polarized* SGs (**PGs**) such SGs. A negative relation node with label $(-r)$ and neighbors $(c_1...c_k)$ expresses that "there is no relation r between $c_1...c_k$" (or if $k = 1$, "c_1 does not possess the property r"); it is logically translated by Φ into the literal $\neg r(e_1...e_k)$, where e_i is the term assigned to c_i. Let us consider the very simple example of figure 1. G describes a situation where there is a pile of three cubes A, B and C; A is blue and C is not blue. Whether B is blue or not is not specified.

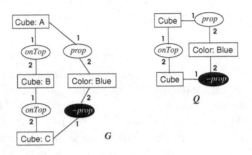

Fig. 1. Atomic negation

Projection on PGs is similar to projection on SGs with a simple extension of the order on relation node labels. The opposite type order is considered for negative labels: we set $-r_1 \leq -r_2$ if $r_2 \leq r_1$.

Definition 4 (Extended order on relation labels). *Given two relation labels l_1 and l_2, $l_1 \leq l_2$ if, either l_1 and l_2 are both positive labels, say $l_1 = (r_1)$ and $l_2 = (r_2)$, and $r_1 \leq r_2$, or l_1 and l_2 are both negative labels, say $l_1 = (-r_1)$ and $l_2 = (-r_2)$, and $r_1 \geq r_2$.*

Since negation is introduced a PG can be inconsistent.

Definition 5 (inconsistent PG). *A PG is said to be inconsistent if its normal form contains two relation nodes $+r(c_1...c_k)$ and $-s(c_1...c_k)$ with $r \leq s$. Otherwise it is said to be consistent.*

Property 1. For any PG G on a support $\mathcal{S}$, G is inconsistent iff $\Phi(\mathcal{S}) \cup \{\Phi(G)\}$ is (logically) inconsistent.

Negation on concept types. Negation in concept labels can be defined in a similar way. A concept node labeled by $-t$ (and a marker) is interpreted as "*there is an entity that is not of type t*", and not as "*there is not an entity of type t*", that is we keep an existential interpretation. Since the universal concept type is supposed to represent all entities, it cannot be negated. Let us point out that, if negation on concept types is interesting from a modeling viewpoint, it does not add expressiveness. Indeed concept types can be processed as unary relation types. More precisely, consider SGs on a support $\mathcal{S}$. Let $\mathcal{S}'$ be the support built by translating all concept types, except the universal type $\top$, into unary relation types (keeping the same partial order). The concept type set of S' is composed of the single type $\top$. Then, SGs on S can be transformed into SGs on S', while preserving projections and logical deduction: each concept node with label $(\sim t, m)$, where $\sim t$ can be positive or negative and $t \neq \top$, is translated into a concept node with label $(\top, m)$ and one neighboring relation node with label $(\sim t)$. A simple and uniform way of processing negation on concepts and relations thus involves applying the transformation sketched above, processing the obtained graphs with algorithms given in this paper and, if needed, applying the reverse transformation to present the results. Another solution is to adapt the algorithms, which is straightforward.

3   Different Kinds of Atomic Negation

In this section we study three ways of understanding negation in relation with the notions of *query* and *answer*.

3.1   Closed-World Assumption

A first way of understanding "not A" is "A is not present in the KB" (and more generally A cannot be obtained from the KB by inference mechanisms). Such a view is consistent with the "closed-world assumption" generally made in databases and the "negation by failure" in logic programming. Although only positive information needs to be represented in the KB, we will not forbid a PG representing facts to contain negative relations. A *completed* PG is obtained from a PG by expliciting in a negative way all missing information about relations. Then a query is not mapped to the original KB but rather to its completed version.

Definition 6 (completed PG). *The* completed PG *of a PG G, denoted by $completed(G)$, defined over a support S, is the only PG obtained from the normal form of G by adding all possible negative relations: for all relation type r of arity k in S, for all concept nodes $c_1...c_k$, if there is no relation $r'(c_1...c_k)$ in $nf(G)$ with $r' \leq r$, add the relation $-r(c_1...c_k)$.*

Definition 7 (CWA-PG deduction problem). *The PG deduction problem with closed-world assumption semantics takes two PGs Q and G as input and asks whether $Q \succeq completed(G)$.*

The mapping to classical logical deduction is obtained via the completed KB:

Property 2. Let Q and G be *PGs* defined on a support S, with G being consistent. $Q \succeq completed(G)$ if and only if $\Phi(S), \Phi(completed(G)) \vdash \Phi(Q)$.

Definition 8 (CWA-PG query answering problem). *Let Q be a (polarized) query and G be a (polarized) KB. The query answering problem asks for the image graphs of Q by all projections to $completed(G)$.*

Obviously the completed PG (or the part of it concerning the negated relations of the query) does not have to be computed in practice. Indeed, let Q^+ be the subgraph obtained from Q by considering concept nodes and solely positive relations. We have to select the projections from Q^+ to G that do not lead to "map" a negative relation in Q to a contradictory positive relation in G.

Definition 9. *A negative relation $-r(c_1...c_k)$ in a PG Q is satisfied by a projection π from Q^+ to a PG G if G does not contain a positive node $+s(\pi(c_1)...\pi(c_k))$ with $s \leq r$.*

Property 3. Let Q and G be two PGs defined over a support $\mathcal{S}$, with G being consistent. There is a bijection from the set of projections from Q^+ to G satisfying all negative relations in Q to the set of projections from Q to *completed(G)*.

Algorithms 1 and 2 take advantage of this property. In algorithm 2, $Ans(Q, \pi)$ is the PG obtained from Image(Q^+,π) by adding negative relations corresponding to negative relations in Q (i.e. for each $-r(c_1...c_k)$ in Q, one adds $-r(\pi(c_1)...\pi(c_k))$ to Image(Q^+,π)). In other words, $Ans(Q, \pi)$ is the image of Q by a projection (extending π) to *completed(G)* and not to G. Indeed, the closed-world assumption cannot be made on answer graphs, as the absence of a relation in an answer graph could come from the absence in the KB but also from its absence in the query. For instance, consider figure 2, which shows the only answer obtained by applying the query Q to the KB G in figure 1; the relation of label $(-prop)$ is added whereas it does not appear in G.

Algorithm 1. CWADeduction

Data: PGs Q and G
Result: true if Q can be deduced from G with CWA, false otherwise
begin

    Compute P the set of projections from Q^+ to G;
    forall $\pi \in P$ **do**
        Good $\leftarrow$ true;

1   **forall** *negative relation* $-r(c_1 ... c_k)$ *in* Q **do**
            if *there is* $s(\pi(c_1) ... \pi(c_k))$ *in* G *with* $s \leq r$ **then**
                Good $\leftarrow$ false ;   // π is not good
                exit *this for loop* ;

2   **if** *Good* **then return** *true*;
    return *false*;
end

3.2   Open-World Assumption

Let us now interpret the example in figure 1 with open-world assumption: nothing is known about the color of the cube B. Seen as a yes/no question, Q asks *whether* there is a blue cube on top of a non-blue cube. Seen as a query, Q asks for *exhibiting* objects having these properties. In both cases, what

---

Algorithm 2. CWAQueryAnswering

Data: PGs Q and G
Result: the set of answers to Q in G with closed-world assumption
begin

>Compute P the set of projections from Q^+ to G;
>$Answers \leftarrow \emptyset$;
>**forall** $\pi \in P$ **do**
>>**Good** $\leftarrow$ true;

1

>>**forall** *negative relation* $-r(c_1 \ldots c_k)$ *in* Q **do**
>>>**if** *there is* $s(\pi(c_1) \ldots \pi(c_k))$ *in* G *with* $s \le r$ **then**
>>>>**Good** $\leftarrow$ false ;   // π **is not good**
>>>>**exit** *this for loop* ;

2

>>**if** *Good* **then** $Answers \leftarrow Answers \cup \{Ans(Q, \pi)\}$;

>**return** $Answers$;

end

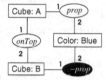

Fig. 2. Single result obtained by applying the CWA-PG query answering in figure 1

should be answered to Q? Let us first point out that spontaneously a non-logician (an end-user for instance) would say that the answer to the yes/no question is *no*. This intuition corresponds to the observation that there is no answer to the query. However, in classical FOL, the answer to the yes/no question is *yes*. Indeed the logical formulas assigned to Q and G by Φ are respectively of form $\Phi(Q) = \exists x \exists y \ (p(x, Blue) \wedge \neg p(y, Blue) \wedge r(x, y))$ and $\Phi(G) = p(A, Blue) \wedge r(A, B) \wedge r(B, C) \wedge \neg p(C, Blue)$ (where $p = prop, r = onTop$ and atoms assigned to concept nodes are ignored). $\Phi(Q)$ can be deduced from $\Phi(G)$ using the valid formula $p(B, Blue) \vee \neg p(B, Blue)$ (every model of $\Phi(G)$ satisfies either $p(B, blue)$ or $\neg p(B, blue)$; $\Phi(Q)$ is obtained by interpreting x and y as B and C if $p(B, blue)$ holds, and as A and B in the opposite case). Classical deduction thus ensures that there is a "solution" to Q but it is not able to *construct* it. Hence, there is no answer to Q as a query. This example leads to the following observations:

- The assertions "Q is (classically) deducible from G" and "the set of answers to Q in G is not empty" might disagree. In other words, deduction and the decision problem associated with query answering are different problems (which was not the case for SGs).
- The difference between the notions of deduction and the existence of an answer is due to the use of the law of excluded middle, which states here that "either B is blue or it is not blue".

Trying to formalize the preceding observations led us to distinguish two semantics for negative relations, namely according to *intuitionistic* logic and to classical
logic. In intuitionistic logic, the law of excluded middle does not hold. In fact,
this logic appears to br completely in line with the notion of answer, as detailed
later: Q is intuitionistically deducible from G if and only if the set of answers to
G is not empty. Note we do not claim that intuitionistic logic is the only logic
suitable to our framework. Another candidate would have been 3-value logic, in
which 3 truth vales are considered instead of 2: *true*, *false* and *undetermined*.

Intuitionistic negation. Intuitionistic logic is a well-established logic belonging to constructive mathematics [Fit69]. It is built upon the notion of *constructive proof*, which rejects the *reductio-ad-absurdum* reasoning. For instance, a
proof of $(A \vee B)$ is given by a proof of A or a proof of B; a proof that the falsity
of $(A \vee B)$ leads to a contradiction does not yield a proof of $(A \vee B)$ since it does
not determine which of A or B is true. Intuitionistic (natural) deduction rules
are those of classical logic except that the absurdity rule (from $\Gamma, \neg A \vDash \bot$ deduce
$\Gamma \vDash A$) does not hold. Clearly each theorem of intuitionistic logic is a theorem
of classical logic but not conversely. Some characteristic examples of classical
logic theorems not provable in intuitionistic logic are $(A \vee \neg A)$, $(\neg\neg A \rightarrow A)$ and
$((A \rightarrow B) \rightarrow (\neg A \vee B))$. We denote by $\Vdash$ intuitionistic deduction (recall that $\vdash$
is classical deduction). The relationship between classical and intuitionistic logic
in the logical fragment of PGs can be expressed as follows:

Property 4. For any predicate r with arity k, let $\mathcal{E}(r)$ be the formula $\forall x_1 \ldots x_k$
$(r(x_1, \ldots, x_k) \vee \neg r(x_1, \ldots, x_k))$. Given a support $\mathcal{S}$, let $\mathcal{E}_\mathcal{S}$ be the set of formulas
$\mathcal{E}(r)$ for all predicates r corresponding to relation types in $\mathcal{S}$. Then: $\Phi(\mathcal{S})$, $\mathcal{E}_\mathcal{S}$,
$\Phi(G) \Vdash \Phi(Q)$ if and only if $\Phi(\mathcal{S})$, $\Phi(G) \vdash \Phi(Q)$.

Let us come back to the example in figure 1. According to intuitionistic logic,
formula $p(B, Blue) \vee \neg p(B, Blue)$ can be considered as true only if it can be
shown that $p(B, Blue)$ is true, or that $\neg p(B, Blue)$ is true. Since none of these
two statements can be proven, Q cannot be deduced; hence the answer to Q as
a yes/no question is *no*, which corresponds to the fact that there is no answer
to Q as a query. Such an interpretation of a yes/no question can be seen as the
query answering problem in its decision form which asks for the existence of
an answer, that is the existence of a projection. This problem is equivalent to
intuitionistic deduction checking, as shown by the next theorem.

Property 5. A polarized SG G defined on a support $\mathcal{S}$ is inconsistent iff $\Phi(\mathcal{S}) \cup$
$\{\Phi(G)\}$ is intuitionistically inconsistent.

Theorem 1. *Let Q and G be two polarized SGs defined on a support $\mathcal{S}$, with
G being consistent. $Q \succeq nf(G)$ if and only if $\Phi(\mathcal{S}), \Phi(G) \Vdash \Phi(Q)$.*

This theorem yields the following property, which shows that intuitionistic negation is completely in line with the notion of answer to a query.

Property 6. Given two PGs Q and G, when Q is deducible from G with classical
negation but not with intuitionistic negation, there is no answer to Q in G.

We are now able to define the intuitionistic deduction problem as well as the query answering problem in terms of projection.

Definition 10 (OWA-PG intuitionistic deduction problem). *The PG intuitionistic deduction problem takes two PGs Q and G as input and asks whether $Q \succeq G$.*

Definition 11 (OWA-PG query answering problem). *The OWA-PG query answering problem takes two PGs Q and G as input and asks for the set of image graphs of Q by all projections to G.*

Classical negation. The classical semantic of negation leads to a case-based reasoning: if a relation is not asserted in a fact, either it is true or its negation is true. We thus have to consider all ways of completing the knowledge asserted by a PG. The next definition specifies the notion of the completion of a PG relative to a support S.

Definition 12 (Complete PG). *A complete PG on a support S is a consistent (normal) PG satisfying the following condition: for each relation type r of arity k in S, for each k-tuple of concept nodes $(c_1...c_k)$, where $c_1...c_k$ are not necessarily distinct nodes, there is a relation $+s(c_1...c_k)$ with $s \leq r$ or (exclusive) there is a relation $-s(c_1...c_k)$ with $s \geq r$. A PG is complete w.r.t. a subset of relation types $T \subseteq T_R$ if the completion considers only elements of T.*

Property 7. If a relation node is added to a complete PG, either this relation node is redundant (there is already a relation node with the same neighbor list and a label less or equal to it) or it makes the PG inconsistent.

A complete PG is obtained from G by repeatedly adding positive and negative relations as long as adding a relation brings new information and does not yield an inconsistency. The so-called completed PG defined for closed-world assumption (cf. section 3.1) is a particular case of a complete PG obtained from G by adding negative relations only. Since a PG G is a finite graph defined over a finite support, the number of different complete PGs that can be obtained from G is finite. We can now define deduction on PGs.

Definition 13 (OWA-PG (classical) deduction problem). *The PG (classical) deduction problem with open-world assumption semantics takes two PGs Q and G as input and asks whether each complete PG G^c obtained from G is such that $Q \succeq G^c$.*

This problem is known to be Π_p^2-complete (Π_p^2 is co-NP^{NP}) whereas deduction on SGs is NP-complete. The following property expresses that the PG deduction is sound and complete with respect to the classical deduction in FOL.

Theorem 2. *Let Q and G be two PGs defined on a support S. G is a consistent PG. Then Q can be (classically) deduced from $nf(G)$ if and only if $\Phi(S), \Phi(G) \vdash \Phi(Q)$.*

Algorithm 3 presents a brute-force algorithm scheme for OWA deduction. Let us recall that the other OWA problems (intuitionistic deduction and query answering) are directly based on projection. An immediate observation for generating the G^c is that we do not need to consider all relations types but only those appearing in Q. The algorithm generates all complete PGs relative to this set of types and for each of them checks whether Q can be projected to it. A complete graph to which Q cannot be projected can be seen as a counter-example to the assertion that Q is deducible from G. Although algorithm improvments are beyond the scope of this paper, let us outline an improved way of checking deduction. Consider the space of graphs $\mathcal{G}$ leading from G to its completions (and ordered by subgraph inclusion). The question "is there a projection from Q to each $G^c \in \mathcal{G}$?" can be reformulated as "is there a set of (incomparable) SGs $\{G_1 , ..., G_k\}$ in this space, which covers $\mathcal{G}$, i.e. each $G^c \in \mathcal{G}$ has one of the G_i as subgraph, and such that there is projection from Q to each G_i?". The brute-force algorithm takes $\mathcal{G}$ as the covering set. A more efficient method consists in building a covering set by incrementally completing G, one relation node after another.

Algorithm 3. OWAClassicalDeduction

Data: PGs Q and G, G being consistent
Result: true if Q can be (classically) deduced from G, false otherwise
begin
    Compute $\mathcal{G}$ the set of complete PG obtained from G w.r.t. relation types in Q;
    forall $G^c \in \mathcal{G}$ **do**
        if *there is no projection from Q to G^c* **then**
            return *false* ;   // G^c is a `counter-example`
    return *true*;
end

---

4   Equality and Difference

In this section we extend previous framework to equality and inequality, also called *coreference* and *difference*. Equality is classically represented in conceptual graphs by a special feature called a coreference link. A coreference link relates two concept nodes and indicates that these nodes represent the same entity. See figure 3: coreference links are represented by dashed lines; in addition there is an implicit coreference link between two nodes with the same individual marker (here c_1 and c_5). Formally, coreference can be defined as an equivalence relation, denoted by *coref*, added to a SG, such that nodes with the same individual marker necessarily belong to the same equivalence class. As most knowlegde representation formalisms, conceptual graphs make the "unique name assumption" (UNA). Consequently, nodes with different individual markers necessarily belong to different equivalence classes. In addition, coreferent concepts must have *compatible* types (see the discussion in [CM04]). Let us point out that equality does not bring more expressiveness to SGs (at least in the context of UNA). Indeed a SG with coreference, say G, is logically equivalent to the normal SG $nf(G)$ obtained by merging all nodes belonging to the same coreference class (see figure 3). We present it for clarity reasons, since difference is naturally seen as the negation of coreference.

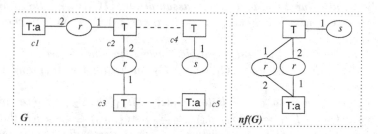

Fig. 3. Coreference: $coref = \{\{c_1,\ c_3,\ c_5\}, \{c_2,\ c_4\}\}$

Let us introduce inequality (or difference) as a special element of the SG syntax, called a *difference link*. A difference link between two nodes c_1 and c_2 expresses that c_1 and c_2 represent distinct entities. See Figure 4: difference links are represented by crossed lines. Due to the unique name assumption, there is an implicit difference link between nodes having distinct individual markers. Formally, difference is added to SGs as a symmetrical and antireflexive relation on concept nodes, called *dif*. In next definitions, we distinguish between the set of explicit coreference and difference links (E_{coref} and E_{dif}) and the relations (*coref* and *dif*) obtained from explicit *and* implicit links.

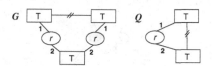

Fig. 4. $PG^{\neq}$: Q is classically deducible from G

Definition 14 ($PG^{\neq}$). *A (polarized) SG with equality and inequality (notation $PG^{\neq}$) is a 6-tuple $(C,\ R,\ E,\ l,\ E_{coref},\ E_{dif})$ where:*

- *(C, R, E, l) is a (polarized) SG;*
- *E_{coref} and E_{dif} are sets of specific edges between distinct nodes of C.*

Definition 15 (*coref* relation). *The relation coref on a $PG^{\neq}$ $G = (C,\ R,\ E,\ l,\ E_{coref},\ E_{dif})$ is the equivalence relation over C defined as the reflexive and transitive closure of the union of:*

- *the symmetrical relation induced by E_{coref} over C;*
- *implicit links due to multiple occurrences of individual markers: $\{\{c, c'\} \mid c, c' \in C,\ marker(c) \neq * \text{ and } marker(c) = marker(c')\}$.*

Before defining *dif*, we introduce a relation Dif on the *equivalence classes* of *coref*.

Definition 16 (Dif **relation**). *The relation Dif on a $PG^{\neq}$ $G = (C, R, E, l, E_{coref}, E_{dif})$ is the symmetrical relation over equivalence classes of coref defined as the union of:*

1. *the symmetrical relation induced by E_{dif}: $\{\{C_1, C_2\} \mid$ there are $c_1 \in C_1$, $c_2 \in C_2$ with $\{c_1, c_2\} \in E_{dif}\}$;*
2. *implicit links due to unique name assumption: $\{\{C_1, C_2\} \mid$ there are $c_1 \in C_1$, $c_2 \in C_2$ with $marker(c_1) \neq *$, $marker(c_2) \neq *$ and $marker(c_1) \neq marker(c_2)\}$;*
3. *implicit links due to incompatible types (in the sense of [CM04]): $\{\{C_1, C_2\} \mid$ there are $c_1 \in C_1$, $c_2 \in C_2$ such that $type(c_1)$ and $type(c_2)$ are incompatible $\}$;*
4. *implicit links due to contradictory relations: $\{(C_1, C_2) \mid$ there are $c_1 \in C_1$, $c_2 \in C_2$ and $r_1 = +t(d_1...d_q)$, $r_2 = -s(e_1...e_q) \in R$ such that $t \leq s$ and for all $k \in \{1..q\}$, one has $\{d_k, e_k\} \in coref$ except for exactly one value of k.*

Sets 2, 3 and 4 could be deduced from set 1 and knowledge in the support but we prefer adding them explicitly in Dif. Set 4 is illustrated by figure 5: the relation nodes have opposite labels and coreferent neighborhood except for c and c'; making c and c' coreferent would lead to an inconsistent graph.

Definition 17 (dif **relation**). *The relation dif on a $PG^{\neq}G = (C, R, E, l, E_{coref}, E_{dif})$ is the symmetrical relation over C defined as the cartesian product of all pairs of coref classes belonging to Dif (i.e. if $\{C_1, C_2\} \in Dif$ then for all $c_1 \in C_1$ and $c_2 \in C_2$, $\{c_1, c_2\} \in dif$).*

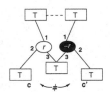

Fig. 5. Implicit dif link

For a $PG^{\neq}$ in normal form, one has $dif = Dif$. A $PG^{\neq}$ is consistent if it does not contain contradictory information, i.e. it is consistent *as a PG* and *coref* and *dif* are not contradictory ($coref \cap dif = \emptyset$). Note that dif and Dif are then antireflexive.

The FOL formula assigned to a $PG^{\neq}$ translates coreference by assigning the same term (variable or constant) to coreferent nodes, or equivalently, by adding an atom $e_1 = e_2$ for each pair of coreferent nodes with assigned terms e_1 and e_2. dif is translated by $\neq$. Every consistent $PG^{\neq}$ possesses a normal form which is obtained by merging all concept nodes of the same *coref* class (note that this is a generalization of the normal form of *SGs*, where *coref* is implicitly defined, two nodes being in the same *coref* class if and only if they have the same individual marker). The obtained $PG^{\neq}$ is logically equivalent to the original one.

Let us come back to (positive) SGs and consider the processing of coreference and/or difference. A projection π from Q to G has to respect coreference: for all nodes c_1 and c_2 of Q, if $\{c_1, c_2\} \in coref_Q$ then $\{\pi(c_1), \pi(c_2)\} \in coref_G$ (note $\pi(c_1)$ and $\pi(c_2)$ can be the same node). As for positive SGs without coreference, completeness is obtained only if the target SG is in normal form. Recall that projection can be replaced by coref-projection to ensure completeness without this condition. If difference is added to SGs, projection has to respect the dif relation as well: for all nodes c_1 and c_2 of Q, if $\{c_1, c_2\} \in dif_Q$ then $\{\pi(c_1), \pi(c_2)\} \in dif_G$. Concerning completeness, the same discussion as for negative relations about the use of the law of excluded middle can be brought as illustrated by figure 4. The formulas assigned to G and Q are respectively $\Phi(G) = \exists x \exists y \exists z \ (r(x, z) \wedge r(y, z) \wedge \neg(x = y))$ and $\Phi(Q) = \exists x \exists y \ (r(x, y) \wedge \neg(x = y))$. $\Phi(Q)$ can be (classically) deduced from $\Phi(G)$, using the law of excluded middle for $x = z$ and/or $y = z$, while there is no projection from Q to G. As for PGs, projection is sound with respect to classical deduction and intuitionistic deduction, and it is complete for intuitionistic deduction only.

The extension to $PG^{\neq}$s of algorithms designed for PGs is easy. In the CWA case, nodes of the KB G not known as being coreferent are considered as being connected by a dif link. Thus a projection π from Q^+ to G has to satisfy: for all nodes c_1 and c_2 in Q, if $\{c_1, c_2\} \in dif_Q$ then $\{\pi(c_1), \pi(c_2)\} \notin coref_G$ (or simply $\pi(c_1) \neq \pi(c_2)$ if G is in normal form). The algorithms 1 (deduction) and 2 (query answering) are extended by insertion of a new step checking this condition between steps 1 and 2. In the OWA case, no changes are to be done for query answering and intuitionistic deduction: the fact that projection preserves coreference and difference is sufficient. Concerning classical deduction, case-based reasoning has to be done as for negative relations. Algorithm 4 is a brute-force algorithm computing all dif-complete $PG^{\neq}$ (i.e. forall c, c' distinct concept nodes, either $\{c, c'\} \in dif$ or $\{c, c'\} \in coref$). Computing completions incrementally during deduction checking would of course be more efficient. Note that case 1 updates $coref$ but may also involve updating dif (due to potential contradictory relations as illustrated in figure 5), while case 2 updates dif only.

Algorithm 4. AllCompleteGraphsForDif

Data: a $PG^{\neq}$ G
Result: the set of all (normal) dif-complete $PG^{\neq}$ obtainable from G
begin
    $CompleteSet \leftarrow \emptyset$;
    CompleteRec(G);
    return $CompleteSet$;
end

---

5 Perspectives

In this paper we study separately three kinds of negation. In practice it may be useful to combine them. An interesting approach in this perspective is that of G.

Procedure CompleteRec(G)

Data: a PG$^{\neq}$ G

Result: this procedure computes recursively the completion of G; it has access
to all data of the main algorithm (alg. 4)

begin

  if $dif \cup coref$ *is complete* **then**

    // all pair of distinct concept nodes are in dif or $coref$

    $CompleteSet \leftarrow CompleteSet \cup \{G\}$;

  else

    Choose two (distinct) nodes c, c' in G such that $\{c,\ c'\} \notin dif \cup coref$;

    // case 1: make them coreferent

    let G_1 *be obtained from* G *by adding* $\{c, c'\}$ *to* E_{coref}

      CompleteRec(G_1);

    // case 2: make them ''different''

    let G_2 *be obtained from* G *by adding* $\{c, c'\}$ *to* E_{dif}

      CompleteRec(G_2);

end

Wagner [Wag03] who, after an analysis of different kinds of negation that can be found in existing systems and languages, as well as in natural language, proposes to distinguish between several kinds of predicates. Predicates are separated into *total* predicates and *partial* predicates that may have "truth value gaps" (that is it may be the case that neither P nor $\neg P$ is true). The law of excluded middle applies to the first ones but not the second ones. Total predicates can be *open* or *closed*, according to the underlying completeness assumption, namely OWA or CWA. A kind of negation corresponds to each kind of predicate. The proposed logic for distinguishing between these three kinds of predicates is a *partial logic* with three truth values (true, false and undefined). Although we do not consider the same logical framework, the above three kinds of predicates correspond to the three cases analyzed in the present paper. Similar to Wagner's proposal, we could combine the three ways of processing negation. If information about a relation type is assumed to be complete, closed-world negation is used. If it is not, the question is whether the law of excluded middle applies or not. If the answer is yes, the negation for this relation type is the classical negation, otherwise it is the intuitionistic negation. Since all mechanisms defined in this paper are based on projection (or coref-projection), combining them is not difficult.

References

BBV97. C. Bos, B. Botella, and P. Vanheeghe. Modeling and Simulating Human Behaviors with Conceptual Graphs. In *Proc. of ICCS'97*, volume 1257 of *LNAI*, pages 275–289. Springer, 1997.

BM02. J.-F. Baget and M.-L. Mugnier. The Complexity of Rules and Constraints. *JAIR*, 16:425–465, 2002.

CM04.   M. Chein and M.-L. Mugnier. Types and Coreference in Simple Conceptual Graphs. In *Proc. ICCS'04*, volume 3127 of *LNAI*, pages 303–318. Springer, 2004.

Dau03.   Frithjof Dau. *The Logic System of Concept Graphs with Negation And Its Relationship to Predicate Logic*, volume 2892 of *LNCS*. Springer, 2003.

Fit69.   M. C. Fitting. *Intuitionistic Logic, Model Theory and Forcing*. North Holland, Amsterdam, 1969.

Ker01.   G. Kerdiles.   *Saying it with Pictures: a Logical Landscape of Conceptual Graphs*.   PhD thesis, Univ. Montpellier II / Amsterdam, Nov. 2001. http://turing.wins.uva.nl/~kerdiles/.

Kli05.   J. Klinger. Local Negation in Concept Graphs. In *Proc. of ICCS'05*, volume 3596 of *LNAI*, pages 209–222. Springer, 2005.

ML05.   M.L. Mugnier and M. Leclère. On Querying Simple Conceptual Graphs with Negation. Research Report LIRMM 05-051, revised version to appear in Data and Knowledge Engineering, 2006, July 2005.

Mug00.   M.-L. Mugnier. Knowledge Representation and Reasoning based on Graph Homomorphism. In *Proc. ICCS'00*, volume 1867 of *LNAI*, pages 172–192. Springer, 2000.

Sow84.   J. F. Sowa. *Conceptual Structures: Information Processing in Mind and Machine*. Addison-Wesley, 1984.

Wag03.   G. Wagner. Web Rules Need Two Kinds of Negation. In *Proc. 1st International Workshop PPSW3*, volume 2901 of *LNCS*, pages –. Springer, 2003. http://tmitwww.tm.tue.nl/staff/gwagner.

WL94.   M. Wermelinger and J.G. Lopes. Basic Conceptual Structures Theory. In *Proc. ICCS'98*, volume 835 of *LNAI*, pages 144–159. Springer, 1994.

A Pattern-Based Approach to Conceptual Clustering in FOL

Francesca A. Lisi

Dipartimento di Informatica, Università degli Studi di Bari, Italy
lisi@di.uniba.it

Abstract. This paper presents a novel approach to Conceptual Clustering in First Order Logic (FOL) which is based on the assumption that candidate clusters can be obtained by looking for frequent association patterns in data. The resulting method extends therefore the levelwise search method for frequent pattern discovery. It is guided by a reference concept to be refined and returns a directed acyclic graph of conceptual clusters, possibly overlapping, that are subconcepts of the reference one. The FOL fragment chosen is $\mathcal{AL}$-log, a hybrid language that merges the description logic $\mathcal{ALC}$ and the clausal logic DATALOG. It allows the method to deal with both structural and relational data in a uniform manner and describe clusters determined by non-hierarchical relations between the reference concept and other concepts also occurring in the data. Preliminary results have been obtained on DATALOG data extracted from the on-line CIA World Fact Book and enriched with a $\mathcal{ALC}$ knowledge base.

1 Introduction

Conceptual clustering is a form of unsupervised learning that aims at determining not only the clusters but also their descriptions expressed in some representation formalism [17]. Related to Conceptual Clustering is *Concept Formation* [8]. The difference between the two is substantially in the methods: The former applies bottom-up batch algorithms whereas the latter prefers top-down incremental ones. Very few works on Conceptual Clustering and Concept Formation in the representation framework of (fragments of) First Order Logic (FOL) can be found in the literature [22, 2, 11, 10, 3, 24, 23, 7]. They differ in the approaches (distance-based, probabilistic, etc.) and/or in the representations (description logics, conceptual graphs, E/R models, etc.) adopted.

This paper presents a novel approach to Conceptual Clustering which follows a recent trend in cluster analysis: using *frequent association patterns* as candidate clusters [26, 25]. A frequent (association) pattern is an intensional description, expressed in a language $\mathcal{L}$, of a subset of a given data set $\mathbf{r}$ whose cardinality exceeds a user-defined threshold (*minimum support*). Each frequent pattern emphasizes a regularity in the data set, therefore it can be considered as the clue of a cluster. The method we propose for conceptual clustering extends the *levelwise search* method [16] for frequent pattern discovery with an additional post-processing step to turn frequent patterns into clusters.

H. Schärfe, P. Hitzler, and P. Øhrstrøm (Eds.): ICCS 2006, LNAI 4068, pp. 346–359, 2006.

The representation framework chosen is the one offered by the *hybrid knowledge representation and reasoning* (KR&R) system $\mathcal{AL}$-log [6] (see also Appendix for a brief introduction). It allows for the specification of both relational and structural data: the former is based on DATALOG [4] whereas the latter is based on the description logic $\mathcal{ALC}$ [20]. The integration of the two forms of representation is provided by the so-called constrained DATALOG clause, i.e. a DATALOG clause with variables eventually constrained by concepts expressed in $\mathcal{ALC}$. Frequent pattern discovery in $\mathcal{AL}$-log has been already investigated in [14]. In this paper we extend [14] towards conceptual clustering under the assumption that frequent patterns can be considered as the clue for clusters.

The paper is organized as follows. Section 2 defines our problem of Conceptual Clustering, first in general then within the KR&R framework of $\mathcal{AL}$-log. Section 3 illustrates our solution method whereas Section 4 presents the algorithm together with a theoretical and an empirical evaluation. Section 5 concludes with final remarks and directions of future work.

2   The Problem Statement

In conceptual clustering each cluster $\mathcal{C}$ defines a new concept therefore it consists of two parts: an *intension* $int(\mathcal{C})$ and an *extension* $ext(\mathcal{C})$. The former is an expression belonging to a logical language $\mathcal{L}$ whereas the latter is a set of objects that satisfy the former.

The conceptual clustering problem we face is a *semi-supervised* learning problem. Indeed we are interested in finding subconcepts of a known concept, called *reference concept*. More formally, given

- a reference concept C_{ref},
- a data set **r**,
- a language $\mathcal{L}$

our *conceptual clustering* problem is to find a directed acyclic graph (DAG) $\mathcal{G}$ of concepts $\mathcal{C}_i$ such that (i) $int(\mathcal{C}_i) \in \mathcal{L}$ and (ii) $ext(\mathcal{C}_i) \subset ext(C_{ref})$. Note that C_{ref} is among both the concepts defined in (the structural part of) **r** and the symbols of $\mathcal{L}$. Furthermore $ext(\mathcal{C}_i)$ relies on a notion of satisfiability of $int(\mathcal{C}_i)$ w.r.t. **r**.

Within the KR&R framework of $\mathcal{AL}$-log, the **data set r** is represented as a $\mathcal{AL}$-log knowledge base.

Example 1. As a running example, we consider an $\mathcal{AL}$-log knowledge base $\mathcal{B}_{\mathrm{CIA}}$ that enriches DATALOG facts[1] extracted from the on-line 1996 CIA World Fact Book[2] with $\mathcal{ALC}$ ontologies. The structural subsystem Σ of $\mathcal{B}_{\mathrm{CIA}}$ focus on the concepts Country, EthnicGroup, Language, and Religion. Axioms like

---

[1] http://www.dbis.informatik.uni-goettingen.de/Mondial/
  mondial-rel-facts.flp
[2] http://www.odci.gov/cia/publications/factbook/

```
AsianCountry ⊑ Country.
MiddleEasternEthnicGroup ⊑ EthnicGroup.
MiddleEastCountry ≡ AsianCountry ⊓ ∃Hosts.MiddleEasternEthnicGroup.
IndoEuropeanLanguage ⊑ Language.
IndoIranianLanguage ⊑ IndoEuropeanLanguage.
MonotheisticReligion ⊑ Religion.
ChristianReligion ⊑ MonotheisticReligion.
MuslimReligion ⊑ MonotheisticReligion.
```

define four taxonomies, one for each concept above. Note that Middle East countries (concept MiddleEastCountry) have been defined as Asian countries that host at least one Middle Eastern ethnic group. Two of the 15 countries classified as Middle Eastern are Armenia ('ARM') and Iran ('IR'). Assertions like

```
<'ARM','Armenian'>:Hosts.
<'IR','Arab'>:Hosts.
'Armenian':IndoEuropeanLanguage.
'Persian':IndoIranianLanguage.
'Armenian Orthodox':ChristianReligion.
'Shia':MuslimReligion.
'Sunni':MuslimReligion.
```

belong to the extensional part of Σ.

The relational subsystem Π of $\mathcal{B}_{CIA}$ consists of DATALOG facts like

```
language('ARM','Armenian',96).
language('IR','Persian',58).
religion('ARM','Armenian Orthodox',94).
religion('IR','Shia',89).
religion('IR','Sunni',10).
```

and constrained DATALOG clauses such as

```
speaks(CountryID, LanguageN)← language(CountryID,LanguageN,Perc)&
        CountryID:Country, LanguageN:Language
believes(CountryID, ReligionN)← religion(CountryID,ReligionN,Perc)&
        CountryID:Country, ReligionN:Religion
```

that can deduce new DATALOG facts when triggered on $\mathcal{B}_{CIA}$.

The **language** $\mathcal{L}$ contains expressions, called $\mathcal{O}$-queries, relating individuals of C_{ref} to individuals of other concepts (*task-relevant concepts*). These concepts also must occur in **r**. An $\mathcal{O}$-query is a constrained DATALOG clause of the form

$$Q = q(X) \leftarrow \alpha_1, \ldots, \alpha_m \& X : C_{ref}, \gamma_2, \ldots, \gamma_n,$$

which is compliant with the properties of *linkedness* and *connectedness* [18] and the bias of *Object Identity* (OI)[3] [21]. The $\mathcal{O}$-query

---

[3] The OI bias can be considered as an extension of the unique names assumption from the semantics of $\mathcal{ALC}$ to the syntax of $\mathcal{AL}$-log. It boils down to the use of substitutions whose bindings avoid the identification of terms.

$$Q_t = q(X) \leftarrow \& X : C_{ref}$$

is called *trivial* for $\mathcal{L}$ because it only contains the constraint for the *distinguished variable* X. Furthermore the language $\mathcal{L}$ is *multi-grained*, i.e. it contains expressions at multiple levels of description granularity. Indeed it is implicitly defined by a *declarative bias specification* which consists of a finite alphabet $\mathcal{A}$ of DATALOG predicate names and finite alphabets Γ^l (one for each level l of description granularity) of $\mathcal{ALC}$ concept names. Note that α_i's are taken from $\mathcal{A}$ and γ_j's are taken from Γ^l. We impose $\mathcal{L}$ to be finite by specifying some bounds, mainly $maxD$ for the maximum depth of search and $maxG$ for the maximum level of granularity.

Example 2. We want to describe Middle East countries (individuals of the reference concept) with respect to the religions believed and the languages spoken (individuals of the task-relevant concepts) at three levels of granularity ($maxG = 3$). To this aim we define $\mathcal{L}_{\text{CIA}}$ as the set of $\mathcal{O}$-queries with $C_{ref} =$ MiddleEastCountry that can be generated from the alphabet $\mathcal{A}=$ {believes/2, speaks/2} of DATALOG binary predicate names, and the alphabets

$\Gamma^1 =$ {Language, Religion}
$\Gamma^2 =$ {IndoEuropeanLanguage, ..., MonotheisticReligion, ...}
$\Gamma^3 =$ {IndoIranianLanguage, ..., MuslimReligion, ...}

of $\mathcal{ALC}$ concept names for $1 \leq l \leq 3$, up to $maxD = 5$. Examples of $\mathcal{O}$-queries in $\mathcal{L}_{\text{CIA}}$ are:

$Q_t=$ q(X) $\leftarrow$ & X:MiddleEastCountry
$Q_1=$ q(X) $\leftarrow$ speaks(X,Y) & X:MiddleEastCountry, Y:Language
$Q_2=$ q(X) $\leftarrow$ speaks(X,Y) & X:MiddleEastCountry, Y:IndoEuropeanLanguage
$Q_3=$ q(X) $\leftarrow$ believes(X,Y) & X:MiddleEastCountry, Y:MuslimReligion

where Q_t is the trivial $\mathcal{O}$-query for $\mathcal{L}_{\text{CIA}}$, $Q_1 \in \mathcal{L}^1_{\text{CIA}}$, $Q_2 \in \mathcal{L}^2_{\text{CIA}}$, and $Q_3 \in \mathcal{L}^3_{\text{CIA}}$.

In our context, a **cluster** has an $\mathcal{O}$-query Q as intension and the set $answerset(Q, \mathcal{B})$ of correct answers to Q w.r.t. $\mathcal{B}$ as extension. Note that $answerset(Q, \mathcal{B})$ contains the substitutions θ_i's for the distinguished variable of Q such that there exists a correct answer to $body(Q)\theta_i$ w.r.t. $\mathcal{B}$. In other words, the extension is the set of individuals of C_{ref} satisfying the intension.

Example 3. The cluster having Q_1 as intension has extension $answerset(Q_1, \mathcal{B}_{\text{CIA}})$ = {'ARM', 'IR', 'SA', 'UAE'}. In particular, the substitution $\theta =$ {X/'ARM'} is a correct answer to Q_1 w.r.t. $\mathcal{B}_{\text{CIA}}$ because there exists a correct answer $\sigma=$ {Y/'Armenian'} to $body(Q_1)\theta$ w.r.t. $\mathcal{B}_{\text{CIA}}$.

The **DAG** $\mathcal{G}$ is structured according to the *subset relation* between cluster extensions.

3 The Method

The conceptual clustering problem stated in Section 2 can be decomposed in two subproblems:

I. discovery of frequent patterns in data

II. generation of clusters from frequent patterns

In particular, the subproblem I is actually a variant of frequent pattern discovery which aims at obtaining descriptions of the data set **r** at different levels of granularity [9]. Here **r** typically encompasses a taxonomy $\mathcal{T}$. More precisely, the problem of *frequent pattern discovery at l levels of description granularity*, $1 \leq l \leq maxG$, is to find the set $\mathcal{F}$ of all the frequent patterns expressible in a multi-grained language $\mathcal{L} = \{\mathcal{L}^l\}_{1 \leq l \leq maxG}$ and evaluated against **r** w.r.t. a set $\{minsup^l\}_{1 \leq l \leq maxG}$ of minimum support thresholds by means of the evaluation function *supp*. In this case, $P \in \mathcal{L}^l$ with support s is frequent in **r** if (i) $s \geq minsup^l$ and (ii) all ancestors of P w.r.t. $\mathcal{T}$ are frequent in **r**.

The method proposed for solving one such decomposed problem extends the *levelwise search* method [16] for frequent pattern discovery with an additional post-processing step to solve the subproblem II. This method searches the space $(\mathcal{L}, \succeq)$ of patterns organized according to a generality order $\succeq$ in a breadth-first manner, starting from the most general pattern in $\mathcal{L}$ and alternating candidate generation and candidate evaluation phases. The underlying assumption is that $\succeq$ is a quasi-order monotonic w.r.t. *supp*. For $\mathcal{L}$ being a multi-grained language of $\mathcal{O}$-queries, we need to define first *supp*, then $\succeq$. The *support* of an $\mathcal{O}$-query $Q \in \mathcal{L}$ w.r.t. an $\mathcal{AL}$-log knowledge base $\mathcal{B}$ is defined as

$$supp(Q, \mathcal{B}) = \mid answerset(Q, \mathcal{B}) \mid / \mid answerset(Q_t, \mathcal{B}) \mid$$

and supplies the percentage of individuals of C_{ref} that satisfy Q.

Example 4. Since $\mid answerset(Q_1, \mathcal{B}_{\text{CIA}}) \mid = 3$ and $\mid answerset(Q_t, \mathcal{B}_{\text{CIA}}) \mid = 15 = \mid$ MiddleEastCountry $\mid$, then $supp(Q_1, \mathcal{B}_{\text{CIA}}) = 20\%$.

Patterns are ordered according to $\mathcal{B}$-*subsumption* [14] which can be tested by resorting to constrained SLD-resolution: Given two $\mathcal{O}$-queries $H_1, H_2 \in \mathcal{L}$, $\mathcal{B}$ an $\mathcal{AL}$-log knowledge base, and σ a Skolem substitution for H_2 w.r.t. $\{H_1\} \cup \mathcal{B}$, we say that H_1 $\mathcal{B}$-subsumes H_2, denoted as $H_1 \succeq_{\mathcal{B}} H_2$, iff there exists a substitution θ for H_1 such that (i) $head(H_1)\theta = head(H_2)$ and (ii) $\mathcal{B} \cup body(H_2)\sigma \vdash body(H_1)\theta\sigma$ where $body(H_1)\theta\sigma$ is ground. It has been proved that $\succeq_{\mathcal{B}}$ is a quasi-order that fulfills the condition of monotonicity w.r.t. *supp* [14].

Example 5. It can be checked that $Q_1 \succeq_{\mathcal{B}} Q_2$ by choosing $\sigma = \{$X/a, Y/b$\}$ as a Skolem substitution for Q_2 w.r.t. $\mathcal{B}_{\text{CIA}} \cup \{Q_1\}$ and $\theta = \emptyset$ as a substitution for Q_1. Similarly it can be proved that $Q_2 \not\succeq_{\mathcal{B}} Q_1$. Furthermore, it can be easily verified that Q_3 $\mathcal{B}$-subsumes the following $\mathcal{O}$-query in $\mathcal{L}^3_{\text{CIA}}$

$Q_4 = $ q(A) ← believes(A,B), believes(A,C) &
                 A:MiddleEastCountry, B:MuslimReligion

by choosing $\sigma = \{$A/a, B/b, C/c$\}$ as a Skolem substitution for Q_4 w.r.t. $\mathcal{B}_{\text{CIA}} \cup \{Q_3\}$ and $\theta = \{$X/A, Y/B$\}$ as a substitution for Q_3. Note that $Q_4 \not\succeq_{\mathcal{B}} Q_3$ under the OI bias. Indeed this bias does not admit the substitution $\{$A/X, B/Y, C/Y$\}$ for Q_4 which would make possible to verify conditions (i) and (ii) of the $\succeq_{\mathcal{B}}$ test.

$\mathcal{AL}$-**CoClust**$(\mathbf{r}, \mathcal{L}, \{minsup^l\}_{1 \le l \le maxG})$
1. $\mathcal{G} \leftarrow \emptyset$;
2. $Q_t \leftarrow \{\text{generateTrivialPattern}(\mathcal{L})\}$;
3. $l \leftarrow 1$;
4. **while** $l \le maxG$ **do**
5. $\mathcal{I}^l \leftarrow \emptyset$;
6. $k \leftarrow 1$;
7. $\mathcal{C}_k^l \leftarrow \{Q_t\}$;
8. $\mathcal{F}_k^l \leftarrow \emptyset$;
9.   **while** $k \le maxD$ **and** $\mathcal{C}_k^l \ne \emptyset$ **do**
10. $\mathcal{F}_k^l \leftarrow \text{evaluateCandidates}(\mathbf{r}, \mathcal{C}_k^l, \mathcal{I}^l, minsup^l)$;
11. $\mathcal{G} \leftarrow \text{updateGraph}(\mathcal{G}, \mathcal{F}_k^l)$;
12. $k \leftarrow k + 1$;
13. $\mathcal{C}_k^l \leftarrow \text{generateCandidates}(\mathcal{F}_{k-1}^l, \mathcal{L}^l, \mathcal{I}^l)$;
14.   **endwhile**
15. $l \leftarrow l + 1$;
16. **endwhile**
return $\mathcal{G}$

Fig. 1. The algorithm $\mathcal{AL}$-CoClust

Due to the features of the levelwise search, the resulting conceptual clustering method is *top-down* and *incremental*. Note that it is not hierarchical because it returns a DAG instead of a tree of concepts.

The choice criterion for cluster descriptions plays a key role in defining the subproblem II. Being the method top-down and incremental, we can adopt a simple yet not trivial heuristic: for each cluster, keep the *first* generated description that fulfills some *user-defined constraints*. This heuristic can cope successfully with cases in which other criterions fail. E.g., a criterion such as 'choose the most-general w.r.t. $\succeq_B$' is not reliable because $\succeq_B$ is not a total order.

4   The Algorithm

The algorithm $\mathcal{AL}$-CoClust reported in Figure 1 implements our extension of the levelwise search method. The search starts from the trivial $\mathcal{O}$-query in $\mathcal{L}$ (procedure generateTrivialPattern() at step 2.) and iterates through the cycle of:

- generation of candidate patterns (procedure generateCandidates() at step 13.);
- evaluation of candidate patterns (procedure evaluateCandidates() at step 10.);
- update of the graph of clusters (procedure updateGraph() at step 11.)

for a number of times up to $maxG$ (step 4.) and $maxD$ (step 9.). Let us describe an iteration of this cycle for a given level l, $1 \le l \le maxG$, of description granularity and a given level k, $1 \le k \le maxD$, of search depth.

The procedure generateCandidates() builds the set $\mathcal{C}_k^l$ of candidate k-patterns starting from the set $\mathcal{F}_{k-1}^l$ of frequent $(k-1)$-patterns and the language $\mathcal{L}^l$

by taking the set $\mathcal{I}^l$ of infrequent patterns at level l into account. It consists of a refinement step followed by a pruning step. The former derives new patterns from patterns previously found frequent by applying the rules of a specialization operator proposed to search $(\mathcal{L}, \succeq_\mathcal{B})$ and proved to be ideal [13]. The pruning step allows some infrequent patterns to be detected and discarded prior to evaluation by testing the following conditions arisen from the monotonicity of $\succeq_\mathcal{B}$ w.r.t. $supp$: A k-pattern Q in $\mathcal{L}^l$, $1 \leq l \leq maxG$, is infrequent if it is $\mathcal{B}$-subsumed w.r.t. an $\mathcal{AL}$-log knowledge base $\mathcal{B}$ by either (i) an infrequent $(k-1)$-pattern in $\mathcal{L}^l$ or (ii) an infrequent k-pattern in $\mathcal{L}^{l-1}$ [14].

The procedure evaluateCandidates() computes the set $\mathcal{F}_k^l$ of frequent k-patterns by starting from the set $\mathcal{C}_k^l$ of candidate k-patterns. It is responsible for filtering out those candidate patterns with insufficient support. Furthermore it exploits the $\succeq_\mathcal{B}$ relations between patterns to make the computation of answer sets more efficient [12].

For each frequent pattern $P \in \mathcal{F}_k^l$ that fulfills the user-defined constraints, the procedure updateGraph() checks whether a cluster $\mathcal{C}$ with extension $ext(\mathcal{C}) = answerset(P)$ already exists in $\mathcal{G}$. If one such cluster is not retrieved, a new node $\mathcal{C}$ with intension $int(\mathcal{C}) = P$ and extension $ext(\mathcal{C}) = answerset(P)$ is added to $\mathcal{G}$. Note that the insertion of a node can imply the reorganization of the DAG to keep it compliant with the subset relation on extents. If the node is already present in $\mathcal{G}$, no action is performed.

4.1   Theoretical Evaluation

The algorithm $\mathcal{AL}$-CoClust terminates bulding a graph $\mathcal{G}$ of subconcepts of C_{ref}. Note that $\mathcal{G}$ can be empty, meaning that no cluster has been detected in **r** with the current parameter settings.

Proposition 1. $\mathcal{AL}$-CoClust *is correct.*

Proof. The algorithm consists of two nested loops. First, we concentrate on the inner loop (steps 9.-14.). Given a fixed l, we prove that the following properties hold invariantly at the beginning of each iteration (step 9.):

R_1: $\mathcal{G}$ *contains nodes having frequent patterns $Q \in \mathcal{L}^l$ up to depth level $k-1$ as intents.*

R_2: $\mathcal{C}_k^l$ *is the superset of frequent k-patterns $Q \in \mathcal{L}^l$.*

R_3: *frequent patterns $Q \in \mathcal{L}^l$ from depth level $k+1$ onwards are specializations of queries in $\mathcal{F}_k^l$.*

R_4: *all patterns in $\mathcal{I}^l$ are infrequent.*

It is easy to verify that R_1, R_2, R_3, and R_4 hold at the beginning of the first iteration, i.e. after initialization (steps 1. and 5.-8.). Let us now assume that R_1, R_2, R_3, and R_4 hold up to the n-th iteration, $n < maxD$ and prove that these relations still hold at the beginning of the $(n+1)$-th iteration.

- *After step 10., R_2 and R_4 are still valid, since the patterns added to $\mathcal{I}^l$ are infrequent, and $\mathcal{F}_k^l$ is equal to the set of frequent k-patterns. Also R_3 still holds, since, due to the monotonicity of $\succeq_\mathcal{B}$ w.r.t. supp, none of the frequent patterns from depth level $k + 1$ onwards are specializations of the infrequent patterns moved to $\mathcal{I}^l$. Finally, R_1 is valid since $\mathcal{F}$ nor k have been modified.*
- *After step 11., $\mathcal{G}$ contains nodes with frequent patterns up to depth level k as intents, and R_2, R_3, and R_4 still hold.*
- *With the increment of k in step 12. R_1, R_2, R_3, and R_4 again hold.*
- *In step 13., the $(k + 1)$-patterns are generated, and since R_3 holds at that moment, $\mathcal{C}_{k+1}^l$ is the superset of $\mathcal{F}_{k+1}^l$.*

The loop terminates if either the depth bound maxD has been reached or $\mathcal{C}_k^l$ is empty. If the latter is the case, then it follows from R_2 and R_3 that language $\mathcal{L}^l$ contains no frequent patterns at depth level k or from level $k + 1$ onwards. Since $\mathcal{G}$ contains nodes with frequent patterns in $\mathcal{L}^l$ up to depth level $k - 1$ as intents, $\mathcal{G}$ at that moment contains all clusters with intents chosen among the frequent patterns in $\mathcal{L}^l$ as required. The depth bound assures termination of $\mathcal{AL}$-CoCLUST in case of infinite languages.

The outer loop (steps 4.-16.) also terminates because step 15. makes l converge to maxG.

The main source of complexity in $\mathcal{AL}$-CoCLUST are the $\mathcal{B}$-subsumption tests during candidate generation and evaluation phases. In [13, 12] it is shown how appropriate algorithmic and implementation choices can help mitigating the computational cost of these phases. More generally, incrementality and top-down direction of search are properties desirable when dealing with FOL formalisms.

4.2 Empirical Evaluation

In this section we report the results of two experiments with the following parameter settings: $maxD = 5$, $maxG = 3$, $minsup^1 = 20\%$, $minsup^2 = 13\%$, and $minsup^3 = 10\%$. Both returned 53 frequent patterns out of 99 candidate patterns. The two experiments differ as for the form of the frequent patterns to be considered as descriptions of candidate clusters.

In the first experiment we have required descriptions to have all the variables constrained by concepts of any granularity level. In this case $\mathcal{AL}$-CoCLUST returns the following 14 clusters:

$C_0 \in \mathcal{F}_1^1$
q(A) ← A:MiddleEastCountry
{ARM, BRN, IR, IRQ, IL, JOR, KWT, RL, OM, Q, SA, SYR, TR, UAE, YE}

$C_1 \in \mathcal{F}_3^1$
q(A) ← believes(A,B) & A:MiddleEastCountry, B:Religion
{ARM, BRN, IR, IRQ, IL, JOR, KWT, RL, OM, Q, SA, SYR, TR, UAE}

$C_2 \in \mathcal{F}_3^1$
q(A) ← speaks(A,B) & A:MiddleEastCountry, B:Language
{ARM, IR, SA, UAE}

$\mathcal{C}_3 \in \mathcal{F}_5^1$
q(A) ← believes(A,B), believes(A,C) &
        A:MiddleEastCountry, B:Religion, C:Religion
{BRN, IR, IRQ, IL, JOR, RL, SYR}

$\mathcal{C}_4 \in \mathcal{F}_5^1$
q(A) ← speaks(A,B), believes(A,C) &
        A:MiddleEastCountry, B:Language, C:Religion
{ARM, IR, SA}

$\mathcal{C}_5 \in \mathcal{F}_3^2$
q(A) ← speaks(A,B) & A:MiddleEastCountry, B:AfroAsiaticLanguage
{IR, SA, YE}

$\mathcal{C}_6 \in \mathcal{F}_3^2$
q(A) ← speaks(A,B) & A:MiddleEastCountry, B:IndoEuropeanLanguage
{ARM, IR}

$\mathcal{C}_7 \in \mathcal{F}_5^2$
q(A) ← speaks(A,B), believes(A,C) &
        A:MiddleEastCountry, B:AfroAsiaticLanguage, C:MonotheisticReligion
{IR, SA}

$\mathcal{C}_8 \in \mathcal{F}_3^3$
q(A) ← believes(A,'Druze') & A:MiddleEastCountry
{IL, SYR}

$\mathcal{C}_9 \in \mathcal{F}_3^3$
q(A) ← believes(A,B) & A:MiddleEastCountry, B:JewishReligion
{IR, IL, SYR}

$\mathcal{C}_{10} \in \mathcal{F}_3^3$
q(A) ← believes(A,B) & A:MiddleEastCountry, B:ChristianReligion
{ARM, IR, IRQ, IL, JOR, RL, SYR}

$\mathcal{C}_{11} \in \mathcal{F}_3^3$
q(A) ← believes(A,B) & A:MiddleEastCountry, B:MuslimReligion
{BRN, IR, IRQ, IL, JOR, KWT, RL, OM, Q, SA, SYR, TR, UAE}

$\mathcal{C}_{12} \in \mathcal{F}_5^3$
q(A) ← believes(A,B), believes(A,C) &
        A:MiddleEastCountry, B:ChristianReligion, C:MuslimReligion
{IR, IRQ, IL, JOR, RL, SYR}

$\mathcal{C}_{13} \in \mathcal{F}_5^3$
q(A) ← believes(A,B), believes(A,C) &
        A:MiddleEastCountry, B:MuslimReligion, C:MuslimReligion
{BRN, IR, SYR}

organized in the DAG $\mathcal{G}_{\text{CIA}}$. They are numbered according to the chronological order of insertion in $\mathcal{G}_{\text{CIA}}$ and annotated with information of the generation step. From a qualitative point of view, clusters $\mathcal{C}_5{}^4$ and $\mathcal{C}_{11}$ well characterize Middle East countries. Armenia, as opposite to Iran, does not fall in these clusters. It rather belongs to the weaker characterizations $\mathcal{C}_6$ and $\mathcal{C}_{10}$. This proves that $\mathcal{AL}$-CoCLUST performs a 'sensible' clustering. Indeed Armenia is a well-known borderline case for the geo-political concept of Middle East, though the Armenian is usually listed among Middle Eastern ethnic groups. Modern experts tend nowadays to consider it as part of Europe, therefore out of Middle East. But in 1996 the on-line CIA World Fact Book still considered Armenia as part of Asia.

The second experiment further restricts the conditions imposed by the first one. Here only descriptions with variables constrained by concepts of granularity from the second level on are considered. In this case the algorithm $\mathcal{AL}$-CoCLUST returns the following 12 clusters:

$\mathcal{C}_0' \in \mathcal{F}_1^1$
q(A) ← A:MiddleEastCountry
{ARM, BRN, IR, IRQ, IL, JOR, KWT, RL, OM, Q, SA, SYR, TR, UAE, YE}

$\mathcal{C}_1' \in \mathcal{F}_3^2$
q(A) ← believes(A,B) & A:MiddleEastCountry, B:MonotheisticReligion
{ARM, BRN, IR, IRQ, IL, JOR, KWT, RL, OM, Q, SA, SYR, TR, UAE}

$\mathcal{C}_2' \in \mathcal{F}_3^2$
q(A) ← speaks(A,B) & A:MiddleEastCountry, B:AfroAsiaticLanguage
{IR, SA, YE}

$\mathcal{C}_3' \in \mathcal{F}_3^2$
q(A) ← speaks(A,B) & A:MiddleEastCountry, B:IndoEuropeanLanguage
{ARM, IR}

$\mathcal{C}_4' \in \mathcal{F}_5^2$
q(A) ← speaks(A,B), believes(A,C) &
        A:MiddleEastCountry, B:AfroAsiaticLanguage, C:MonotheisticReligion
{IR, SA}

$\mathcal{C}_5' \in \mathcal{F}_5^2$
q(A) ← believes(A,B), believes(A,C) &
        A:MiddleEastCountry, B:MonotheisticReligion, C:MonotheisticReligion
{BRN, IR, IRQ, IL, JOR, RL, SYR}

$\mathcal{C}_6' \in \mathcal{F}_3^3$
q(A) ← believes(A,'Druze') & A:MiddleEastCountry
{IL, SYR}

---

[4] $\mathcal{C}_5$ is less populated than expected because $\mathcal{B}_{\text{CIA}}$ does not provide facts on the languages spoken for all countries.

$\mathcal{C}'_7 \in \mathcal{F}^3_3$
q(A) ← believes(A,B) & A:MiddleEastCountry, B:JewishReligion
{IR, IL, SYR}

$\mathcal{C}_8 \in \mathcal{F}^3_3$
q(A) ← believes(A,B) & A:MiddleEastCountry, B:ChristianReligion
{ARM, IR, IRQ, IL, JOR, RL, SYR}

$\mathcal{C}'_9 \in \mathcal{F}^3_3$
q(A) ← believes(A,B) & A:MiddleEastCountry, B:MuslimReligion
{BRN, IR, IRQ, IL, JOR, KWT, RL, OM, Q, SA, SYR, TR, UAE}

$\mathcal{C}'_{10} \in \mathcal{F}^3_5$
q(A) ← believes(A,B), believes(A,C) &
        A:MiddleEastCountry, B:ChristianReligion, C:MuslimReligion
{IR, IRQ, IL, JOR, RL, SYR}

$\mathcal{C}'_{11} \in \mathcal{F}^3_5$
q(A) ← believes(A,B), believes(A,C) &
        A:MiddleEastCountry, B:MuslimReligion, C:MuslimReligion
{BRN, IR, SYR}

organized in a DAG $\mathcal{G}'_{\text{CIA}}$ which partially reproduces $\mathcal{G}_{\text{CIA}}$. Note that two scarsely significant clusters in $\mathcal{G}_{\text{CIA}}$, $\mathcal{C}_2$ and $\mathcal{C}_4$, do not belong to $\mathcal{G}'_{\text{CIA}}$ thanks to the stricter constraints.

5   Conclusions and Future Work

The form of conceptual clustering considered in this paper can be cast as a problem of *Ontology Refinement*. Indeed the method takes an ontology as input and returns subconcepts of one of the concepts in the ontology. This is done by discovering strong associations between concepts in the input ontology. The hybrid FOL formalism $\mathcal{AL}$-log enables the uniform treatment of both structural and relational data. The closest work to ours is Vrain's proposal [24] of a top-down incremental but distance-based method for conceptual clustering in a mixed object-logical representation. Several application areas, notably the *Semantic Web* [1], can benefit from our proposal. Indeed the interpretability of clustering results and the expressive power of FOL are desirable properties in the Semantic Web area. Yet previous works in this area either apply conceptual clustering systems which do not deal with FOL, e.g. [5, 19], or propose clustering methods for FOL which are not conceptual, e.g. [15].

For the future we plan to extensively evaluate the method. Experiments will show, among the other things, how clusters depend on the minimum support thresholds set for the stage of frequent pattern discovery. Also a comparison with other approaches (if available) to conceptual clustering in FOL on a larger hybrid dataset (if available) would be of great interest. Finally we wish to investigate other choice criterions for concept descriptions.

References

1. T. Berners-Lee, J. Hendler, and O. Lassila. The Semantic Web. *Scientific American,* May, 2001.
2. G. Bisson. Conceptual Clustering in a First Order Logic Representation. In B. Neumann, editor, *ECAI 1992. Proceedings of the 10th European Conference on Artificial Intelligence,* pages 458–462. John Wiley & Sons, 1992.
3. I. Bournaud and J.-G. Ganascia. Conceptual Clustering of Complex Objects: A Generalization Space based Approach. In G. Ellis, R. Levinson, W. Rich, and J.F. Sowa, editors, *Conceptual Structures: Applications, Implementation and Theory,* volume 954 of *Lecture Notes in Computer Science,* pages 173–187. Springer, 1995.
4. S. Ceri, G. Gottlob, and L. Tanca. *Logic Programming and Databases.* Springer, 1990.
5. P. Clerkin, P. Cunningham, and C. Hayes. Ontology discovery for the semantic web using hierarchical clustering. In G. Stumme, A. Hotho, and B. Berendt, editors, *Working Notes of the ECML/PKDD-01 Workshop on Semantic Web Mining,* pages 1–12, 2001.
6. F.M. Donini, M. Lenzerini, D. Nardi, and A. Schaerf. $\mathcal{AL}$-log: Integrating Datalog and Description Logics. *Journal of Intelligent Information Systems,* 10(3):227–252, 1998.
7. N. Fanizzi, L. Iannone, I. Palmisano, and G. Semeraro. Concept Formation in Expressive Description Logics. In J.-F. Boulicaut, F. Esposito, F. Giannotti, and D. Pedreschi, editors, *Machine Learning: ECML 2004,* volume 3201 of *Lecture Notes in Computer Science,* pages 99–110. Springer, 2004.
8. J.H. Gennari, P. Langley, and D. Fisher. Models of incremental concept formation. *Artificial Intelligence,* 40(1-3):11–61, 1989.
9. J. Han and Y. Fu. Mining multiple-level association rules in large databases. *IEEE Transactions on Knowledge and Data Engineering,* 11(5), 1999.
10. A. Ketterlin, P. Gançarski, and J.J. Korczak. Conceptual Clustering in Structured Databases: A Practical Approach. In *Proceedings of the First International Conference on Knowledge Discovery and Data Mining,* pages 180–185, 1995.
11. J.-U. Kietz and K. Morik. A polynomial approach to the constructive induction of structural knowledge. *Machine Learning,* 14(1):193–217, 1994.
12. F.A. Lisi and F. Esposito. Efficient Evaluation of Candidate Hypotheses in $\mathcal{AL}$-log. In R. Camacho, R. King, and A. Srinivasan, editors, *Inductive Logic Programming,* volume 3194 of *Lecture Notes in Artificial Intelligence,* pages 216–233. Springer, 2004.
13. F.A. Lisi and D. Malerba. Ideal Refinement of Descriptions in $\mathcal{AL}$-log. In T. Horvath and A. Yamamoto, editors, *Inductive Logic Programming,* volume 2835 of *Lecture Notes in Artificial Intelligence,* pages 215–232. Springer, 2003.
14. F.A. Lisi and D. Malerba. Inducing Multi-Level Association Rules from Multiple Relations. *Machine Learning,* 55:175–210, 2004.
15. A. Maedche and V. Zacharias. Clustering Ontology-Based Metadata in the Semantic Web. In T. Elomaa, H. Mannila, and H. Toivonen, editors, *Principles of Data Mining and Knowledge Discovery,* volume 2431 of *Lecture Notes in Computer Science,* pages 348–360. Springer, 2002.
16. H. Mannila and H. Toivonen. Levelwise search and borders of theories in knowledge discovery. *Data Mining and Knowledge Discovery,* 1(3):241–258, 1997.
17. R.S. Michalski and R.E. Stepp. Learning from observation: Conceptual clustering. In R.S. Michalski, J.G. Carbonell, and T.M. Mitchell, editors, *Machine Learning: an artificial intelligence approach,* volume I. Morgan Kaufmann, San Mateo, CA, 1983.

18. S.-H. Nienhuys-Cheng and R. de Wolf. *Foundations of Inductive Logic Programming*, volume 1228 of *Lecture Notes in Artificial Intelligence*. Springer, 1997.

19. T.T. Quan, S.C. Hui, A.C.M. Fong, and T.H. Cao. Automatic generation of ontology for scholarly semantic web. In S.A. McIlraith, D. Plexousakis, and F. van Harmelen, editors, *The Semantic Web - ISWC 2004*, volume 3298 of *Lecture Notes in Computer Science*, pages 726–740. Springer, 2004.

20. M. Schmidt-Schauss and G. Smolka. Attributive concept descriptions with complements. *Artificial Intelligence*, 48(1):1–26, 1991.

21. G. Semeraro, F. Esposito, D. Malerba, N. Fanizzi, and S. Ferilli. A logic framework for the incremental inductive synthesis of Datalog theories. In N.E. Fuchs, editor, *Proc. of the 7th Int. Workshop on Logic Program Synthesis and Transformation*, volume 1463 of *Lecture Notes in Computer Science*, pages 300–321. Springer, 1998.

22. R.E. Stepp and R.S. Michalski. Conceptual clustering of structured objects: a goal-oriented approach. *Artificial Intelligence*, 28(1):43–69, 1986.

23. K. Thomson and P. Langley. Concept formation in structured domains. In D.H. Fisher, M.J. Pazzani, and P. Langley, editors, *Concept Formation: Knowledge and Experience in Unsupervised Learning*. Morgan Kaufmann, San Francisco, CA, 1991.

24. C. Vrain. Hierarchical conceptual clustering in a first order representation. In Z.W. Ras and M. Michalewicz, editors, *Foundations of Intelligent Systems*, volume 1079 of *Lecture Notes in Computer Science*, pages 643–652. Springer, 1996.

25. H. Xiong, M. Steinbach, A. Ruslim, and V. Kumar. Characterizing pattern based clustering. Technical Report TR 05-015, Dept. of Computer Science and Engineering, University of Minnesota, Minneapolis, USA, 2005.

26. H. Xiong, M. Steinbach, P.-N. Tan, and V. Kumar. Hicap: Hierarchical clustering with pattern preservation. In M.W. Berry, U. Dayal, C. Kamath, and D.B. Skillicorn, editors, *Proc. of the 4th SIAM Int. Conference on Data Mining*, 2004.

Appendix: The KR&R System $\mathcal{AL}$-log

The system $\mathcal{AL}$-log [6] integrates two KR&R systems: Structural and relational. The **structural subsystem** Σ is based on $\mathcal{ALC}$ [20] and allows for the specification of knowledge in terms of classes (*concepts*), binary relations between classes (*roles*), and instances (*individuals*). Complex concepts can be defined from atomic concepts and roles by means of constructors (see Table 1). Also Σ can state both is-a relations between concepts (*axioms*) and instance-of relations between individuals (resp. couples of individuals) and concepts (resp. roles) (*assertions*). An *interpretation* $\mathcal{I} = (\Delta^{\mathcal{I}}, \cdot^{\mathcal{I}})$ for Σ consists of a domain $\Delta^{\mathcal{I}}$ and a mapping function $\cdot^{\mathcal{I}}$. In particular, individuals are mapped to elements of $\Delta^{\mathcal{I}}$ such that $a^{\mathcal{I}} \neq b^{\mathcal{I}}$ if $a \neq b$ (*unique names* assumption). If $\mathcal{O} \subseteq \Delta^{\mathcal{I}}$ and $\forall a \in \mathcal{O} : a^{\mathcal{I}} = a$, $\mathcal{I}$ is called $\mathcal{O}$-*interpretation*. The main reasoning task for Σ is the *consistency check*. This test is performed with a *tableau calculus* that starts with the tableau branch $S = \mathcal{T} \cup \mathcal{M}$ and adds assertions to S by means of *propagation rules* until either a contradiction is generated or an interpretation satisfying S can be easily obtained from it.

The **relational subsystem** Π extends DATALOG [4] by using the so-called *constrained* DATALOG *clause*, i.e. clauses of the form

$$\alpha_0 \leftarrow \alpha_1, \ldots, \alpha_m \& \gamma_1, \ldots, \gamma_n$$

Table 1. Syntax and semantics of $\mathcal{ALC}$

bottom (resp. top) concept	$\bot$ (resp. $\top$)	$\emptyset$ (resp. $\Delta^{\mathcal{I}}$)
atomic concept	A	$A^{\mathcal{I}} \subseteq \Delta^{\mathcal{I}}$
role	R	$R^{\mathcal{I}} \subseteq \Delta^{\mathcal{I}} \times \Delta^{\mathcal{I}}$
individual	a	$a^{\mathcal{I}} \in \Delta^{\mathcal{I}}$
concept negation	$\neg C$	$\Delta^{\mathcal{I}} \setminus C^{\mathcal{I}}$
concept conjunction	$C \sqcap D$	$C^{\mathcal{I}} \cap D^{\mathcal{I}}$
concept disjunction	$C \sqcup D$	$C^{\mathcal{I}} \cup D^{\mathcal{I}}$
value restriction	$\forall R.C$	$\{x \in \Delta^{\mathcal{I}} \mid \forall y \ (x,y) \in R^{\mathcal{I}} \rightarrow y \in C^{\mathcal{I}}\}$
existential restriction	$\exists R.C$	$\{x \in \Delta^{\mathcal{I}} \mid \exists y \ (x,y) \in R^{\mathcal{I}} \wedge y \in C^{\mathcal{I}}\}$
equivalence axiom	$C \equiv D$	$C^{\mathcal{I}} = D^{\mathcal{I}}$
subsumption axiom	$C \sqsubseteq D$	$C^{\mathcal{I}} \subseteq D^{\mathcal{I}}$
concept assertion	$a : C$	$a^{\mathcal{I}} \in C^{\mathcal{I}}$
role assertion	$\langle a,b \rangle : R$	$(a^{\mathcal{I}}, b^{\mathcal{I}}) \in R^{\mathcal{I}}$

where $m \geq 0$, $n \geq 0$, α_i are DATALOG atoms and γ_j are *constraints* of the form $s : C$ where s is either a constant or a variable already appearing in the clause, and C is an $\mathcal{ALC}$ concept. A constrained DATALOG clause of the form $\leftarrow \beta_1, \ldots, \beta_m \& \gamma_1, \ldots, \gamma_n$ is called *constrained* DATALOG *query*. For an $\mathcal{AL}$-*log knowledge base* $\mathcal{B} = \langle \Sigma, \Pi \rangle$ to be acceptable, it must satisfy the following conditions: (i) The set of predicate symbols appearing in Π is disjoint from the set of concept and role symbols appearing in Σ; (ii) The alphabet of constants used in Π coincides with $\mathcal{O}$; (iii) For each clause in Π, each variable occurring in the constraint part occurs also in the DATALOG part. The interaction between Σ and Π allows the notion of *substitution* to be straightforwardly extended from DATALOG to $\mathcal{AL}$-log. It is also at the basis of a model-theoretic semantics for $\mathcal{AL}$-log. An *interpretation* $\mathcal{J}$ for $\mathcal{B}$ is the union of an $\mathcal{O}$-interpretation $\mathcal{I}_{\mathcal{O}}$ for Σ and an Herbrand interpretation $\mathcal{I}_{\mathcal{H}}$ for Π_D (i.e. the set of DATALOG clauses obtained from the clauses of Π by deleting their constraints). The notion of *logical consequence* paves the way to the definition of *correct answer* and *answer set* similarly to DATALOG. Reasoning for an $\mathcal{AL}$-log knowledge base $\mathcal{B}$ is based on *constrained SLD-resolution*, i.e. an extension of SLD-resolution with the tableau calculus to deal with constraints. *Constrained SLD-refutation* is a complete and sound method for answering queries, being the definition of *computed answer* and *success set* analogous to DATALOG. A big difference from DATALOG is that the derivation of a constrained empty clause does not represent a refutation but actually infers that the query is true in those models of $\mathcal{B}$ that satisfy its constraints. Therefore in order to answer a query it is necessary to collect enough derivations ending with a constrained empty clause such that every model of $\mathcal{B}$ satisfies the constraints associated with the final query of at least one derivation.

Karl Popper's Critical Rationalism in Agile Software Development

Mandy Northover, Andrew Boake, and Derrick G. Kourie

Espresso Research Group, Department of Computer Science,
School of Information Technology, University of Pretoria,
Pretoria, 0001, South Africa
mandy.northover@siemens.co.za,
andrew.boake@up.ac.za,
dkourie@cs.up.ac.za

Abstract. Sir Karl Popper's critical rationalism – a philosophy in the fallibilist tradition of Socrates, Kant and Peirce – is applied systematically to illuminate the values and principles underlying contemporary software development. The two aspects of Popper's philosophy, the natural and the social, provide a comprehensive and unified philosophical basis for understanding the newly emerged "agile" methodologies. It is argued in the first four sections of the paper – *Philosophy of Science, Evolutionary Theory of Knowledge, Metaphysics,* and *The Open Society* – that the agile approach to software development is strongly endorsed by Popper's philosophy of critical rationalism. In the final section, the relevance of Christopher Alexander's ideas to agile methodologies and their similarity to Popper's philosophy is demonstrated.

1 Introduction

Karl Popper was one of the leading philosophers of science in the 20th century. His theories have been applied to many different fields of human inquiry and have influenced diverse disciplines. While some writers [1, 2, 3, 4] have applied Popperian concepts successfully to aspects of the software engineering discipline, this paper aims to do so far more systematically. In particular, Popper's ideas are used in an attempt to understand the philosophical basis of the newly emerged "agile" methodologies.

The two aspects of Popper's philosophy, the natural and the social, are unified in a single philosophy – *critical rationalism* – which belongs to the tradition of *fallibilism*[1]. Philosophers in this tradition reject the quest for certainty in

---

[1] Fallibilism emphasizes the uncertainty of knowledge. Philosophers who belong to the fallibilist tradition include Socrates, Kant, Peirce and Popper. Peirce, in particular, anticipates some of Popper's essential ideas: Peirce's "abduction" resembles Popper's falsificationism and his conjectures resemble Popper's hypotheses. Furthermore, as does Popper, Peirce stresses the importance of inquiry – open inquiry – to the development of useful and reliable knowledge and states, "Do not block the way to inquiry - our survival depends on it" [5]. It is also arguable that Peirce's principle of *unfettered inquiry* can only be realised in the open society that Popper envisages. Since Peirce's death in 1914, these ideas, unpublished in his lifetime, have become increasingly influential in various disciplines.

H. Schärfe, P. Hitzler, and P. Øhrstrøm (Eds.): ICCS 2006, LNAI 4068, pp. 360–373, 2006.
© Springer-Verlag Berlin Heidelberg 2006

knowledge and maintain, instead, that knowledge is always hypothetical, always a tentative approximation to the truth. Consequently, the emphasis of Popper's philosophy is on hypothesis formation as the way knowledge advances, and on critical arguments as the way knowledge is controlled. This leads naturally to an evolutionary approach to the progression of knowledge.

In the *natural sciences*, Popper promotes critical rationalism through a series of falsifiable conjectures and attempted refutations. In the *social sciences*, he promotes a piecemeal approach to social reform through the application of critical rationalism to the problems of social life.

This paper argues that Popper's critical rationalism can illuminate the underlying values and principles of the "lightweight" software methodologies that emerged during the Internet era. Traditional software methodologies seem unable to cope with the challenges of this era, which include shortened market time windows, rapid technological change and increasingly volatile business requirements. Arguably, all these challenges can be reduced to the problem of accommodating change. As a result, a number of "lightweight" methodologies have emerged: Adaptive Software Development, Extreme Programming[2], Scrum, Crystal, Feature-Driven Development, Dynamic System Development Method, and "pragmatic programming".

In February 2001, seventeen representatives of these methodologies met to discuss the similarities of their approaches. The meeting resulted in the representatives agreeing to adopt the name "agile" instead of "lightweight" and issue the *Manifesto for Agile Software Development* which includes a statement of the following four central values underlying the agile approach [6]:

> *"Individuals and interactions* over processes and tools,
> *Working software* over comprehensive documentation,
> *Customer collaboration* over contract negotiation,
> *Responding to change* over following a plan."

Later, in order to support these four values, a further dozen principles were formulated which emphasize: prompt customer satisfaction, flexible response to change, frequent delivery of software, daily interaction between customers and developers, motivated individuals, face-to-face communication, measuring progress through working software, sustainable development, technical excellence, simplicity, self-organising teams, and reflection. All this culminated in the formation of *The Agile Alliance*, a non-profit organisation whose stated mission it was to promote these principles and values.

This paper applies Popper's philosophy to deepen an understanding of the values and principles underlying agile software development in four main sections: *Philosophy of Science*, *Evolutionary Theory of Knowledge*, *Metaphysics*, and *The Open Society*. At the end of the paper, the ideas of the philosopher and architect, Christopher Alexander, will be discussed. Alexander had a profound influence on the agile movement: he initiated the patterns movement in

---

[2] Kent Beck's methodology, Extreme Programming (XP), undoubtedly established the popularity of "lightweight" methods, hence the discussion in this paper of certain principles specific to XP.

architecture which inspired the leaders of the design patterns movement in software, many of whom became the leaders of the agile movement. Alexander's philosophy also contains many ideas which have a strong affinity with Popper's.

2   Philosophy of Science

In this section, Popper's principle of falsificationism is transferred from the domain of physical science to the domain of computer science. It is applied to software testing in general and to agile software development in particular.

2.1   Falsificationism

Popper rejects the traditional view of scientific method – the inductive method of verification – in favour of the deductive scientific method of *falsification*. He also rejects the inductivists' quest[3] for firm foundations in science believing, instead, that scientific theories are only tentative approximations to the truth.

Falsificationists argue that general statements, like the laws of physics, can never be conclusively verified but can, however, be conclusively falsified by a single counter-instance. Popper uses the following simple example to indicate the difference between falsification and verification: no number of singular observation statements of white swans, however large, can empirically verify the statement "All swans are white" but the first observation of a black swan can falsify it. If we had attempted to find verifying instances of this statement, we would not have had much difficulty. But this would only have confirmed – but not verified – the general statement.

According to Popper, a theory belongs to science only if it is testable.[4] And a theory is testable only if some imaginable observation can falsify it. Falsifiability, therefore, is Popper's criterion of demarcation between science and non-science.

Popper believed that scientific theories should be formulated as clearly and precisely as possible so as to expose them most unambiguously to falsification. They should also make bold claims because the bolder a theory, the greater its informative and predictive content. Bold theories also have a greater likelihood of being disproven.

Popper's principle of falsificationism requires us to test our theories rigorously, to search for our errors actively, and to eliminate our false theories ruthlessly. It is a process which emphasizes criticism: tests should be designed critically to falsify theories, not to support them. There are three important ways of using criticism to establish a theory as defective:

---

[3] Inductivists believe that knowledge develops as a result of observation and experiment. They base general statements on accumulated observations of specific instances and search for evidence to verify their theories.

[4] Theories which cannot be empirically tested, according to Popper, do still have meaning and can still be critically discussed. These theories are part of *Metaphysics* which is discussed in Section 4.

1. find experimental conditions under which the theory fails;
2. demonstrate that the theory is inconsistent with itself or some other theory;
3. demonstrate that the theory is "unfalisifiable".

Amongst the falsifiable theories that survive repeated attempts at refutation, the best corroborated[5] one is chosen.

2.2 Falsificationism and Software Testing

The susceptibility of software to testing demonstrates its falsifiability, and thus, Popper would no doubt agree, the scientific nature of software development. Software developers, on the one hand, resemble theoretical scientists. They are responsible for establishing an overall theory that guides the development of working programs. Software testers, on the other hand, resemble experimental scientists. Each test case they write is like a "scientific experiment" which attempts to falsify a part of the developer's overall theory.

Coutts, in [1], uses many Popperian ideas to trace the similarities between software testing and the scientific method. He concludes that falsification is just as essential to software development as it is to scientific development: whereas in science the falsification criterion is often used as a demarcation between science and non-science, in software the falsification criterion demarcates the testing discipline from the analysis and development disciplines.

Popper's falsificationism is also useful in providing a philosophical understanding of the view that no amount of software testing can prove a program correct.[6] Consequently, software testers should not strive to prove that programs are error-free. They should rather try to eliminate as many errors as they can by writing critical test cases with the purpose of detecting failures rather than demonstrating the absence of errors. To aid this process of rigorous testing, software developers should write software that better supports falsification.

2.3 Falsificationism and Agile Software Development

This section will show that many agile practices support falsificationism and the related ideas of criticism and error elimination.

One of the fundamental differences between agile methodologies, especially Extreme Programming (XP), and traditional software methodologies is their approach to software testing. Traditional methodologies usually leave testing until

---

[5] Corroboration is a failed attempt at refutation, which is not equivalent to a successful attempt at confirmation, since confirmation aims at proving theories, whereas Popper argues that no matter how well corroborated a theory may be, it remains a tentative approximation to the truth.

[6] The famous mathematician, E.W. Dijkstra, agrees with this view: "Testing can only demonstrate the presence of errors, not their absence.". Some have considered this statement to be a special case of Popper's falsifiability principle [3]. However, despite the apparent similarity between their views regarding testing, Popper and Dijkstra draw radically different conclusions. Dijkstra advocates that programs should be deductively verified whereas Popper rejects verification completely.

the end of a project while agile methodologies advocate *test-driven development*. One of the problems with the traditional approach is evident when projects run late. The testing phase is shortened which invariably results in incomplete or poor-quality testing. Agile developers avoid this problem through their practice of test-driven development. The main focus of this approach is on developers defining test cases before writing any code. A failing automated test case is written which points out lacking functionality. Then the functionality is implemented to ensure the test case passes. This cycle is repeated, in order of priority of the most valuable business requirements, until all the required functionality is implemented. The practice of test-driven development can be understood, from a Popperian perspective, as continuous attempts at falsification. The fact that test-driven development is the basis of the development approach of agile methodologies like XP, means that Popper would consider these methodologies to be pre-eminently scientific.

Another of the agile practices which encourages falsificationism is the *collective ownership of code*. Individual contributions to source code are stored in collectively owned repositories and, through a democratic peer review process, the functioning of the source code in these repositories is thoroughly assessed.[7] Thus, the code is more open to falsificationism.

In line with Popper's falsificationism, agile methodologies acknowledge that people invariably make mistakes. Embracing this fact, they adopted an iterative and incremental approach to development. This approach, which derives from the Rational Unified Process (RUP), is now central to all agile methodologies. It allows for mistakes to be detected early and repaired, by reworking pieces of the software. This is similar to Popper's method of error detection and elimination.

The agile practice involving an *on-site customer* also resembles the role of falsificationism in Popper's method. Agile methodologies, being people-oriented, advocate strongly the importance of continuous and frequent collaboration between customers, developers and stakeholders. Customers form an integral part of the agile team and remain on-site throughout development. They provide feedback which helps the team to detect and eliminate errors as early as possible. Popper would view the transparency of this development process both as scientific and democratic.

Finally, *pair programming*, where two people sit together and co-write a program, also supports and encourages falsificationism because the source code is subjected to a critical peer review process.

It is clear, from the discussion above, that Popper's principle of falsificationism is crucial to optimal software testing and to many agile practices.

3   Evolutionary Theory of Knowledge

In this section, Popper's evolutionary theory of knowledge is applied to agile's iterative development paradigm.

---

[7] This ethic is also central to Open Source Software (OSS) development where competing designs, analogous with Popper's competing scientific hypotheses or theories, are available for extensive peer review.

According to Popper, all knowledge, of which scientific knowledge is the exemplar, begins with problem-solving. He uses the following four-stage model to describe the way scientific knowledge advances:

$$P_1 \rightarrow TS \rightarrow EE \rightarrow P_2 \,. \tag{1}$$

P_1 is the initial problem, TS the proposed trial solution, EE the process of error elimination applied to TS, and P_2 the resulting solution with new problems [7]. This model holds that complex structures can only be created and changed in stages, through a critical feedback process of successive adjustments. It is an evolutionary model concerned with continuous developments over time instead of full-scale development based on an unchanging model or blueprint.[8] According to this model, scientific knowledge is perpetually growing. There is no state of equilibrium because the tentative adoption of a new solution which solves one problem (P_1), invariably exposes a whole set of new problems (P_2).

The order of the four stages suggests that we always start with a problem and then use the method of trial and the elimination of error to solve this problem. Although this problem-solving process is continually iterated, it is not merely cyclic because P_2 is always different from P_1.

Applied to the development of knowledge in the individual, each iteration is to some degree informed by memories of what survived prior iterations. If we trace this process backwards, we find unconscious expectations that are inborn. This shows how Popper's theory of knowledge merges with a theory of biological evolution. What distinguishes the evolution of scientific knowledge from biological evolution is the conscious application by human beings of the critical method. Whereas lower organisms often perish with their false theories, humans can use language to present their theories objectively: "by criticizing our theories we can let our theories die in our stead" [8].

Popper's evolutionary theory of knowledge has been applied to many fields of human inquiry besides science, including music, art, economics, logic and learning.[9] In fact, virtually all forms of organic development can be understood in this way. In the rest of this section, Popper's four-stage model is applied to agile software development.

The advocates of agile methodologies recognise that software development is a non-linear process which requires short cycles and continuous feedback. As a result, they advocate an iterative and incremental approach to development which controls unpredictability and allows for adaptability. Each iteration is like a miniature software project, and includes all the tasks necessary to release a subset of the required functionality in a short time period. During this iterative process, software applications evolve incrementally, in small steps.

Altogether, the four stages in Popper's evolutionary model can be understood, from an agile perspective, as a single development iteration: P_1 corresponds to a subset of the customer's requirements that are to be implemented during the iteration – these are the problems that the developer has to solve; TS is

---

[8] Popper applies this idea to his social philosophy which is described in Section 5.

[9] This model is applied to the field of architecture in the final section of this paper.

equivalent to the solution proposed by the developer and *EE* with the attempt, by the software tester, to eliminate errors by writing critical test cases; finally, P_2 is parallel to the tested software that contains a subset of the customer's requirements – P_2 usually gives rise to unanticipated problems or requirements that must be addressed in one of the future iterations.

Before the start of the following iteration, agile proponents use reflection to make refinements and to reprioritise the project's requirements. At this stage, changes can easily be incorporated into the project. Agile teams also refactor the code, if necessary, at the end of an iteration to eliminate errors and improve the design and quality of software. Both reflection and refactoring further support error elimination in Popper's model. Thus, in much the same way that scientific knowledge advances through this evolutionary process, software approximates more and more closely the required solution over repeated iterations.

4   Metaphysics

In this section, it is argued that Popper's *three worlds epistemology* provides a more adequate framework for knowledge management, especially in agile organisations, than Polanyi's paradigm of "tacit" knowledge which currently pervades the knowledge management discipline.

Popper divides existence and the products of cognition into three ontologically related domains called *Worlds*:

1. *World 1*, the world of physical objects or of physical states, independent of any perceptions;
2. *World 2*, the subjective world of private mental states;
3. *World 3*, the objective but intangible world of products of the human mind.

Popper further distinguishes between two types of knowledge: *knowledge in the subjective sense* and *knowledge in the objective sense*. The first type of knowledge belongs to World 2, the world of subjective belief, and the second type to World 3, the world of objective knowledge.

Worlds 1 and 2 are the familiar worlds of matter and mind. It is World 3 that requires further explanation. Knowledge in World 3 consists of spoken, written or printed statements – critical arguments, artistic and literary works, scientific problems, logical contents of books – that are open to public criticism and can be experimentally endorsed or disputed. Knowledge in World 2, by contrast, is private and barely criticizable. This is the kind of knowledge studied by traditional epistemology. The fact that knowledge in World 3 is objective and open to public scrutiny, establishes the superiority of World 3 knowledge over World 2 knowledge. Furthermore, according to Popper, knowledge in World 3 is essential to the growth of scientific knowledge.

Although World 3 knowledge is a product of the human mind, it is largely autonomous of World 2. The autonomy of World 3 knowledge can be understood if we consider that this knowledge will continue to exist, will remain rediscoverable, even if all knowing subjects are temporarily destroyed. Another unique property of World 3 is the exclusive existence in it of certain types of objects. For example, the fact that calculus was discovered independently by Newton and Leibniz shows that it was not merely created by a subjective mind but that calculus pre-existed in World 3 awaiting discovery. In terms of World 2, Leibniz made the discovery first, but in terms of World 1, Newton was the first to publish the discovery in physical form. The interaction and feedback between Worlds 1, 2 and 3 commonly occur in this way – World 3 indirectly influences World 1 through the mental World 2.

Popper's three worlds epistemology can be used as a framework for managing knowledge in organisations. To date, the knowledge management discipline has focused most of its attention on knowledge in World 2 of Popper's epistemology. This focus is possibly due to the reliance of pioneers in knowledge management, Nonaka and Takeuchi, on the work of scientist and philosopher Michael Polanyi. Polanyi's work focuses primarily on personal or *tacit*[10] knowledge, which Popper places in World 2, and neglects Popper's, more important, World 3 of objective knowledge. Furthermore, Polanyi's paradigm addresses only the one type of knowledge identified by Popper: knowledge in the subjective sense. Polanyi relies solely on this type of knowledge because he regards truth to be what the majority of experts subjectively believe to be true. Consequently, he believes only in subjective truth. He dismisses knowledge in the objective sense and, therefore, objective truth.

Popper disagrees with Polanyi's view of truth. He believes that human knowledge is fallible and argues that in all sciences, even the experts are sometimes mistaken. So we cannot rely merely on subjective knowledge.[11] Furthermore, Popper believes that it can only be on the basis of objective tests and criticism of theories that scientists come to an agreement about the truth.

The fallibility of human knowledge necessitates the rigorous questioning of our knowledge and our assumptions. New knowledge is produced as a result of this questioning process. Therefore, it is crucial to the knowledge management discipline that open questioning is possible. In order to achieve this, we must endeavour to change our subjective, World 2 knowledge into objective, World 3 knowledge where it is open to public scrutiny.

Popper's three worlds epistemology provides a more comprehensive framework for understanding the approach taken by different organisations towards managing knowledge. Organisations that adopt process methodologies like the Capability Maturity Model (CMM), by focussing on the importance of documentation, overemphasize Worlds 1 and 3 while neglecting World 2. They also fail to recognise the dynamic nature of knowledge in World 3. Organisations that adopt Polanyi's paradigm, on the other hand, overemphasize World 2 and dismiss World 3. The rest of this section argues that, while knowledge management

---

[10] The notion of tacit knowledge was introduced by Polanyi to mean "a backdrop against which all actions are understood". Nonaka and Takeuchi use the term differently from Polanyi to denote "knowledge that is intrinsic to individuals and not easily articulated or shared".

[11] In fact, Popper did explicitly raise these arguments against Polanyi in [9].

in agile organisations seems to be adequately explained by Polanyi's paradigm, it does in fact require a more balanced explanation that acknowledges the existence of World 1 and the equal importance of Worlds 2 and 3.

Knowledge creation and sharing is a crucial part of agile software development. Agile teams derive much of their agility by relying on the tacit knowledge embodied in the team, rather than on explicitly writing the knowledge down in documents. Their frequent use of informal communication channels such as co-located teams, customer collaboration, pair programming and regular stand-up meetings, facilitates tacit knowledge transfer and fosters a culture of knowledge sharing in the organisation. The enhanced informal communication between team members compensates for the reduction of documentation and other forms of explicit knowledge. All this seems to suggest that Polanyi's model of tacit knowledge is sufficient to explain the means by which agile organisations manage knowledge.

However, agile organisations do recognise the importance of explicit, World 3 knowledge in certain scenarios [10]. What makes explicit knowledge in agile organisations unique is that it is collectively owned by the team. Collective ownership facilitates the advancement of knowledge because team members share responsibility for keeping the information up-to-date. Collective ownership of information cannot be adequately characterised as subjective, or even inter-subjective, as in Polanyi's model, since it involves knowledge that exists outside and independently of each team member. Instead, this relation is more adequately described in terms of Popper's three worlds epistemology: what is owned are the objective products of the team's collective endeavours, which cannot be reduced to the sum of their individual tacit knowledge contributions.

5   The Open Society

Popper's social philosophy, which is described in this section, is seamlessly interwoven with his philosophy of science: "Rationality, logic and a scientific approach all point to a society which is 'open' and pluralistic, within which incompatible views can be expressed and conflicting aims pursued." [7]. According to Popper, all life is a process of problem-solving and the best societies should support this process by protecting their citizens' freedom of thought and right to criticise.

Popper considers two opposing approaches to social change, *piecemeal social engineering* and *utopian social engineering*. He firmly advocates the former approach and believes it is essential to achieving a state of liberal democracy which he calls *The Open Society*[12]. Popper stresses the importance of openness and transparency in human social systems. He believes all discourse should flourish within open societies, free from forms of totalitarian political control that

---

[12] The Open Society is "an association of free individuals respecting each others' rights within the framework of mutual protection supplied by the state, and achieving, through the making of responsible, rational decisions, a growing measure of humane and enlightened life." [7].

might try to engineer a society's development according to supposedly absolute laws of development.

In what follows, Popper's distinction between utopian and piecemeal social engineering is used to illuminate the differences between traditional and agile methodologies in terms of their divergent approaches to software development.

5.1 Utopian Social Engineering

Utopian social engineering aims at remodelling the "whole of society" according to an idealistic, unchanging plan or blueprint. Utopian societies are controlled centrally by a few, powerful, individuals who believe that social experiments are only valuable if carried out holistically.

The utopian approach, which is based on the following apparently rational reasoning, seems to be very scientific:

- Any action must have a particular aim.
- The action is considered rational if it seeks its aim consciously and determines its means according to this aim or end.
- In order to act rationally, therefore, the first thing to do is to choose the end.
- There must be a clear distinction between the intermediate ends and the ultimate end.

The utopian approach to social engineering develops a blueprint of society before considering ways for its realisation. By assuming that societies can be designed according to an unchanging plan, this approach incorporates the same mistaken notion of certainty as does the traditional, inductivist view of science. Despite the apparently rational reasoning of the utopian approach, several of its premises are questionable and can be criticised as follows:[13]

- Even scientists cannot predict everything, therefore, the idea of a totally planned society is impossible.
- The notion of an unchanging blueprint of society is untenable because change is continuous.
- Since social life is too complicated to predict, unintended consequences are inevitable. The greater the holistic changes, the greater the unintended consequences, thus forcing the holistic engineer, ultimately, into piecemeal improvisation.
- So much is reconstructed at once that it is unclear which measures are responsible for which results. And if the goal is distant, it is difficult to say if the results are in the direction of the goal or not.

There are remarkable similarities between utopian social engineering, as characterised by Popper, and the traditional approach to software development.

The mistaken assumption that underlies many traditional software methodologies is that project requirements are fixed before implementation begins. This is a fondly held assumption because fixed requirements allow a project to be

---

[13] The following critique is adapted from [11].

planned with more certainty from the start. As a result, traditional methodologies advocate comprehensive up-front design as a prerequisite to implementation. It is evident that this approach to design shares many similarities with idealistic blueprint planning. However, software engineering projects are usually too complex to guarantee comprehensive up-front design using a blueprint. Furthermore, project requirements are seldom static which means there will inevitably be deviations from the blueprint.

Much like the utopian engineer's approach to social engineering, traditional methodologists believe in the infallibility of their approach which, if followed strictly, will lead to the success of the project. The tendency towards inflexibility and software bureaucracy in this approach can stifle both creativity and innovation. Moreover, this approach often leads to blind adherence to a process, which is incompatible with the critical attitude that Popper advocates.

5.2   Piecemeal Social Engineering

Agile software development shares many similarities with Popper's piecemeal approach to social engineering and builds on what Christopher Alexander[14] calls "piecemeal growth".

Piecemeal social engineering, in contrast to utopian social engineering, does not aim to redesign society "as a whole". Instead, it tries to achieve its aim in a humane and rational manner by small adjustments and re-adjustments which can be continually improved upon. It is a cautious, critical approach that acknowledges the uncertainty introduced by the "human factor".

The main premise of the piecemeal approach is that "We do not know how to make people happy, but we do know ways of lessening their unhappiness." [7][15]. It draws attention to solving social problems instead of the pursuit of an ideal or utopian society.

Consequently, the task of the piecemeal social engineer is to identify and remedy the most urgent problems in society instead of searching for and imposing on society the elusive ideal of the greatest common good. The piecemeal social engineer makes her way cautiously, step by step, consciously and actively searching for mistakes in her method. She is aware that she can learn from her mistakes and knows that if anything goes wrong, her plan can be terminated or even reversed. The damage would not be too great and the re-adjustment not too difficult.

The most striking similarity between the agile approach to software development and Popper's piecemeal approach to social engineering is their rejection of an overall blueprint for design. Agile methodologies are adaptive rather than anticipatory, which means they place less emphasis on comprehensive up-front design. They do not believe in planning new projects in their entirety nor do they believe in redesigning existing projects from scratch. Instead, agile developers believe that software should be built in a piecewise fashion through a series of short iterations with continuous feedback. Feedback makes the development

---

[14] Alexander's philosophy is discussed in the final section of this paper.

[15] This approach is analogous with the falsification of scientific hypotheses.

process flexible and reversible which means that the developers can accommodate change more easily and correct mistakes as early as possible. This is why agile methodologies embrace, rather than avoid, change. Through this process, software is built in a "bottom-up" fashion, in small steps rather than the large leaps that occur in blueprint planning.

Another similarity between the two approaches is their focus on the present rather than the future. Agile methodologies emphasize simplicity in their designs. To achieve simplicity, they implement only those requirements which are certain. They do not implement functionality for potential future requirements because of the inherent uncertainty of the future. Their focus on the immediate problem is similar to the piecemeal engineer's focus on identifying the most urgent problems in society.

5.3 The Agile Team and the Open Society

The agile team can be described both in terms of a scientific community within the open society and in terms of an open society as a whole.

The practices and principles that regulate an agile team, as well as the environment in which they work, encourage open and critical discussion, qualities which Popper believes are essential to the progression of knowledge in a scientific community. In particular, face-to-face communication, pair programming and white-board discussions encourage the free exchange of ideas as well as creative hypothesis formation. Furthermore, the presence of on-site customers, facilitates openness and courage, which Popper believes are essential to democratic citizens and scientists.

It may seem that by emphasizing the importance of the team over the individual members, agile methodologies contradict Popper's belief in the supremacy of the individual. However, Popper believes that "The objectivity of science ... [is] in the hands not of individual scientists alone, but of scientists in a scientific community." [8].

In terms of the open society as a whole, the manager of the agile team has similar responsibilities to the team members as the state has to its citizens. Teams are not managed by an authoritarian leader. Rather, project responsibility is shared equally between developers and managers: managers ensure that conditions are conducive to productive software development and give developers the responsibility for making all the technical decisions, trusting them to get the work done. Like responsible citizens in an open society, agile team members develop a sense of community and accountability through their shared goals and responsibilities as well as their self-organising nature.

6   Christopher Alexander's Social Engineering Architecture

Many of Popper's ideas are evident in the field of architecture in the approach to design espoused by the architect and philosopher, Christopher Alexander.

Alexander is best known for his observations on the design process and is recognised as the "father" of the patterns movement in architecture. His ideas have proved influential far beyond the field of architecture: he inspired the design patterns movement in software which indirectly influenced the agile movement because many of the current leaders of the agile movement were leaders of the design patterns movement.

Alexander promotes a philosophy of incremental, organic and coherent architectural design by which facilities emerge in an evolutionary fashion into communities of related buildings. Alexander believes architects should be accountable to their clients who should be able to criticise the architect's designs before they are implemented. Inherent in this philosophy are several of Popper's principles including his evolutionary epistemology, and his principles of social openness, democracy, criticism and rationalism.

Alexander's approach to architectural design, which has often been referred to as *social engineering architecture*, also shares certain principles in common with Popper's piecemeal social engineering. Just as Popper rejects the master plan of utopian social engineering in favour of piecemeal social engineering, Alexander rejects a master plan for architectural design. According to Alexander, the development of a master plan for architectural design is the first and fatal deviation from the *Timeless Way of Building*[16]. Instead of a master plan, Alexander proposes a meta-plan which constitutes three parts [13]:

1. a philosophy of piecemeal growth;
2. a set of patterns or shared design principles[17] governing growth; and
3. local control of design by those who will occupy the space.

Under Alexander's meta-plan, communities grow in an evolutionary fashion through a series of small steps, in an ongoing and iterative fashion[18] which results in what Alexander calls *organic order*. Consequently, Alexander would agree with the piecemeal engineer who, according to Popper, recognises that "only a minority of social institutions are consciously designed while the vast majority have just 'grown', as the undesigned results of human actions." [8].

It is significant that many of Popper's ideas appear to be inherent in Alexander's approach to architectural design. For, if Popperian ideas influenced Alexander, then these ideas may have indirectly influenced some of the leaders of the agile movement.

7   Conclusion

This paper has used Popper's critical rationalist philosophy to illuminate the values and principles underlying software development, with a particular emphasis

---

[16] XP founder, Kent Beck, acknowledges Alexander in [12] by entitling one of the chapters "The Timeless Way of Programming".

[17] It is no coincidence that the software design principles in the *The Agile Manifesto* resemble Alexander's shared design principles for architecture.

[18] There are clear links between this process and Popper's evolutionary epistemology.

on the agile approach to software development. In the first section, Popper's criterion of falsificationism provided a scientific justification for rigorous software testing, and it was shown how several agile practices support falsificationism. In the second section, Popper's model of the evolutionary growth of scientific knowledge accounted for the iterative development paradigm of agile methodologies. It was argued, in the third section, that Popper's three worlds epistemology provided a more comprehensive framework for knowledge management than Polanyi's paradigm of "tacit" knowledge. In the fourth and final section, Popper's distinction between utopian and piecemeal social engineering was used to illuminate the divergent approaches to software development of traditional and agile methodologies. The agile team was also favourably described in terms of Popper's notion of an open society. Finally, it was shown that Popper's piecemeal social engineering is endorsed by Alexander's social engineering architecture.

References

1. Coutts, D.: The Test Case as a Scientific Experiment. http://www.stickyminds.com
2. Hall, W.P.: Managing Maintenance Knowledge in the Context of Large Engineering Projects: Theory and Case Study. Journal of Information & Knowledge Management, Vol. 2, No. 2 (2003)
3. Snelting, G.: Paul Feyerabend and Software Technology. Journal of the German Computer Science Society (1998)
4. Moss, M.: Why Management Theory Needs Popper: The Relevance of Falsification. The Journal of Philosophy of Management, Vol. 3, Issue 3 (2002)
5. Firestone, J.M., McElroy, M.W.: The Open Enterprise: Building Business Architectures for Openness and Sustainable Innovation. KMCI Online Press (2003)
6. "Manifesto for Agile Software Development." http://www.agilemanifesto.org
7. Magee, B.: Popper. In: Kermode, F. (ed.): Fontana Press, Glasgow (1973)
8. Popper, K.: A Pocket Popper. In: Miller, D. (ed.): Fontana Press, Glasgow (1983)
9. Popper, K.: All Life is Problem Solving. Routledge, New York (1999)
10. Chau, T., Maurer, F., Melnik, G.: Knowledge Sharing: Agile Methods vs. Tayloristic Methods. IEEE International Workshops on Enabling Technologies, Austria (2003)
11. Popper, K.: The Open Society and its Enemies: Volume 1: The Spell of Plato. Routledge, London (1966)
12. Beck, K., Andres, C.: Extreme Programming Explained: Embrace Change. 2nd edn. Addison-Wesley, Boston (2005)
13. DeMarco, T., Lister, T.: Peopleware: Productive Projects and Teams 2nd edn. Dorset House Publishing Co., New York (1999)

On Lattices in Access Control Models

Sergei Obiedkov, Derrick G. Kourie, and J.H.P. Eloff

Department of Computer Science, University of Pretoria, Pretoria 0002, South Africa

Abstract. Lattices have been extensively used for implementing mandatory access control policies. Typically, only a small sublattice of the subset lattice of a certain alphabet is used in applications. We argue that attribute exploration from formal concept analysis is an appropriate tool for generating this sublattice in a semiautomatic fashion. We discuss how two access control models addressing different (in a sense, opposite) requirements can be incorporated within one model. In this regard, we propose two operations that combine contexts of the form (G, M, I) and (N, G, J). The resulting concept lattices provide most of the required structure.

1 Introduction

Multiuser computer systems must ensure that information they contain is accessible only to those who are authorized to use it. Therefore, choosing and implementing an access control model suitable for a particular computer system is an important part of its development.

Some of the most well-know access control models make heavy use of lattices [1]. They are based on Denning's Axioms formulating assumptions under which an information flow policy can be regarded as a lattice [2]. Due to relative complexity of their creation and maintenance, these models have enjoyed only limited use in practice.

In the access control setting, one speaks about active subjects (such as users or processes) accessing passive objects (such as files or resources). This separation is not absolute: an object such as a program can become a subject when trying to access another object such as a file.

In lattice-based control models, security labels are assigned to entities (subjects and objects). These security labels are partially ordered and, in fact, form a lattice. There are good reasons for it to be a lattice rather than just a partial order: the supremum and infimum operations play a role in, for instance, some versions of the Biba model [3,4]. Information is allowed to flow only in one direction, e.g., from entities with security labels that are lower in the lattice to entities with security labels that are higher. Section 2 describes the lattice-related aspects of the Bell-LaPadula model [5].

A security label is usually a combination of a security level (*secret*, *confidential*, etc.) and a subset of categories (project names, academic departments, etc.). In practice, only a small fraction of all possible labels is used. In Section 3, we argue that attribute exploration from formal concept analysis (FCA) can help effectively identify such useful labels. We describe attribute exploration rather informally

H. Schärfe, P. Hitzler, and P. Øhrstrøm (Eds.): ICCS 2006, LNAI 4068, pp. 374–387, 2006.

(since formal definition is available elsewhere [6]) and, perhaps, more than necessary for the general ICCS audience, but this is to give security experts without FCA background an idea of how they can benefit from using this technique.

A lattice-based access control model typically addresses some particular security concerns, such as confidentiality (the Bell-LaPadula model [5]) or integrity (the Biba model [3]). In terms of formal concept analysis, the two models can be expressed by two formal contexts such that the object set of one context is the attribute set of the other: (G, M, I) and (N, G, J). Some researchers combine the lattices of the two models into one lattice [7] to obtain a unified information flow policy. In Section 4, we discuss the resulting structure, which arises from a combination of the contexts mentioned above. Then, we outline an attribute exploration procedure to get the necessary components of this structure.

2   Lattices in Access Control Models

Definition 1. *Let H denote a totally ordered set of* classifications *or security* levels *(we use standard notation for this order: e.g., $<$ and $\leq$) and let C denote a set of* categories *such as names of departments within a company, project names, etc. Then, a* compartment *is defined as a subset of categories, and a* security label *is a pair (h, D) consisting of a security level $h \in H$ and a compartment $D \subseteq C$.*

Security labels are partially ordered: $(h, D) \leq (g, E)$ if and only if $h \leq g$ and $D \subseteq E$. It is easy to see that this partial order is a lattice: it is the product of two lattices, $(H, \leq)$ and $(\mathfrak{P}(C), \subseteq)$. Such lattices are common in military security models. Sometimes, only a subset of the product is used in practice.

Example 1. Suppose that there are three security levels: *unclassified, secret, top secret*. Let $C = \{a, b, c\}$ contain the names of three different projects in the system. Figure 1 presents a security lattice built from H and C. Subjects with the label *(secret, $\{a\}$)* can access objects with the same label and objects labeled *(unclassified, $\emptyset$)*, but cannot access objects labeled, e.g., *(top secret, $\{a, b\}$)* or *(secret, $\{b\}$)*.

It should be noted that by accessing we mean essentially reading (rather than writing or modifying). Thus, an access control model, such as the one from Example 1, addresses the *confidentiality* requirement [4]:

Confidentiality: Prevention of unauthorized disclosure of information.

One of the most popular models dealing with confidentiality issues is the Bell-LaPadula model [5]. There exist many variants of this model; we concentrate only on its lattice-related aspects and follow the presentation in [1].

Assuming that $\lambda(e)$ is the security label of the subject or object e, the *simple-security property* is formulated as follows:

– Subject s can read object o only if $\lambda(o) \leq \lambda(s)$.

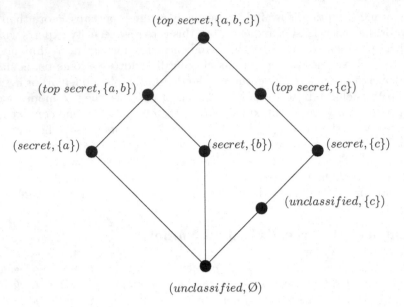

Fig. 1. An example of a security lattice

With security labels defined as above, a subject can read an object only if the subject is at the same security level as the object or higher and the subject has access to all categories associated with the object. However, the simple-security property is not enough: some restrictions are necessary on how subjects can write or modify objects. If there are no restrictions, a secret subject can read a secret document and create an unclassified copy of it, thus providing unclassified subjects with access to secret information. To avoid this, the *-property is introduced:

– Subject s can write object o only if $\lambda(s) \leq \lambda(o)$.

According to these two rules, information can flow only upwards: from less secure objects to more secure subjects and from less secure subjects to more secure objects.

The *-property implies that higher-level subjects cannot send messages to lower-level subjects. This is not always acceptable. To avoid this problem, subjects are sometimes granted *downgrading* capabilities: they are allowed to temporarily change their security label to one that is lower in the lattice. For example, a secret user may be associated with a secret subject and an unclassified subject. Then, the user can write an unclassified document by logging in as an unclassified subject. During this session, the user will not be able to read any secret documents. It is trusted that the user will not betray the information she got during a previous session when logged in as a secret subject. In the case of a subject with downgrading capabilities being a program rather than a human, previous sessions are not an issue.

From Example 1, it is clear that not all possible combinations of security levels and categories make sense as security labels. For instance, Smith describes a lattice from military practice based on four security levels and eight categories, which potentially gives rise to 1024 labels [8]. However, the actual lattice contains only 21 elements: there are no compartments consisting of more than three categories (apart from the compartment associated with the top element) and compartments are used only in combination with two top-most security levels. Hence, development of an access control model would involve identification of meaningful combinations of security levels and categories. In the next section, we argue that attribute exploration from formal concept analysis can help organize this process in a semiautomatic fashion and, in some sense, ensure the validity of its results.

3   Building Access Control Models by Attribute Exploration

First, we recall some basic notions of formal concept analysis (FCA) [6].

Definition 2. *A formal context is a triple* (G, M, I)*, where* G *is a set of objects,* M *is a set of* attributes, *and* $I \subseteq G \times M$ *is the* incidence relation *providing information on which objects have which attributes.*

Formal contexts are naturally represented by cross tables, where a cross for a pair (g, m) means that this pair belongs to the relation I.

Definition 3. *For* $A \subseteq G$*,* $B \subseteq M$*, the following* derivation operators *are defined:*

$$A^I := \{m \in M \mid gIm \text{ for all } g \in A\}$$
$$B^I := \{g \in G \mid gIm \text{ for all } m \in B\}$$

If the relation I *is clear from context, one writes* A' *and* B' *instead of* A^I *and* B^I.

Derivation operators are used to define concepts:

Definition 4. *The pair* (A, B)*, where* $A \subseteq G$*,* $B \subseteq M$*,* $A' = B$*, and* $B' = A$ *is called a* (formal) concept *(of the context* K*) with* extent A *and* intent B*. A concept* (A, B) *is* more general *(less specific, etc.) than a concept* (C, D)*, if* $C \subseteq A$ *(equivalently,* $B \subseteq D$*).*

For $g \in G$ and $m \in M$ the sets $\{g\}'$ and $\{m\}'$ are called *object intent* and *attribute extent*, respectively.

The operation $(\cdot)''$ is a closure operator [6], i.e., it is idempotent ($X'''' = X''$), extensive ($X \subseteq X''$), and monotone ($X \subseteq Y \Rightarrow X'' \subseteq Y''$). Sets $A \subseteq G$, $B \subseteq M$ are called *closed* if $A'' = A$ and $B'' = B$. Obviously, extents and intents are closed sets. Since the closed sets form a closure system [9], the set of all formal concepts of the context $\mathbb{K}$ forms a lattice called a *concept lattice* and usually denoted by $\underline{\mathfrak{B}}(\mathbb{K})$ in FCA literature.

Definition 5. *A* many-valued context *is a quadruple* (G, M, W, I), *where G and M are object and attribute sets, respectively; W is a set of attribute values; and $I \subseteq G \times M \times W$ is a ternary relation satisfying the following condition:*

$$(g, m, w) \in I \text{ and } (g, m, v) \in I \Rightarrow w = v.$$

Thus, every object has *at most* one value for every attribute. For our purposes, it is convenient to assume that every object has *exactly* one value for every attribute.

To apply FCA methods to many-valued contexts, one first needs to transform them into one-valued contexts (i.e., contexts in the sense of Definition 2). A standard transformation technique here is *plain conceptual scaling*. We replace every many-valued attribute m with a set of one-valued attributes M_m by building a one-valued context (G_m, M_m, I_m), where G_m is the set of all possible values of m, i.e., $\{w \mid w \in W \text{ and } \exists g \in G : (g, m, w) \in I)\} \subseteq G_m$. The relation I_m translates every value of m into a subset of M_m. Then, the many-valued context (G, M, W, I) is replaced by a one-valued context

$$(G, \bigcup_{m \in M} \{m\} \times M_m, J),$$

where $(g, (m, n)) \in J$ if and only if there is $w \in W$ such that $(g, m, w) \in I$ and $(w, n) \in I_m$.

Now, we can easily express the Bell-LaPadula model in terms of FCA. The set H of security levels and the set C of categories (see Definition 1) constitute our attribute set. In fact, security level is a many-valued attribute, but the scaling is pretty straightforward: we use the context $(H, H, \geq)$ as a scale. In the case of Example 1, we get the following (ordinal) scale:

	unclassified	*secret*	*top secret*
unclassified	×		
secret	×	×	
top secret	×	×	×

The attribute *unclassified* is clearly redundant; so, in principle, we do not have to include the smallest security level as an attribute in the one-valued context.

The elements of G in our context are subjects and objects and the incidence relation should assign categories and levels to them. Then, security labels correspond to concept intents, and the security label of an entity (subject or object) is the intent of the least general concept covering the entity (i.e., containing the entity in its extent). The concept lattice we get is the reverse of the Bell-LaPadula lattice, since larger intents correspond to less general concepts; the set of concept intents with usual subset order is precisely the Bell-LaPadula lattice[1].

The problem here is that a complete list of subjects and objects is usually unknown at the moment when the access control model is being developed (and,

---

[1] To get the same order in the concept lattice, we can also consider categories as objects of the context and entities (subjects and objects of the access control model), as attributes. Then, security labels are concept extents.

if a system is open to new users or new documents, a complete list never becomes available). To construct the lattice of security labels, the developer of the system must envision all types of potential subjects and objects and describe them in terms of security levels and compartments. It would be good to have a way of verifying that all possible cases have been considered. Besides, it would not make harm to bring some order to the process of selecting these relevant combinations. Attribute exploration is a technique that does precisely this.

Definition 6. *An expression $D \to B$, where $D \subseteq M$ and $B \subseteq M$, is called an (attribute) implication. An implication $D \to B$ holds in the context (G, M, I) if all objects from G that have all attributes from the set D also have all attributes from the set B, i.e., $D' \subseteq B'$.*

There is a connection between implication sets and closure operators.

Definition 7. *An attribute set $F \subseteq M$ respects an implication $D \to C$ if $D \nsubseteq F$ or $C \subseteq F$. A set F respects an implication set Σ if it respects all implications from Σ. If every set that respects an implication set Σ also respects the implication $D \to C$, then we say that $D \to C$ (semantically) follows from Σ.*

All sets that respect some implication set form a closure system (and, hence, there is a closure operator corresponding to every implication set). A minimal (in terms of size) set of implications from which all other implications of a context semantically follow was characterized in [10]. It is called the *Duquenne-Guigues basis* or *stem base* in the literature.

Note that, having the Duquenne-Guigues basis of the context, we are able to construct the lattice of concept intents even without knowing the actual objects of the context. The join-irreducible[2] elements of this lattice correspond to object intents that have to be in the context. Such necessary object intents form the *representation context* of the concept lattice.

The goal of attribute exploration is to find this representation context and construct its lattice. The attribute exploration process is quite standard [6] and, perhaps, does not have to be formally explained here. In its simplest version, it can be outlined as follows. Given some initial (possibly empty) set of objects of a subject domain, which is known to have considerably more (perhaps, an infinite number of) such objects, and their intents, attribute exploration aims to build an implicational theory of the entire domain (summarized by the Duquenne–Guigues basis) and a representation context. Obviously, an object of the representation context must respect all implications from the generated implication basis and provide a counterexample for every implication that does not follow from the basis. It means, in particular, that the concept lattice of the domain is isomorphic to the concept lattice of this relatively small representation context.

The process of attribute exploration is interactive. In the general setting, the process is as follows: the computer suggests implications one by one; the user

---

[2] We are working under the assumption that the order is that of concept generality, i.e., the reverse of the intent subset order. Therefore, join-irreducible intents are those that cannot be presented as intersections of other intents.

(the expert) accepts them or provides counterexamples. Attribute exploration is designed to be as efficient as possible, i.e., to suggest as few implications as possible without any loss in completeness of the result. This is achieved by generating implications from the Duquenne-Guigues basis in the order consistent with the subset ordering of implication premises (from smaller to larger premises). Then, if the user rejects an implication $D \rightarrow B$ at some step, it does not affect implications whose premise is not a superset of D: if such implications were in the basis, they will remain there.

Advanced versions of attribute exploration allow the user to input background knowledge, e.g, in form of implications the user knows to be true. Note that background knowledge is not limited to implications [11]. The presence of background knowledge usually decreases the number of questions the user has to answer, since if the answer to a certain question follows from the background knowledge (combined with already accepted implications), this question is not asked.

Background knowledge is particularly useful in the case of many-valued attributes. Moreover, it is readily available in this case: it can be automatically generated from scales. Indeed, a scale completely specifies all possible combinations of the corresponding new one-valued attributes. If we want to limit ourselves to implicational background knowledge, all that is necessary is to generate the Duquenne-Guigues basis for each scale. A method for generating the complete propositional axiomatization of the scale (with attributes interpreted as propositional variables) also exists [12] and is surprisingly similar to a method for generating the implication basis [13].

In our case, there is only one many-valued attribute: security level. It can be shown that the Duquenne–Guigues basis provides the axiomatization for the whole propositional theory of an ordinal scale, which is the type of the scale we used above for this attribute. If H is the set of security levels and $l \in H$ is the lowest level, the Duquenne–Guigues basis consists of the following implications:

$$\emptyset \rightarrow \{l\}$$

and

$$\{l, h\} \rightarrow \{k \mid k \leq h\}$$

for all $h \in H$ such that $\exists j \in H(l < j < h)$. Therefore, in our case, we can do with only background implications. The basis for the context in Example 1 is as follows:

$$\emptyset \rightarrow \{unclassified\};$$

$$\{unclassified, top\ secret\} \rightarrow \{unclassified, secret, top\ secret\}.$$

By entering these implications as background knowledge we avoid questions on implications between these attributes. In the case of Example 1, we would start with the context $(\emptyset, M, \emptyset)$, where $M = \{top\ secret, secret, unclassified, a, b, c\}$ and the two background implications above. The first question asked by the system would be:

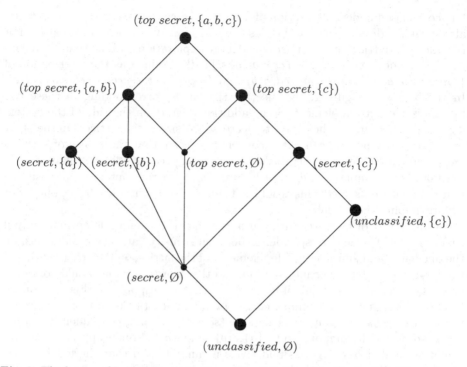

Fig. 2. The lattice obtained by attribute exploration based on Example 1. The order is the subset order of intents. Smaller nodes correspond to nodes absent from the lattice in Figure 1.

Is it true that all objects must have all attributes?

This is the most strong implication, and there are no counterexamples to it in our so far empty context. However, the answer is clearly "no", and we have to enter an object that lacks some attributes. Suppose that we enter an object with the intent {*top secret, secret, unclassified, a, b*}. Since this will be the only object of our context, the system assumes that all objects from the domain are like that and asks if it is true that every entity is labeled as at least (*top secret, a, b*), to which we have to enter another counterexample, say, {*top secret, secret, unclassified, c*}. Then, we may have to enter object intents {*secret, unclassified, a*} and {*unclassified, c*}. The system is not going to ask us whether every object is labeled as at least *unclassified*, as it knows from the background implications that this is so.

It is obvious that the process always stops at some point. In fact, the number of questions we are going to answer is equal to the sum of the sizes of the implication basis and representation context. The latter, however, depends on what objects we enter as counterexamples.

The resulting lattice may (and in the case of Example 1, will) contain more elements than it is necessary, that is, there may be some concepts that do not correspond to any realistic security labels (see Figure 2). This is because the concept lattice is closed under intersection of intents, whereas the security lattice, generally speaking, does not have to be closed under intersection of security labels. For instance, the lattice we obtain from attribute exploration on Example 1 will contain an intent $\{top\ secret, secret, unclassified\}$ obtained as the intersection of $\{top\ secret, secret, unclassified, a, b\}$ and $\{top\ secret, secret, unclassified, c\}$. In the lattice of Figure 1, the meet of the labels corresponding to these two intents is the bottom element. Some additional effort on the side of the system developer is required if they want to decrease the size of the lattice. On the other hand, the existence of such extra concepts suggests that the choice of security levels and categories or their distribution among security labels might not be optimal. For example, all labels in Figure 1 that do not contain c can easily be marked as *unclassified*: this would not affect the information flow policy, but would simplify the model.

As a matter of fact, we can even argue that the labeling is optimal only if there is a one-to-one correspondence between security categories and levels, on the one hand, and join-irreducible[3] elements of the lattice, on the other hand, and the lattice is closed under intersection. In other words, a node should contain a category or a level if and only if it is the one that corresponds to this category or level or if it is above such unique corresponding node. In the case of Example 1, the level *secret* corresponds to the node $(secret, \{c\})$ and, consequently, should not co-occur with compartments $\{a\}$, $\{b\}$, and $\{a, b\}$. Such approach ensures that the number of categories and levels is minimal and that the labels are as small as possible.

4   Combining Confidentiality and Integrity Models

As said above, the Bell-LaPadula model is concerned with confidentiality. Another issue that requires attention is *integrity*:

Integrity: Prevention of unauthorized modification of information.

One of the most common models dealing with integrity is the Biba model [3]. Biba proposed several integrity models, of which the best known is *strict integrity*. This model is dual to the Bell-LaPadula model: again we have a lattice of (integrity) labels, but this time the information is allowed to flow only downwards—from entities with higher integrity to entities with lower integrity—this is to prevent corruption of "clean" high-level entities by "dirty" low-level entities [4]. Assuming that $\omega(e)$ is the integrity level of the entity e, the rules of the model are formulated as follows [1]:

Simple-integrity property: Subject s can read object o only if $\omega(s) \leq \omega(o)$
Integrity ⋆-property: Subject s can write object o only if $\omega(o) \leq \omega(s)$

Of course, it is generally not very important whether the information flows only upwards or only downwards. However, it becomes important if we want to combine the Bell-LaPadula and Biba models in order to address both confidentiality

---

[3] Assuming the subset order as in Figure 1.

and integrity issues. In the combination of these two models, the rules of information flow are as follows [1]:

- Subject s can read object o only if $\lambda(o) \leq \lambda(s)$ and $\omega(s) \leq \omega(o)$.
- Subject s can write object o only if $\lambda(s) \leq \lambda(o)$ and $\omega(o) \leq \omega(s)$.

If the same security labels are used both for Bell-LaPadula and Biba models, then the information flow policy boils down to allowing subjects to read and write only objects from their own security level and compartment. The case when labels are different for the two models is more interesting (and useful).

Let $\mathbb{K}_1 = (G, M, I)$ and $\mathbb{K}_2 = (N, G, J)$ be formal contexts. By using G as the object set of $\mathbb{K}_1$ and as the attribute set of $\mathbb{K}_2$, we address the difference in the information flow direction of the two models. We want to combine these contexts into one structure in a way that preserves both orders of the corresponding concept lattices. The largest possible combination of the concept lattices is their product: $(\mathfrak{B}(\mathbb{K}_1) \times \mathfrak{B}(\mathbb{K}_2), \leq)$, where $(c_1, c_2) \leq (d_1, d_2)$ if and only if $c_1 \leq d_1$ and $c_2 \leq d_2$ (with respect to the usual "generality" order on concepts). The join and meet operations of the lattice $(\mathfrak{B}(\mathbb{K}_1) \times \mathfrak{B}(\mathbb{K}_2), \leq)$ are obvious:

$$(c_1, d_1) \vee (c_2, d_2) = (c_1 \vee c_2, d_1 \vee d_2)$$

$$(c_1, d_1) \wedge (c_2, d_2) = (c_1 \wedge c_2, d_1 \wedge d_2)$$

Here, joins and meets of concept pairs are taken in their respective lattices.

However, the product is too large for our purposes: it reflects the structure of each of the component lattices, but it fails to capture the dependencies between the elements of different lattices. These dependencies are given via the relations between the set G and corresponding sets (M and N) in the two contexts.

So, we are looking for an adequate subset $\mathfrak{B}_\diamond(\mathbb{K}_1, \mathbb{K}_2)$ of $\mathfrak{B}(\mathbb{K}_1) \times \mathfrak{B}(\mathbb{K}_2)$. The first observation is that $\mathfrak{B}_\diamond(\mathbb{K}_1, \mathbb{K}_2)$ must contain concept pairs of the form

$$((\{g\}^{II}, \{g\}^I), (\{g\}^J, \{g\}^{JJ})) \tag{1}$$

for every $g \in G$. That is, every element of G must get exactly the same description as it is given by the two initial contexts.

Then, we have several options for how to proceed. We can adopt a minimalist approach and add to $\mathfrak{B}_\diamond(\mathbb{K}_1, \mathbb{K}_2)$ only those concepts from $\mathfrak{B}(\mathbb{K}_1) \times \mathfrak{B}(\mathbb{K}_2)$ that are required to make our initial set of concepts a lattice. This makes sense if we know that G is the entire object set of our domain (rather than a set that gives rise to representation contexts for $\mathbb{K}_1$ and $\mathbb{K}_2$) and if all we need is to properly order elements from G. In the case of combining access control models, this is indeed all we need (ideally, every concept of the lattice should be a meaningful security label for some subject or object), but, unfortunately, the set G does not necessarily contain all possible entities, but only those enough to get a representation context for each of the two models. Therefore, we choose a different approach.

The (almost) maximalist approach we choose is to take the sublattice of $\mathfrak{B}(\mathbb{K}_1) \times \mathfrak{B}(\mathbb{K}_2)$ generated by concept pairs from (1), i.e., the smallest subset of $\mathfrak{B}(\mathbb{K}_1) \times \mathfrak{B}(\mathbb{K}_2)$ containing all object pairs (1) and closed under the join and meet operations. From now on, $\mathfrak{B}_\diamond(\mathbb{K}_1, \mathbb{K}_2)$ denotes this lattice. This approach ensures, in particular, that every concept of each component has a counterpart in the resulting lattice:

1. For every $c \in \mathfrak{B}(\mathbb{K}_1)$, there is $d \in \mathfrak{B}(\mathbb{K}_2)$ such that $(c, d) \in \mathfrak{B}_\diamond(\mathbb{K}_1, \mathbb{K}_2)$;
2. For every $d \in \mathfrak{B}(\mathbb{K}_2)$, there is $c \in \mathfrak{B}(\mathbb{K}_1)$ such that $(c, d) \in \mathfrak{B}_\diamond(\mathbb{K}_1, \mathbb{K}_2)$;

This is useful if we believe that every concept of the initial lattices corresponds to a (complete w.r.t. the given attributes) description of some entity (for the combination of the Bell-LaPadula and Biba models, to a security label to be attached to some subject or object)—then, we cannot discard any concept.

We consider two important subsets of $\mathfrak{B}_\diamond(\mathbb{K}_1, \mathbb{K}_2)$:

$$\mathfrak{B}_\triangle(\mathbb{K}_1, \mathbb{K}_2) := \{((A^{II}, A^I), (B^{JJ}, B^J)) \mid A \subseteq G, B = \bigcup_{a \in A} \{a\}^J\}$$

and

$$\mathfrak{B}_\triangledown(\mathbb{K}_1, \mathbb{K}_2) := \{((B^I, B^{II}), (A^J, A^{JJ})) \mid A \subseteq G, B = \bigcup_{a \in A} \{a\}^I\}$$

It can be easily seen that each of the sets $\mathfrak{B}_\triangle(\mathbb{K}_1, \mathbb{K}_2)$ and $\mathfrak{B}_\triangledown(\mathbb{K}_1, \mathbb{K}_2)$ is a lattice, and (possibly, without the bottom and top elements, respectively) they are the set of all joins and the set of all meets of subsets of pairs (1), respectively.

We now define two contexts whose concept lattices are isomorphic to the lattices $\mathfrak{B}_\triangle(\mathbb{K}_1, \mathbb{K}_2)$ and $\mathfrak{B}_\triangledown(\mathbb{K}_1, \mathbb{K}_2)$.

Definition 8. *Let $\mathbb{K}_1 = (G, M, I)$ and $\mathbb{K}_2 = (N, G, J)$ be formal contexts. Then,*

$$\mathbb{K}_1 \triangle \mathbb{K}_2 := (G, G \cup M, I \cup I_\triangle),$$

where $I_\triangle = \{(g, h) \mid g \in G, h \in G, \text{ and } \{g\}^J \subseteq \{h\}^J\}$, and

$$\mathbb{K}_1 \triangledown \mathbb{K}_2 := (G \cup N, G, J \cup J_\triangledown),$$

where $J_\triangledown = \{(g, h) \mid g \in G, h \in G, \text{ and } \{h\}^I \subseteq \{g\}^I\}$.

Proposition 1. *The concept lattice $\mathfrak{B}(\mathbb{K}_1 \triangle \mathbb{K}_2)$ is isomorphic to the lattice $\mathfrak{B}_\triangle(\mathbb{K}_1, \mathbb{K}_2)$ and the concept lattice $\mathfrak{B}(\mathbb{K}_1 \triangledown \mathbb{K}_2)$ is isomorphic to the lattice $\mathfrak{B}_\triangledown(\mathbb{K}_1, \mathbb{K}_2)$.*

Proof. We define a mapping $f_\triangle : \mathfrak{B}(\mathbb{K}_1 \triangle \mathbb{K}_2) \to \mathfrak{B}_\triangle(\mathbb{K}_1, \mathbb{K}_2)$ as follows. For $A, C \subseteq G$,

$$f_\triangle((A, A^I \cup C)) = ((A^{II}, A^I), (C^J, C)).$$

To show that this mapping is well-defined we need to prove that, for every concept (A, B) of $\mathfrak{B}(\mathbb{K}_1 \triangle \mathbb{K}_2)$, there is $C \subseteq G$ such that $B = A^I \cup C^{JJ}$. It is obvious that $B \cap M = A^I$. One can see that $B \cap G$ is $(\cdot)^J$-closed, as

$B \cap G = \{h \mid h \in G \text{ and } \forall g \in A(\{g\}^J \subseteq \{h\}^J)\} = (\bigcup_{g \in A}\{g\}^J)^J$. Note that $((A^{II}, A^I), (C^J, C)) \in \underline{\mathfrak{B}}_\triangle(\mathbb{K}_1, \mathbb{K}_2)$, since $C = (\bigcup_{g \in A}\{g\}^J)^J$.

Clearly, $f_\triangle$ is order-preserving. The inverse mapping is as follows:

$$f_\triangle^{-1}((A^{II}, A^I), (B^{JJ}, B^J)) = (\{a \in A^{II} \mid \{a\}^J \subseteq B^{JJ}\}, A^I \cup B^J).$$

Without loss of generality, we may assume that $B = \bigcup_{a \in A}\{a\}^J$. Then, $A \subseteq \{a \in A^{II} \mid \{a\}^J \subseteq B^{JJ}\}$, and it is easy to see that $f_\triangle^{-1}((A^{II}, A^I), (B^{JJ}, B^J))$ is indeed a concept of $\mathbb{K}_1 \triangle \mathbb{K}_2$.

The mapping $f_\triangledown : \underline{\mathfrak{B}}(\mathbb{K}_1 \triangledown \mathbb{K}_2) \to \underline{\mathfrak{B}}_\triangledown(\mathbb{K}_1, \mathbb{K}_2)$ and the inverse mapping are given below:

$$f_\triangledown((A^J \cup C, A)) = ((C, C^I), (A^J, A^{JJ}));$$

$$f_\triangledown^{-1}((B^I, B^{II}), (A^J, A^{JJ})) = (A^J \cup B^I, \{a \in A^{JJ} \mid \{a\}^I \subseteq B^{II}\}).$$

We omit the rest of the proof. □

Let us consider implications in these contexts. How should they be interpreted? An implication of $\mathbb{K}_1 \triangle \mathbb{K}_2$ may contain elements of G, as well as elements of M. An implication $A \to B$ holds in $\mathbb{K}_1 \triangle \mathbb{K}_2$ if and only if, for all $g \in G$, $B \cap M \subseteq g^I$ and $g^J \subseteq (B \cap G)^J$ whenever $A \cap M \subseteq g^I$ and $g^J \subseteq (A \cap G)^J$. In words:

> If (in the two initial contexts) an element of G is related to all elements from $A \cap M$ and no elements from $N \setminus (A \cap G)^J$, then it is related to all elements from $B \cap M$ and no elements from $N \setminus (B \cap G)^J$.

An (object) implication $C \to D$ over $G \cup N$ in the context $\mathbb{K}_1 \triangledown \mathbb{K}_2$ reads as follows:

> If (in the two initial contexts) an element from G is related to all elements from $C \cap N$ and no elements from $M \setminus (C \cap G)^I$, then it is related to all elements from $D \cap N$ and no elements from $M \setminus (D \cap G)^I$.

These implications express the relation between the attributes of M and negations of attributes from N (and vice versa). Note however that the implication system of, e.g., $\mathbb{K}_1 \triangle \mathbb{K}_2$ is different from the implication system of $(G, M \cup N, I \cup (G \times N) \setminus J^{-1})$, the context obtained by combining attributes from M and negations of attributes from N. The difference is due to the fact that, in the case of $\mathbb{K}_1 \triangle \mathbb{K}_2$, our additional attributes are only certain conjunctions of negated attributes from N.

Now, we outline how attribute exploration can be organized in the case when the set G described by two contexts (G, M, I) and (N, G, J) is not completely available. We take the problem of combining Bell-LaPadula and Biba model as an example.

Suppose that the elements of M are confidentiality categories and confidentiality levels (of the Bell-LaPadula model) and the elements of N are integrity categories and levels (of the Biba model).

We start by constructing $\mathbb{K}_1 = (G, M, I)$ and $\mathbb{K}_2 = (N, G, J)$ using standard attribute exploration and object exploration[4], respectively. The user is asked to verify implications over M and implications over N (but not implications over $M \cup N$). However, when providing a counterexample to one of these implications, the user must enter its complete description in terms of both M and N.

Now, we know all confidentiality labels and all integrity labels. What we do not know is what combinations of confidentiality and integrity labels are possible. We construct contexts $\mathbb{K}_1 \triangle \mathbb{K}_2$ and $\mathbb{K}_1 \triangledown \mathbb{K}_2$. At this stage, the concept lattices of these contexts might not be exactly what we need, but they do contain every combination of confidentiality and security labels attached to at least one element from G. To explore other possibilities, we use attribute exploration on $\mathbb{K}_1 \triangle \mathbb{K}_2$ and object exploration on $\mathbb{K}_1 \triangledown \mathbb{K}_2$ to build appropriate lattices.

Questions asked during attribute exploration of $\mathbb{K}_1 \triangle \mathbb{K}_2$ sound more natural than one could imagine by looking at the previously given formulas. If $M = \{a, b, c\}$ and $N = \{d, e, f\}$, we may be asked to verify an implication $\{\neg d\} \rightarrow \{b, \neg e\}$, which can be understood as "if a subject may not write to d, it may not write to e, either, but may read from b instead".

Having built $\mathbb{K}_1 \triangle \mathbb{K}_2$ and $\mathbb{K}_1 \triangledown \mathbb{K}_2$, we can construct $\mathfrak{B}_\Diamond(\mathbb{K}_1, \mathbb{K}_2)$ by applying the join and meet operations to $f_\triangle(\mathfrak{B}(\mathbb{K}_1 \triangle \mathbb{K}_2)) \cup f_\triangledown(\mathfrak{B}(\mathbb{K}_1 \triangledown \mathbb{K}_2))$.

5   Conclusion

We have discussed some lattice-related aspects of access control models. It seems likely that attribute exploration can be useful in their construction. The process may take long, but it is worth the effort in serious applications: attribute exploration explicitly forces the system developer to consider issues that can be overlooked otherwise. Although a lattice produced by attribute exploration can contain more elements than it is necessary for a given set of security labels and categories, it is a better starting point than the lattice of all subsets and it still contains all necessary elements. In fact, the presence of extra elements may indicate that the choice of security labels and categories is not optimal.

We have also shown how a model addressing confidentiality and a model addressing integrity can be combined within one lattice and how this lattice can be obtained with the help of attribute exploration. This is only a step towards formalizing the combined model and further research is necessary to estimate the benefits of the proposed approach and to evaluate other possibilities.

---

[4] Object exploration is a process dual to attribute exploration: the user is asked to confirm an implication between objects (also defined dually to the attribute implication) or enter a new attribute as a counterexample. The only reason we are talking about object exploration is that the set N is the object set of $\mathbb{K}_2$. One can think of object exploration as attribute exploration in the transposed context (where objects and attributes change places).

References

1. Sandhu, R.: Lattice-based access control models. IEEE Computer **26** (1993) 9–19
2. Denning, D.: A lattice model of secure information flow. Comm. ACM **19** (1976) 236–243
3. Biba, K.: Integrity considerations for secure computer systems. Report TR-3153, Mitre Corporation, Bedford, Mass. (1977)
4. Gollmann, D.: Computer Security. John Wiley & Sons Ltd, Chichester (1999)
5. Bell, D., LaPadula, L.: Secure computer systems: Mathematical foundations and model. Report M74-244, Mitre Corporation, Bedford, Mass. (1975)
6. Ganter, B., Wille, R.:   Formal Concept Analysis: Mathematical Foundations. Springer, Berlin (1999)
7. Lipner, S.: Nondiscretionary controls for commercial applications. In: Proc. IEEE Symp. Security and Privacy, Los Alamitos, Calif., IEEE CS Press (1982) 2–10
8. Smith, G.: The Modeling and Representation of Security Semantics for Database Applications. PhD thesis, George Mason Univ., Fairfax, Va. (1990)
9. Birkhoff, G.: Lattice Theory. Amer. Math. Soc. Coll. Publ., Providence, R.I. (1973)
10. Guigues, J.L., Duquenne, V.: Familles minimales d'implications informatives resultant d'un tableau de données binaires. Math. Sci. Humaines **95** (1986) 5–18
11. Ganter, B.: Attribute exploration with background knowledge. Theoretical Computer Science **217** (1999) 215–233
12. Ganter, B., Krausse, R.: Pseudo models and propositional horn inference. Technical Report MATH-AL-15-1999, Technische Universität Dresden, Germany (1999)
13. Ganter, B.: Two basic algorithms in concept analysis. Preprint Nr. 831, Technische Hochschule Darmstadt (1984)

An Application of Relation Algebra to Lexical Databases

Uta Priss and L. John Old

Napier University, School of Computing
u.priss@napier.ac.uk, j.old@napier.ac.uk

Abstract. This paper presents an application of relation algebra to lexical databases. The semantics of knowledge representation formalisms and query languages can be provided either via a set-theoretic semantics or via an algebraic structure. With respect to formalisms based on n-ary relations (such as relational databases or power context families), a variety of algebras is applicable. In standard relational databases and in formal concept analysis (FCA) research, the algebra of choice is usually some form of Cylindric Set Algebra (CSA) or Peircean Algebraic Logic (PAL). A completely different choice of algebra is a binary Relation Algebra (RA). In this paper, it is shown how RA can be used for modelling FCA applications with respect to lexical databases.

1 Introduction

Formal Concept Analysis (FCA) is a method for data analysis and knowledge representation that provides visualisations in the form of mathematical lattice diagrams for data stored in "formal contexts". A formal context consists of a binary relation between what are named "formal objects" and "formal attributes" (Ganter & Wille, 1999). FCA can, in principle, be applied to any relational database. Lexical databases, which are electronic versions of dictionaries, thesauri and other large collections of words provide a challenge for FCA software because the formal contexts of lexical databases tend to be fairly large consisting of 100,000s of objects and attributes. An overview of FCA applications with respect to lexical databases can be found in Priss & Old (2004). That paper also provides pointers to future research, for example, with respect to specific implementations of a construct called a "neighbourhood lattice". In this paper, we are continuing the thread of that research and provide a further mathematical analysis supported by empirical data from Roget's Thesaurus.

For the development of a mathematical analysis of structures in lexical databases it is beneficial to make use of existing relational methods. In particular, this paper elaborates on how methods from relation algebra (RA) can be applied. RA is a perfect companion for FCA because, while FCA facilitates visualisations of binary relations, RA defines operations on binary relations. Thus RA seems a natural candidate for a context-relation algebra as defined in Sect. 3. Section 2 compares RA to other algebras of relations and explains why RA is more suitable for FCA applications than the other methods. Section 4 focuses on the applications of RA/FCA in the area of lexical databases using examples from Roget's Thesaurus.

H. Schärfe, P. Hitzler, and P. Øhrstrøm (Eds.): ICCS 2006, LNAI 4068, pp. 388–400, 2006.

2   Algebras as a Representation of Logical Structures

Logicians usually employ set-theoretic models for the formal semantics of formal languages. But it is also possible to use algebraic structures instead of set-theoretic models because a formal language can be first interpreted as an algebra, which then further can be interpreted as a set-theoretic structure. The advantage of this approach is that additional structures are represented in the algebra. Historically, the development of formal logic has been closely tied to the development of algebras of relations (cf. Maddux (1991)) as is evident in Peirce's work in both fields. The modern field of Algebraic Logic studies the relationship between different types of logics and different types of algebras (cf. Andreka et al., 1997). Surprisingly many structures from logic can be translated into algebraic structures and vice versa. The translations are beneficial because theorems and methods from either field become available in the other.

With respect to relational databases and other structures based on n-ary relations (such as power context families in FCA (Wille, 2002)), traditionally the most popular algebras use n-ary relations as basic units. Codd's (1970) modelling of relational databases with relational algebra (RLA) is well known. With respect to power context families, the algebra of choice is usually Peirce Algebraic Logic (PAL), such as in the modelling suggested by Eklund et al. (2000). Both RLA and PAL and a third algebra called Krasner algebra (cf. Hereth Correia & Pöschel, 2004) are very similar in nature to Henkin & Tarski's (1961) Cylindric Set Algebras (CSA). CSA, RLA, PAL and Krasner algebra all have the expressive power of first order logic (FOL) with equality (cf. Van den Bussche (2001) for CSA and RLA and Hereth Correia & Pöschel (2004) for PAL and Krasner). It is quite possible that they are equivalent or even isomorphic to each other in some sense (for CSA and RLA this is investigated by Imielinski & Lipski (1984); for PAL and Krasner, by Hereth Correia & Pöschel (2004)).

In addition to CSA, Tarski (1941) also studied an algebra which is quite different and goes back to Peirce, De Morgan and Schroeder (cf. van den Bussche (2001) for an overview). This algebra is called relation algebra (RA) – the similarity in name to Codd's relational algebra (RLA) is unfortunate, but these names are established in the literature. The difference between RA and CSA is that RA uses exclusively binary relations, whereas CSA and the other algebras use n-ary relations. At first sight, this appears to be a limitation of RA. But in fact, RA is quite powerful and has the expressive power of FOL with up to three variables. If one adds a form of projection operation to RA, one can obtain what is called a Fork algebra (FRA)[1] which has the expressive power of full FOL, but still only uses binary relations (Frias et al. 2004). In FRA, n-ary relations are encoded as a part-whole structure among binary relations. For example, a ternary relation consists of a binary relation, whose left or right element contains two parts which can be retrieved using the projection operation. While this may sound complicated, the operations of RA are overall much simpler (and easier to implement) than

---

[1] For the purposes of modelling FCA contexts with RA (Priss, 2006), only two elements that contain information about projections are required from FRA but none of the other operations. Thus in the remainder of this paper, the abbreviation RA is used to mean "RA including FRA elements if needed".

the operations of CSA; and RA has many interesting properties that can be calculated on an abstract level (cf. Pratt (1992) and (1993) for an overview).

We believe that RA is currently under-valued among computer scientists, although there has recently been an increased interest in relational methods in computer science as evidenced by the recent creation of a new journal in this field[2]. Part of the reason for RA's lack of popularity may be the fact that any university student who learns some mathematics is likely to learn some linear algebra. But not even every full time mathematics student learns about universal algebra, algebraic logic and lattice theory. With respect to physical space, linear algebra and vector spaces are useful models, but we would argue that, with respect to information spaces, perhaps other algebras than linear algebra can be more useful[3]. This claim is supported by the results of algebraic logic, which demonstrate the close connection between logic and algebra. An elaboration of this claim with respect to the use of non-linear algebra-based methods in information retrieval can be found in Priss (2000).

Apart from the availability of simpler operations in RA, another advantage of RA is the availability of visualisations. There are several types of visualisations for different algebras: Venn Diagrams for set-based Boolean algebras, n-dimensional coordinate systems for vector spaces and CSA-style operations (cf. Andreka et al. 1997), Peirce's Existential Graphs for PAL-style operations. But all of these have the limitation that they become difficult or impossible to draw as the complexity increases. They tend to be suitable for instances of relations (the relationship between a few points in space or between a few Existential Graphs), but not for an overview of a larger system of relations. On the other hand, because RA uses solely binary relations, the visualisation methods of FCA are instantly available (i.e. concept lattices). Although FCA visualisations also have a limit with respect to how much complexity can be represented, that limit is much higher than for n-dimensional vector spaces. For example, it is already difficult to represent 3-dimensional vector spaces on paper, but a concept lattice can easily represent a (Boolean) lattice with up to 5 independent co-atoms (corresponding to 5 dimensions), and many more if they are not completely independent. Methods of zooming and nesting are available for larger systems using the software TOSCANA (Eklund et al., 2000). In that manner, FCA visualisations enable users to obtain an overview of larger sets of data and to explore hidden structures among the data.

Of course, FCA visualisations have always been applied to power context families and many-valued contexts. The strategy that is normally used is to first apply PAL operations to formal contexts and then to extract a binary relation from the n-ary relations, by selecting two columns with or without scaling and applying combination operations to columns before selecting them. But that implies that it is also conceivable to reverse these two steps and to construct a binary encoding of the n-ary relations right away and then to use RA to operate on these formal contexts.

---

[2] www.jormics.org

[3] An anonymous reviewer of this paper remarked that a similar dichotomy can be found in other areas of mathematics: "in topology, Hausdorff spaces are convincing models of real geometries, but T0 spaces are more interesting from a logical viewpoint. Likewise, classical metric spaces are nice generalizations of Euclidian distances, while ultrametrics are often more appropriate to measure the 'distance' between pieces of information".

3 Context-RAs and Context Algebraic Structures

This section provides a brief introduction into how relation algebra can be applied to FCA. A more detailed explanation of this topic can be found in Priss (2006), from which the definitions below are taken. The cross table of a formal context can be considered a Boolean (or binary) matrix in the sense of Kim (1982), for which matrix operations are defined as follows: with $(i, j)_I$ denoting the element in row i, column j in matrix I and $\vee$, $\wedge$ and $\neg$ denoting Boolean OR, AND and NOT: $(i, j)_{I \cup J} := (i, j)_I \vee (i, j)_J$; $(i, j)_{\overline{I}} := \neg(i, j)_I$; $(i, j)_{I \circ J} := 1$ iff $\exists_k : (i, k)_I \wedge (k, j)_J$; $(i, j)_{I^d} := (j, i)_I$. The operations $\cap$ and $\subseteq$ are as usual: $I \cap J := \overline{\overline{I} \cup \overline{J}}$ and $I \subseteq J :\Longleftrightarrow I \cap J = I$. A matrix containing all 0's is denoted by nul; a matrix with all 1's is one, a matrix with 1's on the diagonal and 0's otherwise is dia. A matrix is symmetric if $I = I^d$, reflexive if $dia \subseteq I$, transitive if $I \circ I \subseteq I$.

To distinguish between operations on sets and on matrices, we use typewriter font (e.g. A, B) for sets and italics (e.g. A, B) for matrices. A formal context normally consists of two sets and a matrix: $(\mathrm{G}, \mathrm{M}, I)$. In this paper, all formal contexts of an application are assumed to be defined with respect to a finite, linearly ordered set ACT called an *active domain*. That means that for all sets of objects and attributes: $\mathrm{G}, \mathrm{M} \subseteq \mathrm{ACT}$. A set $\mathcal{A}$ is defined as the set of all Boolean $|\mathrm{ACT}| \times |\mathrm{ACT}|$ matrices together with an interpretation that ensures that, semantically, for $I \in \mathcal{A}$ and $1 \leq n \leq |\mathrm{ACT}|$, the nth row and column in I corresponds to the nth element in ACT. It is then said that I *is based on $\mathcal{A}$* denoted by $I_{\mathcal{A}}$ (although the subscript can be omitted if it is clear which active domain is meant).

Definition 1. *A matrix-RA based on $\mathcal{A}$ is an algebra $(R, \cup, {}^-, one, \circ, {}^d, dia())$ where $one \in R$ is a reflexive, symmetric and transitive matrix; $R := \{I_{\mathcal{A}} | I_{\mathcal{A}} \subseteq one\}$ is a set of Boolean matrices; $\cup, {}^-, \circ, {}^d$ are the usual Boolean matrix operations; and for any set $\mathrm{S} \subseteq \mathrm{ACT}$ and $\mathrm{a(n)}$ denoting the nth element in ACT, $dia(\mathrm{S})$ is defined by $(i, j)_{dia(\mathrm{S})} = 1$ iff $i = j$ and $\mathrm{a(i)} \in \mathrm{S}$ (but only if $dia(\mathrm{S}) \subseteq one$).*

It can be shown that a matrix-RA is an RA and fulfills all the axioms of an RA, such as $(R, \cup, \cap, {}^-, nul, one)$ is a Boolean algebra; $\circ$ is associative and distributive with $\cup$; dia is a neutral element for $\circ$ (but unique inverse elements need not exist); $(I^d)^d = I$; $(I \cup J)^d = I^d \cup J^d$; $(I \circ J)^d = J^d \circ I^d$; and so on (see Priss (2006)).

With $\mathrm{G}, \mathrm{M} \subseteq \mathrm{ACT}$, a formal context can be represented "based on $\mathcal{A}$": the sets can be represented as $dia(\mathrm{G})$ and $dia(\mathrm{M})$ or combined as $sqr(\mathrm{G}, \mathrm{M}) := dia(\mathrm{G}) \circ one \circ dia(\mathrm{M})$. A formal context can then be represented using three matrices: $(dia(\mathrm{G}), dia(\mathrm{M}), I_{\mathcal{A}})$ or using two matrices $(sqr(\mathrm{G}, \mathrm{M}), I_{\mathcal{A}})$, in both cases with $I_{\mathcal{A}} \subseteq sqr(\mathrm{G}, \mathrm{M})$. A formal context without empty rows or columns can be represented by a single matrix: $I_{\mathcal{A}}$. A natural RA can be defined for any set of formal contexts:

Definition 2. *A context-RA based on $\mathcal{A}$ for a set of formal contexts is the smallest matrix-RA based on $\mathcal{A}$ that contains these contexts.*

Instead of representing every formal context using $|\mathrm{ACT}| \times |\mathrm{ACT}|$ matrices, it is also possible to represent formal contexts in the usual way, but in that case, the RA operations need to be modified (see the next definition). The sets $\mathrm{G}, \mathrm{M} \subseteq \mathrm{ACT}$ are assumed to be linearly ordered according to ACT. This linear order ensures that operations can be defined among formal contexts which have different sets of objects and attributes. In this representation, a formal context (G, M, I) *based on* ACT contains a Boolean matrix I of size $|G| \times |M|$ where the ith row corresponds to the ith element in G and the jth column corresponds to the jth element in M. This can be denoted as $I_{G,M}$. But in the remainder of this paper the subscript of I is omitted, if the formal context is denoted using sets of objects and attributes (not matrices). In these cases, it is always assumed that I's dimensions correspond to its sets of objects and attributes.

Definition 3. *For formal contexts* $\mathcal{K}_1 := (G_1, M_1, I)$ *and* $\mathcal{K}_2 := (G_2, M_2, J)$ *based on* ACT *the following context operations are defined:*

$\mathcal{K}_1 \sqcup \mathcal{K}_2 := (G_1 \cup G_2, M_1 \cup M_2, I \sqcup J)$ *with* $gI \sqcup Jm :\Longleftrightarrow gIm$ *or* gJm

$\mathcal{K}_1 \sqcap \mathcal{K}_2 := (G_1 \cup G_2, M_1 \cup M_2, I \sqcap J)$ *with* $gI \sqcap Jm :\Longleftrightarrow gIm$ *and* gJm

$\mathcal{K}_1 \diamond \mathcal{K}_2 := (G_1, M_2, I \diamond J)$ *with* $gI \diamond Jm :\Longleftrightarrow \exists_{n \in (M_1 \cap G_2)} : gIn$ *and* nJm

$\overline{\mathcal{K}_1} := (G_1, M_1, \overline{I}); \quad \mathcal{K}_1^d := (M_1, G_1, I^d).$

Definition 4. *A context algebraic structure (CAS) based on* ACT *is a three sorted algebra* $(R_1, R_2, R_3, \sqcup, ^-, \diamond, ^d, dia(), \mathtt{set}(), (\cdot, \cdot, \cdot))$ *where* R_2 *is a set of subsets of* ACT, R_3 *is a set of Boolean matrices,* R_1 *is a set of formal contexts based on* ACT *and constructed using the partial function* $(\cdot, \cdot, \cdot) : R_2^2 \times R_3 \to R_1;$ $\sqcup, ^-, \diamond, ^d$ *are according to Definition 3;* $\mathtt{set}_G(I) := \{g \in G \mid \exists_{m \in M} : gIm\};$ $\mathtt{set}_M(I) := \{m \in M \mid \exists_{g \in G} : gIm\};$ *and* $dia_G(S_\to)$ *is defined by* $(i, j)_{dia_G(S_\to)} = 1$ *iff* $i = j$ *and for the ith element in* G: $g(i) \in S; dia_M(S_\uparrow)$ *is defined analogously.*

If $G_1 = G_2$, $M_1 = M_2$, then $\cup$ and $\cap$ can be used instead of $\sqcup$ and $\sqcap$. If $M_1 = G_2$, then $\circ$ can be used instead of $\diamond$. Using these definitions, the normal FCA operations can be defined, such as the prime operator (cf. Priss (2006)), using either context-RAs or CAS. For many-valued contexts and power context families, a fork algebraic extension is required (Priss, 2006), but that is not relevant for this paper.

For implementation purposes, it should be noted that it is not suggested that by using matrix-RAs for modelling contexts that these actually have to be represented as giant matrices in a computer. In FCA applications, giant formal contexts are usually sparse matrices or contain many duplicate rows or columns. When dealing with sparse matrices, non-RA-based algorithms might be faster, but RA can be used as an underlying theory for proving theorems. This is shown using examples from lexical databases in the next sections. On the other hand, with respect to small contexts (i.e. less than 100 objects and attributes), RA operations can be directly implemented. RA software already exists[4] and could possibly be combined with existing FCA software. RA operations could then be used as an additional query interface.

4   Lexical Databases

A lexical database is defined here as an organised collection of words in electronic form. That includes dictionaries such as Webster's or the Oxford English Dictionary,

---

[4] RelView: www.informatik.uni-kiel.de/~progsys/relview.shtml

with formal definitions and glosses; semantic lexicons such as Roget's Thesaurus (Roget, 1962) and WordNet (Miller et al. (1990)), organised by conceptual categories and lexical relations; and bilingual dictionaries, where the primary structure is a conceptual mapping between languages. Previously, Priss & Old (2004) developed methods for suitable representations for lexical databases – a set of guidelines for visualising lexical relations using FCA.

Roget's Thesaurus (ROGET), in particular in the form of a lexical database[5], is the source of examples used in this paper. ROGET consists of a conceptual hierarchy, at the bottom of which are words grouped by shared meaning. These groups are referred to as senses, and members of a particular group are commonly referred to as synonyms. The relationship between words and senses can be represented as a formal context where instances (actual entries of particular words in particular senses) are represented by a cross in the context (e.g. Fig. 1). In this paper, the term "entry" is reserved for this relationship between words and senses. A word has usually more than one sense (several entries in ROGET); and each sense is usually represented by (and contains) more than one synonym. The word "over", for example, has 22 senses in ROGET, represented by 22 entries in the thesaurus; and the number of entries in those 22 senses (synonyms sharing a sense with over) ranges from 3 to 37. The question which arises is: which words and senses should be included in a formal context for a given word? A possible answer to this question is provided by neighbourhood contexts and lattices as discussed in the next section.

WORDS ⟶ SYNONYMY

SENSES	over	above	beyond	across	past	in excess of	...
9:12:1	X						
206:24:2	X	X					
206:27:4	X	X	X			X	
227:40:1	X	X					
...	X	X	X	X	X		

POLYSEMY

Fig. 1. Words, senses and entries in Roget's Thesaurus

4.1 A Formalisation of Neighbourhood Lattices Using RA

A mathematical definition of "neighbourhood lattices" of Roget's Thesaurus was originally defined in an unpublished manuscript by Rudolf Wille and can be found in Priss (1998). But, although neighbourhood lattices have been produced for numerous examples over the years, so far the underlying theory has not been further investigated. This section shows how RA can be used to analyse and model the structures of neighbourhood lattices. A neighbourhood lattice starts with a word w, collects all the senses that w

---

[5] The relational database used in this paper is based on Roget (1962), which was converted to electronic format by Dr. W. A. Sedelow and Dr. S. Yeates Sedelow and edited and enhanced by L. J. Old.

has, then collects all other words that also have these senses and so on. Or, alternatively, the process can be started with a sense, collecting all its words and so on. The operation of collecting all objects that an attribute has (or vice versa) is called a "plus operator" (Priss & Old, 2004). Under different names, the idea of the plus operator can be found in many linguistic applications, such as starting with a word in one language, then retrieving all the translations in another language, and so on (e.g. Wunderlich (1980) or Dyvik (1998)). Using RA, the plus operator is defined as follows:

Definition 5. *For a context* (G, M, I), $H \subseteq G$, $N \subseteq M$, *a column matrix H that has a 1 in the position of each element of* H *(i.e.,* $H := diag_G(H_\rightarrow) \circ one_{G,\{\}}$*), and N an analogous row matrix, a plus operator is defined as* $H^+ := H^d \circ I$ *and* $N^+ := I \circ N^d$.

It follows that H^+ is a row matrix and N^+ is a column matrix. Applying the plus operator twice yields $H^{++} = I \circ (H^d \circ I)^d = I \circ I^d \circ H$ and $N^{++} = N \circ I^d \circ I$; three times: $H^{+++} = H^d \circ I \circ I^d \circ I$ and $N^{+++} = I \circ I^d \circ I \circ N^d$, and so on. A neighbourhood context can utilise the plus operator any number of times and can be started with sets of objects or attributes. A typical neighbourhood context is $(set_G(H^{++}), set_M(H^{+++}), I)$. The RA representation of the plus operator implies that the operator is essentially formed by repeated composition of I and I^d. For finite matrices repeated composition leads to a "transitive closure", normally defined as $I^{trs} := I \cup (I \circ I) \cup (I \circ I \circ I) \cup \dots$. It should be noted that the calculation of a transitive closure is not an FOL operation, but an additional operation that cannot axiomatically be derived from the other RA operations.

   If a matrix I is reflexive in all rows that are not empty, then $I^{trs} = I \circ I \circ I \circ \dots$ because $\forall_{x,y} : (xIy \Rightarrow xIx, xIy) \Rightarrow \forall_{x,y}\exists_z : (xIy \Rightarrow xIz, zIy) \iff I \subseteq I \circ I$. The matrix $(I \circ I^d)$ which is used in neighbourhood contexts is reflexive for non-empty rows, i.e. $xIy \Rightarrow x(I \circ I^d)x$, because $x(I \circ I^d)y \iff \exists_z : xIz, yIz$. Thus $(I \circ I^d)^{trs} = (I \circ I^d) \circ (I \circ I^d) \circ \dots$. The matrix is also symmetric because $I \circ I^d = (I \circ I^d)^d$ according to the rules for ^{d}. This implies the following lemma and leads to the definition of a context which uses this matrix. Figure 2 provides an illustration.

Lemma 1. *If I is a matrix of a formal context without empty rows and without empty columns, then* $(I \circ I^d)^{trs}$ *is an equivalence relation on objects and* $(I^d \circ I)^{trs}$ *is an equivalence relation on attributes.*

Definition 6. *With* $I^{dec} := I \circ (I^d \circ I)^{trs}$, *the neighbourhood closure context of a set* H *of objects is defined as* $(set_G((I \circ I^d)^{trs} \circ H), set_M(H^d \circ I^{dec}), I)$ *and of a set* N *of attributes as* $(set_G(I^{dec} \circ N^d), set_M(N \circ (I^d \circ I)^{trs}), I)$. *Its corresponding lattice is called the neighbourhood closure lattice.*

I^{dec} (where "dec" stands for "decomposition") has some interesting properties. For example, it does not matter whether one starts the calculation with objects or attributes. The properties are summarised in the next lemma. A horizontal decomposition of a lattice is a decomposition into components whose horizontal sum (Ganter & Wille, 1999) is the original lattice. If one removes the top and bottom nodes from a lattice, then the remaining connected graphs are the components of the horizontal decomposition.

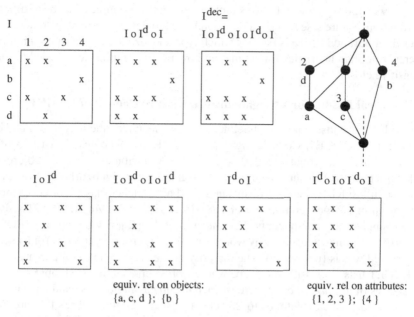

Fig. 2. An example illustrating Lemma 1 and Lemma 2

Lemma 2. *For finite matrices* I:

1. $I^{dec} = (I \circ I^d)^{trs} \circ I = I \circ (I^d \circ I)^{trs}$.
2. $I \subseteq I^{dec}$.
3. I^{dec} *implies a horizontal decomposition of the concept lattice of* I.

Proof: 1) If, for example, $(I^d \circ I)^{trs} = (I^d \circ I)$ and $(I \circ I^d)^{trs} = (I \circ I^d) \circ (I \circ I^d)$ but $(I \circ I^d) \circ (I \circ I^d) \circ I \neq I \circ I^d \circ I$ then this would be a contradiction to $I^d \circ I \circ I^d \circ I = I^d \circ I$ because of $(I^d \circ I)^{trs}$.
2) $xIy \Rightarrow \exists_{z_1, \dots, z_n} : xIz_1, z_2Iz_1, z_2Iz_3, \dots, z_nIy \iff x(I \circ I^d \circ \dots \circ I^d \circ I)y$ with $x = z_i$ for even i and $y = z_i$ for odd i.
3) Because of Lemma 1 and Lemma 2.2.

Thus the neighbourhood closure lattice of an object or an attribute is the component of the horizontal decomposition of the original lattice that contains the object or attribute. This is, of course, what was to be expected given the definition of the plus operator – but the RA modelling provides an interesting representation of how the neighbourhood closure context is computed.

For small formal contexts, it is trivial to calculate neighbourhood closure contexts because one simply needs to decompose the lattice. But for large formal contexts, it is not efficient to first calculate the whole lattice if one is only interested in the neighbourhood of some objects or attributes. In that case, it is much more feasible to calculate only neighbourhood closure lattices because these contain the complete information about an object or attribute. This is especially the case if a lattice contains a few large components and many small components and one is interested in objects or attributes that belong to the small components (see the next section). In some cases, even neighbourhood closure lattices may be too big, in which case simple neighbourhood lattices may be a sufficient approximation. Future research should be conducted to develop heuristics to determine which neighbourhood lattice is best for which type and size of context. It might also be of interest to obtain heuristics for estimating how many iterations are needed to reach the transitive closure.

4.2   Empirical Results for a Neighbourhood Closure Context of ROGET

Using the RA formalisation as a guideline, we have calculated the neighbourhood closure context for ROGET. Considering that the full ROGET context consisting of all words and senses has about 113,000 objects, 71,000 attributes and 200,000 crosses, calculating a transitive closure is not trivial. Calculating the neighbourhood closure context for the full ROGET database results in 26,314 equivalence classes, or components, ranging from one largest component of 138,919 entries (belonging to 38,621 senses), to 22,206 single-entry components. This means that the majority of entries in ROGET (about 70%) are connected, either by words with shared senses (synonymy) or by senses having shared words (polysemy). The majority of the single-entry components derive from ROGET lists, a classification type where words representing such objects as ship parts, species of animal, or capital cities, each occupy a single sense and have no synonyms. The components with 10 to 28 senses typically contain words that are fairly specialised, but still somewhat polysemous, such as belief systems ("freethinker", "nihilist"), occupations ("moneylender"), musical forms ("serenade"), temporal adjectives ("dayly") and countries and capital cities, which happen to occur in more than one list.

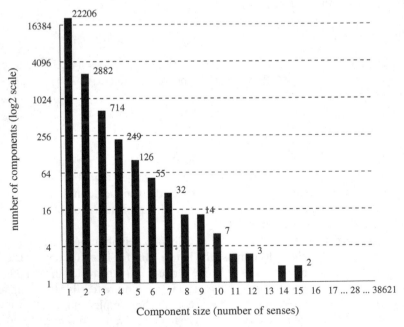

Fig. 3. Exponential distribution of components

The distribution of component sizes is shown in Fig. 3. To display the number of components (y-axis), a log2 function was used. A distribution consisting of one large component, a large number of tiny components and a smaller number of exponentially distributed components in between has been called a "power law distribution" by Strogatz (2001). Power law distributions are a phenomenon found among such diverse areas as word or letter frequency in text, strength of earthquakes, connectivity in the brain and HIV epidemics (Barabasi, 2002), the distribution of forest fires, species group size in biology, and web sites on the Internet. There has been some debate as to the actual significance of power distributions but Barabasi (p. 77) makes the claim that: "Nature normally hates power laws. In ordinary systems all quantities follow bell curves, and correlations decay rapidly, obeying exponential laws. But all that changes if the system is forced to undergo a phase transition. Then power laws emerge – nature's unmistakable sign that chaos is departing in favor of order." Thus it is of potential significance that components in ROGET follow this particular distribution.

4.3 Antonymy

A second example is the treatment of antonymy. Antonyms are generally known as words with opposite meanings. Antonymy is interesting for two reasons. First, antonymy is what is called a "lexical relation" in WordNet (Miller et al., 1990). That means that it is a relationship between senses of words and not between words. For example, "big" and "small" can be considered antonyms. But synonyms of "big" (such as "huge") need not be antonyms of "small" and its synonyms. To model this using RA and the ROGET context from the previous section, it would be necessary to use a Fork relational (FRA) matrix and projections. This is because a binary relation among antonyms would correspond to a relation among word/sense pairs which would need to be encoded using FRA. For the purposes of this paper, we are treating antonymy as a relation among words (similar to synonymy), thus avoiding the need for FRA.

Second, even though antonyms express some form of contrast, negation or duality, they are also usually quite close in meaning. For example, "hot" and "cold" both describe temperature; "up" and "down" both describe direction. Word association studies

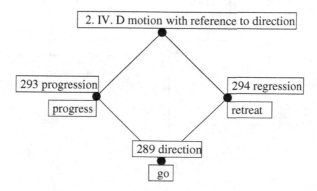

Fig. 4. Antonymous categories share meaning

(where a person is told a word and has to name the first word that comes to mind) show that antonyms are as closely associated in the mental lexicon as are synonyms (Miller et al., 1990).

ROGET does not identify antonyms on a sense-level. The first edition of the book version of Roget's Thesaurus identified antonymous categories by arranging them opposite to each other in the table of contents. A similar structure on the category-level has been included in ROGET. Furthermore, we have included in ROGET some data that is available from word association tests (Nelson et al., 1998) in the form of an antonymy relation among words.

The goal of this section is to detect and analyse this second feature of antonymy (contrast versus shared synonyms) in ROGET. For example (cf. Fig. 4), on a category

Table 1. Antonyms that share synonyms

antonym 1	shared synonym	antonym 2
give	yield	accept
add	compute	subtract
descend	slope	ascend
editor	reviewer	author
after	then	before
white	bleak	black
sweet	brisk	bitter
suck	snuffle	blow
sharp	abrupt	blunt
effect	sequence	result
major	elective	minor
future	sometime	past

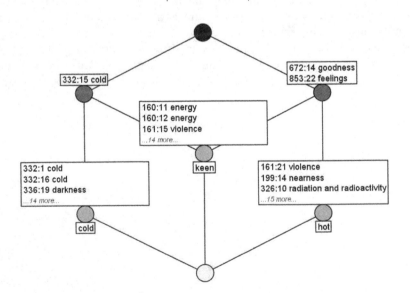

Fig. 5. Overlapping neighbourhoods for antonyms

level, "293 progression" (containing "progress") and "294 regression" (containing "retreat") are antonyms in ROGET. They share the higher-level class of "Motion With Reference To Direction". They also share a synonym "go" in category "289 direction". Table 1 shows a (manually compiled) list of antonyms from word association data, which share a synonym in ROGET. It should be noted that in this case, synonymy is not only evaluated on the sense-level but also on the paragraph level. That means that all words in a paragraph of ROGET are considered synonyms for this purpose.

One approach to studying shared meanings among antonyms is to identify those neighbourhood contexts of antonyms where the sets of attributes are not disjoint. In this case, again it is useful to consider paragraphs as attributes instead of senses because otherwise there would not be much overlap. But because the sets of synonyms at that level are large, the plus operator is used only once. Thus, if H denotes the column matrix which contains exactly one 1 in the position of a word w and J denotes the column matrix which contains one 1 in the position of the antonym of w and if $H^{++} \cap J^{++} \neq \emptyset$ then the neighbourhood context $(\mathtt{set_G}(H \cup J \cup (H^{++} \cap J^{++})), \mathtt{set_M}(H^+ \cup J^+ \cup (H^{++} \cap J^{++})^+), I)$ is formed. Figure 5 shows the example of "hot" and "cold", which share "keen" as a synonym if synonymy is evaluated at the paragraph level. In this case the shared synonym refers to metaphoric senses of the original words. It would be of interest to conduct a more detailed investigation of all such cases in ROGET in future research.

5 Conclusion

This paper demonstrates the applicability of RA formalisations for modelling lexical databases with FCA by using "neighbourhood contexts and lattices". Specific examples are neighbourhood closure contexts of the complete ROGET context and neighbourhood lattices for antonymous words. The examples show that different types of neighbourhood contexts are relevant for different aspects of lexical databases. But the research presented in this paper is not restricted to linguistic applications. Similar structures could be investigated in concept lattices in other application areas.

References

1. Andreka, H.; Nemeti, I.; Sain, I. (1997). *Algebraic Logic*. In: Gabbay (ed.), Handbook of Philosophical Logic, Kluwer.
2. Barabasi, A. L. (2002). *Linked: The New Science of Networks*. Perseus Publishing.
3. Codd, E. (1970). *A relational model for large shared data banks*. Communications of the ACM, 13:6.
4. Dyvik, H. (1998). *A Translational Basis for Semantics*. In: Johansson and Oksefjell (eds.): Corpora and Crosslinguistic Research: Theory, Method and Case Studies, Rodopi, 51-86.
5. Eklund, P,; Groh, B.; Stumme, G.; Wille, R. (2000). *A contextual-logic extension of TOSCANA*. In: Ganter & Mineau (eds.), Conceptual Structures: Logical, Linguistics, and Computational Issues, Springer LNAI 1867, p. 453-667.
6. Frias, M.; Veloso, P.; Baum, G. (2004). *Fork Algebras: Past, Present and Future*. Journal on Relational Methods in Computer Science, 1, p. 181-216.
7. Ganter, B.; Wille, R. (1999). *Formal Concept Analysis*. Mathematical Foundations. Berlin-Heidelberg-New York: Springer, Berlin-Heidelberg.

8. Henkin, L.; Tarski, A. (1961). *Cylindric algebras*. In: Dilworth, R. P., Lattice Theory, Proceedings of Symposia in Pure Mathematics, AMS, p. 83-113.
9. Hereth Correia, J.; Pöschel, R. (2004). *The Power of Peircean Algebraic Logic (PAL)*. In: Eklund (Ed.) Concept Lattices. LNCS 2961, Springer Verlag. p. 337-351.
10. Imielinski, T.; Lipski, W. (1984). *The relational model of data and cylindric algebras*. Journal of Computer and Systems Sciences, 28, p. 80-102.
11. Kim, K. H. (1982). *Boolean Matrix Theory and Applications*. Marcel Dekker Inc.
12. Maddux R. (1991). *The origin of relation algebras in the development and axiomatization of the calculus of relations*. Studia Logica, 5, 3-4, p. 421-456.
13. Nelson, D. L., McEvoy, C. L., & Schreiber, T. A. (1998). The University of South Florida word association, rhyme, and word fragment norms. Available at http://www.usf.edu/FreeAssociation/
14. Miller, G.; Beckwith, R.; Fellbaum, C.; Gross, D.; Miller, K. J. (1990). *Introduction to WordNet: an on-line lexical database*. International Journal of Lexicography, 3, 4, p. 235-244.
15. Pratt, V.R. (1992). *Origins of the Calculus of Binary Relations*. Proc. IEEE Symp. on Logic in Computer Science, p. 248-254.
16. Pratt, V.R. (1993). *The Second Calculus of Binary Relations*. Proc. 18th International Symposium on Mathematical Foundations of Computer Science, Gdansk, Poland, Springer-Verlag, p. 142-155.
17. Priss, U. (1998). *Relational Concept Analysis: Semantic Structures in Dictionaries and Lexical Databases*. (PhD Thesis) Verlag Shaker, Aachen 1998.
18. Priss, U. (2000). *Lattice-based Information Retrieval*. Knowledge Organization, Vol. 27, 3, 2000, p. 132-142.
19. Priss, U.; Old, L. J. (2004). *Modelling Lexical Databases with Formal Concept Analysis*. Journal of Universal Computer Science, Vol 10, 8, 2004, p. 967-984.
20. Priss, U. (2006). *An FCA interpretation of Relation Algebra*. ICFCA'06. Springer Verlag, LNCS (to appear).
21. Roget, Peter Mark (1962). *Roget's International Thesaurus*. 3rd Edition Thomas Crowell, New York.
22. Strogatz, H. (2001). Exploring complex networks. Nature, 410, 268-276.
23. Tarski, A. (1941). *On the calculus of relations*. Journal of Symbolic Logic, 6, p. 73-89.
24. Van den Bussche, J. (2001). *Applications of Alfred Tarski's Ideas in Database Theory*. Proceedings of the 15th International Workshop on Computer Science Logic. LNCS 2142, p. 20-37.
25. Wille, R. (2002). *Existential Concept Graphs of Power Context Families*. In: Priss; Corbett; Angelova (Eds.) Conceptual Structures: Integration and Interfaces. LNCS 2393, Springer Verlag, p. 382-395.
26. Wunderlich, D. (1980). *Arbeitsbuch Semantik*. Königstein/Ts: Athenaeum.

A Framework for Analyzing and Testing Requirements with Actors in Conceptual Graphs

B.J. Smith and Harry Delugach

Computer Science Department
University of Alabama in Huntsville
Huntsville, AL 35899
smithbj@email.uah.edu, delugach@email.uah.edu

Abstract. Software has become an integral part of many people's lives, whether knowingly or not. One key to producing quality software in time and within budget is to efficiently elicit consistent requirements. One way to do this is to use conceptual graphs. Requirements inconsistencies, if caught early enough, can prevent one part of a team from creating unnecessary design, code and tests that would be thrown out when the inconsistency was finally found. Testing requirements for consistency early and automatically is a key to a project being within budget. This paper will share an experience with a mature software project that involved translating software requirements specification into a conceptual graph and recommends several actors that could be created to automate a requirements consistency graph.

1 Introduction

We use software to record TV shows. We use software to search the internet. We use software to write papers. We use software to communicate with other people far away from us. We use software to plan trips. We use software to track our spending. We use software to write software. And when our compilers don't work, we can't do our jobs. We rely on some form of software almost every day of our life.

Eliciting requirements, in some form, is the first step in most software development cycles. If an error were found in the requirements phase, the cost to detect and repair that error is five to ten times less than the cost of detecting and repairing that error in the development phase [1]. It is therefore the purpose of this paper to suggest one way of strengthening requirements gathering procedures. This method involves using conceptual graphs, as popularized by Sowa [2], to build a graph that can be tested automatically for inconsistencies by using actors.

There are several reasons that we care about modeling requirements. In a sufficiently complex system, it's important to know which requirements are related to each other, and how strong those relationships are. If a requirement is considered for changing, the overall impact on cost and schedule need to be analyzed to make sure that that is the right choice to make. By using a tool with well-modeled conceptual graphs of requirements, someone can see the interdependencies of the requirements, and what effects adding to, deleting from, or modifying that requirement has on the entire project.

H. Schärfe, P. Hitzler, and P. Øhrstrøm (Eds.): ICCS 2006, LNAI 4068, pp. 401–412, 2006.

In a sufficiently complex system, requirements don't stop changing until the project's end of life. It's also not possible for one person to completely understand every detail of every requirement. Since multiple people would be working on different portions of the specification, it's conceivable that inconsistencies could be inserted without the other people's knowledge of it. This causes inconveniences if caught early, and expensive complications if caught too late. An automated logical system can catch some inconsistencies as soon as they are developed, saving both time and effort.

Conceptual graphs provide a visual representation of the relationships between details of requirements that a text document does not. Some conceptual graph tools offer the visual representation of concepts that a human can infer from. We would like the tools to do the inferencing themselves, and we're on the verge of that now. With data driven software, we use certain concepts' referents, along with actors, to show consistencies within a requirement and then within the entire graph.

Conceptual graphs have been shown to be helpful in modeling requirements in the past [3]. This work takes a medium scale mature project, and while graphing this project, a way was found to evaluate the consistency of certain parts of the requirements.

2 General Approach

This work modeled the requirements from a medium scale mature project. The software produced by this medium size project acted as a graphical user interface to an aircraft's radio. This graphical user interface constrained the data available to a pilot based on what configuration he was operating his radio. To accurately graph the project software's requirements, we had to locate the original representation. This original representation was an Interface Control Document that detailed what the radio expected in different configurations and a Software Requirements Specification that generalized the configurations of the Interface Control Document. After finding the main software specification and the supporting documentation, we had to parse the document by hand to isolate the important concepts and relations needed for an accurate graph of the system. This was cumbersome and time consuming. The complexity of the specification and supporting document severely limited automated parsing. The requirement that we will detail concerns security. Our requirement is found in the Software Requirements Specification and is as follows:

> *There shall be one and only one Security Level for our system. Valid values of the security level shall be None, Medium, and High.*

From the Interface Control Document, we learn that there are two key indexes associated with the security levels. They are named LP12_Key_Index and LP3_Key_Index. LP12_Key_Index can have values between 0 and 20, inclusive. LP3_Key_Index can have values between 0 and 10, inclusive. There also exists an enumerated control over the indices: LinkProtection1 can have a value between 0 and 3, inclusive, depending upon the Security Level. Fig. 1 represents these facts.

Fig. 1. Each concept has individual constraints as represented above

Furthermore, these keys have constraints with respect to each other as follows:

- When the Security Level is None, the data that it sends has no encryption.
- When the Security Level is Medium, the data that it sends has a level of encryption that is found on the piece of hardware itself.
- When the Security Level is High, the data that it sends has a level of encryption that is supplied by an additional piece of hardware.

Further research into the hardware interface explains that as these security level states change, several other fields are affected.

- When the Security Level is None, the two key indexes, LP12_Key_Index and LP3_KeyIndex, will both have values of 0.
- When the Security Level is Medium, LP12_Key_Index must have values between 1 and 20, inclusive, and LP3_Key_Index must have a value of 0.
- When the security level is High, LP3_Key_Index must have a key index between 1 and 10, inclusive, and LP12_Key_Index must have a key index of 0.

An example of how constraints between concepts and states that are not allowed are graphed using conceptual graphs is presented in Fig. 2.
This graph would read in text form:

- If Security Level is None, then LinkProtectionType1 must be 0, LP12_Key_Index must be 0 and LP3_Key_Index must be 0.
- If Security Level is Medium, then LinkProtectionType1 can be 1 or 2, LP12_Key_Index must not be 0, and LP3_Key_Index must be 0.
- If Security Level is High, then LinkProtectionType1 must be 3, LP12_Key_Index must be 0, and LP3_Key_Index must not be 0.

To check for consistency within the facts asserted in this conceptual graph, another graph needs to be created. This graph uses actors and a text file (called a "database" in Charger[7]) to check for consistency within the graph. Before we discuss the representation of the test graph of this requirement, we propose the invention of at least two new actors and the modification of one existing actor for this graph to be effective as described. A more thorough demonstration of the use of these actors together will be shown later. These two new actors, <2key_lookup> and <counter>, are described, and the modified actor, <lookup> is described later in this paper. The <counter> and <2key_lookup> actors work together to search for all instances of a key inside of a database. This is important because these database files are what hold the data constraints, and we need to find every relationship between the independent and dependent data.

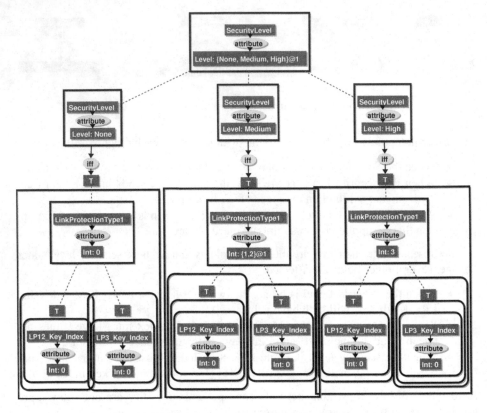

Fig. 2. The constraints of the security level requirement

A <2key_lookup> actor takes three inputs and produces one output.

- Database: file – input – *file* is a tab-delimited file with column headings.
- Input1: Referent1 – Input1 is the name of one of the column headers being searched, and Referent1 is the value searched for under the column header Input1.
- Input2: Referent2 – Input2 is the name of another of the column headers being searched, and Referent2 is the value searched for under the column header Input2.
- Output: Referent3 – the value of Referent3 is
    - Null if there do not exist three different column headers named Input1, Input2, and Output inside of the database file.
    - Null if there do not exist a single row that has all three referents Referent1, Referent2, and Referent3 from the corresponding types Input1, Input2, and Output in it inside of the database file.
    - the value at the location in the Output column that corresponds to the same row as Input1's referent (Referent1) and Input2's referent (Referent2) inside the database file.

An example of the use of a <2key_lookup> actor is shown below in Fig. 3. In this figure we assume that the Input1 column has values of 5 and 5, the Input2 column has values of 6 and 7, and the Output column has values of 1 and 2. Note that in the graph where Input1 is 4 and Input2 is 7, we could have Output equal to 2, since this is unambiguous. However, we have defined this <2key_lookup> actor to require both keys to be present. Further research could create the case to relax this restriction.

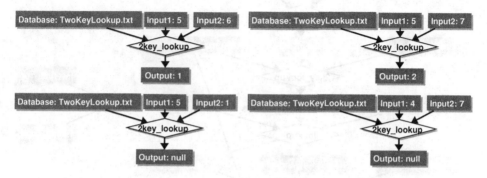

Fig. 3. Four examples using the <2key_lookup> actor

A <counter> actor takes two inputs and produces one output.

- Interval – the referent is the time in between increments of a change to the output. This value should be large enough so that the count does not change until all activity has completed downstream.
- Reset – if the referent is T then the next output will be reset to 1. If the referent is null (or more generally not T), then the next output will be incremented by 1.
- Counter – after time elapses (based on the interval's referent), the Counter's referent increases by 1.

An example of the use of <counter> is shown below in Fig. 4. In this figure we assume that the Input column of the test file has values 1 and 2, and the Output column of the test file has corresponding values of 3 and 4.

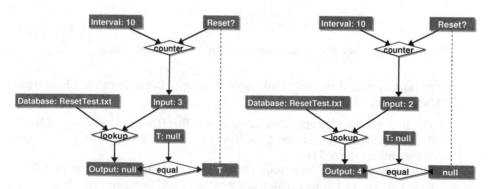

Fig. 4. Two examples using the <counter> actor

The <lookup> actor could be modified to output multiple values, based upon the concepts' names. During the writing of this paper, this feature was instituted into CharGer. This will allow multiple outputs to be populated from one lookup actor, cleaning the graph.

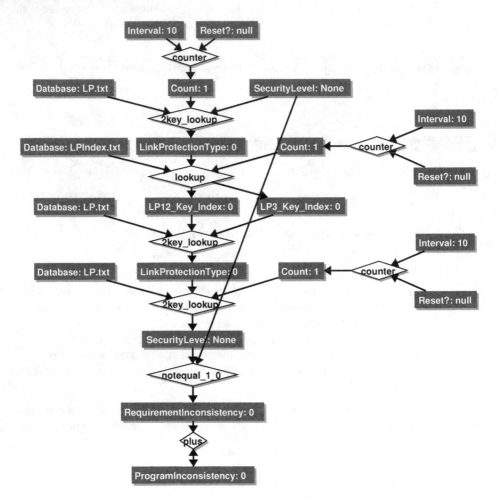

Fig. 5. Requirement checking graph using actors and lookups

The graph represented in Fig. 5 looks up the Security Level from the LP.txt database. The sequence of the graph follows:

1. The first <2key_lookup> actor retrieves the first (Count: 1) instance of Link-ProtectionType that matched SecurityLevel: None. The LinkProtectionType Value that is returned is 0.
2. The <lookup> actor then retrieves the first (Count: 1) instance of LP12_ Key_Index and LP3_Key_Index that matched LinkProtectionType: 0.

3. The second <2key_lookup> actor then retrieves the SecurityLevel that matched both LP12_Key_Index: 0 and LP3_Key_Index: 0.
4. The <notequal_1_0> actor produces RequirementInconsistency: 1 if the SecurityLevel at the start of the graph (from the LP.txt database) does not match the SecurityLevel at the end of the graph (from the LPIndex.txt database).
5. The plus actor adds the referent of RequirementInconsistency to the referent of ProgramInconsistency and put the result in the referent of ProgramInconsistency.
6. After this thread is completed, the second <counter> actor activates and the second part of the graph runs again. The first counter only runs again until all <counter> actors downstream are completed (signified by the Reset? concept being set to T).

This graph could indicate that there is a requirement inconsistency. Further, if there were a requirement inconsistency, then the accumulation of requirement inconsistencies is the total number of program inconsistencies, so that at the end of the run of this graph, we count how many total requirement inconsistencies there were in the entire graph. To demonstrate this, we created two concepts, RequirementInconsistency and ProgramInconsistency. Their values both start at 0. RequirementInconsistency can only have a referent (value) of 0 or 1. RequirementInconsistency can have a integer value greater than or equal to 0. These two referents are populated as follows:

- When the SecurityLevel referent that we expect is not the actual SecurityLevel referent, the <notequal1_0> actor's value becomes (or stays at) 1.
- When the SecurityLevel referent that we expect is the actual SecurityLevel referent, the <notequal1_0> actor's value becomes (or stays at) 0.
- Whenever the <notequal1_0> actor's value changes, that value is added to the ProgramInconsistency referent. (If the two SecurityLevel values are the same, then 0 is added to the ProgramInconsistency value, changing nothing.)

The inputs and output to the accumulating <plus> actor are as follows:

- RequirementInconsistency – can have a value of 0 or 1.
- ProgramInconsistency – can have a integer value greater than or equal to 0.
- ProgramInconsistency – result of adding either 0 or 1 (RequirementInconsistency referent) to ProgramInconsistency's referent.

In the future, we would prefer that all the data that is useful in determining consistency be represented within the conceptual graph itself. However, with current tools, this is even less possible than using actors and databases. More actors are needed to take advantage of sets, instead of using a database. An example implementation is presented in Fig. 6.

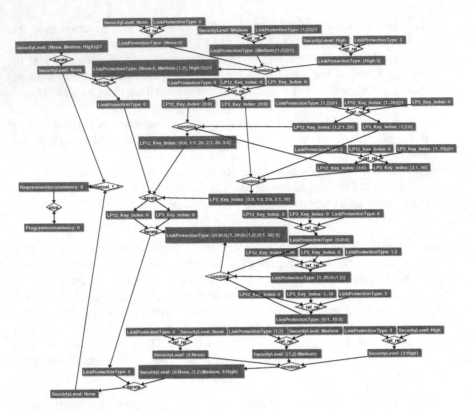

Fig. 6. A graph testing the consistency of requirements without the use of databases

This graph introduces several new actors. These actors are <ref_rel>, <2_ref_rel>, <combine>, and <iterate>. A <ref_rel> actor takes at least two inputs (numbered 1 through N) and produces one less than the number of inputs as outputs (numbered 1 through N-1).

- Input1 – this referent is the pointer to the other inputs' (Input2 through InputN) referents.
- Input2 through InputN – this referent will be the value that the Input1 referent points to.
- Output1 through OutputN-1 – this referent consists of two values inside of its set. The values to the left of the colon (:) are the available pointer values that point to the values to the right of the colon.

An example of the use of <ref_rel> is shown below in Fig. 7.

Fig. 7. Two examples using the <ref_rel> actor

In the graph on the left, we take the two referents, and create a referent relationship between them, to form what could be considered a pointer (None), and the value that it points to (0). The output of the <ref_rel> actor is the Type of the value that the pointer points to. In the graph on the right, we take three referents, and create two referent relationships. The first arc will always be inside of the output of the <ref_rel> actor. The other inputs will be the Types of the outputs of the <ref_rel> actor.

A <2_ref_rel> actor takes at least three inputs (numbered 1 through N) and produce two less than the number of inputs as outputs (numbered 1 through N-2).

- Input1 and Input2 – these referents, together, will be the pointer to the other inputs' (Input3 through InputN) referents.
- Input3 through InputN – this referent is the value that the Input1 and Input2 referent points to.
- Output1 through OutputN-2 – this referent consists of three values inside of its set. The values to the left of the semicolon (:) are the available pointer values that point to the values to the right of the colon. The values to the left of the semicolon (;) are the referents of the Input1 concept, and the values to the right of the semicolon (;) are the referents of the Input2 concept.

A <combine> actor takes any number of inputs (numbered 1 through N) and produces one output.

- Input1 through InputN – input – these referents will each be a different element in the output set.
- Output – output – the type is the same as the inputs' type, and the referent will be the set of all of the inputs' referents.

An example of the use of <combine> actor is shown below in Fig. 8.

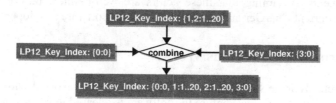

Fig. 8. An example of the use of the <combine> actor

In the graph the combine actor takes any number of inputs, all with the same Type, and create an output whose referent is a set of the referents that are inputs to the <combine> actor.

An <iterate> actor takes at least one input and produce one output. If there is only one input, then it functions as follows:

- Input – to be useful, the referent should be a set.
- Output – the referent of Output takes on each of the values of Input's referent; the referent changes to the next value of the set once the thread completes. The Output's type is the same type as Input.

If there are two inputs, then the actor functions as follows:

- Input1 – the referent of this concept will be used to search Input2
- Input2 – to be useful, the referent should be a set, and the set will consist of ordered pairs. You could consider the ordered pairs to be pointers on the left of the colon, and the value at the address of the pointer to be the value on the right side of the colon.
- Output – if the referent of Input1 matches the value to the left of the colon (":") of Input2, then the Output's referent is the value to the right of the colon; the referent changes to the next value that matches once the thread completes. The Output's type is the same type as Input2.

An example of the use of <iterate> is shown below in Fig. 9.

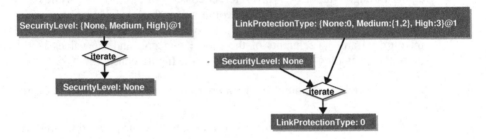

Fig. 9. An example of the use of the <iterate> actor

This section described several actors that can be used in conjunction with requirements in conceptual graphs that will allow some automatic consistency checking of the requirements. Assuming that these actors are well-defined, they can be easily added to the current CharGer architecture and used by software developers.

3 Background (or Previous Work)

With a sufficiently large software project, many people are involved with the creation and approval of project artifacts before the software is released to the customer. Even if some of those people felt comfortable modeling with conceptual graphs, not everyone would understand the full meaning of the graphs, and because of this, requirements are often modeled according to simple descriptions, and if we're lucky, primitive diagrams. Many of these types of requirements are vague and informal; such methods already in wide use to represent requirements are Block Diagrams, Flowcharts, Timing Diagrams and natural languages. A way to translate these requirements into conceptual graphs is covered by Cyre [5]. Other methods used to represent requirements include OMT object diagrams and Data Flow Diagrams. A way to translate these requirements into conceptual graphs is covered by Delugach [6].

Cyre [5] also demonstrates a way to show inconsistencies in conceptual graphs. Their method involves comparing definitions of graphs to the actual graphs, and looking for missing relations and concepts. Our proposed approach of showing inconsistencies lies more with inconsistencies in the data being represented, and less with the topology of the graph's components. For example, if we take a university domain where a freshman has between 0 and 32 credit hours and a sophomore has between 33 and 64 credit hours, Cyre's method would determine that the following graph is consistent, because it assumes definitions are consistent and does not take into account dependency between data.

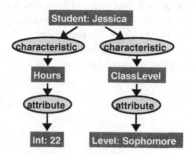

Fig. 10. A simple inconsistent graph

Conceptual graphs are based upon our interpretation of natural language. Sowa [2] has shown that conceptual graphs can be mapped directly to first-order logic. With many commonly used requirement representations able to be mapped to conceptual graphs, our process to test for the consistency of data can be applied widely to existing translations.

4 Conclusion and Summary

We have shown that real requirements consistency checking can be achieved with a few modifications to current tools. One such modification is an <ifthen> actor that fires if its input is true (easily done by comparing a dynamic value to a static value such as IF var == NULL or IF var == Bob). Another addition is an <ifnotthen> actor that fires if its input is NULL (or 0). Further research could address these problems. Without these actors, we would need to duplicate much of a graph.

This approach still does not answer the perhaps more important question of "Are these the requirements that we really want?" This work does not address that issue, except that this supports the human activity of finding inconsistencies, which clearly mean that something was incorrect and needs to be examined by humans.

References

1. Davis, Alan M. Software Requirements: Objects, Functions, and States. Englewood Cliffs, NJ: Prentice-Hall, 1993.
2. J.F. Sowa, Conceptual Structures: Information Processing in Mind and Machine. Reading, Mass.: Addison-Wesley. 1984.
3. Delugach, H. and Lampkin, B., "Acquiring Software Requirements as Conceptual Graphs," Fifth International Symposium on Requirements Engineering, Los Alamitos, California: IEEE Computer Society Press, 2001, pp. 296-297.

4. Ryan, Kevin and Mathews, Brian, Matching Conceptual Graphs as an Aid to Requirements Re-use, in Proc. of the IEEE Sump. On Requirements Engineering, San Diego, January 1993, ISBN 0-8186-3120-1, 112-120.
5. Cyre, Walling R.: Capture, Integration, and Analysis of Digital System Requirements with Conceptual Graphs, IEEE Transactions on Knowledge Data Engineering 9(1): 8-23, 1997.
6. Delugach, Harry S., ``An Approach To Conceptual Feedback In Multiple Viewed Software Requirements Modeling," Proc. Viewpoints 96: Intl. Workshop on Multiple Perspectives in Software Development, Oct. 14-15, 1996, San Francisco.
7. Delugach, Harry S., "CharGer Conceptual Graph Editor," version 3.5b1. http:// sourceforge.net/projects/charger/, accessed January, 2006.

Query-Based Multicontexts for Knowledge Base Browsing: An Evaluation

Julien Tane, Philipp Cimiano, and Pascal Hitzler

AIFB, Universität Karlsruhe (TH)
76128 Karlsruhe, Germany
{jta, pci, hitzler}@aifb.uni-karlsruhe.de

Abstract. In [7], we introduced the *query-based multicontext theory*, which allows to define a virtual space of views on ontological data. Each view is then materialised as a formal context. While this formal context can be visualised in a usual formal concept analysis framework such as Conexp or ToscanaJ, [7] also briefly described how the approach allowed the creation of a novel navigation framework for knowledge bases. The principle of this navigation is based on supporting the user in defining pertinent views. The purpose of this article is to discuss the benefits of the browsing interface. This discussion is performed, on the one hand, by comparing the approach to other Formal Concept Analysis based frameworks. On the other hand, it exposes the preliminary evaluation of the visualisation of formal contexts by comparing the display of a lattice to two other approaches based on trees and graphs.

1 Introduction

Semantic technologies have matured in recent years with many new advances concerning the creation, management and application of ontologies and knowledge bases (see [8]). Diverse paradigms have been proposed to interact with knowledge bases. One of these paradigms uses diagrams representing *concept lattices*[1] (see [5]) to visualise and interact with a knowledge base (related approaches can be found in [1,3,10]). Moreover, a overview of the tool support for Formal Concept Analysis can be found in [12][2]

The use of formal concept lattices for the display and interaction with knowledge has some interesting features and some major drawbacks. Though concept lattices are suitable structures to represent binary relations and hierarchical knowledge, they suffer from mainly two limitations. First, the reading of a concept lattice requires some training. But, the evaluation found in [4] and the one presented in this paper showed that novices could read lattice diagrams with a minimum of training. The second limitation lies in the increasing difficulty of drawing a readable concept lattice diagram for larger lattices.

---

[1] In this paper, we assume that the reader is acquainted with the basic definitions of traditional Formal Concept Analysis such as: formal contexts, formal concepts and concept lattices.

[2] See also Uta Priss' FCA home page: http://www.fcahome.org.uk/

H. Schärfe, P. Hitzler, and P. Øhrstrøm (Eds.): ICCS 2006, LNAI 4068, pp. 413–426, 2006.

In [7], we proposed a new approach to knowledge browsing using concept lattices. This approach relies on the use of queries to define and manipulate views over a knowledge base. In order to give a meaning to the views used, we developped the *Query-Based Multicontext theory*. This theory allows the manipulation of surrogate objects to represent the formal contexts underlying the concept lattices used by the browsing tool. Knowledge browsing in this framework consists in supporting the user in defining relevant views. To help the reader in grasping the principle of this approach, we illustrate it by some examples of views which can be defined. For example, the tool allows a user to define a lattice showing the distribution of the publications of researchers of a given group working on the topics *Formal Concept Analysis*, *logic* or *text-mining*. The concept lattice of this view is presented in Figure 1.

The purpose of this article is to discuss the benefits of our approach in two ways. First, the features of our approach are compared with those of other Formal Concept Analysis based approaches in Section 2. Then Section 4 presents an evaluation we performed with a prototype of the tool. This evaluation compares the selection mechanism of three visualisation paradigms: the lattice view, the tree view and the graph view.

The plan of this article is as follows. In Section 2, we give a short intuitive motivation for semantical views on data, comparing our approach to other Formal Concept Analysis based frameworks. In Section 3, we recapitulate the main aspects of the Query-Based Multicontext approach necessary for a full understanding of the rest of the article. In Section 4, we discuss the results of our evaluation of the visualisation aspects of the Query-Based Multicontext navigation paradigm. We finally draw some conclusions from our work in Section 5.

2   Motivation

In [7], we introduced the *Query-Based Multicontext theory*. This theory is a new approach to interacting with data based on a combination of Formal Concept Analysis and knowledge base querying. In the next section, we reintroduce the necessary formalism to understand this paper, but to give the reader a general idea we expose in this section the general principle in an intuitive way. We also introduce an example which we use for illustration purposes throughout this paper. The main idea is to support and guide the user in defining views on the data to be visualised with a given paradigm.

Suppose Katharina is a researcher interested in the research fields of *Formal Concept Analysis*, *text-mining* and *logic*. While studying related works in her area, she notes that the AIFB Institute has proposed some interesting approaches. She would like to get an overview of the researchers working on this topic at AIFB. The AIFB portal[3] offers the possibility to download a knowledge base representing many aspects of this institution. To give an idea of the content

---

[3] See http://www.aifb.uni-karlsruhe.de/about.html. We refer to this web site for more information on this knowledge base and its associated ontology.

Table 1. Table displaying the kind of elements in the knowledge base

type	example
concepts	persons, projects, publications, research group, research topics...
relations	author, isWorkedOnBy, memberOf, isCarriedOutBy, hasProject
instances	Julien Tane, Pascal Hitzler, FCA, Logic...
relation instances	isWorkedOnBy(FCA, Julien Tane), isWorkedOnBy(Logic, Pascal Hitzler)...

of this knowledge base, we illustrate parts of its content in Table 1. For example, the instance of *research topic* Formal Concept Analysis is worked on by the *PhD Student* Julien Tane. To query and interact with the knowledge base, there exists an ontology API (the current implementation is based on KAON[4]). We showed in [11] that it is possible to create a query language on top of this API.

The portal contains some information on more than 900 publications and around 900 authors. Displaying the formal concept lattice of the binary relation *author* as a Hasse diagram would be hardly practical since this lattice contains 996 formal concepts. Let us consider diverse options to cope with this issue:

- use an iceberg concept lattice
- use conceptual scaling
- select the context objects and attributes to be used for the displaying of the lattice[5]
- define queries to create object and attribute sets and their incidence relation

The first technique is based on frequent item sets. Using the TITANIC algorithm (see [9]), it is possible to construct the lattice of the frequent item sets of the formal context. Setting a threshold on the number of publications would reduce the size of the lattice to be displayed. While this can return interesting results, some relevant authors might be missing from the lattice because they do not have enough publications. Moreover, for large lattices with a small number of layers, this does not ease the visualisation very much.

The second technique uses the traditional technique of *conceptual scaling* (see [6]). Scaling is a powerful technique when the object set can be studied according to orthogonal sets of attributes and where these attribute sets are organised in a lattice. The choice of the scales has a great influence on the form of the lattice. In the present case, several different scales are imaginable. A possible scale could be to consider the type of the authors of the publications: *professor, assistant professor, PhD Student*. Another possible scale could be the research group to which the authors of the publications are affiliated. The scales are useful because they offer a powerful means of factorising the search space but there is in general

---

[4] See http://kaon.semanticweb.org.

[5] Like in the Concept Explorer, where objects and attributes can be selected or deselected using check boxes. You can download and use the Concept Explorer from http://sf.net/projects/conexp.

no reason that this factorisation corresponds to an interesting view on the data. The conceptual scaling lacks the facility for choosing the attribute set flexibly.

The third technique is implemented in the Concept Explorer by Serhiy Yevtushenko. The interface offers the possibility to decide for each attribute and for each object of the context whether it is visible or not. While this technique allows for a fine granular decision of what should be visualised, it is tedious to choose the given elements.

The final option corresponds to our approach. We support the user in the creation of the concept lattice to be displayed. For this, the user can define precisely the three parts of the formal context: the object set, the attribute set and the incidence relation. Section 3.4 shows how a template mechanism helps simplify the creation of complex views.

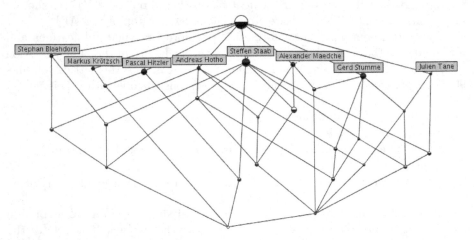

Fig. 1. The lattice representing the distribution of the publications published by members of the *Knowledge Management Group* among the researcher working on *Formal Concept Analysis*, *logic* or *text-mining*

For example, our researcher may be particularly interested in the research papers published by the *Knowledge Management Group* who are working on some topics relevant for his own research: *Formal Concept Analysis*, *logic* or *text-mining*. The resulting concept lattice has only 28 concepts. It can be created using a special kind of query primitive: *the role query* which is introduced in Section 3.2. Figure 1 shows the lattice resulting from the view creation process. Two comments should be made about this diagram. First, it has been edited for this paper in order to make it more readable. The editing took much longer than the creation process which took less than a minute[6]. The editing phase is not essential when using the tool because of the highlighting mechanism. The other observation is that the object labels are not shown in the diagram. The reason for this lies in the higher number of publications than persons. Displaying them

---

[6] This is fairly complex view to create many views require less time.

would make the diagram much less readable. This is not a limitation when using the browser, since the labels can be displayed on demand.

Finally, it is crucial to ensure that the lattice paradigm is suitable for knowledge browsing, we performed the evaluation presented in Section 4. Before that, we reintroduce the basic elements behind our approach in the following sections.

3   Query-Based Multicontext

In this section, we first give a formal definition of the Query-Based Multicontext. Then we illustrate this definition with a very simple example. In this section, we recall the formal definition of the Query-Based Multicontext, and illustrate it using a simple example. Finally, we discuss briefly the knowledge browsing paradigm based on the Query-Based Multicontext theory.

3.1   Definition: Query-Based Multicontext

Each of the views of the system corresponds to a triple p of queries which can be interpreted to a formal context by means of an evaluation function κ.

$$p := (q_1, q_2, q_3) \longrightarrow \mathbb{K}_p := \kappa(p)$$

Each triple p is called a *context index*, whereas the result $\kappa(p)$ of the interpretation of p is called the *realised* (formal) context of p. In the Query-Based Multicontext approach, navigating means going from one realised context to a new one. We reintroduce here a simplified[7] version of the formal definition given in [7]. For this, we call a *query language* a set L together with an *evaluation mapping eval*. The latter is a mapping from L to $\mathcal{P}(\Omega)$ or to $\mathcal{P}(\Omega \times \Omega)$, where Ω is some given set and $\mathcal{P}(X)$ denotes the power set of the set X. By usual abuse of terminology, we will call L itself a query language if the corresponding evaluation mapping is understood.

The following definition formalises the idea of a Query-Based Multicontext:

Definition 1 (Query-Based Multicontext). *Let Ω be a set called universe and let L_1 and L_2 be two query languages with evaluation mappings $eval_1$ for L_1, with $eval_1(q_1) \subseteq \mathcal{P}(\Omega)$, and $eval_2$ for L_2, with $eval_2(q_3) \subseteq \mathcal{P}(\Omega \times \Omega)$. Elements of L_1 are sometimes called set queries, whereas elements of L_2 are called relation queries. Let $\mathbb{P} := L_1 \times L_1 \times L_2$. We call an element p of $\mathbb{P}$ a context index.*

For a context index $p = (q_1, q_2, q_3) \in \mathbb{P}$, we define its induced query-based context as $\mathbb{K}_p := (eval_1(q_1), eval_1(q_2), eval_2(q_3) \cap (eval_1(q_1) \times eval_1(q_2)))$. We call QBMC:$=\{\mathbb{K}_p | p \in \mathbb{P}\}$ a Query-Based Multicontext. We call the mapping from $\mathbb{P}$ in QBMC the context realisation function and we denote it by κ. So, for all $p \in \mathbb{P}$, $\kappa(p) = \mathbb{K}_p$.

---

[7] Only one query infrastructure and one instance of this query infrastructure is considered in the present definition.

Given the choice of the two mappings $eval_1$ and $eval_2$, a context index fully specifies the content of a formal context. Note that the incidence relation is defined as the pairs common to the relation returned by the third query (returning a set of pairs) and to the cross product of the two others. The idea behind this construction is that the relation desired is only between the objects and attributes of the resulting context. The relation between other objects and attributes is not relevant at that point. In other words, only those relation elements are kept which are between the object set and the attribute set. For instance, if the goal of the formal context is to display the relation between professors and their research topics, then the pairs of the relation between PhD Students and research topics are not relevant. The underlying database, however, might not make any difference and return all the pairs.

In [7] we introduced operators on queries and context indices, the next section introduces three operators necessary for the context indices used for the evaluation presented in this paper. However, we refer to [7,11] for other primitives useful when dealing with knowledge bases.

3.2   Query Operators

In order to illustrate the importance of query primitives for flexible context index creation, we introduce three operators[8]: the set operator, the inverse operator and finally the role operators. These are all the query operators needed to construct the context index used in this paper and for the evaluation.

Set Operator. The set operator allows a fine granular selection of instances. For elements $i_1, \ldots, i_n$ in Ω, the *set query expression* "$\{i_1, \ldots, i_n\}$" is a query element. The evaluation of a set expression returns the set itself. Let $Set(\Omega)$ be the set of all set query expressions.

Inverse Operator. The *inverse operator* is an operator for elements of L_2 swapping the components of the answers of a given query. For example, for q_r in L_2 and $eval_2(q_r) = \{(a,b),(c,d)\}$ we have that q_r^{-1} is also a query and $eval_2(q_r^{-1}) = \{(b,a),(d,c)\}$. For a query language L_2, the set of all inverse queries is denoted by $\mathrm{Inv}(L_2)$.

Role Operator. We consider now an important query operator: the role operator. This query operator takes two arguments: a relation query q_r (i.e. an element $q_r \in L_2$, which returns per definition a set of pairs) and a set query q_i (i.e. an element $q_i \in L_1$, which returns per definition a set of singletons). For every q_r and q_i, the expression: "$\exists_I q_r(q_i)$" is called a role query for the elements for the relation q_r with parameters q_i – the subscript I of the existential quantifier serves as a reminder that the operator returns sets of instances[9], defined as follows:

$$eval_1(\exists_I q_r(q_i)) := \{x \in \Omega | \exists y \in eval_1(q_i), (x,y) \in eval_2(q_r)\}.$$

---

[8] The original idea behind these operators comes from description logics. However, they are not restricted to description logics.

[9] This emphasizes the difference with another operator introduced in [7] which returns sets of concepts.

From the mathematical definition, it can easily be seen that the role query selects only the elements verifying the following statement: the first component of the pairs resulting from the evaluation of q_r, where the second component is in the result of the evaluation of q_i. For two query languages L_1 and L_2, the set of relation queries $Role(L_2, L_1)$ is the set consisting of all the role queries which can be constructed using the following recursion rule:

- $\forall q_r \in L_2, \forall q_i \in L_1,$ "$\exists_I q_r(q_i)$"$\in Role(L_2, L_1)$
- $\forall q_i \in L_1, \forall q_{role} \in Role(L_2, L_1),$ "$\exists_I q_{role}(q_i)$"$\in Role(L_2, L_1)$

For example, the role query $\exists_I brothers(\{mymother, myfather\})$ returns my uncles, if the evaluation $\{mymother, myfather\}$ returns my father and mother, and the *brothers* query returns all the brotherhood pairs of my family. The use of the $\exists$ quantifier allows to retrieve the uncles from both sides of the family whereas the $\forall$ quantifier would have returned an empty set since no brother of my mother is a brother of my father.

3.3 Example

We now present an example of Query-Based Multicontext related to the evaluation of this paper.

As a toy example, we consider four concept names: *person, publication, project, research group* and *research topic* as well as binary relations names: *isWorkedOnBy, memberOf, projectInfo, hasProject, isCarriedOutBy* and *author*.

We can specify the languages L_1 and L_2 in the following way, let L_{1C} and L_{2R} be two sets.

- L_{1C} corresponds to concept names:
  $L_{1C} := \{person, research topic, research group, project, publication\}$
- L_{2R} corresponds to relation names:
  $L_{2R} := \{isWorkedOnBy, member, author, projectInfo, hasProject, isCarriedOutBy\}$
- $L_2 := L_{2R} \cup Inv(L_{2R})$
- $L_1 := L_{1C} \cup Set(\Omega) \cup Role(L_2, L_{1C} \cup Set(\Omega))$

L_1 and L_2 can be seen as query languages. We consider the following evaluation functions:

- $eval_1$
  - $eval_1(research group) = \{Chair 1, Chair2, Chair 3, Chair 4 \} \subseteq \Omega$
  - $eval_1(research topic) = \{E\text{-}Learning, Semantic Web, Logic, Genetic Algorithms, Complexity Theory, Petri Nets, \ldots\} \subseteq \Omega$
  - $eval_1(person) = \{Julien Tane, Phillip Cimiano, Daniel Sommer, Andreas Hotho, \ldots\} \subseteq \Omega$
  - $eval_1(project) = \{SEKT, DIP, SESAM, VIROR, \ldots\} \subseteq \Omega$
  - $eval_1(publication) = \{tane\text{-}icfca05, hotho\text{-}icml02, stumme\text{-}fcamerge01, \ldots\} \subseteq \Omega$

– $eval_2$: The evaluation functions of the relations *projectInfo*, *isCarriedOutBy*, *isWorkedOnBy*, *memberOf* and *author* can be understood as formal contexts since they return sets of pairs of elements.

Figure 1 displays the relation between authors of the topic of *text mining*, *logic* and *Formal Concept Analysis*, and the publications of the *Knowledge management group* as discussed in Section 2. The context index corresponding to this view is:

$(\exists_I author(\exists_I memberOf(Chair3)), \exists_I isWorkedOnBy^{-1}$ (text mining, *data mining, knowledge discovery), author*).

Using this Query-Based Multicontext infrastructure, it is possible to define many context indices. But it has mainly been chosen because all the context indices used in the evaluation presented Section 4 can be expressed.

3.4 Knowledge Browsing

In [7], we presented a navigation framework for ontologies based on the use of context indices. Before discussing in the next section the evaluation we performed, we explain the principle behind our navigation mechanism.

In our browsing paradigm, context indices can be seen as the conceptual representation of views. Browsing the knowledge base corresponds to changing the view, that is generating a new context index. In [7], we presented a simple way of creating relevant views by parameterising some function, called constructor, in order to return a pertinent context index. We presented diverse kinds of

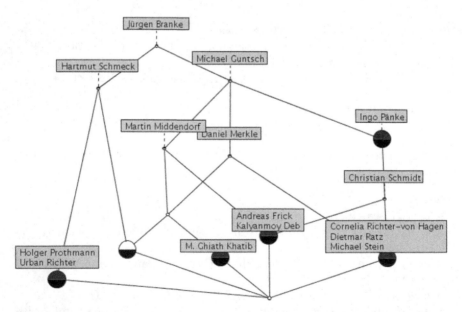

Fig. 2. The lattice representing the author-publication relation of the coauthors of Jürgen Branke

constructors for common types of views to be displayed: relation, joins, subsumption hierarchies, etc.

In order to use a constructor, the parameters of the function need to be set. For example, the *CoRelation* constructor CoR takes two parameters: a set query q_i (in L_1) and a relation query q_r (in L_2). The evaluation of CoR returns a context index (i.e. a triple of queries)

$$CoR(q_i, q_r) := (\exists_I q_r(q_i), \exists_I q_r^{-1}(\exists_I q_r(q_i)), q_r).$$

This constructor has been used at the end of Section 3.3 for the second context index. The realised context of this context index represents the author-publication relation of the coauthors of Jürgen Branke and Figure 2 shows the corresponding lattice. In order to create the context index, the user only needs to set the parameters of the *CoRelation* constructor. In that special case, these parameters are: the instance parameter {Jürgen Branke} and the relation parameter $author^{-1}$.

In the following section, we investigate three different means of presenting the content of a realised context to a user.

4   Evaluation

We now present the results of our evaluation which compared the efficiency of user performance when answering certain type of question on three different paradigms. The rationale behind this is to find criterias as to which view paradigm is the most suitable for certain types of questions. Before describing the evaluation's methodology and results, we briefly describe the visualisation paradigms.

4.1   Visualisation Paradigms

Our evaluation used three kinds of visualisation paradigms. We introduce them briefly. During the evaluation, the three visualisation paradigms displayed the realised context of the context index corresponding to the question.

Lattice View: The lattice view shows the Hasse diagram of a certain context index. Diverse types of interaction are possible. A user can select objects of a node using the double middle click. He can also move nodes and label or display the objects and/or attributes[10] in a specific panel. Figure 2 shows the appearance of the lattice view.

Tree View: The tree view of the questions used in the experiment of Section 4 displays the binary relation of the realised context index. In the case of a normal[11] context index, the tree has only a height of 2^{12}. Figure 3 shows the appearance of the tree view when answering a training question.

---

[10] [7] presented the diverse selection and interaction modes for the lattice.

[11] For a subsumption and subsumption-instance context index, it corresponds more to the usual tree view of the subsumption hierarchies.

[12] The root node does not carry any information and is therefore hidden.

```
┆   ┆ AntAlg – Ameisen Algorithmen   (0)
o┄ ┆ EUCOR Virtuale – EUCOR Virtuale   (1)
o┄ ┆ EVOLEARN – Evolution und Lernen   (5)
φ┄ ┆ ITSAM – IT–Unterstützung für das Asset Management   (2)
    ├─ ┆ multicriteria optimization   (0)
    └─ ┆ portfolio management   (0)
o┄ ┆ KIM – Karlsruher Integriertes Informations–Management   (5)
o┄ ┆ KUBIK – Kooperationsprojekt zur Unterstützung von Bankenlösungen mit Informations– und Kommunikationstechniken   (13)
φ┄ ┆ MULO – Multikriterielle Optimierung   (1)
    └─ ┆ multicriteria optimization   (0)
o┄ ┆ NUKATH – Notebook Universität Karlsruhe (TH)   (1)
o┄ ┆ OC – Organic Computing   (18)
o┄ ┆ OTC – Organic Traffic Control   (11)
o┄ ┆ OptRek – Optimierung auf rekonfigurierbaren Rechensystemen   (10)
φ┄ ┆ P–Opt – Portfoliooptimierung mit komplexen Nebenbedingungen   (4)
    ├─ ┆ computational finance   (0)
    ├─ ┆ multicriteria optimization   (0)
    ├─ ┆ portfolio management   (0)
    └─ ┆ risk management   (0)
o┄ ┆ PEdit – Programmierumgebung für parallele Systeme   (1)
o┄ ┆ QE – Quantitative Emergenz   (11)
o┄ ┆ SCP – Supply Chain Planung   (1)
─────────────────────────────────────────────────
│ Lattice line diagram │ Tree View │
```

Fig. 3. The tree view of the content of a realised context used in the experiment

Graph View: The graph view shows the relations between instances. It has been adapted from the visualisation used in the OIModeller as part the KAON project[13]. For the purpose of the evaluation, we adapted it in the following way: attributes of the realised context are marked in different colours. Figure 4 shows the graph for one of the realised contexts used in the experiment. The graph view offered the possibility to hide nodes or to pin them, that is they do not move with the other elements when these are moved.

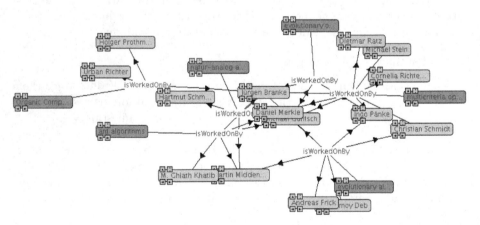

Fig. 4. The graph view of one of the realised context used as training in the experiment

4.2   Evaluation Methodology

The methodology we used for our experiment consists in a comparion of the performance of our test persons on the task of answering certain questions using the lattice against using one of the other paradigms. Due to the small amount of

---

[13] See http://sf.net/projects/kaon.

data, we used the T-test distribution (see [2]) to determine the significance of the difference in testers' performance when using the lattice against using one of the two other paradigms. For each of these tests, we formulate the null-hypothesis as:

The time need by users to answer the given question does not differ in a significant manner between the two paradigms (i.e, lattice and tree or lattice and graphs).

We compare the average time in seconds needed by the diverse views to answer the questions.

The number of users had been chosen in order to ensure the statistical relevance of the test. A group of eighteen testers were chosen. Each tester was asked seven questions: Four training questions and three evaluation questions. Of the four training questions, three were presented to the user as lattice view and the last one as graph view. The idea behind this was to allow the user to familiarise themselves with the diverse paradigms. No training was used for the tree view, because it was assumed that users were already acquainted with the paradigm. This small training was introduced in the evaluation after we found out in a preliminary evaluation that users without training had difficulties with the lattice paradigm and usually performed much better with the tree view.

Each tester was asked the three evaluation questions in a different paradigm. For each question and each paradigm, six testers were presented that question with that paradigm. For example, to person1 first the lattice view was presented, then the tree view and finally the graph view. Five other testers were presented this combination.

The test individuals were all academic people, who had already heard of Formal Concept Analysis, but were not used to reading concept lattices. A very short crash course in reading concept lattices was given to them.

We now give the original wording of the three questions used in the evaluation as well as the corresponding context indices:

- **Question 1.** The panel above displays the relation between the projects and research groups of the Institute.
  Task: Please select the projects carried out by the two groups "Efficient Algorithms" and "Knowledge Management" at the same time.
  Context Index: $(\exists_I isCarriedOutBy(\text{research group}),$ research group,
  $$isCarriedOutBy^{-1})$$
- **Question 2.** This panels shows the persons working on the text mining field at the institute as well as their publications.
  Task: Please select the publications of the author who did not share any publication with any other of this group of authors.
  Context Index: $(\exists_I author(\exists_I isWorkedOn^{-1}(\{text\text{-}mining\})),$
  $$\exists_I isWorkedOn^{-1}(\{text\ mining\}), author)$$
- **Question 3.** The panel above displays the distribution of the publications of the SEKT project among the members of the *Knowledge Management* group.

Task: Select all the publications where at least two of the following persons are authors:

- Arthur Judson Brown
- Roger Wilson
- Arthur Lehning

Context Index: $(\exists_I\, projectInfo(\{SEKT\}), \exists_I\, memberOf(Chair3), author)$

All the three evaluation questions required the user to select some objects of the context.

4.3   Evaluation Conclusions

Diverse conclusions could be drawn from this evaluation. First, as shown in Table 2, the users performed in average quicker with the lattice paradigm than with the other paradigms. We computed the t-value for each question, and determined the corresponding significance level α of the Null hypothesis. The corresponding results are found in Table 3. The low significance level (inferior to 0.10) for all tests shows that the differences in the time needed to answer the questions are unlikely to be due to chance.

Moreover, this performance also proves that the training restricted to four preliminary questions is enough for user to perform the tasks. Observe that without the training phases, users tended to be slower than the tree view.

Table 2. Time in s needed for the three evaluation questions in the three paradigms

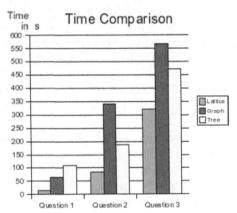

Another important result is that there were more errors using the two other paradigms. This is mainly due to the cumbersome nature of the chosen tasks for the tree and graph paradigms. For all questions, the amount of interaction needed to answer the question with the lattice view was much smaller than with the other paradigms. The evaluation confirms the intuition that users should perform better with the lattice paradigm if the number of elements to select is

Table 3. T-values and significance levels of each of the lattice-tree and lattice-graph comparisons

Question	Tree	Graph
Question 1	3.94 ($\alpha = 0.05$)	3.69 ($\alpha = 0.05$)
Question 2	1.99 ($\alpha = 0.10$)	2.29 ($\alpha = 0.05$)
Question 3	1.86 ($\alpha = 0.10$)	2.27 ($\alpha = 0.05$)

large or the number of elements which have to be examined for a given selection is large.

Note that for the diverse tasks to be performed, the performance of the selection using the lattice could have been greatly improved if an additive highlighting mechanism had been available. Therefore, there seems to be still room for improvement. Finally, it should be made clear that the questions asked are not representative of all the possible tasks occuring when visualising a view. However, for these kinds of tasks, the lattice approach is more advantageous.

5   Conclusion

In this article, we discussed the benefits of the Query-Based Multicontext approach to knowledge browsing. In Section 2, we first discussed the added value of the approach compared to other Formal Concept Analysis based knowledge browsing solutions. Then we presented a comparison of the visualisation of realised contexts for diverse paradigms. The evaluation has shown that novice users could answer questions using the lattice paradigm more quickly than with the other paradigms, as long as they had been given some preliminary training in Formal Concept Analysis. However, our experiment remained quite limited, and supplementary tests should be performed.

Acknowledgements

We would like to thank our colleagues for their participation in the evaluation and the reviewers for many critical comments which helped us improve the quality of the paper greatly. Finally, we acknowledge support by the German Federal Ministry for Education and Research (BMBF) under the SmartWeb project.

References

1. Peter Becker, Jo Hereth, and Gerd Stumme. ToscanaJ: An open source tool for qualitative data analysis. In V. Duquenne, B. Ganter, M. Liquiere, E. M. Nguifo, and G. Stumme (eds.), editors, *Advances in Formal Concept Analysis for Knowledge Discovery in Databases. Proc. Workshop FCAKDD of the 15th European Conference on Artificial Intelligence (ECAI 2002). Lyon, France, July 23, 2002*, 2002.

2. Jürgen Bortz. *Statistik für Human- und Sozialwissenschaftler.* Springer-Lehrbuch. Springer Verlag, ISBN: 3-540-21271-X 2005.
3. Richard Cole and Gerd Stumme. CEM - a Conceptual Email Manager. In Bernhard Ganter and Guy W. Mineau, editors, *Proc. ICCS 2000*, volume 1867 of *LNAI*, pages 438–452. Springer, 2000.
4. Peter W. Eklund, Jon Ducrou, and Peter Brawn. Concept Lattices for Information Visualization: Can Novices Read Line-Diagrams? In Peter W. Eklund, editor, *ICFCA*, volume 2961 of *Lecture Notes in Computer Science*, pages 57–73. Springer, 2004.
5. Berhard Ganter and Rudolf Wille. *Formal Concept Analysis – Mathematical Foundations.* Springer Verlag, Berlin – Heidelberg, 1999.
6. Bernhard Ganter and Rudolf Wille. Conceptual Scaling. In *F. S Roberts (eds): Applications of combinatorics and graph theory to the biological sciences*, pages 139–167, New York, 1989. Springer Lecture.
7. Julien Tane. Using a Query-Based Multicontext for Knowledge Base Browsing. In *Formal Concept Analysis, Third International Conf., ICFCA 2005-Supplementary Volume*, pages 62–78, Lens, France, 2005. IUT de Lens, Universite d'Artois.
8. Rudi Studer and Steffen Staab, editors. *Handbook on Ontologies in Information Systems.* Springer Verlag, Berlin – Heidelberg, 2003.
9. Gerd Stumme, Rafik Taouil, Yves Bastide, Nicolas Pasquier, and Lotfi Lakhal. Computing iceberg concept lattices with TITANIC. *Data Knowl. Eng.*, 42(2):189–222, 2002.
10. Gerd Stumme and Rudolf Wille. A geometrical heuristic for drawing concept lattices. In Roberto Tamassia and Ioannis G. Tollis, editors, *Graph Drawing*, volume 894 of *Lecture Notes in Computer Science*, pages 452–460. Springer, 1994.
11. Julien Tane. Query Based Multicontext based browsing: a technical report. Technical report, Research Unit Knowledge and Data Engineering, University of Kassel, Germany, September 2004. http://www.kde.cs.uni-kassel.de/tane/techreport.
12. T. Tilley, R. Cole, P. Becker, and P. Eklund. A survey of formal concept analysis support for software engineering activities. In Gerd Stumme, editor, *Proceedings of the First International Conference on Formal Concept Analysis - ICFCA'03*. Springer-Verlag, February 2003.

Representation and Reasoning on Role-Based Access Control Policies with Conceptual Graphs

Romuald Thion and Stéphane Coulondre

LIRIS: Lyon Research Center for Images and Intelligent Information Systems,
Bâtiment Blaise Pascal (501), 20, avenue Albert Einstein,
69621 Villeurbanne Cedex, France
firstname.surname@liris.cnrs.fr

Abstract. This paper focuses on two aspects of access control: graphical representation and reasoning. Access control policies describe which permissions are granted to users w.r.t. some resources. The Role-Based Access Control model introduces the concept of role to organize users' permissions. Currently, there is a need for tools allowing security officers to graphically describe and reason on role-based policies. Thanks to conceptual graphs we can provide a consistent graphical formalism for Role-Based Access Control policies, which is able to deal with specific features of this access control model such as role hierarchy and constraints. Moreover, once a policy is modeled by CGs, graph rules and inference procedures can be used to reason on it; This allows security officers to understand why some permissions are granted or not and to detect whether security constraints are violated.

1 Introduction

Opening our information systems to the world-wide web is really seducing, but highlights a problem which becomes more and more crucial: security. Nowadays, every on-line computer must be equipped with security update agent, firewall, anti-virus software and even anti-spyware, otherwise within a few minutes information system can become the target of automated scanning scripts, hackers, malicious websites, etc.: security is now of main importance. In this paper, we are dealing with a particular aspect of security process: access control.

1.1 Reasoning on Access Control Policies

Access control denotes the fact of determining whether a *user* (process, computer, human user, etc.) is able to perform an *operation* (read, write, execute, delete, etc.) on an *resource* (a tuple in a database, a table, an object, a file, etc.). An operation right on a resource is called *permission*. Access control policies define the users permissions in order to enforce security of an organization.

As a matter of fact, a large part of flaws in information systems are due to administration mistakes or security misconceptions. The number of users is increasing, and rules become more complex. Moreover, in order to deal with border-line cases and organizational peculiarities, security constraints have been

H. Schärfe, P. Hitzler, and P. Øhrstrøm (Eds.): ICCS 2006, LNAI 4068, pp. 427–440, 2006.

introduced. Security constraints define states in which policies are inconsistent. As policy engineering is considered to be of high pratical importance [3], there is a need for tools facilitating design and maintenance of security policies, which can be a though job with large constrained policies. According to [4], we do think that such tools cannot be designed without a proper formal framework. Policy engineering tools need to meet several requirements:

- have an appropriate graphical interface,
- be able to capture access control model mechanisms and constrained policies,
- be able to check consistency of policies,
- be able to answer queries for particular permissions or relation holdings in the policies,
- have comprehensible inference mechanism, even by non-logicians.

1.2   Why Conceptual Graphs?

Rather than tailoring a dedicated fragment of first-order logic [9](FOL) or using a pure logic-based approach [4,3] with traditional resolution methods (such as Robinson principle), we focused on Conceptual Graphs (CGs) and their dedicated inference procedures. CGs has been proposed as a mathematically well founded knowledge and reasoning model [15,6,1]. The CGs framework meets the requirements described in section 1.1:

- CGs is a formal system (the function Φ gives a straight equivalence in FOL [17]),
- the CGs support can model complex structures such as hierarchies,
- individual markers and graph rules are expressive enough to capture authorization mechanisms of access control models,
- chaining procedures [12] allow direct graphical-based inferences on graphs.

The following section presents fundamentals on CGs and Role-Based Access Control policies. Section 3 presents how to use the CGs framework to capture RBAC concepts, relations, constraints and authorization mechanisms. Last section survey related work and presents some perspectives.

2   Fundamentals

2.1   Role Based-Access Control Policies

The Role-Based Access Control (RBAC) model is an access control model in which permissions are associated with roles (roles can be seen as collections of permissions), and users are made members of appropriate roles. The definition of a role is quoted from [13]: "A role is a job function or job title within the organization with some associated semantics regarding the authority and responsibility conferred on a member of the role".

Figure 1 is the common representation of the RBAC model. In this figure, "URA" is a short for "User-Role Assignment" and "PRA" for "Permission-Role

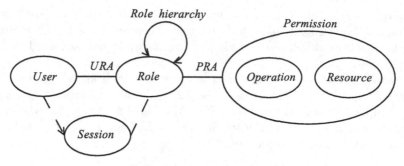

Fig. 1. RBAC Model from [13]

Assignment". In the RBAC model, permissions are not directly assigned to users. Thus, a user is authorized to perform an operation on a resource only if he/she endorses a role which is granted this permission (possibly through inheritance). This authorization principle is referred to as the *core rule* of RBAC. The role hierarchy is a way of minimizing the number of roles in RBAC policies: a role specializing another one inherits all its permissions.

As the major part of access control decisions is based on user functions or jobs, the introduction of roles greatly simplifies policies management. Since roles in an organization are usually stable with respect to users turnover and tasks reassignment, RBAC provides a powerful mechanism for reducing complexity and cost of security management [4]. RBAC was found to be among the most attractive solutions for providing access control in e-commerce, e-government or e-health, and is also a very active research field [8].

However, organizations may involve thousands of roles and administration of RBAC policies can still be complex. Graphical based modeling and reasoning is a step beyond easy-to-use policy administration interfaces and better comprehension of policies.

2.2   Representation and Reasoning with Conceptual Graphs

The following definitions are quoted from [12,1].

We call a *support* the structure that represents the ontological knowledge for a specific domain application. The support is composed of a concept type lattice, a relation type set, a set of invidual markers and a mapping from marker to concept type. A simple graph G defined over a support S is a finite non oriented bipartite graph of concepts and relations, not necessarily connected.

A rule $R: G_1 \Rightarrow G_2$ is composed of two simple CGs G_1 and G_2, respectively called *hypothesis* and *conclusion*. There may be co-reference links between concepts of G_1 and G_2, which can be set only between vertices having the same type. A rule express knowledge of form "If G_1 is present, then G_2 can be added to the knowledge base". In this paper we do not use the colored graph representation of rules (one color for the hypothesis and another one for the conclusion) but the arrowed representation. A *knowledge base* is composed of a set of simple CGs defined over a support S (facts) and a set of rules.

The deduction problem asks whether there is a sequence of rules applications enriching the knowledge base such that a goal Q can be reached. Two reasoning strategies has been developed for this problem: forward and backward chaining of graph rules.

Forward chaining is typically used in order to enrich knowledge base with new facts, implicitly present in the base [12]. It is used to prove a goal graph G_{goal} is implied by a knowledge base KB. Main idea is to calculate the closure of the set of facts of KB by the set of rules of KB to infer all the deductible knowledge. Basic outline of procedure is:

- choose a rule R from KB,
- let be G a fact from KB. If G fulfils the hypothesis of R, then the conclusion of R can be added to KB, if it is not already present (to avoid redundant information),
- repeat until no new facts can be produced,
- if there exists a projection from G_{goal} to facts from KB, then G_{goal} is implied by the knowledge base.

This knowledge addition principle is the graph dual of bottom-up resolution approach in first-order logic. This procedure is not goal-directed, thus generates information that may not be useful. This procedure is sound and complete if G_{goal} and KB are in normal form and is semi-decidable: all positive answers can be computed in finite time, but not for all negative answers.

Backward chaining is typically used to find a linear resolution of a goal (a request). Whereas forward chaining acts from facts to goal, backward chaining is top-down: from conclusion to hypothesis. The goal G and a rule R from KB are analyzed in order to detect if R may have produced G. If an unification is found, the procedure constructs a new goal G'. If the new goal is deductible from KB, then G is deductible too. This basic operation is called *piece unification* [12]. The basic outline of this procedure is:

- set G_{goal} as the current goal G (initialization step),
- choose a rule R from KB such as its conclusion is piece unifiable with G,
- remove the unified conclusion of R from G,
- add the specialized rule hypothesis of R to produce a new subgoal G',
- repeat until the empty subgoal is obtained.

The philosophy of this procedure is close to the classical SLD-resolution (Selected, Linear, Definite), though it is more complex. Like the forward chaining, this procedure is sound and complete and semi-decidable.

3   Conceptual Graphs for Graphical Representation and Inference on Role-Based Policies

In section 2, we described the RBAC model and we outlined two inference procedures. In this section we present how a knowledge base can model a RBAC policy.

3.1   Modeling Basic Concepts of RBAC

The core elements of RBAC models are described in figure 1. For sake of readability, we do not include sessions (which lead to more complex authorization rules) in this paper. Sessions express which roles are currently endorsed by users. According to [2], it is possible to model sessions by adding a new relation between users and roles.

The support must include the following concepts and relations:

- *Role* concept type,
- *Resource* concept type,
- *Operation* concept type,
- *User* concept type,
- *ura* of arity 2 (User,Role) relation type,
- *pra* of arity 3 (Role,Operation,Resource) relation type.

Once the RBAC model is translated into a conceptual graph support, we need to add the different roles, users, operations and resources involved in the policy:

- for each resource within the policy, add a conformity relation *resource_id* : *Resource*,
- for each role within the policy, add a conformity relation *role_id* : *Role*,
- for each operation within the policy, add a conformity relation *operation_id* : *Operation*,
- for each user within the policy, add a conformity relation *user_id* : *User*.

Once the basic concepts and relations are created, the next step towards a complete modeling of the RBAC model (figure 1) is to translate user-role and

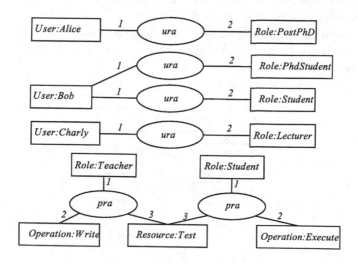

Fig. 2. A graph of user-role and role-permission assignments

permission-role assignment. The assignments are translated into the knowledge base by a set of simples CGs, called *facts*. The facts describe which roles are granted to users, and which operations are granted to roles. The figure 2 is an example of these assignments. In this graph, the user *Bob* is made member of roles *Student* and *PhdStudent* (multiple user-role assignment), the user *Charly* is made member of the role *Lecturer*, and the user *Alice* is made member of the role *PostPhD*. The role *Teacher* is granted *Write* access on resource *Test* and the role *Student* is granted *Execute* access on resource *Test*.

Thus, we need to add some facts to the knowledge base:

- for each permission-role assignment in the policy, add to the knowledge base a *pra* relation between the role, the operation and the resource,
- for each user-role assignment in the policy, add to the knowledge base a *ura* relation between the user and his/her roles.

3.2   Modeling Role Hierarchy

Figure 3 is a sample role hierarchy. For example, a user assigned to a role *Se-niorLecturer* will be granted every permissions granted to role *Lecturer*, *Teacher* and *Researcher*. Graph rules are used to represent the inheritance relation properties. Direct inheritance (e.g. *Lecturer* directly inherits permissions from both *Teacher* and *Researcher* roles) is modeled by an annex relation *seniorDirect*. The complete inheritance relation is modeled with the *senior* relation (reflexive and transitive closure of the *seniorDirect* relation).

To model the RBAC inheritance in conceptual graphs, we need to:

- add the relation type *seniorDirect* of arity 2 (Role,Role) to the support,
- add the relation type *senior* of arity 2 (Role,Role),
- add an order relation *seniorDirect* $\leq$ *senior*,
- add a graph rule (figure 4) settling that the inheritance relation is reflexive,
- add a graph rule (figure 5) settling that the inheritance relation is transitive,
- add direct inheritance relations between roles to the knowledge base (e.g. to model edges from the figure 3).

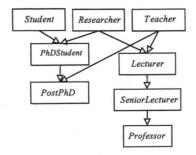

Fig. 3. A sample role hierarchy, arrow is from least to most privileged role

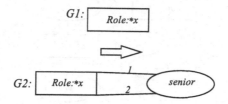

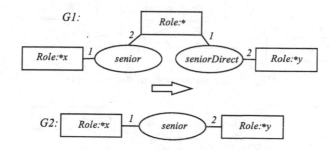

Fig. 4. A role inherits itself

Fig. 5. Role inheritance is transitive

3.3 Modeling Authorization Decision Mechanism and RBAC Constraints

The last step for modeling RBAC policies is to create a rule in the knowledge base for authorization decision. A relation *permitted* of arity 3 (User, Operation, Resource) is added to the support. The graph rule given in figure 6 expresses how users are granted access on resources. This graph rule can be referred to as the *core rule* of the RBAC model (section 2.1).

In the RBAC model, constraints have been introduced to reflect peculiarities of organizations [8]. Examples of such constraints are *mutually exclusive roles* (e.g. no user can be assigned both *Student* and *Lecturer*), *mutually exclusive operations* (e.g. figure 7) or *prerequisites* (e.g. if a user is granted *Write* access on a resource, he must be granted *Read* access too). Other organizational constraints can be expressed with CGs. For example, we can add a rule stating that a *Root* role inherits all other ones. We have chosen to model these constraints by graph rules producing an *Inconsistency* concept. Thus, checking the policy is consistent is to infer if an *Inconsistency* can be produced through chaining of graph rules. We do not use the notion of constraints from [1] because they implies definition of multiple constraints when only one rule may be sufficient (e.g. for symmetric organizational constraints). For example, to capture that *"Execute and Write* are mutually exclusive operations on resource *Test"* (figure 7), we must set two constraints: "if a user is granted *Execute* operation then he cannot be granted *Write* operation" and "if a user is granted *Write* operation then he cannot be granted *Execute* operation".

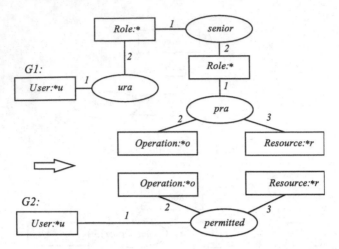

Fig. 6. The authorization decision rule

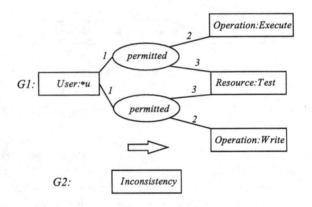

Fig. 7. No user can be permitted both *Execute* and *Write* operations on *Test*

3.4   Reasoning on Policies

The knowledge base includes:

- the set of RBAC concepts,
- the set of RBAC relations, used to model hierarchy, assignments and authorizations,
- the set of rules, modeling properties of role inheritance, authorization core rule and constraints,
- the set of facts, modeling the users, roles, operations, resources and inheritance relations present in the policy.

Given a query, graph rules inferences allow Security Officers (SOs):

- to know granted permissions,
- to query the policy on more complex statements (e.g. figure 8),

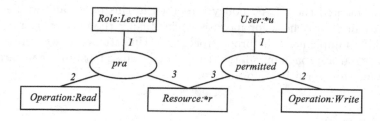

Fig. 8. A sample query

- to find inconsistencies,
- to graphically follow the authorization mechanism,
- to check if settling new assignments or organizational constraints do not introduce inconsistencies.

Let us consider a SO working on the university toy RBAC policy. Role hierarchy is described in figure 3, facts are pictured in figure 2, rules governing role inheritance and authorization mechanism are included in the knowledge base too. Let us suppose the SO is defining a constraint on the RBAC policy, to fit as best as possible the policy to his organization. In order to avoid any student from enter his/her own mark on a test, the SO adds the exclusion rule "no user can be both granted *Execute* and *Write* access on *Test*" (figure 7). He then runs an inference procedure on his policy to know if this new constraints produces an inconsistency . . . And it does!

Following the steps of the inference, the SO is able (without specific requirements on resolution principle) to understand why this constraint can not be added to the policy. We illustrate here the backward chaining inference procedure, which starts with a graph goal modeling the inconsistency. The procedure is outlined as follows (the complete illustration can be found in annex):

- there is an inconsistency if a user is both granted *Execute* and *Write* access on tests (step 1 to 2),
- a user is granted access through his/her roles (step 2 to 4),
- role *Student* is *Execute* granted access on *Test* (step 4 to 5)
- role *Teacher* is *Write* granted access on *Test* (step 4 to 5),
- a role inherits permissions of its ancestors,
- role *PostPhD* inherits from both roles *Student* and *Teacher* (step 5 to 9),
- *Alice* is granted role *PostPhD* (step 9 to 11),
- thus, as the empty subgoal is obtained, the policy is not consistent according to the new constraint.

This example points out how an adequate graphical tool for RBAC policy can be useful for modeling purposes. Without such an inference tool, an SO can happen to add some constraints which can be violated but not detected, until he receives a complain from *PostPhD* users who may not login in the system anymore. We have validated our approach with the CoGITaNT toolkit [7]: We have implemented the RBAC policy presented in this paper and we have used forward chaining to check for inconsistencies (using the closure operator from the toolkit to implement the proof procedure). Unfortunately, we were not able to validate our approach on a large real policy because most of organizations do not provide such sensitive information.

4   Discussion

In this paper, we have shown how to use the CGs framework to design and infer on RBAC policies. Such inferences can be used to check policies consistency and to answer queries in order to formally test or debug policies. CGs have an appropriate graphical interface useful for specifying policies and especially for explaining inferences to security engineers.

4.1   Graphical Representation of Role-Based Access Control Policies

In [11], the authors point out that in order to avoid inconsistencies, it is appropriate to have a precise modeling of access control rules (their work is based on RBAC examples), and a visual modeling to make administration tasks easier.

Their paper compare readability and checking ability of three approaches. The first approach is based on *UML*. The second model studied is *Alloy*: a structural modeling language based on first-order logic, for expressing complex structural constraints and behaviour. Finally, the third model is *Graph Transformations*. It is a graph model where transformation rules, which are rewriting rules, are used to deduce all possible knowledge.

When a new constraint (e.g. mutually exclusive roles) is added to a given policy, these models can verify if the resulting policy is inconsistent, and give a counterexample. Nevertheless, the paper mainly focused on constraints: those tools do not answer questions such as *which roles are allowed to execute this operation on this object*, or other more complex queries. Indeed, UML is only a semi-formal model, Allow Constraint Analyzer is a consistency checker, and Graph Transformations do not have FOL logical semantics.

In [16], the authorization schema is described using Dynamically Types Access Control (DTAC) graphical formalism. The authors aim at graphically represent RBAC schemas. However they do not propose an inference method in order to ask queries or to check for inconsistencies.

4.2   Reasoning on Role-Based Access Control Policies

In [2], constraint logic programming (first order predicate calculus + constraints, e.g. Prolog + CLP library) is used to express and infer on RBAC policy. The authors describe access control programs able to deal with roles hierarchy and objects hierarchy.

The authors of [9] describe a fragment of FOL which is tractable and sufficiently expressive to capture policies for many applications. This work is really interesting and points out tractability and complexity results on their logic. Unfortunately, the authors did not investigate deeply some emerging problems when dealing with complex access control models involving abstract container between users and permissions (e.g.: groups, roles, tasks, etc.) and their hierarchies.

We do agree with the authors statement about the use of logic programming by non-logicians, but we do not agree that a "filling the blank on English sentences" interface is sufficient for security administrators. We think that administrators must have a computer-aided software engineering (CASE) interface to design and check policies built on a consistent graphical model. Moreover such CASE should provide a comprehensible reasoning trace, which can be provided through graph operations (e.g. projection of a query onto the set of facts)

4.3   Modeling Extensions of RBAC Models

For sake of clarity the example exposed in this paper does not include sessions. We are investigating the interest of chase procedure to check RBAC policies involving sessions. For example, using chase procedure we might answer queries like "Are the policies consistent for all possible sessions ?". Moreover, incorporating the model for administration of roles exposed in [14] is promising for distributed policies verification purpose.

The RBAC family includes some extensions, for example GTRBAC [10] which provide a temporal framework for specifying an extensive set of temporal constraints. Such extensions cannot be taken in considerations in the present CGs model. Indeed, it may require the expression of built-in (i.e. whose semantics is hard-coded) concepts and relations such as linear arithmetic constraints, set constraints, etc. Therefore we plan to study the extension of the CGs graphical formalism and inferences to arbitrary built-in elements.

4.4   Modeling Role Hierarchy Using Concept Lattice

One could notice that the conceptual graphs concept lattice might be useful to model the RBAC role hierarchy. While it is indeed well suited for role hierarchy modeling, some problems arise when it comes to describing relations between roles, resources and users. For example, a graph rule is not well-suited for modeling the fact that if a user is assigned two different roles, then an inconsistency holds. Indeed, it may require a coreference link between two generic markers within the same hypothesis, which adds additional graphical elements. We think is it not desirable, however it is still possible.

References

1. Baget, J.-F., Mugnier, M.-L.: Extensions of Simple Conceptual Graphs: the Complexity of Rules and Constraints. Journal of Artificial Intelligence Research (JAIR) **16** (2002) 425–465
2. Barker, S., Stuckey, P. J.: Flexible access control policy specification with constraint logic programming. ACM Trans. Inf. Syst. Secur. **6** (2003) 501–546

3. Bertino, E., Catania, B., Ferrari, E., Perlasca, P.: A logical framework for reasoning about access control models. ACM Trans. Inf. Syst. Secur. **6** (2003) 71–127
4. Bonatti, P. A., Samarati, P.: Logics for authorization and security. Logics for Emerging Applications of Databases. (2003) 277–323
5. Chein, M., Mugnier, M.-L.: Représenter des connaissances et raisonner avec des graphes. Revue d'Intelligence Artificielle. **10** (1996) 7–56
6. Chein, M., Mugnier, M.-L.: Conceptual Graphs: Fundamental Notions Revue d'Intelligence Artificielle. **6-4** (1992) 365–406
7. Knowledge Graphs Research Group: Conceptual Graphs Integrated Tools allowing Nested Typed graphs (CoGITaNT). http://cogitant.sourceforge.net/
8. Ferraiolo, D.F., Kuhn, D.R., Chandramouli, R.: Role Based Access Control. Artech House (2003)
9. Halpern, J. Y., Weissman, V.: Using first-order logic to reason about policies. CSFW. (2003) 187–201
10. Joshi, J., Bertino, E., Latif, U., Ghafoor, A.: A Generalized Temporal Role-Based Access Control model. IEEE Transactions on Knowledge and Data Engineering. **17** (2005) 4–23
11. Koch, M., Parisi-Presicce, F.: Visual specifications of policies and their verification. FASE. Lecture Notes in Computer Science **2621** (2003) 278–293
12. Salvat, E., Mugnier, M.-L.: Sound and complete forward and backward chaining of graph rules. ICCS. Lecture Notes in Computer Science **1115** (1996) 248–262
13. Sandhu, R. S., Coyne, E. J., Feinstein, H. L., Youman, C. E.: Role-Based Access Control models. IEEE Computer. **29** (1996) 38–47
14. Sandhu, R. S., Munawer, Q.: The ARBAC99 model for administration of roles. ACSAC. (1999) 229–239
15. Sowa, J. F.: Conceptual Structures: Information Processing in Mind and Machine. Addison Wesley (1984)
16. Tidswell, J., Potter, J.: A graphical definition of authorization schema in the DTAC model. ACM Symposium on Access Control Models and Technologies. (2001) 109–120
17. Wermelinger, M.: Conceptual Graphs and First-Order Logic. Third International Conference on Conceptual Structures. LNAI **954** (1995) 323–337

Annex: A Graphical Backward Reasoning

We here show the backward inference steps of section 3.4. The dotted lines show the pieces which are removed in every step by application of backward chaining.

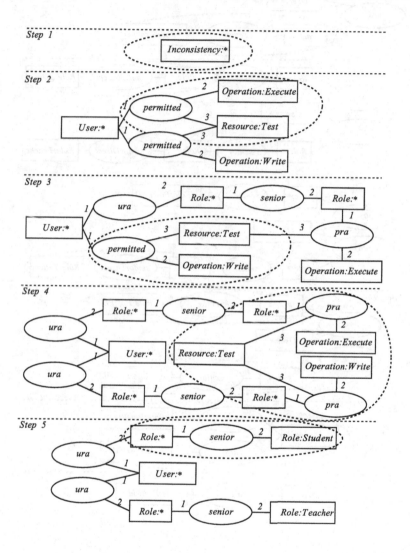

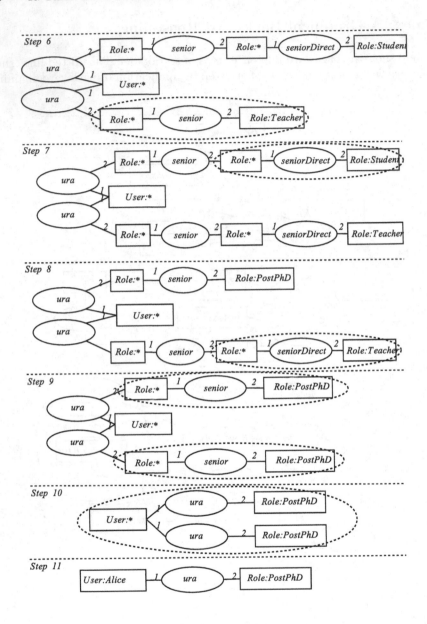

Representing Wholes by Structure*

Yang Yu, Ji Wang, Ting Wang, and Huowang Chen

National University of Defense Technology
Changsha, Hunan, P.R. China 410073
{yangyu, jiwang, tingwang, hwchen} @nudt.edu.cn

Abstract. The influence of Part-Whole relationship (*partof*) on ontology languages is weak because the special semantics of *partof* is missed in both representations and interpretations. In this paper, we first exemplify the complexity and limitation of using *partof* for representing wholes. We then investigate the special semantics of *partof* by analyzing the properties of parts and wholes. The conclusion is that a part should be regarded as something internal to certain whole and a whole is a structured thing which is comprised of some internal things. The special semantics are achieved by introducing some new constructs which are *whole structure, part type, role type, Ontology, defined-in* relation and *play* relation. Because the whole structure possesses inherent modularity and local semantics, the representation is very natural and simple. We show the generality and applicability of the presented approach in terms of the problems pointed out in the paper.

1 Introduction

Ontology is a conceptualization of a domain and plays a crucial role in many communities. Typically, ontology consists of entities, attributes, relationships and axioms [6]. Ontology languages are, different from logic languages, generally based on the understanding of the world and provide intuitive syntax and constructs. Without exceptions, all ontology languages provide the *is-a* relationship by which a domain can be organized based on hierarchical structures.

Another relationship that has received considerable attention is *part-whole* relationship (*partof*) [1,3,7,4,10,12,17]. Psychological experiments [17] have shown that parts play an important role in the thought process of human and some researchers supposed that *partof* should be given the "*first class*" dignity [1]. However, so far it ranks far behind the concept of *is-a* [10]. The use of *partof* can not make knowledge representation evidently intuitive and simple.

As an ubiquitous phenomena in real world, part-whole relations have not been abstracted in the powerful constructs in ontology and conceptual languages. One of the reasons is that traditional languages are short of an important construct

---

* This research is supported by the National Natural Science Foundation of China (60403050) and (90612009), and the National Grand Fundamental Research Program of China under Grant No. 2005CB321802.

H. Schärfe, P. Hitzler, and P. Øhrstrøm (Eds.): ICCS 2006, LNAI 4068, pp. 441–454, 2006.

to embody the special semantics of part-whole relation. We consider this construct is corresponding to the so called *Thirdness* in the trichotomy principle of Peirce (see 4.1). In *conceptual graphs* (CGs), the construct context is provided for representing *Thirds* as noted by Sowa [14]. In another notable language UML2.0 [11], a similar construct called composite structure is newly introduced. However, these constructs are both from syntax but not semantics.

In this paper, we first exemplify the complexity and limitation of using *partof* for representing wholes. Then the special semantics of part-whole relation is investigated by analyzing the interpretations and the properties of *parts* and *wholes*. The important finding is that a part should be regarded as something *internal* to certain whole and a whole is a *structured* thing which is comprised of some internal things. To achieve this, we shift the emphasis to the *structure of whole*. The special semantics of part-whole relation is embodied by introducing some new constructs: *whole structure, part type, role type, Ontology, defined-in* relation and *play* relation. The result is a novel language by extending the traditional ones with these constructs. In this language, the structure of whole possesses inherent modularity and local semantics, whereby the representation is natural and simple. We show the generality and applicability of the introduced approach in terms of the problems pointed out in the paper.

2   Definitions and Terms

There are some characteristics [3,7] related to *partof* relations. We only introduce some relatively important ones used in this paper.

The first one is transitivity. In the formal theory mereology, *partof* is defined as a partial ordering relation; *proper partof* is defined as a strict ordering relation. However, it is well known that the transitivity does not always hold [17]. The other definitions and terms used in this paper are shareability, existential dependence, essential dependence, configuration and role.

Definition 1 (Shareability). *Shareability is a property of partof relation and denotes the ability of the part belonging to two or more wholes between which there are no partof relations at the same time. The parts possessing the ability are called shareable parts (Spart), else exclusive parts (Epart).*

Definition 2 (Existential dependence [13]). *Let P and Q be object types. P is existence dependent on Q if and only if the life of each occurrence p of type P is embedded in the life of one particular and always the same occurrence q of type Q. p is called the dependent object, and is existence dependent on q. The parts possessing the property of P are called existential dependence parts (Dpart)*

Definition 3 (Essential dependence). *Let P and Q be object types. P is essentially dependent on Q if, in order the occurrence p of type P to exist, an occurrence q of type Q has to exist.*

Definition 4 (Configuration, Integrity constraints). *Configuration is some constraints among the parts of a whole for characterizing its integrity.*

Another related notion is role. The nature of roles and the way of representing them have been discussed for a long time in different fields. In the field of knowledge representation, Sowa distinguished between natural types "that relate to the essence of the entities" and role types "that depend on an accidental relationship to some other entity" [14]. Developing Sowa's ideas further, Guarino[5] presented an ontological distinction between role and natural types: a role type must be an anti-rigid concept whose instances can enter and leave the extent of the concept without losing their identity; a natural type is a rigid concept whose instances cannot drop this type without losing its identity. As noted by Loebe [8], most approaches consider roles as determined by some external entities. Loebe tries to characterize different approaches on the basis of the ontological nature of the contexts that determine roles, and he individuates and analyzes in detail three kinds of role: relational roles, processual roles, and social roles. In [9], Masolo et al. establish a formal framework for representing social roles based on definitional dependence. The definition of definitional dependence can be formulated as follows: "a property is definitionally dependent on a property if, necessarily, any definition of ineliminably involves". There are some other characteristics of roles discussed in other literature. In [16], Steimann individuates 15 fundamental characteristics of roles at both the class and the instance levels. These characteristics can be used to evaluate a role model. For a detailed discussion of various approaches the reader is referred to [15,8,9,16]. Summing up, roles are commonly regarded as external things to context, and roles are dependent on context and their players.

3   Motivation

In this section, we point out some problems on the representation of parts and wholes. They are our motivation to focus on the structure of whole.

3.1   Some Problems on the Representation of Parts and Wholes

Problems on Representing Shareable Parts. We show that it is not adequate to express some facts with partof relation only. For example, the Father and Child in a family are disjoint. It perhaps be wrongly expressed as formula (1).

$$Family(x) \land P(y,x) \land P(z,x) \land Father(y) \land Child(z) \rightarrow y \neq z \qquad (1)$$

In this formula, P is the abbreviation for *partof*. Substituting b for x, a for y and z, we have the inference:

$$Family(b) \land P(a,b) \land P(a,b) \land Father(a) \land Child(a) \rightarrow a \neq a.$$

The premise is satisfiable because a person a can be both the father of family b and the child of another family. Formula (1) is wrong because partof relation cannot fix the roles played by a part. In fact, the meaning of (1) is that every person cannot be both a father and a child.

To solve the problem, we can introduce a special relation called *partnameof* which is comprised of a part (or role) name and "of". Intuitively, *partnameof* relation specializes from partof by binding the role played by part. With part-nameof relations we can correctly express the constraint as:

$$Family(x) \land Fatherof(y, x) \land Childof(z, y) \to y \neq z.$$

It seems that there are no general finite classifications of partof relation because of the necessary of using partnameof relations.

Definitional Dependence between Parts and Wholes. We exemplify definitional dependence between parts and wholes by the example of car. For simplicity, we regard car as a composite comprised of one engine and four wheels in which there are two front wheels (FW) and two rear wheels (RW). For some kind of cars (called Car_1), their front wheels are driving wheels (DrW) which connect with their engines; for some others (called Car_2), their rear wheels are driving wheels.

It is impossible to define part concepts without using whole concepts. For example, we can say little about FW besides being a wheel. To make things more clear, we give two constraints related to FW and RW expressed in FOL:

$$FW(x) \land P(x,y) \land Car(y) \to DrW(x), RW(x) \land P(x,y) \land Car(y) \to DrW(x).$$

It is apparent that whether FW or RW is DrW is dependent on Car_1 and Car_2. So the definitions of part types is not fixed. The problem pointed out above also holds for sharable parts. For example, the definitions of student in different universities are different from each other.

Exclusiveness and Sharability. Another general property related to Dpart and Epart is that any instance of them belongs to just one whole without considering transitivity. Because different types of wholes can own the same type of parts, it is difficult to express this kind of constraints: for instance, to express that an engine can be part of just one instance of Car, Truck, Cropper, and others, we have to enumerate all of them. Note that this constraint is dependent on whic concepts are defined in the knowledge base.

A Spart can belong to more than one whole. However, this view will lead to some difficulties. For example, if a person works for several companies simultaneously, he has several ID and Salary attributes; we cannot use a person instance to collect all these attributes due to *property conflicts*.

Complexity of Representing Integrity Constraints. When defining a whole concept, we should not only specify which parts its instance should contain, but also specify the configuration among parts to characterize its integrity. These integrity constraints are often expressed in lengthy formulae. For example, the configuration between Engine and Driving Wheel is partially expressed as:

$$DrW(x) \land P(x,y) \land Car(y) \to \exists z(Eng(z) \land P(z,y) \land Car(y) \land Connect(x,z)).$$

In this formula, we must use $Car(y)$, $P(x,y)$, and $P(z,y)$ to bind $DrW(x)$ and $Eng(z)$ in order to ensure that this constraint can be applied only to a car and its parts. The complexity mainly results from integrity constraints.

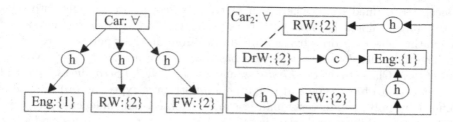

Fig. 1. Representing whole with CGs

3.2   Representing Wholes with CGs

"Conceptual Graphs (CGs) are an extension of C. S. Peirce's existential graphs with features adopted from linguistics and AI" [15]. In a conceptual graph, the boxes are called concepts, and the circles are called conceptual relations. On the left of each concept box is a type field, which contains a type label. On the right is a referent field, which may contain a name, a quantifier like $\forall$, or a plural specification {*} which may be followed by a qualifier to indicate the count of elements in the set (we write number in brackets for simplicity). If no other quantifier is specified, the default quantifier for the variable is the existential $\exists$.

An outstanding capability of CGs is that it can represent logical formulae in a more natural, simple, and readable way: i) they eliminate variables used in logic formulae; ii) they can remarkably simplify the representation, especially when using context; iii) they can be automatically translated into first order logic (FOL). The simplicity mainly results from the construct context. For example, the configuration between driving wheels and engine mentioned in the previous section is dramatically simplified as the right graph shown in Fig. 1.

However, there are still three questions related to context unclear: i) what context is used to represent; ii) when the context should be used; and iii) how the context is used. In CGs, the usage of context is dependent on the things expressed. For example, if we do not need to express the integrity constraint mentioned above, we can represent Car as the left graph shown in Fig. 1; or else we have to use context construct because the constraint cannot be correctly expressed without context in CGs. Another problem is that the simplicity is lost when CGs are translated into FOL.

4   Understanding Wholes and Parts

4.1   The Principle of Peirce

The underlying categories we adopt are based on Peirce's trichotomy principle which is used by Sowa in [15]. The principle can be summarized with three basic categories called *Firstness*, *Secondness*, and *Thirdness*: "*First* is the conception of being or existing independent of anything else. *Second* is the conception of being relative to, the conception of reaction with, something else. *Third* is the conception of mediation, whereby a first and a second are brought into relation" [15]. As noted by Sowa, "Peirce's principle is a metalevel distinction for generating new categories by viewing entities from different perspectives".

The terms used in this paper are based on the principle of Peirce. We use *object type* (Class) to denote *Firstness*, and *role type* (Role) *Secondness*. It should be noted that object type and role type are both pure types: an object type is defined by only its inherent qualities, and an object can independently exist without other things; a role type is defined by only extrinsic qualities, and a role instance is existentially dependent on its player and context. From class and role we derive another important type called *part type* (Part) which is a hybrid from object type and role type. A part type, for instance, RearWheel is defined by not only inherent qualities but also extrinsic properties.

Thirdness is often denoted with context although what exactly is a context still remains to be clarified. Because the ontological natural of different context is quite heterogeneous, the classification of context is necessary. In this paper, we use three kinds of contexts called *relation*, *collaboration* and *composite object type* which are roughly corresponding to the context of three kinds of roles presented by Loebe respectively, i.e., *relational* roles, *processual* roles, and *social* roles [8].

4.2   Are Parts Internal or External to Wholes?

A fundamental question worth more attention is whether parts are internal or external to wholes. It is widely acknowledged that Dpart, Epart and Features are internal to their owner. However, this property is not directly entailed and embodied by traditional representation and interpretations.

The difference between *"externalness"* and *"internalness"* is illustrated in Fig. 2: the cup c is external to the table t and vice versa; the wheel w_1, w_2, and body b are internal to the car c_2. However, this difference cannot be directly embodied by the interpretations in FOL. For instance, the interpretations of the example in Fig. 2 are: $\triangle^I =\{$ Cup: c; Table: t; Car: c_1; Wheel: w_1, w_2, w_3, w_4; Body: $b\}$; $on(c,t)$, $partof(w_1,c_1)$, $partof(w_2,c_1)$, $partof(w_3,c_1)$, $partof(w_4,c_1)$, $partof(b,c_1)$. Because partof relation has no special status in set theory, the meaning of "externalness" is missed. In fact, every individual in the domain can be regarded as an external thing to the others. This kind of interpretations is not natural because a whole individual is divided into several individuals. For example, when we say there is a car, there must be a car, a body and four wheels in the domain.

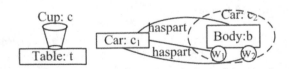

Fig. 2. Difference between "externalness" and "internalness"

It is obvious that the special semantics of partof pertains to the property of "internalness". To achieve this semantics, a part should be regarded as something internal to certain whole and a whole as a structured thing which is comprised of some internal things. The difference between a structured individual and a traditional individual is similar to the one between c_1 and c_2 shown in Fig. 2.

There still remains a question whether roles are internal to wholes. To the best of our knowledge, all related works on role in knowledge representation [5,15,9] advocate that roles are external to whole. It should be noticed that all these works regard role as a hybrid from Firstness and Secondness, and this view of role is similar to Spart.

In this paper, we regard role type as a pure Secondness and as internal property of Thirdness. The main reason is that the connection between role and context is tighter than that one between role and object. For instance, student is an external property to human because the later can exist without the former. Furthermore, the definition of student is independent from the one of human. On the other hand, the definition of student is dependent on and determined by university. In addition, the definition of university is also dependent on student because there is no university without students.

Another reason comes from the principle of unification. Our strategy is to apply the successful principle as long as it provides benefits. By this principle, we can have a uniform view on parts, roles and wholes. As shown in the remainder sections, the unification advantages outweigh the loss (if any), whether from a theoretical or ontological point of view.

4.3   Understanding Wholes

The axiom relating mereology to the theory of dependence asserts that "a whole is strictly dependent on its parts [12]." Furthermore, a whole exists only if its parts exist. "It makes sense to say that a whole is nothing over and above its strict part [4]." In common, a whole is necessarily dependent on its parts.

Intuitively, a whole is comprised of some parts which are interconnected in term of some *integrity constraints*. The intuitive understanding of wholes means that a whole is not a set of parts because of integrity constraints. Just because of falling short of expressing the integrity constraints of wholes, the formal theory mereology cannot tell an integral whole from a scattered sum of disparate entities.

The immediate question is how to represent whole. Because a whole is not a set, we turn to regard a whole as a structured thing called *whole structure*. Whole structure is similar to algebra structures such as *Group*, *Ring* and so on. Typically, a structure is comprised of a set of constituents, some relations and some axioms among these constituents. Then, the special semantics of partof relation is implicitly embodied by a whole structure and its constituents. Because inherent qualities are also "internalness" but not parts, we substitute *defined-in* relation for partof relation as shown in Fig. 3.

Whole structure is a construct corresponding to *Thirdness*. The ontology natural of whole structure is determined by its type (called whole type) which

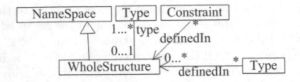

Fig. 3. The metamodel of whole structure

is corresponding to certain kind of Thirds. The metamodel of whole structure is expressed as a UML class diagram in Fig. 3. Its meaning is that a whole structure is a namespace (which is a named element also) with some types and can define some other types and constraints in it: for instance, in the declaration *Composite Class Car* {RearWheel}, the Car is a namespace and at the same time the name of whole structure, the Composite Class is the type of Car, and RearWheel is defined in the structure of Car.

4.4   Understanding Part

Intuitively, a part must be a constituent of some whole. In other words, parts are essentially dependent on wholes at least. A Dpart is existentially dependent on a whole; and all of its external properties are obtained from or by this whole only. Differing from Dpart, an Epart has the freedom of dropping its Epart type and becoming an object by departing from a whole, that is, Epart possesses the property of separability [10]. Being quite different from Dpart and Epart, a Spart has further freedom of obtaining external properties from other wholes.

At first glance, all parts should be "internalness" of wholes. To check this view, we should first consider the properties of "internalness". Intuitively, the internalness mean that they are exclusively owned by others. We call this property of "internalness" exclusivity. This property is proper to Dpart, Epart and Features but fails to Spart (hybrid from class and role). Some examples regarding Dpart, Epart and Spart are shown in Tab. 1. From these examples, we can draw a conclusion that only Spart does not satisfy the exclusivity property. Another problem is also revealed: the Dpart Brain has no corresponding Firstness concept. We leave this problem aside here and discuss it in section 5.

Table 1. Some examples on internal parts and external parts

Contex	Class/Player	Part	Inter/Exter
Car	Wheel	RearWheel	Epart, Internal
Company	Human	Employee	Spart, External
Body	?	Brain	Dpart, Internal
Trade	Company	Buyer, Seller	Spart, External
hasChild	Human	Mother, Child	Spart, External

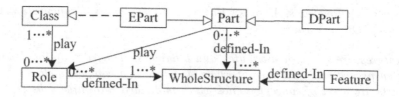

Fig. 4. The metamodel of class, role type and part type

To distinguish Spart from the other two, we resort to role type. We separate the internal properties (i.e. object) of Spart from wholes and leave its external properties (i.e. role) in wholes. A Spart is denoted by several roles which are existentially dependent on wholes and are played by the same object outside these wholes. In another words, we use role type as a substitute for Spart. By this, any *Second* is internal to a certain *Third*.

The metamodel of these types are expressed as a UML class diagram in Fig. 4. We use defined-in relationship to connect the "internalness" with their owner. The defined-in relationship is essential for the integrated representation and will be discussed further in the next section. Note that *Feature* is widely acknowledged as "internalness", and thereby added in Fig. 4. Another important thing is the difference between Epart and Dpart. As pointed out above, Epart possesses the property of separability. This property is embodied by the *is-a* relation denoted with dashed line[1] between Epart and Class.

5   Whole Structure

The previous sections have introduced quite a few concept types and relations. Here, we mainly focus on the whole structure and defined-in relation.

5.1   The Definition of Whole Structure

Before defining whole structure, we first define a crucial concept called Ontology.

Definition 5 (Ontology). *Ontology is a concept type. An instance of Ontology serves as a root concept in which some other concepts are defined.*

Ontology plays a crucial role for representing whole structure (see section 5.2). The instance of *Ontology*, for example, *O* is used as a root to define other concepts. All concepts except *O* should be defined in some whole structure or *O*. All classes defined in *O* must satisfy the property that their instances can exist independently. The inference from this is that all features and *Seconds* must be defined in whole structure.

Definition 6 (Whole Structure). *A whole structure is a description of some type which is a Third[2]. A whole structure consists of: {Features, Parts:⟨ DParts,*

---

[1] The dashed line *is-a* relation is used because Epart is not Class (pure *Firstness*).
[2] Currently, the types of Thirdness are only relation, collaboration, and composite class.

Eparts, Roles⟩, *Relations, Axioms:*⟨ *HasParts, Dependences, Configuration, Qualifications* ⟩, *Others*} *where*

1. *Features are a set of attributes and behaviors. They are optional and are often needed when the type of whole structure is a composite class.*
2. *Parts are a set of concept names which is divided into three subsets.*
   (a) *DParts are a set of concept names whose type is Dpart.*
   (b) *EParts are a set of concept names whose type is Epart.*
   (c) *Roles are a set of concept names whose type is role type.*
3. *Relations are a set of names whose type is relation type. Every relation has a nonnegative integer n called its valence and n is not less than 2.*
4. *HasParts is a set of restrictions which describe how many instances of the concepts of Parts may exist in the whole.*
5. *Dependences are a set of restrictions which describe the dependence relations between the properties of the whole and the properties of the parts.*
6. *Configuration is a set of integrity constraints of the whole.*
7. *Qualifications are a set of restrictions which specify the concepts outside the whole from which the concepts of EParts are specialized or by which the concepts of Roles are played.*

The above definition of whole structure is incomplete and extensible. What a concrete whole structure consists of is determined by its type. Generally speaking, we should elaborate the concrete whole structure for any Thirdness.

There are still some issues things should be noted. Firstly, we regulate that all concepts and constrains defined in a whole structure are local to this structure and make sense only in it. By this, whole structure possesses inherent modularity and local semantics. Secondly, all axioms except the ones in Qualifications can only use the features, concepts and relations defined in the same whole structure. Last but not least, we define whole structure without explicitly using defined-in relation. We can implicitly embody it with the syntax elements brackets (see section 5.2).

Some Constraints on Whole Structure. In order to give a formal language, we should first formalize the new constructs introduced in this paper. In this subsection, we only informally give the constraints with defined-in relation.

The constraints related to whole structure, part type and role type include:

1. If a concept w has a whole structure, then its type is one of composite class, relation, and collaboration (or other introduced types)
2. If a concept c is defined in concept w, then w has a whole structure.
3. Every concept whose type is part type or role type must be defined in another concept that is not an instance of Ontology.

The constraints related to generalization include:

1. If there is a generalization relationship between two concepts, then all the concepts and constrains defined in the super are also defined in subconcept.

2. If a concept c is both defined in concept w_1 and w_2, then there is a generalization relationship between w_1 and w_2.
3. If concept c_1 which is defined in w_1 is a generalization of concept c_2 which is defined in w_2, then w_1 must be a generalization of w_2.

The constraints related to part type or role type include:

1. For the instance x of concept c, if the type of c is part or role, then there is only one instance y of concept w whose type has a whole structure, satisfying $defined\text{-}in(c, w)$ and $partof(x, y)$.
2. For an object or part instance x of concept c and a role instance y of role concept r which is defined in w, if $play(x, y)$, then c is defined outside w.

For any concept w, its upper concepts, denoted with $Upper(w)$, is defined as: $Upper(w) \doteq \{c|defined\text{-}in * (w, c)\}$, where defined-in* denotes the transitive closure of defined-in. The outer concepts of w, denoted with $Outer(w)$, is defined as: $Outer(w) \doteq \{c|defined\text{-}in * (c, s) \land s \in Upper(w)\}$. That c is defined outside w means that c is an outer concept of w.

5.2   Representation and Interpretations

In this section, we briefly show the applicability of whole structure by some examples. For clarity, we use curly brackets to denote defined-in relation without explicitly using it: every concept which occurs in a pair of curly brackets is defined in the direct outer concept which is followed by a colon mark. An Epart example is shown below.

```
O: { Class: Wheel, Engine,
  Car:{Eparts: PEngine(1,1), RW(2,2), FW(2,2), DrW(2,2);
      Config: DrW(x) → ∃y(PEngine(y) ∧ Connect(x,y)),
            DrW(x) → RW(x) ∨ FW(x);
      Qualif: PEngine(x) → O.Engine(x); };
   }
```

In the representation of Car, $PEngine(1, 1)$ means that every car has at least one and at most one engine as part. The others are similar to this.

In the example, the axiom $PEngine(x) \rightarrow O.Engine(x)$ ensures that PEngine is separable from Car. In CGs, the above example can be represented as shown in Fig. 5. It should be noted that part types must occur in whole types and partof relations are eliminated. In addition, when translating them into formal language the structures are preserved and the same as the expressions above. The instance of Car is similar to CGs as shown in Fig. 5.

From the definition of Car, we can define its subconcept Car_1 as: $is\text{-}a$ $(Car_1, Car) \land Car_1\{Configuration : FW(x) \rightarrow DrW(x)\}$, that is, the definition of Car_1 is obtain by adding the above constraint to the one of Car. Similarly, we can define Car_2 as: $is\text{-}a(Car_2, Car) \land Car_2\{Configuration : RW(x) \rightarrow DrW(x)\}$. These two constraints hold in Car_1 and Car_2 respectively but not in the both.

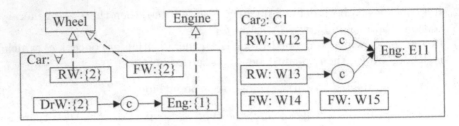

Fig. 5. An example of Epart

Another example is about the Dpart Brain which is defined as:

$$O : \{Class : Body : \{Heart, Brain\}; Heart\}.$$

For Brain, there is no concept C defined in O satisfying that $Brain(x) \rightarrow O.C(x)$. This means that Brain is unseparable from Body. These two examples adequately show the significance of *Ontology* in the language.

The last example is about the Spart Employee which is defined as:

```
O: { Class: Person, Man, Woman, Company:
```
$\qquad$ {Sparts: $Employee(3, *), Manager(1, 1), Secretary(1, 2)$;
$\qquad$ Config: $Mng(x) \vee Sec(x) \rightarrow Emp(x)$;
$\qquad$ Qualif: $Emp(x) \rightarrow \exists! y(O.Person(y) \wedge play(y, x))$,
$\qquad\qquad$ $Mng(x) \wedge Play(y, x) \wedge Obj(y) \rightarrow \neg play(y, z) \wedge Sec(z)$; };
$\qquad$ Axiom: $Person(x) \leftrightarrow Man(x) \vee Woman(x), Man(x) \rightarrow \neg Woman(x)$;} .

Now, we briefly discuss the interpretations of wholes by the Spart example shown above. Our main goal is to present appropriate interpretations which can embody wholes as integrated things. Remind that a whole is comprised of some internal things, the internal structure of whole is determined by itself. Furthermore, because any "internalness" is exclusively owned by a certain whole, the extensions of the former can be determined by the extensions of the later. In terms of this understanding, we allow wholes to have their own interpretations. The key point is that any internal thing can not occur outside its context, and thereby any thing in the outermost domain is external to the others.

For example, the interpretations of the above Spart example are: $\triangle^I = \{$Person: Tom, Bob, Alice; Company: $C_1\}$; further, the company C_1 has its own interpretations: $\triangle^I_{C_1} = \{$ Employee: E_1, E_2, E_3; Manager: E_2; Secretary: E_3; playedby: $(E_1, Bob), (E_2, Tom), (E_3, Alice)\}$. The interpretations of *play* relation embody the dependence between Company and Person. It is even simpler to apply the interpretations to Epart and Dpart because *play* relation are not used.

The benefits obtained from the language are summarized as following:

1. The representation is natural because context can be immediately identified and dependence between concepts is directly embodied by whole structure. Especially, the "internalness" is explicitly represented and embodied.

2. The problematic relation partof is almost eliminated from representation language. We need not use the subcategories of partof and partnameof.
3. This kind of representations possesses inherent modularity, i.e. all concepts and constraints defined in a whole structure make sense only in it. By this, we can overcome the main drawback of the representation in FOL, namely, numerous flatten axioms. Secondly, the expression of the configuration of a whole is simplified because all parts are implicitly qualified by the same whole. Thirdly, the different definitions of the same part concept are separated into and defined in different wholes.
4. The merit of this kind of representation is preserved in formal language, although the goal has not been completely achieved in this paper.

6 Conclusions and Future Work

In this paper we present a novel approach to describing conceptual structures. CGs is the original inspiration for our work. An outstanding capability of CGs is that it can keep representation natural and simple, which lies in the construct context which implicitly employs the property of exclusivity. In order to explicitly embody this intuition, we substitute role type for Spart to achieve this property and further introduce a new construct called whole structure to represent contexts (owners). For the integrated representation of whole structure, some other constructs *part type, Ontology, defined-in and play* relation are presented. The result is a novel language which extends the traditional ones with these new constructs. Whole structure possesses inherent modularity and local semantics, and can be viewed as a fine-grained context, whereby the representations and interpretations of context and parts are natural and simple.

The ongoing work is to formalize the *play* and *defined-in* relations, and give a formal syntax and semantic of the language presented. Another interesting work is to investigate which sort of ontological dependence can be embodied by whole structure.

References

1. A. Artale, E. Franconi, N. Guarino, and L. Pazzi. Part-Whole relations in object-centered systems: an overview. *Data and Knowledge Eng.*, 20(3): 347-383, 1996.
2. F. Baader, D. Calvanese, D. McGuinness, D. Nardi, and P. Patel-Schneider, editors. *The description logic handbook.* Cambridge University Press, January 2003.
3. F.Barbier, B.Henderson-Sellers, A.Le Parc-Lacayrelle, J.-M.Bruel. Formalization of the whole-part relationship in the Unified Modeling Language. *IEEE Trans. on Software Engineering*, 29(5): 459-470, 2003.
4. R.Casati, Varzi. Parts and places: the structures of spatial representation. *Cambridge/MA.*, MIT Press, 1999.
5. N. Guarino. Concepts, attributes and arbitrary relations. *Data and Knowledge Engineering*, 8: 249-261, 1992.
6. N. Guarino and P. Giaretta. Ontologies and knowledge bases: towards a terminological clarification. *Toward Very Large Knowledge Bases: Knowledge Building and Knowledge Sharing*, IOS Press, Amsterdam, 1995.

7. B. Henderson-Sellers and F. Barbier. Black and white diamonds. In *Proc. Second International Conf: Unified Modeling Language (UML'98)*, pp. 550-565, 1999.
8. F. Loebe. An analysis of roles: toward ontology-based modeling. *Master's Thesis*, University of Leipzig, 2003.
9. C. Masolo, L. Vieu, E. Bottazzi, C. Catenacci, R. Ferrario, A.Gangemi, N. Guarino. Social roles and their descriptions. In *Pro of the Ninth International Conference on the Principles of Knowledge Representation and Reasoning*, pp. 267-277, 2004.
10. R. Motschnig-Pitrik and J. Kaasboll. Part-Whole relationship categories and their application in object-oriented analysis. *IEEE Trans. Knowledge and Data Eng.*, 11(5): 779-797, 1999.
11. UML 2.0 Superstructure, 3rd Revision, *OMG document*, Object Management Group, 2003.
12. Simons. Parts-A Study in Ontology. London: Routledge, 1987.
13. M.Snoeck, G. Dedene. Existence dependency: the key to semantic integrity between structural and behavioural aspects of object types, *IEEE Trans. Software Engineering*, 24(24): 233-251, 1998.
14. J. Sowa. Conceptual structures: information processing in mind and machine. *AddisonWesley Publishing Company*, New York, 1984.
15. J. Sowa. Knowledge representation: logical, philosophical, and computational foundations. *Brooks/Cole Publishing Co.*, Pacific Grove, CA, 2000.
16. F. Steimann. On the representation of roles in object-oriented and conceptual modelling. *Data and Knowledge Engineering*, 35(1):83-106, 2000.
17. M. Winston, R. Chaffin, and D. Herrmann. A taxonomy of part-whole relations. *Cognitive Science*, 11: 417-444, 1987.

Author Index

Lecture Notes in Artificial Intelligence (LNAI)

Vol. 3849: I. Bloch, A. Petrosino, A.G.B. Tettamanzi (Eds.), Fuzzy Logic and Applications. XIV, 438 pages. 2006.

Vol. 3848: J.-F. Boulicaut, L. De Raedt, H. Mannila (Eds.), Constraint-Based Mining and Inductive Databases. X, 401 pages. 2006.

Vol. 3847: K.P. Jantke, A. Lunzer, N. Spyratos, Y. Tanaka (Eds.), Federation over the Web. X, 215 pages. 2006.

Vol. 3835: G. Sutcliffe, A. Voronkov (Eds.), Logic for Programming, Artificial Intelligence, and Reasoning. XIV, 744 pages. 2005.

Vol. 3830: D. Weyns, H. V.D. Parunak, F. Michel (Eds.), Environments for Multi-Agent Systems II. VIII, 291 pages. 2006.

Vol. 3817: M. Faundez-Zanuy, L. Janer, A. Esposito, A. Satue-Villar, J. Roure, V. Espinosa-Duro (Eds.), Nonlinear Analyses and Algorithms for Speech Processing. XII, 380 pages. 2006.

Vol. 3814: M. Maybury, O. Stock, W. Wahlster (Eds.), Intelligent Technologies for Interactive Entertainment. XV, 342 pages. 2005.

Vol. 3809: S. Zhang, R. Jarvis (Eds.), AI 2005: Advances in Artificial Intelligence. XXVII, 1344 pages. 2005.

Vol. 3808: C. Bento, A. Cardoso, G. Dias (Eds.), Progress in Artificial Intelligence. XVIII, 704 pages. 2005.

Vol. 3802: Y. Hao, J. Liu, Y.-P. Wang, Y.-m. Cheung, H. Yin, L. Jiao, J. Ma, Y.-C. Jiao (Eds.), Computational Intelligence and Security, Part II. XLII, 1166 pages. 2005.

Vol. 3801: Y. Hao, J. Liu, Y.-P. Wang, Y.-m. Cheung, H. Yin, L. Jiao, J. Ma, Y.-C. Jiao (Eds.), Computational Intelligence and Security, Part I. XLI, 1122 pages. 2005.

Vol. 3789: A. Gelbukh, Á. de Albornoz, H. Terashima-Marín (Eds.), MICAI 2005: Advances in Artificial Intelligence. XXVI, 1198 pages. 2005.

Vol. 3782: K.-D. Althoff, A. Dengel, R. Bergmann, M. Nick, T.R. Roth-Berghofer (Eds.), Professional Knowledge Management. XXIII, 739 pages. 2005.

Vol. 3763: H. Hong, D. Wang (Eds.), Automated Deduction in Geometry. X, 213 pages. 2006.

Vol. 3755: G.J. Williams, S.J. Simoff (Eds.), Data Mining. XI, 331 pages. 2006.

Vol. 3735: A. Hoffmann, H. Motoda, T. Scheffer (Eds.), Discovery Science. XVI, 400 pages. 2005.

Vol. 3734: S. Jain, H.U. Simon, E. Tomita (Eds.), Algorithmic Learning Theory. XII, 490 pages. 2005.

Vol. 3721: A.M. Jorge, L. Torgo, P.B. Brazdil, R. Camacho, J. Gama (Eds.), Knowledge Discovery in Databases: PKDD 2005. XXIII, 719 pages. 2005.

Vol. 3720: J. Gama, R. Camacho, P.B. Brazdil, A.M. Jorge, L. Torgo (Eds.), Machine Learning: ECML 2005. XXIII, 769 pages. 2005.

Vol. 3717: B. Gramlich (Ed.), Frontiers of Combining Systems. X, 321 pages. 2005.

Vol. 3702: B. Beckert (Ed.), Automated Reasoning with Analytic Tableaux and Related Methods. XIII, 343 pages. 2005.

Vol. 3698: U. Furbach (Ed.), KI 2005: Advances in Artificial Intelligence. XIII, 409 pages. 2005.

Vol. 3690: M. Pěchouček, P. Petta, L.Z. Varga (Eds.), Multi-Agent Systems and Applications IV. XVII, 667 pages. 2005.

Vol. 3684: R. Khosla, R.J. Howlett, L.C. Jain (Eds.), Knowledge-Based Intelligent Information and Engineering Systems, Part IV. LXXIX, 933 pages. 2005.

Vol. 3683: R. Khosla, R.J. Howlett, L.C. Jain (Eds.), Knowledge-Based Intelligent Information and Engineering Systems, Part III. LXXX, 1397 pages. 2005.

Vol. 3682: R. Khosla, R.J. Howlett, L.C. Jain (Eds.), Knowledge-Based Intelligent Information and Engineering Systems, Part II. LXXIX, 1371 pages. 2005.

Vol. 3681: R. Khosla, R.J. Howlett, L.C. Jain (Eds.), Knowledge-Based Intelligent Information and Engineering Systems, Part I. LXXX, 1319 pages. 2005.

Vol. 3673: S. Bandini, S. Manzoni (Eds.), AI*IA 2005: Advances in Artificial Intelligence. XIV, 614 pages. 2005.

Vol. 3662: C. Baral, G. Greco, N. Leone, G. Terracina (Eds.), Logic Programming and Nonmonotonic Reasoning. XIII, 454 pages. 2005.

Vol. 3661: T. Panayiotopoulos, J. Gratch, R. Aylett, D. Ballin, P. Olivier, T. Rist (Eds.), Intelligent Virtual Agents. XIII, 506 pages. 2005.

Vol. 3658: V. Matoušek, P. Mautner, T. Pavelka (Eds.), Text, Speech and Dialogue. XV, 460 pages. 2005.

Vol. 3651: R. Dale, K.-F. Wong, J. Su, O.Y. Kwong (Eds.), Natural Language Processing – IJCNLP 2005. XXI, 1031 pages. 2005.

Vol. 3642: D. Ślęzak, J. Yao, J.F. Peters, W. Ziarko, X. Hu (Eds.), Rough Sets, Fuzzy Sets, Data Mining, and Granular Computing, Part II. XXIII, 738 pages. 2005.

Vol. 3641: D. Ślęzak, G. Wang, M. Szczuka, I. Düntsch, Y. Yao (Eds.), Rough Sets, Fuzzy Sets, Data Mining, and Granular Computing, Part I. XXIV, 742 pages. 2005.

Vol. 3635: J.R. Winkler, M. Niranjan, N.D. Lawrence (Eds.), Deterministic and Statistical Methods in Machine Learning. VIII, 341 pages. 2005.

Vol. 3632: R. Nieuwenhuis (Ed.), Automated Deduction – CADE-20. XIII, 459 pages. 2005.

Vol. 3630: M.S. Capcarrère, A.A. Freitas, P.J. Bentley, C.G. Johnson, J. Timmis (Eds.), Advances in Artificial Life. XIX, 949 pages. 2005.

Vol. 3626: B. Ganter, G. Stumme, R. Wille (Eds.), Formal Concept Analysis. X, 349 pages. 2005.

Vol. 3625: S. Kramer, B. Pfahringer (Eds.), Inductive Logic Programming. XIII, 427 pages. 2005.

Vol. 3620: H. Muñoz-Ávila, F. Ricci (Eds.), Case-Based Reasoning Research and Development. XV, 654 pages. 2005.

Vol. 3614: L. Wang, Y. Jin (Eds.), Fuzzy Systems and Knowledge Discovery, Part II. XLI, 1314 pages. 2005.

Vol. 3613: L. Wang, Y. Jin (Eds.), Fuzzy Systems and Knowledge Discovery, Part I. XLI, 1334 pages. 2005.

Vol. 3607: J.-D. Zucker, L. Saitta (Eds.), Abstraction, Reformulation and Approximation. XII, 376 pages. 2005.